New Technology and Conflicting Information

Assessing Consumers' Willingness-to-Pay for New Foods

New Technology and Conflicting Information

Assessing Consumers' Willingness-to-Pay for New Foods

Editors

Wallace E. Huffman
Iowa State University, USA

Jill J. McCluskey
Washington State University, USA

World Scientific

NEW JERSEY • LONDON • SINGAPORE • BEIJING • SHANGHAI • HONG KONG • TAIPEI • CHENNAI • TOKYO

Published by

World Scientific Publishing Co. Pte. Ltd.
5 Toh Tuck Link, Singapore 596224
USA office: 27 Warren Street, Suite 401-402, Hackensack, NJ 07601
UK office: 57 Shelton Street, Covent Garden, London WC2H 9HE

Library of Congress Cataloging-in-Publication Data
Names: Huffman, Wallace E., editor. | McCluskey, Jill J., editor.
Title: New technology and conflicting information : assessing consumers' willingness-to-pay for new foods / [edted by] Wallace E. Huffman, Iowa State University, Jill J. McCluskey, Washington State University.
Description: New Jersey : World Scientific, [2020] | Includes bibliographical references.
Identifiers: LCCN 2020026848 | ISBN 9789811216718 (hardcover) | ISBN 9789811220722 (ebook)
Subjects: LCSH: Genetically modified foods--Public opinion. | Consumer behavior. | Consumers--Attitudes. | Food industry and trade--Technological innovations. | Agricultural innovations--Economic aspects. | New products.
Classification: LCC TP248.65.F66 N488 2020 | DDC 664--dc23
LC record available at https://lccn.loc.gov/2020026848

British Library Cataloguing-in-Publication Data
A catalogue record for this book is available from the British Library.

For any available supplementary material, please visit
https://www.worldscientific.com/worldscibooks/10.1142/11726#t=suppl

Desk Editor: Lum Pui Yee

Typeset by Stallion Press
Email: enquiries@stallionpress.com

Dedication

Wally Huffman passed away in the final stage of publishing this book. He always loved to tell me about his daughter Nadia, who is the same age as my daughter. Lauren was taken from us too soon. I dedicate this book to Nadia and Lauren, two intelligent and beautiful young women.

Contents

1. Introduction and Overview of Book
(Wallace E. Huffman and Jill J. McCluskey) 1

Part I Overview of Research Methods and Effects of Information 15

2. Using Stated Preference Techniques and Experimental Auction Methods: A Review of Advantages and Disadvantages for Each Method in Examining Consumer Preferences for New Technology.
(Wallace E. Huffman and Jill J. McCluskey) 17

3. Labeling of Genetically Modified Foods.
(Wallace E. Huffman and Jill J. McCluskey) 47

4. Public Acceptance of and Benefits from Agricultural Biotechnology: A Key Role for Verifiable Information.
(Wallace E. Huffman and Abebayehu Tegene) 81

5. Media Coverage, Public Perceptions, and Consumer Behavior: Insights from New Food Technologies.
(Jill J. McCluskey, Nicholas Kalaitzandonakes, and Johan Swinnen) 103

Part II New Traits without Added Nutrition: GM Input Traits, Organic, and All Natural Foods **137**

6. Consumer Willingness to Pay for Genetically Modified Food Labels in a Market with Diverse Information: Evidence from Experimental Auctions. (Wallace E. Huffman, Jason F. Shogren, Matthew Rousu, and Abebayehu Tegene) 139

7. Who Do Consumers Trust for Information: The Case of Genetically Modified Foods? (Wallace E. Huffman, Matthew Rousu, Jason F. Shogren, and Abebayehu Tegene) 175

8. Effects and Value of Verifiable Information in a Controversial Market: Evidence from Lab Auctions of Genetically Modified Food. (Matthew Rousu, Wallace E. Huffman, Jason F. Shogren, and Abebayehu Tegene) 191

9. The Effects of Prior Beliefs and Learning on Consumers' Acceptance of Genetically Modified Foods. (Wallace E. Huffman, Matthew Rousu, Jason F. Shogren, and Abebayehu Tegene) 235

10. Willingness-to-pay for Natural, Organic, and Conventional Foods: The Effects of Information and Meaningful Labels. (Jonathan R. McFadden and Wallace E. Huffman) 259

Part III Enhanced Nutrition and Food Safety Achieved Using New Technology **307**

11. Improving the Nutrient Content of Food through Genetic Modification: Evidence from Experimental Auctions on Consumer Acceptance. (Gregory J. Colson, Wallace E. Huffman, and Matthew C. Rousu) 309

12. Consumer Valuation of Information about Food Safety Achieved Using Biotechnology: Evidence from New Potato Products.
(Jonathan R. McFadden and Wallace E. Huffman) 343

13. Consumer Demand for Low-Acrylamide-Forming Potato Products: Evidence from Lab Auctions.
(Jonathan McFadden and Wallace Huffman) 381

14. Consumer Demand for Potato Products and Willingness-to-Pay for Low-Acrylamide, Sulfite-Free Fresh Potatoes and Dices: Evidence from Lab Auctions.
(Katie Lacy and Wallace E. Huffman) 413

15. Consumer Response to Information About a Functional Food Product: Apples Enriched with Antioxidants.
(Armenak Markosyan, Jill J. McCluskey, and Thomas I. Wahl) 451

Part IV Consumer Response to New Technology in Other Cultures 475

16. Russian Consumer Willingness to Pay for Genetically Modified Food.
(Anthony R. Delmond, Jill J. McCluskey, Mirzobobo Yormirzoev, and Maria A. Rogova) 477

17. Consumer Attitudes to Genetically Modified food in Norway.
(Kristine M. Grimsrud, Jill J. McCluskey, Maria L. Loureiro, and Thomas I. Wahl) 509

18. Consumer Response to Genetically Modified Food Products in Japan.
(Jill J. McCluskey, Kristine M. Grimsrud, Hiromi Ouchi, and Thomas I. Wahl) 531

19. Consumer Attitudes Toward Genetically Modified Foods in Beijing, China. (Quan Li, Kynda R. Curtis, Jill J. McCluskey, and Thomas I. Wahl) 553

20. Consumer Response to Information and Second-Generation Genetically Modified Food in India. (Alok Anand, Ron C. Mittelhammer, and Jill J. McCluskey) 571

Part V Conclusion and Recommendations 595

21. Conclusion and Recommendations. (Wallace E. Huffman and Jill J. McCluskey) 597

Index 605

Chapter 1

Introduction and Overview

Wallace E. Huffman and Jill J. McCluskey

The aim of this book is to make the authors' scholarly research in the area of consumers' preferences for new foods that have controversial attributes easily accessible to scholars, researchers, graduate students, food policy-makers, and food marketing specialists. It addresses issues that arise in a market for private goods where conflicting information from interested parties, scientific sources, and the media is present. It begins with a discussion of research methods, stated preference (SP) surveys and experimental auctions, and information issues. The empirical results include how consumers respond to food products that are produced with new technology, e.g., made from genetically modified (GM) or organic ingredients. GM technology may lower farmers' costs of production, enhance nutrition and food safety for consumers, or add quality and variety to consumers' food choices. Organic production is a process-based technology that has only minimal impact on nutrient content, but in some cases might provide fresher produce, which can matter to some consumers.

The authors have undertaken numerous projects to collect unique data in a series of laboratory experiments across various U.S. sites and consumer surveys worldwide, and to undertake novel empirical analyses. Data collected in these projects generally include sociodemographic attributes of subjects, and their revealed willingness to pay (WTP) from auctions of experimental foods and food products under randomly assigned food labels and information treatments and contingent-valuation surveys.

The book is divided into five parts. Part I provides an overview of research methods (consumer surveys and lab auctions) and how information affects public perceptions. Part II discusses new food traits that are the result of the application of science and technology, as well as long-term but rapidly growing organic and natural products. In Part III, research on new foods focuses on those products that possess enhanced nutrition or food safety achieved using new technology, especially biotechnology.[1] The empirical studies presented in Parts II and III focus on empirical evidence from research undertaken in the United States. Part IV addresses consumer response to new foods and technology in other cultures, i.e., studies completed in countries other than the United States. Part V summarizes what we have learned, makes policy recommendations, and offers suggestions for future research.

In Part I, Chapter 2 discusses the two primary research methods for assessing consumers' WTP and willingness to accept (WTA) new products — SP techniques and revealed preference (RP) methods. Multiple variants of these methods exist, but one common approach is discrete choice continent valuation (CV), where respondents are asked (yes or no) whether they are willing to pay a specific price for a specific good. Another approach is the discrete choice conjoint analysis in which respondents are asked to indicate their preferences from among two or more multi-attribute alternatives. McFadden (1980, 1986) has been a pioneer in the development of econometric discrete choice models of consumers' choice among products.[2] However, a strong set of assumptions are required in order to obtain parameter estimates in complex discrete choice models (e.g., McFadden and Train, 2000).

[1]Biotechnology is a technology that utilizes biological systems and/or living organisms to develop or make different products, e.g., cheeses, wines, beers, cured meats and fish, and genetically modified organisms (GMOs). GMOs are those products modified by the use of genetic engineering, e.g., gene splicing or gene editing technologies. These include GM corn and cotton plants that are resistant to a subclass of insects, and corn, cotton, canola, sugar beets, and canola plants that are herbicide tolerant. Biotechnology is a broader set of techniques than genetic modification, GM.

[2]Many SP methods utilize random utility models (Manski, 1997).

Direct surveys for SP research can be conducted with in-person interviews or questionnaires, telephone, mail, email, or online. Because no purchase is involved, these responses are non-binding and hypothetical. However, since the cost of conducting surveys is relatively low, it is possible to have a larger and more representative sample. Subjects can complete the surveys at home, so travel to a central location is not required, and normally surveys are less time-consuming to complete — usually 15–20 minutes. Subjects are paid only a small amount for their participation in surveys. A second advantage is that hypothetical choice experiments can use products that are not usable in an auction. For example, bread made from GM wheat could not be used in an auction in Russia because GM wheat is not approved for sale there. Thus, hypothetical choice experiments can sometimes provide information when RP techniques cannot.

RP techniques, such as experimental auctions and hedonic price analysis, use actual consumer decisions on products in an auction or food purchases from a store setting to model consumer preferences.[3] We focus on lab auctions where randomly chosen subjects from a metropolitan area come to a central location and participate in an experiment that may take 80–100 minutes of their time. Subjects are paid a non-trivial amount, depending on the time requirement for participating. Activities include learning the auction mechanism, e.g., the Vickery second-price auction, where the highest bidder pays the second highest price (Vickery, 1961), but more frequently a random nth-price auction (Shogren *et al.*, 2001). In the latter auction, if a session group size is 20 subjects, and the random n equals five, all subjects bidding strictly higher than the fifth subject are winners and pay the fifth highest bid. Information effects and treatments can be incorporated into these experiments to examine how subjects respond to such stimuli. An advantage of lab auctions is that it is in the best interest of each subject to state his or her true WTP for the experimental good. However, new product development and regulatory approval must be far enough along that the actual

[3]Smith (1976) is credited with first designing a set of experiments that incorporated private information revealing features of experimental auction markets.

products can be put on display in the lab for the experimental auction, i.e., the subjects are bidding on truthfully labeled products.[4]

Raw materials, made possible by genetic engineering, became widely available starting in the mid-1990s. These products used gene-splicing methods for inserting genes, especially across species. For example, herbicide-resistant soybeans and insect-resistant corn were developed by transferring genes from soil bacteria into these plants. Enhanced nutritional content of some foods has also been accomplished, e.g., golden rice has enhanced levels of vitamin A, vegetables may contain enhanced levels of antioxidants and vitamin C, and potatoes have dramatically reduced potential to produce acrylamide when cooked at high temperatures.[5]

Chapter 3 examines the welfare effects of information communicated by interested parties on the decisions of producers of agricultural products, potential users of biotechnology inputs, and consumers of food biotechnology products.[6] Interested parties are considered private agricultural biotechnology companies whose profits depend positively on the sale of GM inputs to agricultural producers and (anti-GM) environment groups whose utility depends positively on the quantity and quality of environmental stocks. The agricultural biotechnology companies provide or distribute information primarily in advertising, news releases, information brochures, and websites. The environmental groups provide or distribute information primarily in news releases, social media, informational web sites, and demonstrations. Final consumers of agricultural products are assumed to be approximately "neutral parties" in the genetically modified organism (GMO) communication debate, but they face decisions relating to their welfare. In this chapter, the hypothesis is tested that verifiable information can play a useful role in a market for new goods where there is a conflicted information environment

[4]Rousu *et al.* (2015) and Colson *et al.* (2016) discuss various dimensions of deception that may arise in lab experiments and their likely seriousness. List (2008) concludes that a strict adherence to a no-deception policy is contrary to the goal of conducting and publishing research on important new topics.

[5]A gene-silencing technique, an early type of gene editing, was used to develop the low-acrylamide potato.

[6]Milgrom and Roberts (1986) were the first to lay out the relevant information theory.

provided by the biotech industry and anti-GM environmental groups. This is the information that is verified by knowledgeable third parties, and its existence may force interested parties to be more revealing in their information dissemination process (Milgrom and Roberts, 1986).

Chapter 4 considers the role of the media in informing consumers about new foods, the impact of the media on public perceptions, and consumer behavior toward food that utilizes new technologies. Despite assurances from scientists, governments, and industry that new food technologies are safe, consumers are often skeptical. These negative consumer attitudes are sometimes attributed to consumers being poorly educated or informed. However, consumers from highly educated and affluent segments are often the most negative about food made from new technology (Curtis *et al.*, 2008).[7] This chapter reports that information about new technology is often provided by media organizations that have their own incentives to slant the information they provide in particular ways. A key constraint is that, because there are diminishing returns to time spent processing information and there are increasing amounts of information, consumers may still be rationally ignorant about new technology (McCluskey and Swinnen, 2004) — and hence, are less likely to spend the time to become well informed about the benefits of such new technologies.

Commercial mass media have thus had an increasingly important influence on public opinion. Extensive media coverage affects consumer perceptions of products and risks, but unfortunately, not benefits more generally, including reduced use of environmentally hazardous products, and consequently consumer demand for goods

[7]There is a surprising anti-science movement in the U.S. and some other developed countries. It is an attempt by a vocal minority to modify school curricula, use uncertainty tactics, and discredit the scientific method. Any person or organization that promotes their ideology over scientifically verified evidence is a partisan in favor of the anti-science position. If a position or theory is pro-science, or "science," it will follow the scientific method, be potentially refutable, peer-reviewable, reproducible, and open to change if the position comes in conflict with new confirmed hypotheses or evidence. Anti-science positions violate one or more of these thresholds. However, anti-science tends to be fragmented, e.g., attempts to ban vaccines, label or ban GMOs, promote or deregulate alternative medicine. However, these same people may support findings from some other branches of science (Rational Wiki, 2019).

and services. The media are criticized for sensationalistic and biased reporting, focusing on sound bites, and ignoring careful analyses. There is also concern that the increased commercialization of the media produces a dumbed-down version of the news, e.g. less in-depth reporting and more "infotainment," leading to decreased quality of coverage of complex issues. Competitive pressures are associated with cutbacks in reporting and editorial quality. The emergence of 24-hour news cycles may have actually weakened journalism, and news reports include an increasing number of factual errors. The consolidation of ownership of the media in the U.S. has undoubtedly reduced the competitiveness among media outlets.

Chapter 5 addresses special issues that arise in labeling GM food. There are three policy options for labeling GM foods: a labeling ban, voluntary labeling, and mandatory labeling. A policy of voluntary labeling works well in a niche market, where consumers who prefer GM-free products are willing to pay a premium to obtain them. In the United States, organic food, which is process based, and has a low probability of GM contamination, is the GM-free option.

If only a small share of consumers care about genetic modification, then mandatory labeling serves the same purpose as voluntary labeling but at a higher cost because the entire market must be segregated and labeled. Mandatory labeling becomes more attractive with increases in the share of consumers who care whether their food has been genetically altered. For mandatory labeling to be welfare enhancing, policy-makers must determine whether consumers place enough value on knowing whether their food is GM-free for the benefits of labeling to exceed its costs.

Supporters of mandatory labeling argue that it is needed for informed consumer choice. Critics of mandatory labeling argue that it carries the potential for information overload, and that consumers face both time and knowledge constraints when interpreting labels and making food purchases. Biotechnology companies generally have opposed mandatory labeling, but in a few cases, they have found that labeling for a new consumer trait that is achieved using genetic modification can be profitable, e.g., GM content associated with positive consumer traits.

Part II reports empirical research on U.S. consumers' WTP for new food products made possible by applications of biotechnology, and application of production-based technology — organic and all natural foods. The first four chapters (Chapters 6–9) use data from the same set of lab auctions. The research project was designed to incorporate the private information-revealing feature of experiment auction markets (a random nth-price auction), the randomized treatments of statistical experiment design, and random sampling methods from survey design. The primary purpose of the project was to identify the effects of GM food labels and different perspectives on genetic modification on consumers' WTP for food items that (at the time) might be GM — russet potatoes, vegetable oil (made from soybeans), and tortilla chips (made from yellow corn). The project developed three novel perspectives on biotech food: the biotech industry perspective, the natural food industries perspective, and the environmental group perspective, and created information treatments containing one to three of these information perspectives. When two or more perspectives were presented in a package, the order was randomized. These experiments were conducted in April 2001, in two metropolitan areas — Des Moines, Iowa, and St. Paul, Minnesota.

Chapter 6 tests the effects of food labels: whether conventional or GM label affects WTP for the three food products. Chapter 7 investigates responses of subjects to the question: "Who do you trust for information on biotechnology?" Chapter 8 also tests the novel hypothesis that verifiable information is a moderating force on subjects' WTP when they receive diverse and conflicting information — the industry's perspective on biotechnology or an environmental group's perspective. It also reports estimates of the value of verifiable information in these experiments. Chapter 9 estimates the effects of subjects' prior beliefs about biotechnology and learning from information injected into the lab auctions on consumer WTP for GM and conventional food products.

Vague food labels and distorted product claims have persisted in the "natural" food industry, while organic foods can be certified by the USDA. Chapter 10 examines consumers' WTP for foods carrying the organic, all natural, and conventional food labels under five different information treatments. The data for this study were collected from 106 subjects who participated in lab auctions (random *n*th price) conducted in April 2013, in Des Moines, Iowa. The subjects bid on delicious red apples, fresh broccoli, and large brown eggs. Novel information treatments were constructed from a no-information baseline, the natural food industry's perspective on all natural foods, the organic industry's perspective on organic foods, an independent perspective on all natural foods, and an independent perspective on organic food.

Part III examines U.S. consumers' WTP for new food products that have enhanced nutritional content or food safety. Chapter 11 examines consumers' WTP for foods that are enhanced with consumer attributes — vitamin C and antioxidants — delivered by transgenic (cross species) and intragenic (within species) GM methods controlling for randomly assigned information treatment effects. An interesting feature of intragenic is that scientific techniques have great similarity to conventional plant breeding. The participants were selected randomly from the Harrisburg, Pennsylvania, and Des Moines, Iowa, metropolitan areas. The experimental commodities were fresh broccoli, beefsteak tomatoes, and fresh russet potatoes. The study uses a random *n*th-price auction for revealing WTP. Critical information about the technology used to create the experimental foods was presented in food labels and then pre-packaged information was injected into the experiment. Novel information treatments were constructed from the following perspectives: a biotech-industry perspective on GM; an environmental group perspective on GM; and an independent third-party perspective on GM. When a subject received two perspectives in a treatment, the order was randomized. A Bayesian updating model of bid prices was utilized that simultaneously controls for bid censoring, commodity fixed effects, bidding-round fixed effects, and correlation of unobserved effects across rounds of bidding. The studies also show significant effect of food labels and packaged information on WTP for new goods.

Chapters 12–14 examine consumers' WTP for fresh potatoes that dramatically reduce the potential for forming acrylamide, a likely cancer-causing agent and neurotoxin, when cooked at high temperatures (over 250°F), which is required to make French fries, potato chips, and hash browns. Conventional plant breeding and new potato storage practices are unable to reduce significantly the acrylamide-forming potential of potatoes, but biotechnology methods have been successful. In April and May 2014, experiments were conducted with randomly selected subjects in cities in three distant regions of the country: Boston, Des Moines, and Los Angeles. There were four different starting times at these experimental sites; the first three were focused on fresh potatoes, french fries, and chips.

Chapters 12 and 13 exploit these observations to examine consumers' WTP for enhanced food safety. These studies focus on the effects of food labels and pre-packaged information on consumers' valuation (bids in a random nth-price auction) of low-acrylamide biotech potato products. Novel information treatments were constructed from three different perspectives: an industry perspective on low-acrylamide potatoes using biotechnology, a scientific perspective on acrylamide exposure, and an environmental group perspective on biotechnology. Each information treatment consisted of one or two perspectives, and when the subject received two perspectives, the order was randomized. Subjects bid first on the set of three commodities without information, and then a second time after receiving one randomly assigned information treatment. The empirical analysis is then of bid price differences: *ex ante* versus *ex post* information. Scientific and/or industry information was shown to significantly increase WTP for the new low-acrylamide potato products.

At the fourth starting time, there were only two products: fresh potato and fresh cut potato dices. Conventional fresh-cut potato dices contain sulfites, which reduce discoloration of the product but increase consumer health risks. Chapter 14 examines consumers' WTP (random nth-price auction) for conventional and biotech white potatoes and potato dices. These products have low acrylamide-forming and browning/bruising potential due to biotech advances. The study also examines the effect of food labeling and pre-packaged information on WTP for new potato products. The information perspectives for this study are an empirical perspective on biotechnology and a company perspective on low bruising and low acrylamide potential using biotechnology. Subjects engage in a random nth-price auction. The empirical analysis is of bid price differences by commodity pre-information versus post-information treatment for both the fresh potatoes and fresh cut potato dices.

Chapter 15 reports consumers' WTP for apples enriched with an antioxidant coating. Although functional food products provide health benefits beyond basic nutrition, some consumers may reject them because they utilize new technology. In 2006, consumers in Spokane and Seattle, Washington, completed face-to-face consumer surveys with contingent valuation questions. An abbreviated information treatment stating: "Fruit enhanced with natural antioxidants will improve its health benefits by helping to prevent cancer, cardiovascular and other diseases" was randomly assigned to half of the subjects. The information treatment had a positive and significant effect on the treated subjects' WTP for the enhanced apples. Other findings include that the Seattle consumers were less likely to pay premium for enhanced apples than those from Spokane, which is surrounded by large acreages of wheat. Previous studies have found that consumers living in close proximity to conventional agriculture are more accepting of "innovative" value-added agricultural production (McCluskey *et al.*, 2003), and this effect may contribute to their WTP for antioxidant apples.

Consumer perception and acceptance of GM foods varies across cultures and countries. Part IV addresses other (non-U.S.) consumers' WTP for new GM food products. These empirical studies were conducted in Russia, Norway, Japan, China, and India. Chapter 16 reports on a study of Russian consumers' WTP for bread made with GM wheat. GM foods are banned in Russia, and thus an SP approach is the only feasible research method to obtain information about Russian consumers' WTP for a hypothetical GM food product. The authors used a survey-based contingent valuation approach to elicit consumers' WTP for bread made from GM wheat, relative to preexisting non-GM alternatives. In 2015, the authors surveyed 300 consumers in several grocery stores near the city center in Perm, Russia, a city of about 1 million residents, in the European part of Russia. The authors examined how economic and social factors affect the mean consumer's WTP for bread containing GMOs. They found that their Russian sample was overwhelmingly negative toward the GM bread with 80% of respondents stating that they were unwilling to purchase GM bread even at a discount of up to 50%.

Chapter 17 analyzes factors that induce Norwegian consumers to choose bread made with GM wheat versus bread made with conventional wheat and estimate the discount required for them to choose the bread made with GM wheat. In January 2002, data were collected from in-person interviews using a contingent value survey of 400 shoppers in the Oslo region of Norway. The study finds that the Norwegian sample were, on average, only willing to purchase GM bread with a large discount relative to the corresponding non-GM product.

Chapter 18 identifies and analyzes factors that induce Japanese consumers to choose noodles made with GM wheat relative to noodles made from conventional wheat. In June 2001, 400 consumers at a Seikyou, a grocery store-like setting, in Matsumoto City, Japan, participated in in-person surveys that included contingent valuation questions for GM noodles. This was an early economic study on consumers' WTP for GM foods, and it took place soon after Japan's mandatory GM labeling policy was enacted. Of the 400 respondents, 80% were not willing to purchase the GM noodles even with the discount.

Chapter 19 reports on a study conducted in Beijing, China, to assess consumers' WTP for oil from GM soybeans and GM rice versus non-GM alternatives. In 2002, in-person contingent value surveys were conducted for 599 individuals in four separate locations: a supermarket, two outdoor markets, and a shopping area. The respondents were provided with information that GM rice ("Golden rice") contains additional vitamins not found in conventional rice. The estimated mean WTP for this Chinese sample was a 38% premium for GM rice and a 16.3% premium for GM soybean oil. The estimated premium for GM rice is expected because of the product-enhancing attribute. However, the premium for GM soybean oil is unexpected. However, the study found that a majority of Chinese respondents were positive about the use of biotechnology in foods.

The last chapter (Chapter 20) of Part IV reports on consumers' WTP for chapatti made with GM wheat in India. The dataset used in this study was collected through 600 in-person surveys conducted at eight grocery stores in New Delhi (400 respondents) and Patna (200 respondents), India, in 2004. Two information treatments were applied, relative to a third no-information benchmark group. One third of the sample was provided with a short statement about positive environmental benefits from the GM wheat due to reduced herbicide usage and production costs associated, and one third was given negative private benefit information that there may be long-term negative health consequences associated with consuming GM wheat (post-elicitation, the subjects were debriefed that there is no scientifically demonstrable evidence to support any negative claims). After responding to CV questions about chapatti made from GM wheat, all respondents were asked questions about their WTP for a hypothetical GM wheat that is heart healthy. One can compare this case with the one where positive public benefit information about GM wheat was given, the latter being an example of a "producer friendly" or "first generation" GM product.

Different types of information had significant effects on mean WTP for GM wheat in India. The no-information group was WTP a premium of 7.3%, and the positive public-benefit information (reduced herbicide usage and production costs) respondents were WTP a premium of 10.4%. In contrast, when negative information about potential health effects was provided to subjects, the average respondent required a discount of 139%, meaning that he or she was not willing to buy at any price. This is consistent with many studies, including Tegene *et al.*, (2003), who observed that individuals place a greater weight on negative information than on positive information. The positive private-benefit information respondents were, on average, willing to pay a 23.7% premium for a hypothetical GM wheat that is "heart healthy."

Part V consists of only one chapter (Chapter 21). It summarizes what we have learned, policy recommendations, and what remains to be investigated, including suggestions for future research.

References

Colson, G.J., J.R. Corrigan, C. Grebitus, M.L. Loureiro and M.C. Rousu (2016). Which Deceptive Practices, If Any, Should be Allowed in Experimental Economics Research? Results from Surveys of Applied Experimental Economists and Students. *American Journal of Agricultural Economics* 98, 610–621.

List, J.A. (2008). Informed Consent in Social Science: Response. *Science* 322(5902), 672.

Manski, C.F. (1997). The Structure of Random Utility Models. *Theory and Decision* 8, 229–254.

McCluskey, J.J., K.M. Grimsrud, H. Ouchi and T.I. Wahl (2003). Consumer Response to Genetically Modified Food Products in Japan. *Agricultural and Resource Economic Review* 32, 222–231.

McCluskey, J.J. and J.F.M. Swinnen (2004). Political Economy of the Media and Consumer Perception of Biotechnology. *American Journal of Agricultural Economics* 86(5), 1230–1237.

McFadden, D. (1980). Models of Probabilistic Choice Among Products. *The Journal of Business* 53, S13–S29.

McFadden, D. (1986). The Choice Theory Approach to Market Research. *Marketing Science* 5, 275–297.

McFadden, D. and K. Train (2000). Mixed MNL Models for Discrete Response. *Journal of Applied Econometrics* 15, 447–470.

Milgrom, P. and J. Roberts (1986). Relying on the Information of Interested Parties. *R and Journal of Economics* 17, 18–32.

Rational Wiki. (2019). *War on Science.* Available at: https://rationalwiki.org/wiki/War_on_Science.

Rousu, M.C., G. Colson, J.R. Corrigan, C. Grebitus and M.L. Loureiro (2015). Deception in Experiments: Towards Guidelines on Use in Applied Economics Research. *Applied Economic Perspectives and Policy* 37, 524–536.

Shogren, J.F., M. Margolis, C. Koo and J. List (2001). A Random *n*th Price Auction. *Journal of Economic Behavior and Organization* 46, 409–421.

Smith, V.L. (1976). Experimental Economics: Induced Value Theory. *American Economic Review* 66, 274–279.

Tegene, A., W.E. Huffman, M. Rousu and J.F. Shogren. Affects of information on consumer demand for biotechnical foods: Evidence from experimental auctions USDA, ERS, Technical Bulletin No. 1903.

Vickrey, W. (1961). Counter Speculation, Auctions and Competitive Sealed Tenders. *Journal of Finance* 16, 8–39.

Part I

Overview of Research Methods and Effects of Information

Chapter 2

Using Stated Preference Techniques and Experimental Auction Methods: A Review of Advantages and Disadvantages for Each Method in Examining Consumer Preferences for New Technology*

Wallace E. Huffman and Jill J. McCluskey

Iowa State University and Washington State University

1. Introduction

In the 2001, biotechnology was being touted as the future of agriculture. However, the business leaders, producers, and scientists involved in biotech did not foresee how controversial genetically modified (GM) foods would be with consumers, particularly for export markets such as Japan. In fact, between 1997 and 2005, Monsanto conducted field trials for GM wheat in sixteen U.S. states, with greater than 30 trials in the State of Washington (Ledford, 2013). McCluskey *et al.* (2003) used contingent valuation (CV) data from a survey they conducted in Japan in 2001 to analyze Japanese consumers' willingness to pay (WTP) for noodles made with GM wheat. Stated preference (SP) data were the only option because GM wheat was not an approved product in Japan or anywhere else at the time. McCluskey *et al.* (2003) found that eighty percent of

*This article originally appeared in the **International Review of Environmental and Resource Economics (2017), 10**, 269–297. This is published by permission of **now publishers inc**.

the Japanese consumers who participated said that they would not buy the noodles made with GM wheat even with a discount. The implications were clear: GM wheat could not be profitably sold in Japan. They recommended that Washington growers, who export up to 90% of their production each year, should not switch to GM wheat.

The above account is an example of how new product evaluation methods can provide valuable information for product development and marketing decisions for a wide range of consumer goods, especially products that utilize new technology. These evaluation methods are primarily of two types: revealed preference (RP) and SP techniques. RP techniques, such as experimental auctions and hedonic price analysis, use actual consumer decisions on products in an auction or food purchases from a store setting to model consumer preferences. The experimental auctions that are used to evaluate consumer preferences are designed to be incentive compatible, i.e., it is in the best interest of each subject to state his or her true willingness-to-pay for one unit of a good. In a random n^{th} price auction, all subjects have a positive probability of winning the auction, and some fraction of winners must exchange money for won product. Thus, their decisions have real financial consequences. Hence, RP techniques exploit the fact that these consumer decisions reveal preferences for goods, in both market and non-market contexts.

SP techniques, such as CV, contingent ranking, and choice experiments, ask people a series of questions that are intended to elicit their preferences for a good, but because no purchase is involved, these responses are non-binding and hypothetical. However, the cost of conducting surveys is relatively low, making it possible to have a larger sample size. Subjects can complete the surveys at home, so no travel to a central location is incurred, and normally surveys do not take long to complete — usually 15 to 30 minutes. Subjects are paid only a token amount for participating in the SP surveys. The relative cost savings of surveys relative to experiments depend on the format of the survey. McMaster *et al.* (2017) found that the per-respondent cost of paper-only surveys is 40% greater than web-push surveys.

Surveys are generally easy to administer. For example, there are some standing data bases of adults available for taking surveys. Hence, it is relatively easy to achieve greater geographical representativeness of subjects in SP than for lab experiments. In addition, sometimes it will only be possible to collect SP data. This will be the case for products that are so new that display samples are not approved for human consumption or do not even exist, such as a new type of food product or other product. If evidence of cross validation of approaches exists, then SP data can provide support for policy decisions that are based solely on results from SP data (see Loureiro *et al.*, 2003; Chang *et al.*, 2009). Hence, there are a number of tradeoffs that come to play in making a decision between RP and SP techniques.

In the remainder of this article, we elaborate and discuss the SP methods and the RP method of experimental auctions including cost, representativeness, and generalizability. We discuss issues and concerns for each method, including the issue of deception, which has been raised in a leading agriculture economics journal a few years ago. Note that this article focuses on *private goods* associated with new technologies rather than public goods, such as environmental quality and recreation, which are the major focus of much of the non-market valuation literature. Private goods provide different incentives than public goods — they are rival and excludible while public good are not (Cornes and Sandler, 1996). Public goods are often provided by referendum, while private goods are typically provided by markets. Thus, there are implications for incentive compatibility of the elicitation method used in research. In our context, incentive compatible means that giving a truthful response to the question or providing a bid that is equal to one's true valuation is an optimal strategy for the participant (Carson and Groves, 2007). We emphasize that while neither method is perfect, both can provide useful information for decision makers. We develop a few recommendations for future research.

2. Background on Methods

SP and RP methods have features that are important to our review. First, SP approaches are purely hypothetical, meaning that they do not include actual purchases or require any behavioral changes on the part of the consumer. They directly ask consumers if they would be willing to pay a specific price for a change in a product characteristic or ask consumers to choose or rank products with different characteristics and prices. In contrast, experimental auctions offer an actual product that the consumer has an opportunity to bid on. Actual money is exchanged if the participant "wins" the auction. The difference between bids for the good with the new characteristic and the good with the traditional characteristic provide a straightforward estimate of willingness to pay (WTP) for that characteristic (Corrigan *et al.*, 2009).

There are tradeoffs to weigh in using these approaches. The criticisms of SP techniques center on the hypothetical nature of the approach. Hypothetical bias (HB) is the chief concern. HB refers to situations in which the WTP elicited from hypothetical formats diverge from the WTP elicited from non-hypothetical formats. HB is caused by violations of incentive compatibility. Primarily in the context of environmentally friendly goods, most of the literature suggests that HB is in the form of an overstatement of WTP (List and Gallet, 2001). However, in the context of goods made with new technology, HB may take the form of an overstatement of the required discount for the food products made with new technology, such as genetically modified (GM) foods.

It is recognized in the SP literature that it is difficult to construct incentive compatible valuation surveys even for private goods, which are the focus of this article, for value estimates (demand) and potentially for marginal value estimates (Carson and Groves, 2007 and 2011; Carson *et al.*, 2014; Vossler *et al.*, 2012; and Vossler, 2016). Carson and Groves (2007) argue that introducing a new private good provides incentives to respondents to say "yes — would purchase" as long as the good is potentially desirable. Their argument is that a positive response will encourage the production of the good and the participant can decide later whether actually to purchase it at the market price. Saying "yes" effectively increases the participant's choice set. It is interesting to consider for private goods made with new technology, such as GM foods, many people want to prevent the production and distribution of the good. Thus, the argument is turned on its head for the segment of consumers who want to block GM foods. They should say "no — would not purchase" at any price. In fact, this outcome sometimes occurs in GM SP valuation studies (for example, Grimsrud *et al.*, 2004).

If subjects are familiar with the product used and the prevailing market price for them, HB for conventional products is likely to be lower, but this is not necessarily the case for familiar products made with new technology. In these circumstances, researchers must weigh the costs and benefits of the assessment method by considering a research method that either aligns subjects' incentives with the researchers' or imposes some cost on subjects for offering responses that deviate from their true preferences relative to the feasibility and costs of implementing that method. Hence, interpreting hypothetical results should be taken with caution, since study respondents were only asked hypothetical bidding questions without having to actually execute them.

There are many possible sources of HB. The first is that it is possible that subjects might provide strategic responses that do not correspond to his and her true preferences. That is, the subject may provide extreme responses with the intent to affect the outcome of the study. For example, if the consumer subject holds anti-technology views, he or she might exaggerate the discount required for him or her to purchase a product made with GM ingredients. If subjects believe that their responses are consequential, i.e., their collected responses will be used to inform future business strategies or public policies, that may impact their responses. In such cases, the subject may take the decision-making exercise more seriously or they may provide non-truthful responses to survey questions in an attempt to influence the outcome of the study, e.g., to send messages to the investigator. With this latter scenario, subject's hypothetical values will be biased. Another source of HB is social desirability bias (Kleckner *et al.*, 2002), in which the respondent is influenced by trying to provide responses that he or she thinks the interviewer wants to hear. This might apply to new technology products because people might perceive a wider societal sentiment against GM foods (see Lusk, 2013).

A mitigating factor is that SP methods typically offer prices/bids to which consumers respond that they will buy or not, i.e. they are dichotomous choice (DC) rather than open-ended questions. This limits the impact of renegade subjects who might try to influence the outcomes of the study by responding with outlier bids. With an open-ended question, such as, "How much are you willing to pay for a pound of non-browning GM apples?" A pro-GM respondent could respond with a bid of $100 per pound in order to influence marketing decisions. In contrast, in a DC question format, the respondent would be asked, "Would you pay $3.99 per pound for apples that have been genetically modified not to brown after slicing? Please respond with yes or no." This limits the impact of outlier responses that might be provided for strategic reasons.

On the other hand, if the participants in a SP exercise believe that their responses are inconsequential, e.g., the information gained will not be used in future business strategies or public policies, they may view one response being as good as another. All responses have the same effect on a subject's utility. Hence, these individuals have little incentive to expend cognitive effort on decisions involving hypothetical stated preferences about ratings and rankings of goods. This makes these hypothetical values relatively "noisy" and would not be expected to yield much useful information when analyzed by the researcher (Lusk and Shogren 2007). However, some topics lend themselves to SP surveys of consumers when other methods are not possible.

Some researchers try to avoid HB in SP valuation through the inclusion of "cheap talk" scripts (Taylor and Cummings, 1999) and other framing devices before or after elicitation. As Vossler (2016) discusses, the use of cheap talk scripts may actually exacerbate the problem of HB. The argument is that a cheap talk script can "emphasize excessively that choices and policies are simply hypothetical, [which] dampens the already weak incentives for truthfulness with likely undesirable effects (Vossler, 2016, p. 39). Thus, SP researchers should avoid a typical cheap talk script telling respondents that hypothetical valuations are likely too high.

Even with the significant disadvantages, SP methods have many advantages in consumer research. Advantages include that surveys can be developed and implemented at relatively low cost. When conducting a consumer survey, subjects do not need to come to a central location to participate, and they are typically compensated a small amount for their participation. In contrast, the cost of setting up an experimental auction at a location with adults from a locality is significant. Subjects must come to central locations to participate in the experiments — normally take place one to three cities — are targeted and the total number of participants is in the range of 150 to 300. An outside party, such as a marketing or survey research firm, is generally paid to identify willing subjects. In addition, subjects incur commuting costs to the site, and experiments generally require 1 to 1.5 hours of a subject's time. Hence, subjects are also paid a substantial amount for their participation. There are also travel costs of the investigators and assistants to the experimental site, display products, and materials for displaying and hiding products. Hence, the average cost per subject participating in one of these experimental auctions is significant.[1] In the authors' experience, the ratio of costs per respondent of experimental auctions to CV can be ten to one.

The high costs of experimental auctions can have implications for representativeness and generalizability if they are a binding constraint. Representativeness and generalizability are important considerations in the value of the information that comes from either method. Many experiments have used university student subjects, which as we discuss later, are not representative of most populations

[1]Decisions on the sample size should be made with the aid of a statistician who specializes in statistical experimental design. This may involve statistical power calculations to justify sample size. A population of undergraduates is not generally appropriate for testing new food products to be sold in grocery stores and super markets because they will not represent the population to whom the product will be marketed.

in which researchers are interested. Thus, results from student experiments are often not generalizable. Further, although the researchers may attempt to choose subjects by randomly sampling a list of residents in a city, it is difficult to achieve a representative sample of participants for a large area, e.g., the United States or a major region.

A second advantage associated with SP techniques is that they can be used for products that do not yet exist or product that would be difficult to sell in an auction. An example of this is a biofuel that is made with new technology, which is still being developed. This is a major benefit because SP techniques can provide valuable information in directing the choices for research on new technologies and/or firms' marketing decisions.

A third advantage of SP techniques is that they often more closely describe products in situations that consumers face when buying food, and thus participants may be better able to provide an accurate answer. In virtually all U.S. retail grocery stores, consumers are presented with a fixed price per unit at which they can purchase as many units of a product as they want. In contingent valuation, participants are presented with a price and asked if they would purchase one unit of the product at that price. In choice-based conjoint experiments, participants are presented with two or more goods including a fixed price for each good and asked to choose the one they would buy. In contrast, experimental auctions require subjects to bid on one unit of real products, which is a practice that is not used in U.S. grocery stores and super markets. However, in the U.S. adults frequently have had experience with antique, estate, car, art, livestock, and land auctions. Opportunities to participate in these auctions have increased with the growth of internet auctions.

Experimental auctions are a RP method using non-hypothetical incentive-compatible methods to obtain information from subjects on their valuation of a product. In particular, experimental auctions use real products and real money. They are designed to be incentive compatible, meaning that they induce each bidder to submit a bid that sincerely reflects his or her value for one unit of the good(s) being auctioned (Lusk and Shogren, 2007). Such auction mechanisms include the Vickrey second-price auction (Vickrey, 1961), n^{th}-price auctions, and the random n^{th} price auction (Shogren *et al.*, 2001). These auctions separate the payment of the subject for participating in the experiment from the decision on how much to bid for a product. This can be achieved by paying participants in cash before the auction starts (Rousu *et al.*, 2007).[2] An additional important feature is that the price that a subject bids for a product is not the amount that he or she pays, even if she is a "winner" (Lusk and Shogren, 2007). As such, these auction mechanisms eliminate the winner's curse that may arise in an English auction where the highest bidder is the winner and pays her winning bid. The winner's curse arises in the English auction because as a bidder increases her bid, she increases the probability of being the winner but also increases the probability of paying too much. In a second-price auction, the highest bidder pays the second highest price-bid.[3] If bidders try to send signals to the investigator in their bids that deviate from their true value, they are made worse off.

Results from lab experiments have sometimes been criticized for lack of realism or relevance. As Corrigan *et al.* (2009) mention, experimental auction participants are presented with the unfamiliar problem of being asked to provide the highest price they would be willing to pay for a fixed quantity of a novel good. The difficulty in this exercise is then compounded if the subject has not participated in an auction before. A mitigating factor is that all lab auctions provide training for subjects in the auction mechanism by having

[2] This payment in cash eliminates an endowment effect (that might occur if individuals were given an experimental commodity at the beginning of an experiment and then asked to pay to upgrade to another product). Paying cash up front in an experiment is a type of framing of the events in the auction that add credibility to the institutional structure of the auction. In experiments where there is a settling of accounts at the end of the experiment, subjects are more likely to bid insincerely if they frame decisions as if they are using "house" rather than their "own" money.

[3] In a random n^{th}-price auction, the winner is chosen by the session monitor by drawing a number between 2 and m, where m is the total number of subjects in a session. Subjects who bid strictly more that the nth price are the winner and they must execute the winning bid by purchasing one unit of the product at this price.

subjects bid on practice items first and then showing them how the auction mechanism determines winners. After questions are resolved, the session monitor moves on to the auction of target commodities (e.g., see Rousu *et al.*, 2007, Colson *et al.*, 2011).

Adamowicz, Louviere, and Williams (1994) criticize RP methods that assume models of behavior, e.g. the chosen functional form and included attributes that are regressed on bids, is a maintained hypothesis about the structure of preferences that may not be testable. With RP methods, it is important to have an attribute list that is not large because the attributes are likely to be correlated and create problems with identification of marginal impact of each factor on choices. Also, with many choices subjects take short-cuts to decisions.

Adamowicz *et al.* (1997) argue that it is possible to undertake research that uses both SP and RP methods. When the results are consistent between these approaches for a given research problem, confidence in the results is higher than with SP alone. However, von Haefen and Phaneuf (2008) and Jeon and Herriges (2010) find that responses using the RP and SP methods are frequently inconsistent in magnitude. Loureiro *et al.* (2003) use an economic experiment with a cooperating grocery store in conjunction with a survey to analyze whether consumers' hypothetical WTP responses are effective predictors of actual market behavior. They match individual's survey responses with his or her actual purchase decisions and find that consumers who stated that they were WTP a premium had a higher likelihood of actually buying the product in question. A take-away message is that while HB exists in SP methods, there can still be value in the information they provide.

3. Discussion of Stated-Preference Methods

In SP techniques of product attributes valuation, the participating individuals do not actually make any behavioral changes. The two major types of SP approaches are contingent valuation (CV) and choice-based conjoint (CBC) analysis. We provide more details about both of these methods.

3.1. Contingent valuation (CV) method

The CV method is a commonly used approach to elicit consumer's WTP through a market-like DC questioning format. CV is based on a DC model in which individual WTP can be estimated based on responses to market-type questions (Kanninen, 1993). The basic idea is that the interviewer describes a hypothetical market for the target product and asks people whether they would be willing to pay a specific amount to obtain the targeted product. WTP can be estimated with a limited dependent variable model.

There are typically two types of CV models: the single-bounded model and the double-bounded model. In the single-bounded model, the consumer is asked one question about whether he or she would be paying to pay a specific price (bid) for the item in question. For example, "Would you pay $3.99 per pound for apples that have been genetically modified not to brown after slicing? Please respond with yes or no." The single-bounded model partitions the WTP distribution into two segments. From the participant's response to the example question, we know that the consumer's WTP is either less than $3.99 per pound or greater than or equal to $3.99 per pound. The double-bounded CV model is an extension of the single-bounded model and improves statistical efficiency over the single-bounded method by engaging respondents in two bids instead of one. The second bid question is contingent on the response to the first question. If the initial bid is accepted (i.e., the respondent says yes), a bid question is asked that offers a premium price; whereas if the initial offer is rejected (i.e., the respondent says no), a bid question is asked that offers a discount price. The values of the premium and discount bids are randomly assigned to subjects and calibrated based on the results of a pre-survey. Based on responses to the sequential bidding questions, one can place boundaries on the consumer's WTP, which is depicted in Figure 1.

Let us consider more carefully the choice on GM apples using the double-bounded CV questions: "Would you pay $3.99 per pound for apples that are genetically modified not to brown? Please respond with yes or no. If yes, ask, "Would you pay $ [random premium price]?" and if no, ask, "Would you pay $ [random discount

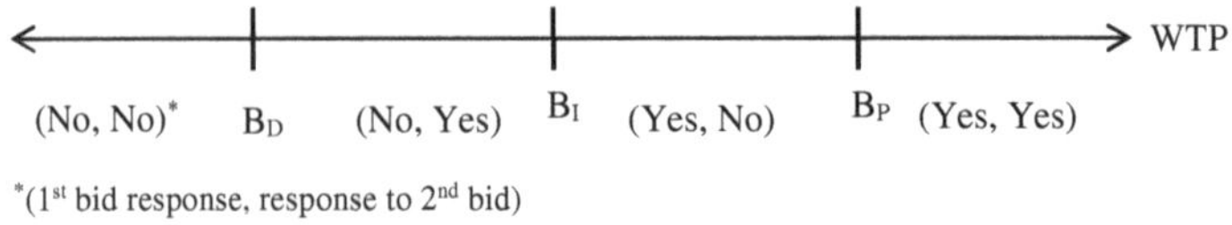

Fig. 1. Partitioning the WTP distribution with the double-bounded question format.

price]?" The researcher should create different versions of the survey with different random bid amounts intended to cover the WTP distribution for GM apples. Thus, most subjects should say "no" to the highest random premium bid, and most should say "yes" to the lowest discount bid. The surveys are to be randomly assigned to subject in the project. Answers to the double-bounded bidding questions can then be sorted into four intervals: $(-\infty, B_D)$ when the first and second answers are both "no", (B_D, B_I) when a discount offer is accepted at the second bid, (B_I, B_P) when a premium is rejected, and $(B_P, +\infty)$ when both answers are "yes".

Hence, the subject's WTP for the product can be separated into four sets of observable outcomes, which is an aid to developing an econometric model of demand. The objective is recover information about key parameters about demand while controlling for demographic information and location effects, which can affect consumers' WTP for the product (Hanemann, Loomis, and Kanninen, 1991; Hanemann, 1984).

There is disagreement over the best number of iterations to include in the bidding procedures used in the CV method. Cameron and Quiggin (1994) consider the problem of anchoring or starting-point bias, a type of path dependence, with iterations of bids. There is some bias with the double-bounded model, primarily due to inconsistencies which may be present between the consumers' first and subsequent bids (Hanemann and Kanninen, 1999). Hanemann, Loomis, and Kanninen (1991) conclude that the gain in efficiency largely prevails over the loss from bias, which tends to be moderate. However, Carson and Groves (2007) discuss that most recent analyses have avoided double-bounded approaches because of the changing nature of preferences that occur in such a two-step (bargaining) approach. It is not common practice to use greater than two iterations in the bidding process.

There are many concerns with the CV method. Concerns include starting point biases from anchoring and "yea saying," (see Chien, Huang and Shaw, 2005 and Herriges and Shrogen, 1996). Misspecification is prevalent when individuals are confronted with unfamiliar goods with an uncertain value, which is less of a problem with choices being made on food products than on nonmonetary valuation of never experienced environmental goods, e.g., polar bears in Alaska.

3.2. *Conjoint analysis*

Although conjoint analysis can be either an SP or RP technique, it has been mostly used in SP approaches. It is used to isolate the effect of product attributes on consumer preferences. The following represents an example of a choice experiment question containing three product profiles:

> "If you were planning to buy milk today, and the following alternatives were available, please check the box under your most preferred option."

Product Attributes	***Option 1***	***Option 2***	*Option 3*
Milk fat	**3.5%**	**2%**	I would
Pasteurized	**Yes**	**No**	Not buy
Price/liter	**$2.99**	**$3.29**	Either product
	□	□	□

The conditional probability of an individual choosing the mth alternative can be expressed in a limited dependent variable model. If price is one of the attributes, then WTP can be estimated.

Conjoint analysis has an advantage of allowing subjects to make decisions about products based on several product attributes. The choice-based approach involves respondents making one choice from each of several sets of stimuli derived from an experimental design (Louviere, 1993). Adamowicz *et al.* (1997) argues that the choice-based conjoint analysis has several advantages over typical CV methods. In particular, since the SP version of this approach mimics what people do in the real world, its results are comparable to consumers' revealed preferences (Adamowicz *et al.*, 1994). This mitigates the potential problem of HB.

The theoretical foundation for an economic interpretation of the results from conjoint analysis is based on Lancaster's approach to consumer theory (Lancaster, 1966), which posits that consumers obtain utility from a product based on the products' attributes. To measure utility, a random utility model is typically used, which assumes that the utility $U_{ij} = v_{ij} + \varepsilon_{ij}$ for individual i and product j is composed of the sum of systematic and random components. It is typically assumed that the systematic component $v_{ij} = X_j\beta + \alpha_i$ is observable and a linear function of the product attributes and individual characteristics.[4] The random component ε_{ij} includes the unobservable factors. X_j is the attribute vector of the jth alternative, β is the coefficient vector representing the weight of attribute in the valuation of alternative j. The parameter α_i is an individual-specific component, and ε_{ij} is stochastic and reflects the idiosyncrasies of this individual in tastes for alternative j (McFadden, 1974). The individual-specific component can be interpreted to represent the individual's economic status (see Adamowicz and Swait, 2011).

One can use choice-based conjoint analysis to estimate $v_{ij} = X_j\beta + \alpha_i$ based on data from hypothetical choice experiments. The approach is to construct a number of hypothetical alternatives and let a sample of respondents make preference judgments. Consider an individual faced with a set of alternatives from which to choose, each of which consists of a different combination of levels of a set of multiple attributes. Suppose individual i faces J alternatives, indexed $j = 1, \ldots, J$ and described by vectors of attributes X_j.

[4]Ladd (1976) provides a thoughtful model of consumer goods characteristics and consumer demand. His approach suggests a different direction that the attributes model can take for explaining consumer choice in SP or RP models. Michael and Becker (1973) use the household production framework to create duo-attribute commodities that are the source of direct household utility. Each commodity is produced using a good purchased in the market and some of a households own time. Hence, the relative goods (time) intensity of commodities varies.

If respondents are randomly chosen, the unknown individual specific component can be interpreted as a random disturbance term.

The probability of an individual choosing the mth alternative is

$$P(m|C,\beta) = P(X_m\beta + \varepsilon_{im} > X_j\beta + \varepsilon_{ij} \quad \forall j \in C \& j \neq m), \qquad (1)$$

where C denotes the choice set. In the case of independently and identically distributed extreme value disturbances, the probability of an individual choosing the mth alternative can be expressed as the following:

$$\mathrm{P}(m|\mathrm{C},\beta) = \frac{\exp(X_m\beta)}{\sum_{j\in C}\exp(X_j\beta)}.^{5} \qquad (2)$$

In this setup, price is considered to be a product characteristic, and WTP estimates are derived by finding the price difference that is needed to make the consumer indifferent between products with different characteristics. Thus, the WTP for a product attribute j can be calculated as:

$$WTP_j = -\frac{\beta_j}{\beta_p} \qquad (3)$$

where β_j is the estimated marginal utility for product attribute j; β_p is the estimated marginal utility for price. In this model interactive variables are used in conjunction with the price variable; and the WTP in equation (3) represents the baseline individual (see Lusk and Schroeder, 2004).

3.3. *Practical considerations for surveys*

All surveys should be carefully designed and pre-tested. Discussing the topic and understanding the distribution of WTP with focus groups will improve the quality and effectiveness of the survey. The researcher should make sure that the issue in question is a factor in

[5]This distributional assumption is common in this area, but since estimation is by maximum likelihood, the results are likely to be sensitive to deviations of the distribution form the assumed one.

the consumers' purchase decision and can check whether bids/prices will likely cover the WTP distribution. Undertaking a pilot study to pretest the survey is also highly recommended. The pretest survey does not need to necessarily be a representative sample. Rather, it can address several potential concerns upfront before the actual survey is conducted. One should assess whether the number of questions and length of the survey optimal? If a survey is too long, survey fatigue will set in and subjects will stop trying to answer questions carefully or make good choices. Also a pretest survey can check for errors or misunderstandings in the directions or questions. The pretest survey can also help the researcher to calibrate the bids or set the offered prices in the conjoint-based choice experiment.

In the CV method, the researcher should calibrate the bids to cover the WTP distribution that was presented in Figure 1. For example, in the double-bounded CV data, there should be consumers who fit into each of the partitioned segments of the WTP distribution. With GM foods in certain parts of the world, the vast majority of people will reject even highly discounted bids. For example, McCluskey *et al.* (2003) found that over 80% of their Japanese sample were in the "no, no" partition of the WTP distribution. Based on the data, the mean required discount to be willing to purchase their sample was outside of the highest discount offered (50%). Since mean WTP estimates are obtained from an extrapolation of the WTP function, the level of uncertainty increases as one gets further outside the data. With an extrapolation, the confidence intervals around the mean WTP estimate must be wider than otherwise would be necessary. Ideally, when calibrating random discounts (premiums), the largest discount (premium) should be large enough so that most consumers will accept (reject) the bid offered.

It is good practice where possible to set the initial bid equal to the average price of the corresponding or similar product that is available in local retail stores. This would provide an anchoring effect that might be useful. For example, if one were evaluating the WTP with GM potatoes, the initial bid could be the average price for 5 pounds (a standard unit of sale) at the area stores. The follow-up bid, which is a premium or discount contingent up the response to the initial bid, can be based on percentage increases or decreases to this initial bid. It is good practice to have at least four versions of random discounts or premium bids, but it can be estimated with fewer.

In planning for the survey, the researcher needs to choose a sample that is representative of population under study. The potential bias of greatest concern in this study is the population choice bias, in which the population chosen does not adequately correspond to the population that will purchase the product. The researchers should collect demographic data and compare to census data for the population they are trying to represent.

4. Discussion of Lab Experiments

Lab experiments provide a controlled environment to test hypotheses about consumer behavior. Investigators have the opportunity to design the institutional structure, develop a special protocol for the experiment, choose their sample, and determine the allocation of subjects to control groups and treatment groups so as to identify behavior that might inform policy (List, 2009).

In addition, the researcher must balance subject control and reality of the context (Lusk and Shogren, 2007). *Control* here means the researcher has control over the environment, or institutional structure, such that no unwanted external force affects subjects' choices. *Context* implies that subjects have some contextual cues about why their decisions might matter in a bigger world. For example, experimental auctions that collect elicited values for real goods in lab settings are context rich and have good subject control (Lusk and Shogren, 2007, p. 15). In contrast, field experiments have very good context but are weak on subject control.

4.1. *Treatment effects in experiments*

Laboratory experiments offer researchers options that are not available to those who use SP or large secondary data sets. With lab experiments, unique data can be collected in private information revealing experimental auction markets. Participants typically report WTP under both the "treatment" and "control," i.e., a within-person statistical experimental design where each subject bids on a set of products without product information (the control) and a second time after information about product attributers have been injected into the experiment (the treatment). Such within-subject comparison is impossible in scanner or survey data. This is especially important when researchers are attempting to evaluate products that are new or that have one or more new attributes, e.g., food safety, genetically modification, and organic, relative to conventional versions of the products.

List (2008a, 2009) emphasizes that randomization of assignment of subjects to the control and treatment (including cases of multiple treatment levels) becomes the instrumental variable (IV) that identifies causal effects across subjects (or trials where there are multiple treatment levels). To see this, consider a given commodity, the (information) treatment effect for the ith participant is $\tau_i = E$ $(WTP_{iT} - WTP_{iC})$, where T refers to treatment and C to control and E is the expectation operator. The average treatment effect (ATE) over all participants is then $\hat{\tau} = \overline{WTP_T} - \overline{WTP_C}$, where $\overline{x}$ represents the sample mean of x over all participants. With random assignment of treatment (control) to subjects, unobservable characteristics of subjects are balanced between subjects assigned to the treatment and control. It is the random assignment of subjects to treatment that becomes the instrumental variable permitting identification of causal treatment effects in subjects' WTP for food products. This experimental design generates data on the counterfactual, and makes the treatment effect easy to estimate — the gold standard in experimental designs (List 2009).

In non-experimental settings, estimating treatment effects is more difficult because the subjects assigned to the treatment and control groups are different. Moreover, the two groups can be expected to differ in both observable and unobservable characteristics, which lead to selection bias. Selection bias on observables can be modeled and controlled using propensity score matching (Rosenbaum and

Rubin, 1983) or Heckman's two-step estimator, which creates an instrumental variable that controls for selection into treatment (Heckman, 1979, 1980). Correction for selection on unobservable characteristics is more difficult, but it sometimes can be achieved by differencing a linear model to eliminate individual fixed or random effects. For example, consider a linear model of subject i's WTP including an unobservable subject-specific fixed effect a_i that does not differ across trials or time for subject i. Now the model for WTP_i pre-information treatment can be subtracted from the model for WTP_i post-information treatment. Since a_i is the same in these two equations, differencing removes them. Then the change in WTP can be regressed on the change in information (see for example, Colson *et al.*, 2011 and McFadden and Huffman, 2017).

Results from lab experiments have been criticized for sometimes using undergraduate students from a college class (List, 2009) when the population of interest is, for example, adults 18 to 65 years of age. Experiments involving new food products have been conducted using college students as subjects (Buhr *et al.*, 1993; Fox *et al.*, 1994; Hayes *et al.*, 1995; Fox *et al.*, 2002). However, college students were not the primary population of interest. Thus, the results are not useful to extrapolate for potential marketing strategies of a wider consumer population. Findings have greater credibility when participants are drawn from a random sample of the population of interest. For example, Menkhaus *et al.* (1992); Noussair *et al.* (2004); Rousu *et al.* (2007); Colson *et al.* (2011), Grebitus *et al.* (2013); and McFadden and Huffman (2017) chose their subjects from a random sample of adult consumers in a study area.

Lab experiments allow researchers to examine how contextual frames affect behavior, identify and measure breakdowns from rationality, explore how people construct preferences and beliefs, and determine reactions to new information. Hence, lab results represent real evidence about how certain people will behave in a given economic environment. The expectation is that this information is valuable in predicting behavior in the market and under new policy regimes.

4.2. *Regulation of human subjects and deception in experiments*

A major advantage of lab experiments, e.g., auctions, relative to stated preference surveys is they contain incentives for individuals to reveal their true preferences for goods. Winners in auctions must execute winning bids and stated preference evaluations are hypothetical. However, interesting research questions may arise before new products have been developed, tested, and completely de-regulated. A challenge that arises is how to minimize deception in obtaining subjects' evaluation of new products. A concern is that an institutional review board (IRB) might not approve the experiments, or even if they approve them, a journal reviewer might use deception as a justification for rejecting a manuscript.

Rousu *et al.* (2015) examine the cases for and against deception in applied economics research, especially economic experiments. They list examples of published papers by agricultural economists that have used some form of deception, and furthermore, they describe some of the ways in which deception might be used successfully in research. In addition, Colson *et al.* (2016) report on a survey of scholars engaged in applied economics research to obtain rankings of severity of deception of various forms. They identify the most severe forms of deception as cases where subjects: 1) are deceived in a way that causes physical or physiological trauma, 2) are promised payments for participating in an experiment but receive nothing, or 3) leave an experiment with a different product than what they thought that they had won. Deception of lesser consequences are identified as: 1) providing subjects with no or misleading information about the main purpose of a study; 2) failing to inform subjects that they are part of a study; 3) using confederates or bogus devices[6]; and 4) providing subjects with incomplete information about a product for sale in an experiment.

[6]Confederates are subjects who appear to be participating in a study but are actually following the experimenter's instructions. Bogus devices do not function as one would expect.

List (2008b) emphasizes that a strict adherence to a no-deception policy is contrary to being able to conduct and publish research on important and interesting topics. For example, in some cases the research can only be undertaken if deception is used, e.g., testing for discrimination by employers against resumes of individuals by gender- or ethnic-associated names. He concludes that the benefits and costs of deception (and informed consent) should be carefully considered in each situation and blanket rules against deception in economic experiments should be avoided. His conclusion notwithstanding, the *American Journal of Agricultural Economics* (*AJAE*) imposed a policy on deception in 2011 — a strict ban on publishing papers that showed even faint deception. However, in 2015, the *AJAE* editors dramatically changed the policy to the following. Authors are now asked to describe the extent and nature of deception involved in submitted papers and to provide justification for deception. The editor then decides whether the deception is warranted or excessive given the research justification.

The U.S. Office of Human Research Protection (2009) does not forbid deception in research, and in some other fields, deception in experiments is common and widely accepted. Psychologists have regularly used deception in their laboratory experiments for more than 100 years (e.g., Solomons, 1987; Sieber *et al.*, 1995), and the use of deception is widely accepted in this field of science. Hertwig and Ortman (2008) report that more than half of the papers published in top social psychology journals in 2002 were based on research that involves some form of deception. Other fields including marketing, nutrition, and medicine publish results regularly from studies that use deceptive (e.g., Nunes and Boatwright, 2004; Wansink, Painer, and North, 2005; Wansink and Cheney, 2005). Also, the Editor at the journal *Food Policy* takes a dim view of a reviewer's claims of deception as a valid argument for rejecting a manuscript (McFadden and Huffman, 2017). When significant physical (and psychological) harm of subjects or non-payment of subjects for project participation are not an issue, a strong case can be made for publishing papers in agricultural and resource economics that have other forms of deception, i.e., the hint of deception should not be used as a reason for rejecting an otherwise scholarly contribution.

A new human subjects' policy for federally funded low-risk project, especially in the social sciences, is going to take effect in January 2018 (Shweder aned Nisbett, 2017). This past January, the federal government issued a memorandum giving universities the opportunity to deregulate vast portions of research in the social sciences, law, and humanities. In particular, the reform of human subjects' policy limits the scope of institutional review boards (IRBs) by exempting low-risk research using human subjects from IRB review. The exempted research activities includes surveys, interviews, and other forms of free communication between researchers and adults, aptitude testing, the observation and recording of verbal behavior in public places, benign behavioral interventions, including ordinary psychological and economic experiments, and other low-risk projects and research procedures. Moreover, the decision as to whether a project is exempt is left to the investigator, but she will most likely be guided using a plainly defined self-administered exemption tool. University research administrators are at liberty to exempt all low-risk research, not just federally funded research.

5. Conclusions

There are tradeoffs to using either stated preference (SP) or experimental auctions, a revealed preference (RP) method, to estimate consumers' preference for goods made with new technologies. The criticisms of SP techniques center on the hypothetical bias (HB). The major advantages include that SP techniques are relatively inexpensive, they can offer situations that more closely resembles purchase situations, they can be used to assess the market potential products that do not yet exist, and deception is not needed. This means that even with HB, the results from SP methods can provide timely and valuable information. Population choice is extremely important if the results are meant to be extrapolated to a larger population. Hence, understanding the tradeoffs can be useful in choosing the best method for a particular research project.

Both SP and RP methods have a role in evaluating new food products. In particular, SP surveys offer a low-cost, timely method to obtain information consumers' preferences for new food products that are on the drawing board. However, the protocol must be carefully developed and pre-tested in order to provide useful and generalizable information. Although subjects who have too much freedom in expressing their preferences for or against a new food product tend to express divergent views, SP researchers should not use cheap talk script telling subjects that hypothetical valuations are likely to be too (low) high. Lusk *et al.* (2004), McCluskey *et al.* (2003) and Vossler *et al.* (2012) provide methods and examples of the value of SP approaches in food research.

RP experiments provide an ideal method for greater in-depth information on consumers' acceptance and WTP for new food products. However, it is necessary that the new food product must be at a more advanced than for SP surveys, i.e., at a stage where at least prototype products are available in sufficient quantity for displace as in an auction market experiments. The greater time commitment of subjects to lab experiments makes RP experiments much more expensive to undertake, but they provide the opportunity for investigators to collect data on social-demographic attributes of subjects and to inject randomly assigned information into the experimental setting. The incentive compatible bids from these experiments can be differenced in a way to eliminate individual fixed or random effects, thereby revealing higher quality information on consumer's preference for new food products, e.g., see Huffman *et al.* (2007), Rousu *et al.* (2007), and McFadden and Huffman (2017).

There is perhaps no more important question for researchers working with SP and RP methods than whether they actually predict behavior in the field, i.e., grocery store shopping behavior. Chang *et al.* (2009) examined data from hypothetical choices and auction experiments to see which more accurately reflected shopping behavior in a grocery store. They concluded RP methods are a better approximation of "true" preferences of shoppers than are SP methods. One reason why research results from both methods might not reproduce real shopping decisions is that the subjects in the

RP and SP research may differ significantly from actual grocery store shoppers. Another reason is that the breadth of concerns of grocery store shoppers extends far beyond that of any well designed research project. The context is rich for in-store shopping, but there is little-to-no control over the environment or attention of the individuals. However, new research projects that attempt to calibrate the coincidence of evidence from SP and RP methods for new food products seem a misplacement of resources and overlook the comparative advantage of different methods.

However, results from economic experiments are more likely to uncover anomalies in consumer behavior. The economic profession has had insightful scholars who have attempted to explain them by making thoughtful modifications to traditional models, e.g., the work of Becker (1992). In addition, a new field of behavior economics has spring up to form a more new perspective on some of the anomalies. However, the likely long-term impact of this field on SP and RP research remains to be seen.

Federal human subjects' policy developed in the early 1970s was to protect individuals from risks associated with biomedical and pharmaceutical experiments, but not for the social sciences, law and humanities. However, institutional review boards on university campus took the opportunity to expand greatly its reaches over the next forty years. However, in January, 2018, low-risk research will be deregulated by the federal government and not subject to review by the IRB. This will be a great relief to many researchers in the social sciences, including economics.

Acknowledgments

The authors are C.F. Curtiss Distinguished Professor of Agriculture and Life Sciences and Professor of Economics, Iowa State University, and Distinguished Professor of Sustainability, School of Economic Sciences, Washington State University. We thank two reviewers and the Editor for assistance. We acknowledge support from the Iowa and Washington State Agricultural Experiment Stations.

References

Adamowicz, W.L., J. Louviere and M. Williams (1994). Combining Revealed and Stated Preference Methods for Valuing Environmental Amenities. *Journal of Environmental Economics and Management* 26, 271–292.

Adamowicz, W.L. and J.D. Swait (2011). "Discrete Choice Theory and Modeling." In J.L. Lusk, J. Roosen, and J.F. Shogren (Eds.), *Oxford Handbook of The Economics of Food Consumption and Policy.* Oxford, England: Oxford University Press, pp. 119–152.

Adamowicz, W.L, J.D. Swait, P. Boxall, J. Louviere and M. Williams (1997). Perceptions vs. Objective Measures of Environment Quality in Combined Revealed and Stated Preference Models of Environmental Valuation. *Journal of Environmental Economics and Management* 32, 65–84.

Becker, G.S. (1992). The Economic Way of Looking at Life. Nobel Lecture, Stockholm, Sweden. Available at: http://www.nobelprize.org/nobel_prizes/economic-sciences/laureates/1992/becker-lecture.pdf

Buhr, B.L., D.J. Hayes, J.F. Shogren and J.B. Kliebensterin (1993). Valuing Ambiguity: The Case of Genetically Engineered Growth Enhancers. *Journal of Agricultural and Resource Economics* 18, 175–184.

Cameron, T. and J. Quiggin (1994). Estimation Using Contingent Valuation Data from a 'Dichotomous Choice with Follow Up'? *Journal of Environmental Economics and Management* 27, 218–234.

Carson, R.T. and T. Groves (2007). Incentive and Informational Properties of Preference Questions. *Environmental and Resource Economics* 37, 181–210.

Carson, R.T. and T. Groves (2011). "Incentive and Information Properties of Preference Questions: Commentary and Extensions." In J. Bennett (Ed.), *International Handbook of Non-Market Environmental Valuation, Northampton.* MA: Edward Elgar.

Carson, R.T., T. Groves and J.A. List (2014). Consequentiality: A Theoretical and Experimental Exploration of a Single Binary Choice. *Journal of the Association of Environmental and Resource Economists* 1, 171–207.

Chang, J.B., J.L. Lusk and F.B. Norwood (2009). How Closely do Hypothetical Surveys and Laboratory Experiments Predict Field Behavior. *American Journal of Agricultural Economics* 91, 518–534.

Chien, Y., C.J. Huang and D. Shaw (2005). A General Model of Starting Point Bias in Double-bounded Dichotomous Choice Contingent Valuation Surveys. *Journal of Environmental Economics and Management* 50, 362–377.

Colson, G., J.R. Corrigan, C. Grebitus, M.L. Loureiro and M.C. Rousu (2016). Which Deceptive Practices, If any, Should be Allowed in Experimental Economics Research? Results from Surveys of Applied Experimental Economists and Students. *American Journal of Agricultural Economics* 98, 610–621.

Colson, G.J., W.E. Huffman and M.C. Rousu (2011). Improving the Nutrient Content of Food through Genetic Modification: Evidence from Experimental Auctions on Consumer Acceptance. *Journal of Agricultural and Resource Economics* 93, 358–363.

Cornes, R. and T. Sandler (1996). *The Theory of Externalities, Public Goods and Club Goods*, 2nd Edn. New York, NY: Cambridge University Press.

Corrigan, J.R., D.P.T. Depositario, R.M. Nayga, Jr., X. Wu and T.P. Laude (2009). Comparing Open-Ended Choice Experiments and Experimental Auctions: An Application to Golden Rice. *American Journal of Agricultural Economics* 91(3), 837–853.

Fox, J.A., D.J. Hayes, J.B. Kliebenstein and J.F. Shogren (1994). Consumer Acceptability of Milk from Cows Treated with Bovine Somatotropin. *Journal of Dairy Science* 77, 703–707.

Fox, J.A., D.J. Hayes and J.F. Shogren (2002). Consumer Preferences for Food Irradiation: How Favorable and Unfavorable Descriptions Affect Preferences for Irradiated Pork in Experimental Auctions. *Journal of Risk and Uncertainty* 24, 75–95.

Grebitus, C., J.L. Lusk and R.M. Naga (2013). Effect of Distance of Transportation on Willingness to Pay for Food. *Ecological Economics* 88, 67–75.

Grimsrud, K.M., J.J. McCluskey, M.L. Loureiro and T.I. Wahl (2004). Consumer Attitudes toward Genetically Modified Food in Norway. *Journal of Agricultural Economics* 55(1), 75–90.

Grimsrud, K.M., J.M. McCluskey and M.L. Loureiro (2004). Policies and Attitudes toward Genetically Modified Foods in Norway. *EuroChoices* 3, 38–45.

Hanemann, W.M. (1984). Welfare Evaluations in Contingent Valuation Experiments with Discrete Responses. *American Journal of Agricultural Economics* 66, 332–341.

Hanemann, W.M., J. Loomis and B.J. Kanninen (1991). Statistical Efficiency of Double Bounded Dichotomous Choice Contingent Valuation. *American Journal of Agricultural Economics* 73, 1255–1263.

Hanemann, W.M. and B. Kanninen (1999). "The Statistical Analysis of Discrete-Response CV Data." In I.J. Bateman and K.G. Willis (Eds.) *Valuing Environmental Preferences: Theory and Practice of the Contingent Valuation Method in the US, EU, and Developing Countries.* Oxford: Oxford University Press, pp. 302–442.

Hayes, D.J., J.F. Shogren, S. Youll Shin and J.B. Kliebenstein (1995). Valuing Food Safety in Experimental Auction Markets. *American Journal of Agricultural Economics* 77, 40–53.

Heckman, J.J. (1979). Sample Selection Bias as a Specification Error. *Econometrica* 47, 153–161.

Heckman, J.J. (1980). "Sample Selection Bias as a Specification Error with an Application to the Estimation of Labor Supply Functions." In J.P. Smith (Ed.), *Female Labor Supply: Theory and Estimation.* Princeton, New Jersey: Princeton University Press, pp. 206–248.

Herriges, J.A. and J.F. Shogren (1996). Starting Point Bias in Dichotomous Choice Valuation with Follow-up Questioning. *Journal of Environmental Economics and Management* 30, 112–131.

Hertwig, R. and A. Ortman (2008). Deception in Experiments: Revising the Arguments and Its Defense. *Ethics and Behavior* 18, 59–92.

Huffman, W.E., M. Rousu, J.F. Shogren and A. Tegene (2007). The Effects of Prior Beliefs and Learning on Consumers' Acceptance of Genetically Modified Foods. *Journal of Economic Behavior and Organization* 63, 193–206.

Jeon, Y. and J.A. Herriges (2010). Convergent Validity of Contingent Behavior Response in Models of Recreation Demand. *Journal of Environmental and Resource Economics* 45, 223–250.

Kanninen, B.J. (1993). Optimal Experimental Design for Double-Bounded Dichotomous Choice Contingent Valuation. *Land Economics* 69, 138–146.

Kanninen, B.J. and M.S. Khawaja (1995). Measuring Goodness of Fit for the Double-Bounded Logit Model. *American Journal of Agricultural Economics* 77, 885–890.

Kleckner, N.S., C.G. Leggett, K.J. Boyle, J. Duffield and R.C. Mitchell (2002). "Evidence of Social Desirability Bias in Contingent Valuation Survey Administered Through In-Person Interviews." Paper presented at the 2nd World Congress of Environmental Economics, Monterey.

Ladd, G.W. and V. Suvannunt (1976). A Model of Consumer Goods Characteristics. *American Journal of Agricultural Economics* 58, 504–510.

Lancaster, K. (1996). A New Approach to Consumer Theory. *The Journal of Political Economy* 74, 132–157.

Ledford, H. (2013). Hunt for Mystery GM Wheat Heats UP. *Nature* 499, 263–264.

List, J.A. (2008a). *Homo Expereimentalis* Evolves. *Science* 321, 207–208

List, J.A. (2008b). Informed Consent in Social Science: Response. *Science* 322, 672.

List, J.A. (2009). An Introduction to Field Experiments in Economics. *Journal of Economic Behavior and Organization* 70, 439–442.

List, J.A. and C.A. Gallet (2001). What Experimental Protocol Influence Disparities between Actual and Hypothetical Stated Values? *Environmental and Resource Economics* 20, 241–254.

Loureiro, M.L., J.J. McCluskey and R.C. Mittelhammer (2003). Are Stated Preferences Good Predictors of Market Behavior? *Land Economics* 79(1), 44–55.

Louviere, J.J. (1993) "Conjoint Analysis." In R. Bagozzi (Ed.), *Handbook of Marketing Research.* New York: Academic Press.

Lusk, J.L. (2013). *Food Police: A Well-Fed Manifesto about the Politics of Your Plate.* New York: Crown Forum.

Lusk, J.L. and J.F. Shogren (2007). *Experimental Auctions: Methods and Applications in Economic and Marketing Research.* New York, NY: Cambridge University Press.

Lusk, J.L. and T.C. Schroeder (2004). Are Choice Experiments Incentive Compatible? A Test with Quality Differentiated Beef Steaks. *American Journal of Agricultural Econonomics* 86, 467–482.

McCluskey, J.J., K.M. Grimsrud, H. Ouchi and T.J. Wahl (2003). Consumer Response to Genetically Modified Food Products in Japan. *Agricultural and Resource Economics Review* 32, 222–231.

McFadden, J.R. and W.E. Huffman (2017). Consumers' Acceptance of Food Safety Achieved Using Biotechnology: Evidence from New Potato Products. *Food Policy* 69, 214–232.

McFadden, D. (1974). "Conditional Logit Analysis of Qualitative Choice Behavior." In P. Zarembka (Ed.), *Frontiers in Econometrics*. New York: Academic Press, pp. 105–142.

McMaster, H.S., C.A. Leard, S. Mann and D.A. Dillman (2017). An Experimental Comparison of Web-Push vs. Paper-only Survey Procedures for Conducting an in-depth Health Survey of Military Spouses. *BMC Medical Research Methodology* 17, 73.

Menkhaus, D.J., G.W. Bordern, G.D. Whipple, E. Hoffman and R.A. Field (1992). An Empirical Application of Experiment Economics in Marketing Research. *Journal of Agricultural and Resource Economics* 17, 44–55.

Michael, R. and G.S. Becker (1973). On the New Theory of Consumer Behavior. *Swedish Journal of Economics* 75, 378–395.

Noussair, C., S. Robin and B. Ruffieux (2004). Do Consumers Really Refuse to Buy Genetically Modified Food? *Economic Journal* 114, 102–120.

Nunes, J.C. and P. Boatwright (2004). Incidental Prices and Their Effect on Willingness to Pay. *Journal of Marketing Research* 41, 457–466.

Office for Human Research Protection (2009). "Protection of Human Research Subjects," Code of Federal Regulation: Title 34, Part 46. Available at: http://www.hhs.gov/ohrp/regulations-and-policy/regulations/45-cfr-46/index.html

Rosenbaum, P.R. and D.B. Rubin (1983). The Central Role of the Propensity Score in Observational Studies for Causal Effects. *Biometrika* 70, 41–55.

Rousu, M.C., G. Colson, J.R. Corrigan, C. Grebitus and M.L. Loureiro (2015). Deception in Experiments: Towards Guidelines on use in Applied Economics Research. *Applied Economic Perspectives and Policy* 37, 524–536.

Rousu, M., W.E. Huffman, J.F. Shogren and A. Tegene (2007). Effects and Value of Verifiable Information in a Controversial Market: Evidence from Lab Auctions of Genetically Modified Food. *Economic Inquiry* 45, 409–432.

Schultz, T.W. (1975). The Value of the Ability to Deal with Disequilibria. *Journal of Economics Literature* 13, 827–846.

Shogren, J.F., M. Margolis, C. Koo and J. List (2001). A Random n-th Price Auction. *Journal of Economic Behavior and Organization* 46, 409–421.

Shweder, R.A. and R.E. Nisbett (2017). Long-Sought Research Deregulation is Upon Us. Don't Squander the Moment. *The Chronicle of Higher Education.*

Sieber, J.E., R. Iannuaao and B. Rodriguez (1995). Deception Methods in Psychology: Have They Changed in 23 Years? *Ethics & Behavior* 5, 67–85.

Solomons, L. (1897). Discrimination in Cutaneous Sensations. *Psychological Review* 4, 246–250.

Taylor, L.O. and R.G. Cummings (1999). Unbiased Value Estimates for Environmental Goods: A Cheap Talk Design for the Contingent Valuation Method. *American Economic Review* 83, 649–665.

Vickrey, W. (1961). Counter Speculation, Auctions and Competitive Sealed Tenders. *Journal of Finance* 16(1), 8–39.

Von Haefen, R.H. and D.J. Phaneuf (2008). Identifying Demand Parameters in the Presences of Unobservales: A Combined Revealed and Stated Preference Approach. *Journal of Environmental Economics and Management* 56, 85–105.

Vossler, C.A. (2016). Chamberlin meets Ciriacy-Wantrup: Using Insights from Experimental Economics to Inform Stated Preference Research. *Canadian Journal of Agricultural Economics* 64, 33–48.

Vossler, C.A., M. Doyon and D. Rondeau (2012). Truth in Consequentiality: Theory and Field Evidence on Discrete Choice Experiments. *American Economic Journal: Microeconomics* 4, 145–171.

Wansink, B. and M.M. Cheney (2005). Super Bowls: Serving Bowl Size and Food Consumption. *Journal of the American Medical Association* 293, 1727–1728.

Wansink, B., J.E. Painter and J. North (2005). Bottomless Bowls: Why Visual Cues of Portion Size May Influence Intake. *Obesity Research* 13, 93–100.

Chapter 3

Labeling of Genetically Modified Foods*

Wallace E. Huffman† and Jill J. McCluskey‡

†*Department of Economics, Iowa State University, Ames, Iowa, USA*

‡*Resource Economic Division, ERS, USDA, Washington, DC, USA*

1. Introduction

The introduction of genetically modified (GM) crops to world markets has created a new division between the crop trading countries. Despite claims from developers, scientists and official institutions that GM products are safe, consumer attitude towards GM foods are largely negative in many developed countries, especially in the European Union and Japan (Lusk *et al.*, 2005). Consumer skepticism is usually attributed to the unknown environmental and health consequences of genetically modified crops. Such consequences include, but are not limited to, unanticipated allergic responses, the spread of pest resistance or herbicide tolerance to wild plants, and inadvertent toxicity to wildlife. Further, many are concerned with the ethical dimensions of biotechnology. The United States is an exception to this rule, where consumers are largely indifferent about GM foods (Nelson, 2001; Chern *et al.*, 2002). Studies conducted

*This article originally appeared in the **Handbook on Agriculture, Biotechnology and Development (2014), 467–485**. This is published by permission of **Edward Elgar Publishing**.

We thank our former graduate students for the collaboration that they provided in the study of consumer responses to biotechnology and information. We also thank the Iowa State Agricultural Experiment and the Washington State Agricultural Experiment Station for Financial Support.

in lesser developed countries (LDCs) find that consumer attitude towards GM foods are less negative and in many cases positive (Li *et al.*, 2002; Subrahmanyan and Cheng, 2000).

The agbiotech industry cites numerous benefits of GM crops, including food availability, nutrition enhancement and economic advantages; importantly, not all of these advantages are attributed to the same crop. The current world population is approximately 6 billion and growing at a rapid pace. There is a need to ensure an adequate food supply for this burgeoning population in the future. GM crops have beneficial characteristics that can be exploited to meet the growing food demand. They can be pest resistant, disease resistant, can tolerate drought and salinity, and can also possess industrial, nutritional or medicinal qualities. Environmental groups host websites, hold demonstration and hand out fliers that claim that GM food is 'FrankenFood', that it is generally harmful to the environment and may be harmful to human health (Greenpeace International, 2001). The media, looking for sound bites, sometimes carries stories about the benefits and risks of GM foods. The result is that the agbiotech industry and environmental groups inject information into the discussion of and markets for GM crops, creating markets with greatly conflicted information. To the extent that consumers are aware of this information, it makes good decision-making more difficult. Food labels may carry information about foods. Owing to a tendency for the food industry to exaggerate claims about their products, the US government has established guidelines on the nature of claims that can be made and the placement of information on food labels for food sold in retail stores.

Developing countries have potentially more to gain from GM foods, since they have more urgent needs in terms of food availability and food security. If there is decreasing marginal utility in calorie consumption per capita, the marginal utility of the last calorie consumed is higher in most developing countries. For example, the yield-enhancing qualities associated with many GM crops are important for countries with large and growing populations, such as china. These governments must recognize that if they are going to continue to feed their own people, they must find more efficient agricultural production methods. Further, in developing countries, small-scale farmers are likely to experience greater yield increases because many currently do not use chemical pest control (Qaim and Zilberman, 2003). Even so, the introduction of GM crops into developing countries has also been controversial (Pinstrup-Andersen and Schiøler, 2001).

Initially, most GM products contained 'first-generation' genetic modifications, which address production considerations and have input traits that may result in public-good benefits, such as reduced pesticide usage. As more 'second-generation' GM products, which include quality traits or product-enhancing attributes such as nutritional benefits, enter the market, consumers may be more accepting of biotechnology in food because of the inherent private benefits. With benefits directly accruing to the consumers, those consumer-friendly GM products could be accepted without discounts or could receive a premium. Golden rice, which by genetic modification produces beta-carotene, a precursor to vitamin A, is one of the emblematic cases of second-generation agbiotech products.

In research aimed at explaining the differences between the United States and Europe, Nelson (2001) concludes that European consumers generally focus on the unknown risks associated with genetically modified products rather than the benefits, whereas US consumers focus neither on the risks nor on the benefits. Nelson argues that European consumers treat the potential harm of GM foods as certain until proven otherwise. This approach is basically the 'Precautionary Principle', which dominates European GM labeling policy. The Precautionary Principle calls for preventive measures to be taken when an activity raises threats even if a direct cause–effect relationship has not been scientifically proven. In its strongest and most distinctive forms, the principle imposes a burden of proof on those who create potential risks, and it requires regulation of activities even if it cannot be shown that those activities are likely to produce significant harms (Sunstein, 2005).

2. Asymmetric Information and Information Quality

GM foods are the result of new developments in crop biotechnology. Uncertainties always exist with new technologies, and agbiotech is no different. Initially some groups raised concerns about the long-term impact of growing GM crops on the environment, including biodiversity. Others raised health and ethical issues. In contrast, the biotechnology industry advertised these crops as the best thing that had happened in food production since hybrid corn.

The markets for new GM crops and foods are ones with contradictory information. The agbiotech industry generally provides information bestowing the benefits of the technology while environmental groups claim a wide range of potential harm to the environment and human health. However, these two groups are interested parties in the debate over genetically modified organisms (GMOs). Since the release of GM crops in the mid-1990s, additional research has been undertaken to identify the benefits and costs, including some of the risks. Information from this research has reduced the uncertainties about GM crops and foods. The presence of GMOs in the US food system since the mid-1990s with no significant health impacts reveals that short-term harm or damages are not significant. The distribution of information about GM crops has been aided by advances in electronic information networks, which transmit information at approximately zero marginal cost.

In contrast to public information which is freely available to everyone, private information, which is held by a particular individual or group, is the source of asymmetric information (Molho, 1997). In two-party interactions in which one party possesses private information, the informed party can be expected to exercise an informational advantage whenever it can expect to gain from it (Molho, 1997). When experienced parties develop intuition about situations potentially leading to opportunistic behavior, asymmetric information can destroy opportunities for welfare-improving exchanges (Akerlof, 1970).

Many decision makers often rely on information provided by individuals or groups who are affected by their decisions. Furthermore, these decision makers may not know all of the alternatives available and have no control over the information provided to them by interested parties. Furthermore, these interested parties may manipulate by distorting or concealing information. For example, farmers rely on information, including advertising, provided by biotechnology companies about the expected performance of new GM crop varieties, which is almost certainly selective. However, these biotechnology firms are constrained by US regulatory policies on GMOs (Vogt and Parish, 1999) and by their interest in repeat sales. Even so, it is widely believed that their communications do not accurately reflect the complete picture. Likewise, communications by anti-GM groups may exaggerate the potential damages to the environment, human health and organic producers, and ignore the benefits, including the dramatic reduction in use of environmentally toxic insecticides.[1]

Environmental groups and organic producers are interest groups. Individuals self-select into these groups because of a common interest or goal focused on the environment, and achieving the group's goal is a public good to its members. Each of these groups has resources — largely members' time, financial contributions and voting power — and their impact is affected by organizational efficiency. By choosing relatively narrow objectives, these groups reduce coordination and decision making costs over organizations that have diverse goals. Free-riding is, however, a major organizational problem (Olson, 1965; Sandler, 1992; Cornes and Sandler, 1996). Free-riding arises when an individual relies on the efforts of others to attain a goal, and, in the extreme, no one exerts any effort toward that goal.

Advances in communication and information technologies have greatly reduced the organizational costs of interest groups and have undoubtedly increased their productivity. They are now able to construct low-cost internet sites for displaying their objectives, news releases, short articles, and other information. For example, during the late 1990s the website of Friends of the Earth was used to display the locations or addresses of field testing of GM crops in England (Huffman and Tegene, 2002). These groups can also use email to rapidly distribute communications among members, for

[1]Over the past decade, organic producers in the United States have become increasingly opposed to GM agriculture and would like to see a ban on GM crops (Huffman and Strzok, 2012).

example dealing with demonstrations, or letters opposing particular GMO issues and policies.

Consider the situation where there are two interested parties providing information in an attempt to sway the decisions of producers and consumers of GM foods. The particular capsule of information is a signal, which reflects the self-interest and private information of the sending party. Sending information is inexpensive because it requires little action on the part of the sender, and new information technologies, such as email and websites, have reduced the cost and greatly increased the speed of delivery, although there remains some modest fixed cost of preparing a communication. Thus, the marginal cost of distributing information has become approximately zero.

Misinformation can be distributed as easily as positive information: computer viruses are just one example of misinformation. The diversity of human interactions has also been greatly expanded with these new information technologies. Individuals now communicate with others whom they do not know personally and with whom they have greatly different norms and values. Hence, new issues arise in accessing information quality. Since signaling with communications is so cheap, one possible outcome is that they degrade the expected quality of information to the point that consumers ignore communications from interested parties.

Although the marginal cost of distributing information is approximately zero, it remains costly for decision makers to interpret this information, especially contradictory information. Consumers and agricultural producers, however, differ in the long-term consequences of using incorrect information. Consumers maximize utility subject to a resource constraint. When they use wrong information in their decision making, their welfare decreases, but this does not generally cause them to die. In contrast, producers, who can be described as long-run profit maximizers, will suffer reduced profits when they use misinformation. If profits become negative over the long term, they will go bankrupt and be forced to exit the industry. This is one dimension on which producers face stronger incentives to scrutinize information about GM technology compared with consumers. For consumers and producers, the ability to process information and make good decisions is a valuable skill in the sense that it can be welfare or profit increasing, and this ability seems most likely to be related to their years of schooling and accumulated experience as decision makers (Huffman, 2001). If decision makers recognize their vulnerability to a particular source of private information, they may choose to collect information from several interested parties possessing different points of view in order to reach a more informed decision. There are potential costs and benefits of these actions.

Milgrom and Roberts (1986) examine decision making in an environment where only private information is available to decision makers, such as consumers. They have shown that it is possible for decision makers to make fully informed decisions when they rely on one interested party for information but the necessary conditions seem quite restrictive and are unlikely to be fulfilled. For example, the interested parties' preferences must be known to the decision maker, the information must be costlessly verifiable, the decision maker must know the facts about which interested party has information to be able to detect situations in which information is being withheld, and the decision maker must be able to draw accurate inference when information is withheld (sometimes referenced as being sophisticated). These conditions are demanding, but perhaps they are not impossible attributes for successful producers to possess. They may exceed the capacities of most food consumers. The implication is that if producers or consumers rely only on information provided by biotechnology supply firms or environmental groups, this is unlikely to lead to fully informed decisions. Hence, good reasons exist for society to be skeptical of claims about GMOs made by both agbiotech firms and environmental interest groups.

Rather than obtaining information from only one interested party, Milgrom and Roberts (1986) show that there are advantages for decision makers from having competition among interested parties who provide information. In this situation, the number of restrictions or assumptions needed to attain good decision making are greatly reduced. However, the interested parties must be able to convey their information to the decision maker, and the decision maker must listen to all interested parties who want to convey information. In essence, there must be an opportunity for different interests to come out or to be revealed.

Agbiotech firms, environmental groups and other interested parties differ in their ability and effectiveness in conveying information that they have. This ability might be associated, among other things, with training, communication skills, personalities, organizational philosophies, and information technologies. For the last eight years rapid advances in communication and information technologies have occurred. With the associated dramatic fall in the cost of sending messages and storing information through new networks, accessibility to these technologies seems minimally constrained by capital or credit. Biotechnology companies are primarily private companies interested in long-term profits associated with the sales of their products and the value of their company, and the information that they distribute can be expected to be consistent with this long-term objective and to be constrained by it. Environmental groups, in contrast, are primarily pursuing non-monetary goals which seem less constraining on their actions and on the objectivity of their information.

With competition among interested parties in providing information, decision makers can be unsophisticated, having little or no idea of available options and issues bearing on the decision or preferences of the interested parties. He or she must, however, be able to process the information effectively that he or she receives, and the information must be verifiable. Verifiability means that the truthfulness of information can somehow be checked against a reliable source. Under these conditions, fully informed decisions are possible.

Much information is being distributed about GM food, but it is not verifiable. Although knowledge about biotechnology has increased significantly over the past fifteen years, some effects and impacts remain unknown. A coalition of anti-GM interests, for example, Greenpeace, Friends of the Earth, Action id, remains in place, more strongly in Europe than in the United States, but they continue to have the objective of slowing the acceptance of GM technology. These groups continue to raise new questions about short- and long- term effects on the environment from using GMOs and consuming GM food. Some of the activities of the anti-GM groups are not beyond disrupting GM field experiments that could lead to advances in the stock of knowledge about GMOs. In response, UK scientists have sent an open letter to these NGO activists asking them not to destroy their GM wheat field trials (Science Media Center, 2012).

Second-generation GM products, including functional foods, are becoming available. They are primarily credence goods (that is consumers cannot detect GM ingredients even after consuming the product). In addition, the food industry may find itself inclined to make health claims about these foods into a marketing device. Policy makers then face significant challenges arising from new claims because of information asymmetries. Issues considered in developing policies might include factors influencing consumer preferences for these products and validity of health claims (Veeman, 2002).

When information is not verifiable, the reliability of information provided by an interested party depends on the congruity between the objectives of the decision maker and those of the interested party. When objectives diverge, optimal decision-making is difficult, and these complex problems have been the topic of a range of incentive schemes in "agent" or agency theory literature. These models are, however, well suited only to decision problems with few players.

In these circumstances, communications by interested parties might lead to unduly restrictive public policies being adopted or it might degrade the information content to the extent that sophisticated decision makers ignore it. This will be generally welfare reducing relative to fully informed decision making. For example, social cost-reducing inputs for crop production might not be used by farmers, or socially beneficial GM foods might not be consumed. More generally, long-term delays in adopting GM technologies may occur when it is time consuming to verify or refute claims by the environmental groups about agricultural-biotechnology products. For example, the EU has chosen to apply the 'precautionary principle' of proceeding very slowly in approving GM crops. This will reduce the expected social and private pay-off to research and development (R&D) in this area. This has implications for where the private sector places its future R&D investments.

When a large number of decision makers need the same verified information, research to produce this information provides a public good which may have great social value. After the information is generated, the knowledge is non-rival, in that use by one decision maker does not affect the quantity or quality available to others. Since the opportunity cost of an added user of the information is essentially zero and each user's value of the information is private information, private decision making on how much to produce will lead to under-provision. Unless some organizational device is used to internalize these unrealized externalities, tendencies to free-ride will lead to sub-optimal outcomes (Huffman and Tegene, 2002). The price system is of no aid in extracting information on the social value of verified information or a system to manage it. Hence, consistent with consumer education, independent third-party or verifiable information about agbiotech may have considerable value if available and disseminated to consumers.

Many studies have examined how information affects valuation. For example, Frewer *et al.* (1999) investigated British consumers' response to a number of messages that all stressed the benefits of GM foods but varied in persuasiveness. None of the messages had a significant impact on consumer attitudes. Scholderer and Frewer (2003) tested communication strategies for changing consumers' attitude towards GM foods in an experimental study involving consumer samples from Denmark, Germany, Italy and the United Kingdom. Different information strategies were tested against a control group for their ability to change consumer attitudes. They found that no attitude change occurred. Unlike attitudes, however, consumers' product choices were sensitive to provision of information. All information materials significantly decreased the probability of consumers choosing GM products.

The scientific consensus is that first-generation GM foods are equivalent to their conventional counterparts. On average, as we discuss in later sections, consumers want a discount in order to choose the GM foods over the conventional product. This reminds us that it is the public's *perception* of risks, rather than scientifically proven risks, that directly affect markets.

3. Food Labels

Food products in the United States can include claims on their labels about nutrition, safety, organic status, and whether they contain GM ingredients. The 1990 Nutrition Labeling and Education Act (NLEA) provided new guidelines for nutrition labels on packaged foods sold in US supermarkets; unpackaged foods, for example fresh fruits and vegetables, are not affected (Balasubramanian and Cole, 2002). This law requires packaged foods to display nutrition information prominently in a new label format, namely the Nutrition Facts Panel (NFP). It also standardizes to some degree serving size, health claims (that link a nutrient to a specific disease), and descriptive terms (for example, 'low fat') on food packages. The goal of this legislation was to improve consumer welfare by providing nutrition information that would assist consumers in making healthy food choices.

3.1. *Costs of effective food labeling*

As an indication of the costliness of effective nutrition labeling, it is estimated that the US food industry has spent $2 billion in order to comply with the 1990 Nutrition Labeling Act (Silverglade, 1996). Some attributes, such as enhanced calcium and vitamins A and C, are viewed positively by consumers, in that more is better in the case of positive consumer attributes. Other attributes, such as sodium, fats and pesticide residues, are negative, so less is better. Prior to the Nutrition Labeling Act, the emphasis on negative labeling was more muted. Balasubramanian and Cole (2002) suggest that this tendency can be explained by consumers having an asymmetric value function, weighing a dollar of loss more heavily than a dollar of gain, which illustrates one of the key points of prospect theory (Tversky and Kahneman, 1981).

NLEA policies tend to emphasize negative rather than positive labeling. First, permissible health claims are those that associate specific nutrients with reduced risk of specific diseases. Of the seven health claims approved by the US Food and Drug Administration (FDA) at the onset of the new Nutrition Labeling Act, three linked negative attributes exclusively with deadly diseases, such as dietary fat with cancer, sodium with hypertension, and dietary saturated fat with high cholesterol and heart disease, and only one claim featured a positive attribute, namely calcium and osteoporosis. Later claims have, however, been more balanced. In general, regulations on nutrient-content claims tend to focus more heavily on negative attributes (calories, sugar, sodium, fat, fatty acids and cholesterol) than on positive attributes such as fiber and vitamins.

Clearly, with foods made currently from crop varieties that contain GM input traits, adding a label for GM content would be an example of labeling a negative food attribute. GM content has not been proven scientifically to have human health consequences, except for the transport of some known allergens to new locations. Hence, GM food labels today would not meet the Nutrition Labeling Law requirement of a proven nutrient intake leading to a better health outcome.

GM products used for food do have to pass a food safety test. In 1992, the FDA announced its landmark decision that GM food and food products will be regulated in the same way as those created by conventional means. This policy allows new GM foods to be treated as conventional foods as long as they meet three conditions: their nutritional value has not been lowered; they incorporate new substances that are already a part of the human diet; and they contain no new allergenic substances. In January 2001, the FDA issued a Guidance for Industry reaffirming this policy, which stated that the only GM foods that need to be labeled are foods that have different characteristics from the non-GM version, such as elevated vitamin A levels. In the United States, labeling food for GM content is not otherwise required. Firms must notify the FDA at least four months before putting a new GM food product on the market, and the scientific description of the product is posted on the internet for review during this time (Just *et al.*, 2006). Only minor changes have been made in these guidelines since 2001.

The GM labeling policy in the United States can be classified as being voluntary, but the European Union (EU) imposed mandatory labeling for all foods containing more than 0.9 per cent GM ingredients. In the United States, if a voluntary label is affixed, the FDA has mandated that it cannot use the phrase 'genetically modified'. The FDA defines the term 'genetic modification' as all forms of breeding, genetic engineering and conventional modifications (US FDA, 2001). The FDA prefers the phrase 'genetically engineered' or 'made through biotechnology'. In contrast, the European Commission GM food label policy, adopted in 1997, requires each member country to enact a law requiring labeling of all new products containing substances derived from GM organisms. Japan, Australia and many other countries have also passed laws requiring GM labels for major foods.

Regardless of the approach, effective GM labeling involves costs, especially the costs of testing for the presence of GM content, segregating GM and non-GM products, variable costs of monitoring for truthfulness of labeling and enforcement of the regulations that exist, and risk premiums for being out of contract (Wilson and Dahl, 2005; Roe and Sheldon, 2007).

Effective GM labeling policies require effective segregation of GM from non-GM commodities. If, however, identity preservation through the production, marketing and processing chain were required, this system would be substantially more costly (Wilson and Dahl, 2005). To the extent that there is a market for non-GM products, buyers would be expected to specify in their purchase contracts some limit on GM content and/or precise prescriptions regarding production, marketing and handling processes. One can envision a marketplace of buyers with differentiated demand according to their aversion to GM content. To make this differentiation effective, additional costs and risks are incurred. Additional testing involves costs of conducting the tests, for which there are several technologies of varying accuracy. The risk is that GM products will be commingled with non-GM products, so the detection system must test to see that customers' shipments are within contract limits for GM contamination. This is a serious economic problem, as agents seek to determine the optimal strategy for testing and other risk mitigation strategies. 'Tolerances' are an important issue in segregation and identity preservation. GM tolerance refers to the maximum impurity level for GM content that is tolerated in a product that still carries the non-GM label. There are two levels where tolerances apply: one is defined by regulatory agencies such as the FDA, and the other is commercial tolerance. Individual firms can and seem likely to adopt different tolerance levels, subject to any regulation. Moreover, different countries are likely to have different tolerance levels, and this increases the risks and costs of segregation or identity preservation.

While private sector handlers routinely segregate and blend grains and beans as a primary function of their business, new risks arise when handling GM and non-GM products, due to the added risk of adventitious commingling. If GM products are considered to be the inferior product, growers and handlers of GM products have an incentive to mix GM with non-GM products. For US grains, Wilson and Dahl (2005) suggest that this risk may be about 4 per cent at the grain elevator level. Farmer–processor contracting in horticultural or specialty crops, however, could reduce this margin by specializing in the product being delivered, such as non-GM or a positive GM trait. Another source of risk is testing, since no test is 100 per cent accurate. Testing risk, however, varies with the technology, tolerance and variety of the products handled, and seems likely to fall over time, as the technology of testing advances.

3.2. *GM labeling policy*

There are three policy options for GM food labeling: a labeling ban, voluntary labeling, and mandatory labeling. A policy of voluntary labeling works well in a niche market where consumers who prefer GM-free products are willing to pay a premium to obtain them. In the United States, organic food, which is process based and has low probability of GM content (due to adventitious presence), is the low GM-content option.

Caswell (1998) argues that mandatory labeling serves the same purposes as voluntary labeling but at a higher cost because the entire market must be segregated and labeled even though only a portion of consumers care that the product is GMO-free. Mandatory labeling becomes more attractive with increases in the number of people who care whether their food has been genetically altered. For mandatory labeling to be welfare enhancing, policy makers must determine whether consumers place enough value on knowing whether their food is GM-free for the benefits of labeling to exceed its costs.

Supporters of mandatory labeling argue that it is needed for informed consumer choice. Critics of mandatory labeling argue that it carries the potential for information overload, and that consumers face both time and knowledge constraints when interpreting labels. Biotech companies are against mandatory labeling. They argue that labeling will unfairly stigmatize products that contain GM ingredients and hurt sales. The most convincing critique of mandatory labeling is the cost of preserving the identity of all foods that could potentially be genetically modified.

The debate over public labeling policy may turn out to be moot because of the expansion of private standards. Public standards do not adjust to changes in consumer preferences or technology as efficiently as private standards. As consumer tastes and preferences continue to change, the private sector has responded to ensure consumer satisfaction. The consolidation in the food retail sector has given the major food retailers the bargaining power to define the private standards at the industry level. Britain's food giants Tesco, Sainsbury's, Morrisons, Marks and Spencer, Somerfield, Aldi and Co-op have maintained a decade-long ban on GM foods.

How the provision of information about the inclusion of GM ingredients affects consumers' food choices and whether consumers value information are important issues for firms and policy makers. Firms typically have more information about the quality of their products than consumers, creating a situation of asymmetric information. It is prohibitively costly for most consumers to acquire information independently of firms.

4. Empirical Studies of Consumer Responses to GM Labels

There is a large body of literature that investigates consumer demand and the use of GM labels. In this section, we first discuss findings of a set of studies that utilize consumer surveys and stated consumer responses to GM labels. We then discuss studies that utilize economic experiments with an emphasis on the effects of information provision.

Consumers' preferences over types and formats of information, their ability to analyze information and time cost of decision making differ. Hence, tendencies to use various types of information vary across consumers (McCluskey and Swinnen, 2004). Labels are likely to be effective when they address specific informational needs and can be processed and used by their target audience (Verbeke, 2005). In this section, we first discuss findings of a set of studies that utilizes consumer surveys and stated consumer responses to GM claims and labels. The survey-based studies reviewed here include those that measure consumers' willingness to pay (WTP) for foods with various labels, including GM content. The second set of articles reviewed and discussed in this section is based on non-experimental data and focuses on estimating consumer GM label valuations on the basis of revealed preferences through consumers' actual purchase behavior. Here the empirical strategy involves reduced-form hedonic empirical models as well as structural demand analyses and estimation of consumers' WTP for foods with a particular label. The last set of papers reviews empirical findings of consumers' WTP for foods using experimental data sets.

4.1. *Survey-based studies*

Advantages and disadvantages exist to using survey-based evidence to assess consumers' preferences and willingness to pay for GM products. A major disadvantage is that a respondent to a survey question may not truly reflect what he/she would do in a real-world setting. This is the 'hypothetical bias', which represents any differences that may exist between responses from real and hypothetical valuation questions (Cummings *et al.*, 1995). However, in the right setting survey-based results have advantages and are useful. In particular, one can obtain information on consumers' characteristics, such as age and gender, and ask questions about consumers' attitude towards health, their beliefs about relevant information or product attributes, among others. Further, researchers can elicit consumer valuations for hypothetical products that do not yet exist. This is especially useful in evaluating consumers' WTP for new products developed using new technologies.

The mostly negative consumer response to food products that contain GM ingredients with 'first-generation' traits is well documented by many researchers. Lusk *et al.* (2005) provides a comprehensive review and a meta-analysis of 25 studies that in total report 57 valuations for GM foods. They found that 87 per cent of the variation in existing value estimates for genetically modified food can be explained by an econometric model that controls characteristics of the sample, elicitation method and characteristics of the food being valued. Given that results are sensitive to elicitation method, a difficult decision is how to inform public policy. To date consumer surveys with valuation questions have been conducted in many countries, and the findings suggest that cultural factors affect consumer acceptance of biotechnology.

A number of examples show how the methods work and the kind of results they are delivering. Some are studies conducted in the United States, where consumers are more accepting of new technology compared with European countries and Japan. Li *et al.* (2004) examined US consumers' value for GM corn-fed beef products. Using dichotomous choice contingent valuation (CV) methodology, they found that respondents who received information about the science and possible consumer benefits of GM, were willing to pay more for GM corn-fed beef products. Respondents who were more environmentally-conscious and more educated were more likely to choose GM corn-fed beef when they received GM information. However, a mean discount of 8 per cent was required across information treatments to get consumers to choose the GM product. This is small compared to findings from other studies in Europe and Japan.

In another US study, Kaneko and Chern (2005) found that most consumers are not knowledgeable about biotechnology or they are indifferent between GM or non-GM foods. They faced consumers with several options: an indifference option in the CV question for vegetable oil, cornflakes and salmon. Respondents were asked to chose either the non-GM food item or GM alternative at the market price. When the respondents choose the non-GM (GM) item, the price of the non-GM (GM) item was randomly discounted. If the respondent was indifferent between these two foods, then he/she randomly received a discount on the GM alternative or non-GM alternative and was asked whether he/she was still indifferent. The authors found that the WTP to avoid GM salmon was the highest in comparison with GM vegetable oil, GM cornflakes and GM-fed salmon.

Heiman and Zilberman (2011) investigate the effects of positively (negatively) framed statements on perceptions about GM and hypothetical choices of Israeli consumers for genetically modified vegetables. Their survey was conducted among students, and the respondents were randomly divided into negative, positive and control groups. The authors suggest that negative framing of the properties of biotechnology not only affected perception but also increased the weight assigned to health and decreased the weight assigned to taste. The main effects of knowledge and non-negative perceptions of the contribution of GM products to health increased consumers' willingness to purchase GM bell peppers. Environmental and moral consideration were found to have little impact on GM selection. The actual choice between GM versus non-GM peppers was significantly affected by price: larger discounts increased demand for GM peppers. Negative framing tended to reduce the likelihood of choosing GM and positive framing tended to increase it, suggesting that wording and presentation of a GM trait matter.

McCluskey *et al.* (2003) use a dichotomous choice CV model to examine the preferences of Japanese consumers. They found that the discount needed for Seikyou consumers to purchase noodles made with GM wheat and tofu made with GM soybeans is positively affected, that is a greater discount is required, when consumers report higher levels of risk toward GM food, higher levels of concern about food safety and the environment, and higher self-reported knowledge about biotechnology, education levels and income. The mean discount required for the Japanese consumers to purchase the GM products was greater than 50 per cent, so one can infer from that a transformation of Japanese consumers' perceptions and attitudes are needed for GM food products to successfully enter the Japanese market.

Grimsrud *et al.* (2004) conducted a consumer survey of 400 shoppers in the Oslo region of Norway in 2002 to identify the factors that affect the discount required for Norwegian consumers to purchase bread made with GM wheat. The Norwegian consumers they sampled were on average willing to purchase GM bread with approximately a 50 per cent discount compared to non-GM bread, indicating that there is strong skepticism toward GM foods in Norway. The strict Norwegian GM food policies are consistent with the consumer consensus antipathy against the inclusion of GM foods in the Norwegian market.

Antonopoulou *et al.* (2009) conducted a survey in Greece to investigate the impact of political factors on consumer attitudes towards GM food. Findings include that rejection of globalization is strongly related with increased probability of being against GM foods, supporting a ban on GM foods, mandatory labeling on allowed foods, and refusing to purchase them at lower or discount price.

Given the skepticism of European Union consumers toward GM foods, Grunert *et al.* (2004) investigated the effects of exposing subjects to positive sensory experiences that can only be obtained by using genetic engineered flavor-enhancing bacteria. The subjects were adults in Denmark, Finland, Norway and Sweden, and the food products were hard yellow cheeses, with one labeled as 'produced using GMOs'. Subjects were first asked to taste the cheeses and to rate them for 'liking' and their willingness to buy over two separate sessions. The cheeses were labeled 'conventional' and 'GM', either without or with a health benefit. The two cheeses were picked from the first session so that the cheese labeled as conventional was rated neutrally among the eight alternatives and the GM cheese was a liked option. They found that price and starter bacteria had the largest impact on subjects' purchase intentions. The non-GM starter has a positive value, whereas the GM starter has a negative value, being larger if GM material is still viable in the final product. Subjects who were in the group that received the cheese labeled 'GM' had a less negative attitude toward GM in food products than the control group. Also, the starter used (GM or conventional) had a smaller impact on subjects' buying intentions than for the control group. Specifically, those subjects who thought that they had tasted a GM cheese attached less weight to the type of starter culture in the rating step, while those that believed that they had tasted a GM cheese with a health benefit (low fat) attached still less weight to it. The study emphasizes that consumers are less negative to GMOs if they provide consumer benefits.

Li *et al.* (2002) conducted a survey of 600 consumers in Beijing, China. Although the majority of surveyed consumers self-reported that they had no knowledge of biotechnology, their attitude toward GM foods was generally positive. Using a CV methodology, they find that Beijing consumers' WTP for GM soybean oil was positively affected by respondents' positive opinion toward GM foods for both products and by higher levels of self-reported knowledge for soybean oil. A few years later, Lin *et al.* (2006) provided an assessment of Chinese consumers' acceptance of biotech foods based on a large-scale survey. They found that about three-quarters of urban consumers in their sample had never heard of biotech foods. In spite of this low level of awareness, most of the Chinese subjects responded with favorable or neutral attitude towards biotech foods.

In Taiwan, Chiang *et al.* (2012) used the CV method to examine consumers' WTP for GM foods and their prior purchase intentions. They found that participants' risk perception, education level and age have significant effects on Taiwanese consumers' prior purchase intentions. The results of their analysis show that Taiwanese consumers would prefer to buy non-GM salmon and they would be willing to pay a higher premium if they were old, or if they deem GM soybean-fed salmon to be risky to human health.

On the African continent, Kimenju and Groote (2008) investigated Nairobi consumers' acceptance of GM food through surveys. They found that 38 per cent of the 604 respondents were aware of GM products and the majority of them were people in higher education and income groups. Even so, the authors found that 68 per cent of the respondents would be willing to buy GM maize meal at the same price as their favorite brand. However, the authors reported that consumers were concerned about possible side-effects, particularly on the environment and biodiversity.

A subset of stated-preference studies considers consumer valuation of the benefits of second-generation GM food products. Li *et al.* (2002) concluded that Chinese consumers in their sample, on average, are willing to pay a 38 per cent premium for Golden Rice. Further, Boccaletti and Moro (2000) reported a decrease in the percentage of consumers who refuse to buy GM food products, from 17 per cent to 12 per cent, when confronted with products that directly benefit them, instead of generic GM food products in Italy. They found that consumers better accept GM products with increased nutritional and organoleptic characteristics, and longer shelf life. Anand *et al.* (2007) found that consumers' acceptance of GM wheat in India changes according to whether its benefits are portrayed as producer-friendly (cost-reducing) or consumer-friendly (disease-fighting).

As more functional foods have been offered on the market, researchers have examined how the benefits associated with functional foods affect consumer acceptance. West *et al.* (2002) reported that Canadian respondents were willing to pay higher prices for functional food products such as anti-cancer tomatoes, and heart-healthy potatoes and chicken. Maynard and Franklin (2003) reported a similar tendency with 'cancer-fighting' dairy products. Grunert *et al.* (2004) investigated the effect of sensory experience of Nordic consumers with GM cheeses and their attitudes and purchase intentions. They reported that the functional benefits of the products, such as fewer calories, less saturated fat, more calcium, and more zinc, generally had small effects. Hartl and Herrmann (2009) found that German consumers are less likely to reject GM foods with consumer-oriented benefits than foods produced using farmer traits.

4.2. *Revealed-preference-based studies*

In Europe and other developed countries, anti-GMO sentiment has been expressed in numerous surveys. Polling of European consumers consistently indicates a high degree of hostility to the presence of GMOs in the food supply. The aversion to GMOs is based on both private considerations, such as health risks or a preference for natural foods, and a social dimension, such as concerns about environmental effects and ethical aspects. While anti-GMO sentiment in surveys could reflect more of a concern for the social dimension of GMOs, private considerations could still dominate actual purchase decisions.

It is well known that individuals' decisions can differ greatly between contingent value studies or other surveys, which are hypothetical, and auctions where there is a real commitment to purchase (Cummings *et al.*, 1995; List and Shogren, 1998; List and Gallet, 2001). Furthermore, given that most surveys do not inquire about actual purchase decisions at specific prices, it has been shown that small contextual cues or small changes in information provided to survey respondents may change results dramatically. In particular, the problem is most severe for willingness-to-pay for public goods. It can be argued that surveys place respondents in the role of citizens, who make judgments from society's point of view, rather the private consumer, who makes optimizing decisions for his or her household (Noussair *et al.*, 2004).

4.3. *Laboratory experiments*

Experimental economics, which is the application of experimental methods to study economic questions, offers new insights into consumer attitudes. Experiments may be conducted in the lab, such as a classroom or real lab setting, or in the field, that is, a real-world setting, such as a grocery store or farmers' market. As the description suggests, many economic experiments are designed to mimic market behavior. For lab experiments, this requires that experiments be conducted under institutional rules that support the information discovery process. The field of experimental economics was launched in the 1960s and 1970s by Vernon Smith (1962 and 1976), Kagel and Roth (1995) and others. The goal in experimental economics is to develop and implement a controlled economic experiment on subjects. Good experiments make empirical analysis of economic data much easier.

A number of economic experiments have been performed to assess what value is placed on commodities that contain GMOs, under different information and labeling treatments. In these experiments, participants bid in an auction setting on real consumer goods in a laboratory session under an incentive-compatible, information-revealing mechanism. Auctions provide several advantages. First, they provide a common homogeneous unit, money, to measure strength of preferences by any one individual for a commodity or across commodities and are free of endowment effects. For example, $1 has a common meaning to respondents, whereas in a survey, 'strongly disagree' and 'agree', are likely to be interpreted differently across respondents. Second, money is free of endowment effects. Endowment effects arise when an individual in an experiment is endowed with a commodity, say a hamburger sandwich, and asked to upgrade to another hamburger sandwich containing different attributes, such as one containing lower bacteria levels. Third, in an auction, an individual is committing to an actual purchase, unlike in a poll where there is no commitment. Fourth, under a demand-revealing mechanism, the dominant strategy is for individuals to indicate their true value for a commodity. In principle, this allows WTP to be directly measured, rather than inferred, as in surveys. The existence of a dominant strategy also simplifies calculation of a participant's best strategy because it is independent of own-risk attitudes and beliefs about other participants. Fifth, the bid submitted in an auction is a summary of an individual's valuation of the various attributes of the good, including GM content. In contrast, in a survey an individual may express a strong preference for non-GM products, but this information does not accurately reveal the weight of this attribute relative to other attributes, such as appearance, size, color and taste. In addition, a survey response might be expected to give greater weight to public dimensions of a good, such as negative externalities that result from widespread use of the product, than a private value auction (Noussair *et al.*, 2004).

One of the earliest economic auctions of GM foods was conducted by Noussair *et al.* (2004). They found that information about the benefits of GM technology significantly reduced the compensation that consumers demanded to consume a GM biscuit (or cookie), and the effect varied by type of information and location. Participants in their experiments were a demographically representative sample of consumers in the Grenoble area of France in July 2000. In the main part of the experiment, participants were asked to place bids on biscuits (or cookies) of the type that are typically available in local stores. The products looked similar and were close substitutes. At the start of the auction, each participant received a sample of each of the four products to taste and was asked to mark down how much she/he liked the product. Each of the four products was simultaneously auctioned over five rounds. In each round, information was incrementally revealed about some or all of the products. No sale prices were drawn until the last (fifth) round. The incrementally revealed information included whether products contained GMOs, the percentage content of GM ingredients, and a four-page handout containing background information about GMOs.

Noussair *et al.* (2004) found that, in contrast to the generally negative public stand of the French against GMOs in 2000, 65 per cent of participants were willing to purchase a biscuit containing GMOs if the price was low enough. Nearly 25 per cent of participants showed no decrease in their WTP in response to learning that a product contained GMOs. Labeling for 0.1 per cent and 1 per cent thresholds for GMO content generated significantly different bids and were perceived as being significantly different. Thus, participants revealed that their demand for biscuits is decreasing in GMO content. Eight-nine per cent of participants were willing to purchase a product satisfying the 1 per cent threshold, the maximum content that the EU exempts from labeling. In essence, injection of information about GMOs into the experiments had little impact on the bidding behavior of these French consumers.

Lusk *et al.* (2004) conducted incentive-compatible auctions to test the value of information (including environmental, health, and benefits to lesser developed countries) across locations in England, France and the United States in 2002. Each subject was endowed with a non-GM chocolate chip cookie and then asked to bid the minimum amount they had to be paid to exchange this cookie for a GM cookie with full knowledge that the consumption of the cookie was required at the end of the auction. The subjects participated in ten rounds of bidding with a random number generated to determine the binding round. Winners were chosen by the fifth price auction. Four information treatment effects were implemented: (i) no information; (ii) potential environmental benefits of GM technology; (iii) potential health benefits; and (iv) potential benefits from greater abundance of food.

They found that average willingness to accept (WTA) a payment to exchange their cookie was lowest in the US locations, \$0.20 higher in England, and \$1 higher in France. This is interpreted as heterogeneity of tastes for GM products across consumers in these developed countries (Lusk *et al.*, 2006). The findings on information effects include that the environmental benefit information resulted in lower impact on WTA among US and English subjects but higher impact on WTA impact for French subjects. Subjects who received the health information had a lower WTA in all areas but France. Subjects only in Texas and England had a lower WTA after receiving information about the potential benefit of greater world food abundance.

Rousu *et al.* (2007) conducted economic experiments to examine the impact on WTP of GM and plain food labels and diverse information with participants from Des Moines, Iowa, and St Paul, Minnesota in 2001. The researchers used a random n^{th}-price auction. Participants were asked to bid simultaneously on three commodities: a bottle of vegetable oil, a bag of tortilla chips, and a bag of Russet potatoes. For each product, there were two food labels: (a) a plain label giving the generic type of food, its weight, and date of preparation in the case of tortilla chips; and (b) a label containing the information in (a) plus the statement 'This product is made using genetic modification (GM)'. Statements about GM foods representing perspectives of the agricultural biotechnology industry, environmental groups, and third-party sources were then packaged into five information treatments: agricultural biotechnology perspective only; environmental group perspective only; agricultural biotechnology and environmental perspectives; agricultural biotechnology and third party perspectives; environmental group and third-party perspectives; and all three perspectives. Food labels and information treatments were randomly assigned to particular lab sessions.

They found that, on average, bidders discounted foods with GM labels by 14 per cent relative to the same food item having a plain food label. The average WTP for a GM product was highest when participants received (only) the industry's perspective about GMOs and the lowest when they received (only) the environmental group's perspective. Across these two information treatments and across the three commodities, the average differential was 64 per cent. Third-party information tended to moderate bidding when it was combined with the agricultural biotechnology industry perspective and the environmental group perspective, having the largest and statistically significant moderating force when participants received the environmental group perspective. The study concludes that third-party information has large social welfare value in this market with conflicting information.

Huffman *et al.* (2007) examine the role of consumers' prior beliefs and new information on their WTP for foods that might be genetically modified. They apply a Bayesian-updating model in which prior beliefs can be updated when the decision maker encounters new information. The study used data from Rousu *et al.* (2007). They found that participants who had informed prior beliefs discounted GM labeled food products more highly than those who had uninformed prior beliefs. Uninformed participants were especially susceptible to information from interested and third parties. In contrast, informed participants were generally not affected significantly by new information. These results show how both skeptics and proponents of new technologies (interested parties) might try to manage information to achieve private objectives. This is most likely to occur when much is unknown scientifically about the impact of new technologies or when third-party information is limited or unavailable. Opponents of a new technology may try to target people who are relatively uninformed about the technology. Proponents of the technology may try to target people who have informative prior beliefs for maximum effectiveness.

Colson *et al.* (2011) assessed consumers' acceptance of GM nutritionally enhanced vegetables using a series of auction experiments administered to a random sample of adult consumers. They were investigating consumer rather than farm traits. The experiments were conducted in 2007 in Des Moines, Iowa, and Harrisburg, Pennsylvania. Commodities chosen for these experiments were Russet potatoes, beefsteak tomatoes andbroccoli. A total of seven different food labels for each product were used in the experiments: (i) plain label giving generic product name and weight; (ii) information in (i) plus the statement 'GM Free Product'; (iii) information in (i) plus the statement 'Intragenic GM Product'; (iv) information in (i) plus the statement 'Transgenic GM Product'; (v) information in (i) plus statements 'Enhanced levels of Antioxidants and Vitamin C' and 'GM product'; (vi) information in (i) plus statements 'Enhanced levels of Antioxidants and Vitamin C' and 'Intragenic GM Product'; (vii) information in (i) plus statements 'Enhanced levels of Antioxidants and Vitamin C' and 'Transgenic GM Product'. Transgenic GM products are made from raw products that contain genes that were from another species, for example, soil bacteria. Intragenic GM products are made from raw products that contain genes from only one species but some genes have been moved long distances using genetic engineering. Following Rousu *et al.* (2007), three perspectives on GMOs were prepared. The information treatments were: (i) No information provided, which served as a baseline; (ii) agricultural biotechnology industry perspective (only); (iii) environmental group perspective (only); (iv) agricultural biotechnology industry and environmental group perspectives; and (v) bio-tech industry, environmental group and third-party perspectives. Food labels and information treatments were randomly assigned to particular lab sessions.

Compared to the baseline of no information, they found that consumers' bids were higher for intragenic GM than for transgenic foods when they received the agricultural biotechnology industry's perspective. When the environmental group perspective was injected, the average bid price by GM technology was reduced below the no information case. When commodities were labeled as enhanced with high levels of anti-oxidants and vitamin C, the average bid prices were higher across all three commodities than when no information on nutrient content was provided. Hence, the increase in nutrient content was shown to have positive economic value. Under the agricultural biotechnology industry perspective and with a label indicating enhanced nutrient content using intragenic GM technology, consumers' bids were significantly higher than for transgenic or a plain label. These results suggest that genetic engineering can be used to improve nutrient content of food, which is a consumer attribute, and that consumers will pay more for that food. Moreover, as these foods come on the market, they may improve US consumers' perspective of the biotech industry.

5. Discussion and Future Research Directions

The scientific consensus is that first-generation GM foods are equivalent to their conventional counterparts. However, on average, consumers want a discount in order to choose first-generation GM products over conventional products. Thus, it is the public's *perception* of risks, rather than scientifically proven risks, that directly affect markets. This brings up the issue of scientific versus consumer sovereignty (Roberts, 1999). Although the scientific consensus is that GM foods are completely safe for consumption aside from potential allergens, it may still be the case that a majority of the population in some countries prefer to avoid GM foods. The evidence shows so far that information provision affects valuations and that the information source also matters.

As more 'second-generation' GM products, which include quality traits or product-enhancing attributes, such as nutritional benefits, enter the market, consumers appear to be more accepting of biotechnology in food because of the inherent private benefits. With benefits directly accruing to the consumers, those consumer-friendly GM products could be accepted without discounts or even receive a premium. As technologies evolve, the lines between GM, functional foods, and other 'enhanced foods' are expected to blur. At the same time, there is a growing segment of consumers who value 'natural' or organic foods. The establishment of the USDA's organic standard in 2002 provided a standardized reference point for those interested in these products. Fifteen years ago, organic products were marketed largely in farmers' markets and specialty food stores, but now organic foods are carried by large supermarkets and big box stores (megastores). Products made from GM raw materials are also carried in these same large supermarkets and big box stores, so organic and GM food products are increasingly coming into direct competition. Although the USDA's organic standard is strictly process based, some producers and consumers believe that organic foods are fresher, safer, more nutritious, environmentally friendly and produced by small, local farmers. However, this characterization is broadly untrue. Hence, research could be undertaken to identify the effects of objective information about the attributes of organic food on consumers' willingness to pay. This seems to provide new research opportunities for assessing consumers' demand for food.

We expect that in the future researchers will focus more on new sources of data, such as retail scanner data and experiments, both in the lab and field. In particular, scanner data can be increasingly individualized, as consumers use grocery store 'loyalty' cards. We also expect researchers to adopt more interdisciplinary or multidisciplinary methods to better understand household food choices. This includes insights from sensory sciences about the factors affecting taste and perceptions and how the experience affects perceptions of technology and the product. Psychology and neuro-economics will probably be integrated into studies of how values, social norms and personality affect choice of food made with new technologies. We expect research to draw increasingly upon findings of behavioral economics either to learn about consumer choice or to overcome cognitive limitations.

Ultimately, the new research agenda may consider how increasing affluence of consumers, increasing health concerns, and increasing expectations for customized products, for example, all the possible combinations offered at Starbuck's and the increasing number of stock-keeping units at grocery stores, affect consumers' food choices. This type of research will be critical in determining how to examine the increased differentiation and variety that exists in today's food markets.

References

Akerlof, G. (1970). The Market for Lemons: Uncertainty and the Market Mechanism. *Quarterly Journal of Economics* 84, 488–500.

Anand, A., R.C. Mittelhammer and J.J. McCluskey (2007). Consumer Response to Information and Second-Generation Modified Food in India. *Journal of Agricultural & Food Industrial Organization* 5, Article 8.

Antonopoulou, L., T. Papadas and A. Targoutzidis (2009). The Impact of Socio-Demographic Factors and Political Perceptions on Consumer Attitudes Towards Genetically Modified Foods. An Econometric Investigation. *Agricultural Economics Review* 10(2), 89–103.

Balasubramanian, S.K. and C. Cole (2002). Consumers' Search and Use of Nutrition Information: The Challenge and Promise of the Nutrition Labeling and Education Act. *Journal of Marketing* 66, 113–127.

Boccaletti, S. and D. Moro (2000). Consumer Willingness to Pay for GM Food Products in Italy. *AgBioForum* 3(4), 259–267.

Caswell, J.A. (1998). Should Use of Genetically Modified Organisms be Labeled? *AgBioForum* 1(1), 22–24.

Chern, W.S., K. Rickertsen, N. Tsuboi and T. Fu (2002). Consumer Acceptance and Willingness to Pay for Genetically Modified Vegetable Oil and Salmon: A Multiple-Country Assessment. *AgBioForum* 5(3), 105–112.

Chiang, J., C. Lin, T. Fu and C. Chen (2012). Using Stated Preference and Prior Purchase Intention in the Estimation of Willingness to Pay a Premium for Genetically Modified Foods. *Agribusiness* 28(1), 103–117.

Colson, G., W.E. Huffman and M. Rousu (2011). Will Consumers Pay More for Product Enhanced Attributes: Evidence from Food Experiments. *Journal of Agricultural and Resource Economics* 36, 343–364.

Cornes, R.C. and T. Sandler (1996). *The Theory of Externalities, Public Goods and Club Goods*, 2nd Edn., New York: Cambridge University Press.

Cummings, R.G., G.W. Harrison and E.E. Rutstrom (1995). Homegrown Values and Hypothetical Surveys: Is the Dichotomous Choice Approach Incentive-Compatible? *American Economic Review* 85(1), 260–266.

Frewer, L.J., C. Howard, D. Hedderley and E. Shepherd (1999). Reactions to Information About genetic Engineering: Impact of Source Characteristics, Perceived Personal Relevance and Persuasiveness. *Public Understanding of Science* 8, 35–50.

Greenpeace International (2001). We Want Natural Food!, available at: http://www.greenpeace.org/~geneng/.

Grimsrud, K.M., J.J. McCluskey, M.L. Loureiro and T.I. Wahl (2004). Consumer Attitudes to Genetically Modified Food in Norway. *Journal of Agricultural Economics* 55(1), 75–90.

Grunert, K.G., T. Bech-Larsen, L. Lähteenmäki, Ø. Ueland and A. Åström (2004). Attitudes Towards the Use of GMOs in Food Protection and Their Impact on Buying Intention: The Role of Positive Sensory Experience. *Agribusiness* 20, 95–107.

Hartl, J. and R. Herrmann (2009). Do They Always Say No? German Consumers and Second-Generation GM Foods. *Agricultural Economics* 40, 551–560.

Heiman, A. and D. Zilberman (2011). The Effects of Framing on Consumers' Choice of GM Foods. *AgBioForum* 14(3), 171–179.

Huffman, W.E. (2001). "Human capital: education and agriculture." In Bruce L. Gardner and Gordon C. Rausser (Eds.), *Handbook of Agricultural Economics*, Vol. IA. Amsterdam: Elsevier Science/North-Holland, pp. 333–381.

Huffman, W.E. and J. Strzok (2012). 'The economics of organic and GMO farming systems: interactions and how they might co-exist', Iowa State University, Department of Economics.

Huffman, W.E. and A. Tegene (2002). "Public acceptance of and benefits from agricultural biotechnology: a key role for verifiable information." In V. Santaniello, R.E. Evenson and D. Zilberman (Eds.), *Market Development for Genetically Modified Foods.* Wallingford: CABI International, pp. 179–189.

Huffman, W.E., M. Rousu, J.F. Shogren and A. Tegene (2007). The Effects of Prior Beliefs and Learning on Consumers' Acceptance of Genetically Modified Foods. *Journal of Economic Behavior and Organization* 63, 193–206.

Just, R.E., J.M. Alston and D. Zilberman (2006). *Regulating Agricultural Biotechnology: Economics and Policy.* New York: Springer Science + Business Media.

Kagel, J. and A. Roth (eds.) (1995). *Handbook of Experimental Economics.* Princeton, NJ: Princeton University Press.

Kaneko, N. and W.S. Chern (2005). Willingness to Pay for Genetically Modified Oil, Cornflakes, and Salmon: Evidence from a US Telephone Survey. *Journal of Agricultural and Applied Economics* 37(3), 701–771.

Kimenju, S.C. and H.D. Groote (2008). Consumer Willingness to Pay for Genetically Modified Food in Kenya. *Agricultural Economics* 38, 35–46.

Li, Q., J.J. McCluskey and T.I. Wahl (2004). Effects of Information on Consumers' Willingness to Pay for GM-Corn-Fed Beef. *Journal of Agricultural and Food Industrial Organization* 2(2), Article 9.

Li, Q., K.R. Curtis, J.J. McCluskey and T.I. Wahl (2002). Consumer Attitudes Toward Genetically Modified Foods in China. *AgBioForum* 5, 145–152.

Lin, W., A. Somwaru, F. Tuan, J. Huang and J. Bai (2006). Consumer Attitudes Toward Biotech Foods in China. *Journal of International Food and Agribusiness Marketing* 18(1/2), 177–203.

List, J. and C. Gallet (2001). What Experimental Protocol Influence Disparities Between Actual and Hypothetical Stated Values? *Environmental & Resource Economics* 20(3), 241–254.

List, J.A. and J.F. Shogren (1998). Calibration of the Difference Between Actual and Hypothetical Valuations in a Field Experiment. *Journal of Economic Behavior & Organization* 37(2), 193–205.

Lusk, J.L., M. Jamal, L. Kurlander, M. Roucan and L. Taulman (2005). A Meta-Analysis of Genetically Modified Food Valuation Studies. *Journal of Agricultural and Resource Economics* 30(1), 28–44.

Lusk, J.L., L.O. House, C. Valli, S.R. Jaeger, M. Moore, B. Morrow and W.B. Traill (2004). Effect of Information About Benefits of Biotechnology on Consumer Acceptance of Genetically Modified Food: Evidence from Experimental Auctions in the United States, England, and France. *European Review of Agricultural Economics* 31, 179–204.

Lusk, J.L., W.B. Traill, L.O. House, C. Valli, S.R. Jaeger, M. Moore and B. Morrow (2006). Comparative Advantage in Demand: Experimental Evidence of Preferences for Genetically Modified Food in the United States and European Union. *Journal of Agricultural Economics* 57, 1–21.

Maynard, L.J. and S.T. Franklin (2003). Functional Foods as a Value-Added Strategy: The Commercial Potential of 'Cancer-Fighting' Dairy Products. *Review of Agricultural Economics* 25, 316–331.

McCluskey, J.J. and J.F.M. Swinnen (2004). Political Economy of the Media and Consumer Perceptions of Biotechnology. *American Journal of Agricultural Economics* 86(5), 1230–1237.

McCluskey, J.J., K.M. Grimsrud, H. Ouchi and T.I. Wahl (2003). Consumer Response to Genetically Modified Food Products in Japan. *Agricultural and Resource Economics Review* 32(2), 222–231.

Milgrom, P. and J. Roberts (1986). Relying on the Information of Interested Parties. *Rand Journal of Economics* 7, 18–32.

Molho, I. (1997). *The Economics of Information.* Malden, MA: Blackwell Publishing.

Nelson, C.H. (2001). Risk Perception, Behavior, and Consumer Response to Genetically Modified Organisms. *American Behavioral Scientist*, 44(8), 1371–1388.

Noussair, C.N., B. Ruffieux and S. Robin (2004). Do European Consumers Really Refuse to Buy GM Food? *Economic Journal* 114, 102–120.

Olson, M. (1965). *The Logic of Collective Action.* Cambridge, MA: Harvard University Press.

Pinstrup-Andersen, P. and E. Schiøler (2001). *Seeds of Contention: World Hunger and the Global Controversy over GM Crops.* Baltimore, MD: Johns Hopkins University Press.

Qaim, M. and D. Zilberman (2003). Yield Effects of Genetically Modified Crops in Developing Countries. *Science* 299, 900.

Roberts, D. (1999). Analyzing Technical Trade Barriers in Agricultural Markets: Challenges and Priorities. *Agribusiness* 15(3), 335–354.

Roe, B. and I. Sheldon (2007). Credence Good Labeling: The Efficiency and Distributional Implications of Several Policy Approaches. *American Journal of Agricultural Economics* 89, 1020–1033.

Rousu, M., W.E. Huffman, J.F. Shogren and A. Tegene (2007). Effects and Value of Verifiable Information in a Controversial Market: Evidence from Lab Auctions of Genetically Modified Food. *Economic Inquiry* 45, 409–432.

Sandler, T. (1992). *Collective Action.* Ann Arbor, MI: The University of Michigan Press.

Scholderer, J. and L.J. Frewer (2003). The Biotechnology Communication Paradox: Experimental Evidence and the Need for a New Strategy. *Journal of Consumer Policy* 26(2), 125–157.

Science Media Center (2012). 'UK Scientists call on anti-GM group to reconsider threats', available at: http://www.sciencemediacentre.co.nz/2012/05/02/uk-scientists-call-for-dialogue-with-anti-gm-group/.

Silverglade, B. (1996). The Nutrient Labeling and Education Act: Progress to Date and Challenges for the Future. *Journal of Public Policy and Marketing* 15, 148–156.

Smith, V.L. (1962). An Experimental Study of Competitive Market Behavior. *Journal of Political Economy* 70, 111–137.

Smith, V.L. (1976). Experimental Economics: Induced Value Theory. *American Economic Review* 66, 274–279.

Subrahmanyan, S. and P.S. Cheng (2000). Perceptions and Attitudes of Singaporeans Toward Genetically Modified Food. *The Journal of Consumer Affairs* 34(2), 269–290.

Sunstein, C.R. (2005). *Laws of Fear: Beyond the Precautionary Principle.* New York: Cambridge University Press.

Tversky, A. and D. Kahneman (1981). The Framing of Decisions and the Psychology of Choice. *Science* 211, 453–458.

US Food and Drug Administration (FDA) (2001). 'Guidance for industry: Voluntary labeling indicating whether foods have or have not been developed using bioengineering', available at: http://vm.cfsan.fda.gov/~dms/biolabgu.html.

Veeman, M. (2002). Policy Development for Novel Foods: Issues and Challenges for Functional Food. *Canadian Journal of Agricultural Economics* 50(4), 527–589.

Verbeke, W. (2005). Agriculture and the Food Industry in the Information Age. *European Review of Agricultural Economics* 32(3), 347–368.

Vogt, D.U. and M. Parish (1999). 'Food biotechnology in the United States: science, regulation, and issues', *US Department of State*, available at: http://fpc.state.gov/6176.htm, accessed January 2012.

West, G.E., C. Gendron, B. Larue and R. Lambert (2002). Consumers' Valuation of Functional Properties of Foods: Results from a Canada-Wide Survey. *Canadian Journal of Agricultural Economics* 50(4), 541–558.

Wilson, W.W. and B.L. Dahl (2005). Costs and Risks of Testing and Segregating Genetically Modified Wheat. *Review of Agricultural Economics* 27(2), 212–228.

Chapter 4

Public Acceptance of and Benefits from Agricultural Biotechnology: A Key Role for Verifiable Information*

Wallace E. Huffman† and Abebayehu Tegene‡

†*Department of Economics, Iowa State University, Ames, Iowa, USA*

‡*Resource Economic Division, ERS, USDA, Washington, DC, USA*

The on-going genetically modified (GM) food and genetically modified organism (GMO) controversy threaten to destroy the near-term market for agbiotech food and inputs and alter greatly the net social benefits that are potentially attainable from agricultural biotechnology. For example, consider:

1. In February 2000 Greenpeace filed a lawsuit challenging the US Environmental Protection Agency's decision to allow the release of GM insect resistant (Bt) crops. In their news release they stated 'The EPA should stop [GE] polluters before the environment is threatened' (Greenpeace, 2000a).
2. During the week of 23 March 2000 Greenpeace joined over 50 other organizations in a petition to the US Food and Drug Administration (FDA) calling on the agency to remove genetically engineered (GE) foods from the market because it failed to require safety testing or labelling (Greenpeace, 2000b).

*This article originally appeared in the **Market Development for Genetically Modified Foods (2002), 179–189**. This is published by permission of **Centre for Agriculture and Bioscience International**.

3. In 1998, the European Union set out to update Council Directive 90/220 covering the deliberate release of GMOs. In the spring of 2000, the EU decided not to approve any more releases until the directive is revised.
4. During 1999, more than 50% of the crop biotech field experiments in the UK were disrupted by anti-GMO activists.

Nettleton (1999) states that the anti-biotech activists have achieved a masterful feat in communication, subverting the purpose of biotechnology, whipping up public alarm and feeding political agendas to protect agricultural markets.

Advances in science enable new technologies and advances in technology increase the demand for science. Advancing science and technology are uncertain and costly activities (Holmstrom, 1989; Huffman and Evenson, 1993; Huffman and Just, 2000). Although some new technologies have benefited society greatly, much uncertainty surrounds most new technologies. For example, little accurate information or knowledge exists about the attributes, including effects, of new ag-biotechnologies, and some of the knowledge/information that does exist is public information and some private.[1] Additional research can be undertaken to increase the knowledge about the beneficial and harmful effects of new technologies, some of which will reduce the uncertainty over future irreversible catastrophes. Advances in communication and information networks make possible rapid worldwide dissemination of public scientific discoveries and other information.

Private information is the source of asymmetric information, and it leads to an informational advantage to the party possessing the information. In two party interactions with one party possessing private information, the informed party can be expected to exercise their informational advantage whenever they can expect to gain from it, and the other party loses. When experienced parties develop intuition about situations potentially leading to opportunistic behaviour of

[1]See Frey (2000) for a discussion of some of the prospects and problems associated with the use of biotechnology in plant breeding.

others, asymmetric information can destroy trade/exchange between parties where the potential gains from trade/exchange are large.

The objective of this chapter is to examine the welfare effects of information from communication (by interested parties) on the decisions of producers of agricultural products and potential users of biotechnical inputs and on consumers of agrobiotechnical products. Interested parties are considered to be private biotechnical companies whose profits depend positively on sales of GM inputs to agricultural producers and 'environmental groups' whose utility depends positively on the quantity and quality of environmental stocks. The ag-biotech companies provide or distribute information primarily in advertising, news releases, informational brochures and web sites, personal contacts and demonstrations. The environmental groups provide or distribute information primarily in news releases, informational web sites and demonstrations. Final consumers of agricultural products and agricultural producers that might use GM inputs are assumed to be approximately 'neutral parties' in this communication process, but they face important decisions relating to their own welfare.[2] We acknowledge that other possibly more trusted sources of information exist for producers and consumers, and the chapter focuses on the importance of these sources to good decision making.[3] We will show that verifiable information plays a central role in socially good decision making and that an independent agency should store and make freely available verified information. This agency might also engage in research needed to refute or confirm claims made by interested parties and others.

---

[2]Hoban (1997, 1999) shows that food safety concerns from biotechnology are rated relatively low by consumers in the USA, Western Europe and Japan, relative to microbial contamination, pesticide residuals and antibiotics or hormones.

[3]Although Hoban (1996) reported that ag-biotech companies and activist groups are not ranked high by US consumers as a trusted information source, this does not mean that they never use any of the information in decision making. In The Netherlands, a survey of the general public showed that environmental and consumer organizations were seen as the most reliable sources of information on biotechnology (Heijs and Midden, 1996). In the UK, Martin and Tait (1992) found that in a local community reaction to a GMO release, the public chose to support the perspective provided by Greenpeace.

1. The Model

Many decision makers must or choose to rely on information provided by individuals or groups who are affected by their decisions. Furthermore, these decision makers may not know the alternatives available and have no control over the information provided to them by interested parties. These interested parties may manipulate by distorting or concealing information. For example, farmers rely on information, including advertising, provided/distributed by biotech companies about the expected performance of new biotech inputs, which is quite selective. These firms are constrained somewhat by an interest in repeat sales, but their communication may not reflect accurately all known impacts. Consumers rely on information and advertising distributed by food companies and environmental groups which seem likely to be tinged with self-interest. For example, communications by Greenpeace and Friends of the Earth opposing GM foods may exaggerate the potential harm to the environment and distract from other important issues.

Greenpeace, Friends of the Earth and other environmental groups are interest groups. Individuals self-select into these groups because of a common interest or goal focused on the environment, and achieving the group's goal is a public good to its members. Hence, free-riding by one member on the efforts of other members is a major organizational problem (Olson, 1965; Sandler, 1992; Cornes and Sandler, 1996: 324–326). Each of these groups has resources — largely members' time and financial contributions — and their impact is affected by organizational efficiency. By choosing relatively narrow objectives, these groups reduce coordination and decision making costs over organizations that have diverse goals. Advances in communication and information technologies have greatly reduced organizational costs of interest groups and have undoubtedly increased their productivity. They are now able to construct low cost web sites for displaying their objectives, news releases, short articles and other information. For example, the web site of Friends of the Earth has been used to display the locations or addresses of large-scale plantings of GMO crops in England (Friends of the Earth, 2000). These groups can also use e-mail to rapidly distribute communications among members, for example dealing with demonstrations, and others, such as letters opposing GMO use and policies.[4]

In our model, the two interested parties provide information in the form of communications attempting to affect agricultural producers' and consumers' decisions. The communications are signals which reflect the self-interest and private information of each party. Communication is cheap because it requires little action on the part of the sender, and new information technologies, such as e-mail and web sites, have reduced its cost and greatly increased the swiftness of delivery. There remains some modest fixed cost of preparing a communication, but the marginal cost of distributing it has become approximately zero (Shapiro and Varian, 1999: 19–51). Misinformation can be as easily distributed as useful information. Computer viruses, like 'I Love You', are one example of misinformation which can be sent swiftly around the world and clog the information network. In particular, the new information technology has greatly expanded the possibility of individuals communicating with others whom they do not know personally and from whom they may have greatly different norms and values. Hence, new problems with assessing the quality of information obtained from web sites and e-mail have arisen (World Bank, 1999: 72–81). Because signalling with communications is so cheap, one possible outcome is that they degrade the quality of information to the point that communications from interested parties are ignored.

Although the marginal cost of distributing information is approximately zero, it remains costly for decision makers to interpret this information, especially contradictory information. Consumers and agricultural producers, however, differ in the long-term consequences of using bad information. Consumers maximize utility subject to a resource constraint. When they fail to use objective information in decision making, their utility or well-being decreases, but this does not generally cause them to exit the economy (except when the consequence is death). Producers on the other hand can be described

[4]Furthermore, advances in communication and information technologies may have facilitated coalition formation among interest groups wishing to oppose ag-biotechnology for whatever reason (see Greenpeace, 2000b and the comment by Nettleton, 1999).

as long-run profit maximizers. If they do not use good information, their profits are reduced. If they become negative over the long term, most likely they will be forced to exit the industry. Thus, there is selectivity operating among farmers that is generally different from that of consumers. For consumers and producers, the ability to process information and make good decisions is a valuable skill in the sense that it can be welfare or profit increasing, and this ability seems most likely to be related to their years of schooling and accumulated experience as decision makers (Huffman, 2002). Given some outlay on information, a decision maker might choose to rely on only one interested party in making their decisions, or they might choose to rely on several interested parties possessing different points of view. There are potential costs and benefits of each of these actions.

Although Milgrom and Roberts (1986) have shown that it is possible for a decision maker to make fully informed decisions when the decision maker relies on one interested party for information, the necessary conditions seem quite restrictive and are unlikely to be fulfilled. For example, the interested party's preferences must be known to the decision maker, the information must be freely verifiable, the decision maker must know the factors about which the interested party has information to be able to detect situations in which information is being withheld, and the decision maker must be able to draw the appropriate inference when information is withheld (i.e. he or she must be a sophisticated, sceptical decision maker). These are demanding, but perhaps not impossible attributes for successful agricultural producers to possess, but they seem to exceed the attributes of most consumers of food. The implication is that for agricultural producers (or consumers) to rely only on information provided/disseminated by biotech supply firms or environmental groups is unlikely to lead to fully informed decisions. Hence, good reasons exist for society to be sceptical of claims made by both the suppliers of ag-biotech inputs and the environmental interest groups.

Adding (having or allowing) competition among interested parties in providing information greatly reduces the restrictions or assumptions necessary for good decision making. The interested parties must be able to convey their information to the decision maker, and the decision maker must listen to all interested parties who want to convey information, that is, there must be an opportunity for the different interests to come out. Ag-biotech firms, environmental groups and other interested parties seem likely to differ in their ability and effectiveness in conveying information that they have. This ability might be associated, among other things, with training, communication skills, personalities, organizational objective/philosophy and information technologies.[5] Rapid advances in communication and information technologies are widely available today, and with the dramatic fall in the cost of sending messages and storing information through new networks (Shapiro and Varian, 1999; World Bank, 1999), accessibility to technologies seems minimally constrained by capital or credit. Biotech companies are primarily private companies interested in long-term profits associated with the sales of their products and the value of the company, and the information that they distribute can be expected to be consistent with this long-term objective and to be constrained by it. Environmental groups are pursuing non-monetary goals which seem likely to be less constraining on their actions and possibly on the objectivity of the information they distribute.

However, the decision makers can now be unsophisticated, having little or no idea of available options, of issues bearing on the decision or preferences of the interested parties. He or she must, however, be able to process the information that he or she receives, and the information must be verifiable. Under these conditions, fully informed decisions are possible. The implication is that agricultural producers and consumers can make better decisions when they use information from diverse and possibly interested parties, provided the information is verifiable.

Much information being distributed these days about ag-biotechnology, however, is not currently verifiable. First, biotechnology is advancing rapidly so many effects and impacts of new

[5]In dealings with strangers of unknown credibility and no binding contract, reputation based on providing accurate and reliable information is valuable. See Sobel (1985) for one perspective on how a concern about credibility can be expected to affect the quality of information provided to a decision maker over time.

products are unknown. Second, a coalition of anti-biotechnology interests has been formed to slow the acceptance of ag-biotechnology. These groups have raised new questions about both the short-term and the long-term effects on health and the environment of using ag-biotechnology and consuming GM foods. Third, some of the activities of the anti-biotech groups seem to be focused on disrupting the experiments that might lead to important and useful advances in the stock of knowledge about ag-biotechnology.

When information is not verifiable, the reliability of information provided by an interested party (or parties) depends on the congruity between the objectives of the decision maker and those of the interested party (parties). When objectives diverge, decision making is difficult, and these complex problems have been the topic of optimal incentive schemes in principal-agent or agency theory literature (e.g. see Holmstrom, 1979; Holmstrom and Milgrom, 1987; Gibbons, 1998). These models are, however, well suited only to decision problems with few, e.g. two, players.

When information is not verifiable, communications by interested parties might lead to unduly restrictive public policies being adopted (e.g. banning GM food production or imports) or it might degrade the information content to the extent that sophisticated decision makers ignore it.[6] This will, however, be generally welfare reducing relative to fully informed decision making. For example, social cost-reducing inputs for crop production might not be used by farmers, or socially beneficial GM foods might not be consumed.[7] More generally, long-term delays in adopting GM technologies because of the time required to verify or refute claims by the environmental groups

[6]The 4-year moratorium on processing patent applications on transgenic plants and animals by the European Patent Office, which ended in December 1999 with a decision by the enlarged board of appeals, seems to have been a matter of interpretation of European Union directives and not of verifiable information (Scheermeier and Dickson, 2000).

[7]As an indication of benefits from agricultural biotechnology, Falck-Zepada *et al.* (2000a) estimate the first-year worldwide economic surplus from the introduction of Bt cotton in the USA was US$240 million. The economic surplus to herbicide tolerant soybeans is larger (see Moschini *et al.*, 1999; Falck-Zepeda *et al.*, 2000b). In the surplus computations, biotech inputs are treated as having neutral effects on human health and the environment. See Alston *et al.* (1995) for more information on methods of social cost-benefit analysis dealing with research.

about ag-biotechnical products will reduce the expected social and private payoff to R&D in this area. This has implications for where the private sector places its future R&D investments (Frey, 2000), meeting future food needs and for economic growth.

When a large number of decision makers can use or need the same verified information, research to produce this information produces a public good, which may be of great social value. The knowledge once produced is non-rival, that is, use by one decision maker does not affect the quantity or quality available to others, and it is not (or may not be) economically feasible to exclude users. Because the opportunity cost of an added user of the information is zero and each user's valuation of the information is private information, private decisions will lead to under provision unless some organizational device is used to internalize externalities associated with free-riding. The price system is of no (or little) aid in extracting information on the social value of verified information or a system to manage it.

2. Institutionalizing a Verified Information System

An ideal verified information system for ag-biotechnology must provide a mechanism for disclosing private information (i.e. information or knowledge that exists but is not available to everyone), establish a process for refuting or confirming claims of interested parties, and advance the stock of knowledge about short- and long-term effects of biotechnology.[8] The size of the problem is large because ag-biotechnology is a global intergenerational public good. Biotech knowledge is non-rival and non-excludable on a global basis, although particular techniques and products have been converted into impure public goods through the institution of patenting and international patent agreements. Furthermore, the impacts (benefits and hazards) are multi-generational — in terms of both potential benefits and hazards.

[8]Avery *et al.* (2000) are concerned with a much narrower issue. They propose a computerized market for the collection and distribution of subjective evaluations of a product of uncertain quality purchased by consumers.

Although most of the currently available ag-biotechnologies have been developed for large-scale agriculture and high income consumers, the potential exists for ag-biotechnology to help low income countries to meet future food needs (Serageldin, 1999; National Research Council, 2000) and many of the low income countries want access to or to have the opportunity to use ag-biotechnology to help meet their future food needs (OECD, 2000). Hence, reliable information on ag-biotechnology is a public good with potentially large global value.

Because new ag-biotechnology has the potential to produce benefits and realize hazards over the long term, it has an intergenerational dimension. This means that sequencing of generations becomes important in setting policies and determining benefits; that is, both equity and efficiency dimensions are important. The current generation has a 'first-mover' advantage because it can choose an agenda or pathway that best serves its own purposes, even though these purposes may be at odds with later generations (Sandler, 1997). This is an especially important issue for actions that are irreversible; for example, reducing the earth's biodiversity, and for some transfers of genes across unrelated species. Farsighted decisions can be promoted by including individuals/representatives distributed over wide age and standard of living ranges in decision making on ag-biotech policy and by evaluating costs and benefits of biotechnology in real terms without discounting.[9]

Some mechanisms for revealing private information lead to better decisions than others. When a product's 'quality' is at issue, 'untruthful' advertising tends to lead to a breakdown in the market for the good and no trade occurs because of adverse selection; for example Akerlof's 'market for lemons' (Akerlof, 1970; Molho, 1997: 19–26). With 'truthful' advertising, monitored effectively by an independent body, good market performance is obtained. Privately

[9]Discounting seems to be a questionable practice for social cost-benefit analysis that spans several generations because with discounting at any positive rate, the current weight to a distant generation's disaster is very small (and frequently negligible). See Sandler (1997: 62–69).

provided information on the route to reputation building and repeat sales is less effective for simple 'quality' issues (Molho, 1997: 52–53), and it can be expected to be quite deficient in the biotech area where the scientific issues are frequently complex, the quality dimensions are frequently changing and the stock of knowledge is steadily advancing.

Private information is best revealed publicly through an independent agency which has the authority and responsibility to independently verify information from interested parties. An independent agency can go about making objective assessments of information and claims made by interested parties and others. These assessments and evaluations are costly to make because they use scarce resources, but once verified the information is a pure public good. New information and communication technologies have greatly reduced the cost of storing and rapidly transferring this information and greatly increased the potential accessibility to a global scale.

Advances in the stock of knowledge are important to a successful information verification system. It expands the topics, issues or dimensions of the knowledge base on which verification can be made. The primary contributors to this activity seem likely to be scientists employed by commercially independent institutions and funded primarily by the public sector, that is, scientists in 'open universities' and possibly government agencies. An open university is one where the direction and funding of research is not driven primarily by commercial interests or any other narrow interest group and where high scientific-control standards are in place (David, 2000; Huffman and Just, 2000). Universities where the research agenda and/or funding has been captured by a single interest group, or one or more large private companies (e.g. a life science company), does not meet this criterion. Also, if the direction of the research programme of a government agency is driven heavily by commercial considerations or has low scientific-control standards, it will not meet the criterion of open and objective science either. For scientific discoveries to be highly supportive of the information verification system, they must originate from an institutional process that signals openness and objectiveness to disinterested parties.

Good research requires considerable time to undertake and to verify itself. When the frontiers of biological and related sciences are advancing rapidly, a significant period may exist where considerable scientific uncertainty exists about outcomes, effects, impacts or the quality of the information (see, e.g. Frey, 2000: 61–79). Interested parties may attempt to exploit this information lag which can be to the detriment of producers and consumers of agricultural products and to society generally. An efficiently functioning knowledge generation and verification system can, however, shorten this lag. By doing so, it creates an environment where interested parties have a strong incentive to reveal voluntarily more (rather than less) of the private information that they possess and where verified information is freely available and easily accessible. The cooperation of interested parties is achievable primarily because sophisticated decision makers would infer even worse outcomes (Milgrom and Roberts, 1986).

No supernational body is likely to be created to provide verifiable information which is a global public good. Nations have been unwilling to empower such bodies with the authority to collect taxes for such purposes (Sandler, 1997). For an individual country to provide the good, the social benefits must exceed the costs. The greater public good nature of verified information created by advances in information technologies means savings for some countries whose decision makers free-ride and weaker incentives for any one country to undertake the verification activity. Only one good biotech verifiable information system is needed; or given one system, the marginal product of another system is zero. In the public economics literature, this has sometimes been labelled as 'best shot' technology of public good supply aggregation (Sandler, 1998). Thinking about a global coordination game across countries for provision of verifiable information, only one country needs to act. The country which has the largest expected net social benefit from action, or largest stake, can be expected to provide a biotech verifiable information system. This seems likely to be a country with a strong research system, a large ag-biotech industry, a technically advanced large-scale agriculture, large population, high price of time (for acquiring and evaluating information) and high income.

3. Trust in Public Institutions

Because a verifiable biotechnology information system must be financed by public tax collections, it most likely will be operated by one or more government agencies.[10] For this institutional framework to be successful, it must have the trust of the public. Currently agencies of national governments dealing with similar issues in Western developed countries vary in the amount of trust or confidence that the public places in information that they provide.

In the USA, public trust is high, but in Western European countries public trust is low. Gaskell *et al.* (1999) reports on a survey of the US public showing a high level of trust in information dealing with the safety of biotechnology provided by the USDA and FDA. Also, Hoban (1997) reports on a 1994 survey of the US public where respondents were asked to rank 15 different sources of information on agricultural biotechnology for trustworthiness. The National Institutes of Health and Food and Drug Administration ranked very high: second and third.[11] (The USDA and EPA were not included in the reported rankings.)

[10]The exchange of information and clearing house mechanism under the 1993 Convention on Biological Diversity have a somewhat different focus. The 1992 Convention on Biological Diversity has as objectives the conservation of biological diversity, the sustainable use of its components and the fair and equitable sharing of the benefits arising out of the utilization of genetic resources. The Convention went into effect in December 1993 with 168 signing countries (the USA being absent). Article 17 requires that members facilitate the exchange of information relevant to the conservation and sustainable use of biodiversity, including results from technical, scientific and social research and other information. Article 18 requires the establishment of a clearing house mechanism to promote and facilitate technical and scientific cooperation (Convention on Biological Diversity, 1992; www.biodiv.org/chm/conv/cbd_text_e.html). Both of these articles are to facilitate access to and exchange of information on biodiversity around the world, especially information needed to implement provisions of the Convention. Although the exchange of information and clearing house activities facilitate voluntary sharing of information on biodiversity, they are not actively involved in quality control or verification. The Cartagena Protocol on Biosafety was adopted in January 2000 to address environmental (but not food safety) impacts of bio-engineered products that cross international borders. The Protocol establishes an Internet-based Biosafety Clearing-House to help member countries exchange scientific, technical, environmental and legal information about living modified organisms.

[11]However, in 1999, US environmental activists and some consumer groups intensely criticized and demonstrated against the FDA's GM food policies (Macilwain, 1999).

For Europe, Gaskell *et al.* (1999) report on a survey of the public in 17 European countries showing low trust in national public bodies 'to tell the truth about GM crops grown in fields'. European governments have accumulated a bad record with the general public on food safety issues because of their past experiences dealing with governments that mishandled information on UK BSE meat and dioxin contamination of dairy and poultry products in Belgium and The Netherlands. Decision making by producers and consumers is made more difficult when government agencies cannot be trusted as a verifiable information source.

Current public information systems in place in the USA and Europe dealing with information on agricultural biotechnology are primarily focused on regulation for environmental and health safety and secondarily focused on proving information to consumers through labelling. In the USA, the regulation of biotechnology products is through the Coordinated Framework established in 1986 (National Research Council, 2000a) which ties together the US Department of Agriculture (USDA), Food and Drug Administration (FDA) and Environmental Protection Agency (EPA). The framework is based on the principle that techniques of biotechnology are not inherently risky and that biotechnology should not be regulated as a process, but rather that the products of biotechnology should be regulated in the same way as products of other technologies. Responsibility and jurisdiction over transgenetic products were assigned as follows: (i) plants came under the jurisdiction of the Federal Plant Pest Act (FPPA) administered by the USDA; (ii) food and feed under the Federal Food, Drug, and Cosmetic Act (FFDCA) administered by the FDA; and (iii) microorganisms and substances used for pest control under the jurisdiction of the Federal Insecticide, Fungicide, and Rodenticide Act (FIFRA) and part of FFDCA administered by the EPA. As new biotech products have been developed, environmental and consumer groups have expressed concerns that EPA rules do not adequately cover all the relevant risk issues (e.g. oral toxicity, potential for allergenicity) and the USDA should examine more thoroughly for risks of new plants out crossing with wild relatives to produce unusually hardy weeds or adversely affect biodiversity (National Research Council, 2000a).

Labelling for GMO content has not been adopted in the USA, but indirectly information will be provided in a new USDA standard for 'organic food' labelling (Golan *et al.*, 2000). For crops, the standard means that the use of genetic engineering, irradiation and sewage sludge in the production or processing stages is prohibited. Although all organic farmers and handlers would be expected to abide by the standard, it remains to be seen whether the standard can be effectively enforced.

For countries in the European Union (EU), the EU has established polices as directives to member nations on environmental legislation. The EU Directive on the Contained Use of Genetically Modified Organisms (Directive 90/219/EEC) and on Deliberate Release (Directive 90/220/EEC) attempt to establish a system for controlling the use of GM organisms (European Commission, 2000). The directives were modelled after the EU's chemical notification directives. Since these EU directives are not implemented uniformly across member countries and no central monitoring authority exists, the system is somewhat loosely controlled.

Directive 90/219/EEC provides common rules throughout the EU for the use of GM microorganisms in research laboratories and industrial facilities and provides appropriate measures to protect human health and the environment from any risks arising from activities using GM microorganisms. The Directive outlines appropriate procedures for risk management. Microorganisms and activities using them are to be classified by their potential for risk and to containment and control measures. Each country must designate 'competent authorities' to receive information from commercial companies and research institutes. These authorities must organize inspection and other control measures. They must also examine the conformity of notifications received with the requirements of the directives. Effective risk management is expected, and it means that a careful risk assessment of contained use must be made, the appropriate level of containment must be exercised and suitable preventive measures must be taken (European Commission, 2000).

Directive 90/220/EEC covers deliberate release of GM organisms into the environment for research and development purposes and the placement on the market of products containing GMOs. The directive takes a preventive approach, emphasizing prior assessment and approval. The main elements are: (i) an environmental risk assessment must be carried out before any experimental or commercial releases of GMOs into the environment; (ii) no release can be carried out without the consent of the competent authority; (iii) an approval procedure by a nation's competent authorities should limit experimental releases to at most 90 days; and (iv) EU Community approval is required for commercial releases of GMOs (European Commission, 2000).

The latter procedure has been implemented as follows (Maurer and Harl, 2000). On receipt of the notification, the competent authority in a member country has 90 days to either forward the notification dossier to the European Commission with a favourable opinion or inform the notifier that the proposed commercial release does not fulfil the requirements of the directive. After receiving a notification dossier, the Commission immediately forwards it to the competent authorities of all other member states. If no objection is raised by the competent authorities of these states within 60 days after the commission forwards the notification, the competent authority that first received the notification issues a written consent to the applicant and informs the other member countries and the Commission of the consent. When member countries do not reach agreement on a notification, the Commission draws on relevant scientific committees for information and opinions. After weighing this information, the Commission makes a decision which is binding on all EU members.[12] Individual countries are also urged to have effective penalties for improper release of GMOs, but none has been established.

[12]For a company attempting to obtain commercial approval for selling GMOs, this is a relatively complicated and time consuming procedure in comparison to having one central EU agency make a decision.

Public confidence in EU GMO policies has been undermined by recent information that unapproved GMOs have been sold in some EU countries and planted by farmers; for example, oilseed rape in several countries and cotton in Greece and Spain (Greenpeace, 2000c,d). The establishment of governmental policies that are unenforceable or weakly enforced does not build public confidence or trust in government. Directive 90/220/EEC is currently under revision.

GMO labelling was adopted by the European Commission in 1997. It requires each member country to enact a law requiring labelling of all new products containing GM organisms approved under Directive 90/220/EEC (i.e. where there are safety reasons). Under this policy, a GMO is defined as 'an organism in which the genetic material has been altered in a way that does not occur naturally by mating and/or natural recombination'. Also, mixtures with non-GMOs would indicate the possibility that GMOs may be present.

The EU's labelling policy has been controversial. First, proven safety risk evidence is generally lacking. Second, the scientific meaning of 'none' has been controversial in an era where rapid advances in scientific instrumentation has made it possible to detect ever smaller units, and accidental physical mixing of GMO and non-GMO commodities can easily occur. Third, the requirements for restaurants and caterers is so low as to be virtually useless to interested consumers.

4. Conclusions

This chapter has addressed the economics of information as it affects the acceptance of and benefits from agricultural biotechnology. We have shown that the producers and consumers can make good decisions on acceptance and use of GM products if there is freely accessible, verifiable information and competition in the provision of information by interested parties. However, when information is not verifiable and decision makers must rely on interested parties, achieving good decisions is much more difficult. Ag-biotech companies, environmental groups and others seem likely to have interests that diverge from those of consumers and producers and to use private information strategically, especially when the supply of new GM products is advancing rapidly. Furthermore, rapid advances in communication and information technologies have greatly reduced the cost and increased the speed with which information can be distributed. The groups' private information can cause the market for GM products in one way or another to collapse. This may be at considerable loss in social welfare.

We have argued that the services provided by an institutionalized information verification system operated by an independent body would be a mechanism for producing good public information services that would have large social value. This institution would reveal private information, establish a process for refuting or confirming claims of interested parties and advance the stock of knowledge on short- and long-term effects of biotechnology. Advances in the stock of knowledge are important to a successful information verification system because it can expand the topics, issues or dimensions of the knowledge base on which verification can be made. The primary contributors to this activity seem likely to be scientists employed by commercially independent institutions, funded primarily by the public sector and working to meet scholarly and scientific standards.

The relevant categories of information would cover topics consistent with a broad range of income and intergenerational interests. To obtain and maintain public trust, it must be 'consumer driven' and broader than pure scientific issues, although they would be one important component. It would, however, include scholarly presented and summarized information on ethical, social, economic, environmental, food safety, scientific and trade issues dealing with GMOs.

Verified information is costly to produce, but once provided, it has international public good attributes. We have argued that the advances in information and communication technologies have increased the free-rider problem by weakening the incentives for any one country to undertake such activities. We suggest that the provision will most likely be by some large country that stands currently to receive a significant share of the net social benefits from a verifiable information system for ag-biotechnology.

Currently, public institutions dealing with regulating GMOs and labelling in the USA have relatively high public trust but none fulfils our conceptual view of a verifiable information system. Public bodies in Europe are experiencing low public trust as an information source, and this is undoubtedly making good decision making more difficult. An effective information system dealing with biotechnology must be managed well in order to provide large social benefits.

Acknowledgments

The authors are Professor of Economics, Iowa State University, and Economist, Resource Economics Division, ERS, USDA, respectively. Matt Rousu provided research assistance, and the Economic Research Service (cooperative agreement no. 43-3AEL-8-80125) and Iowa Agriculture and Home Economics Experiment Station provided financial support. Views presented in this chapter are the authors' and do not represent ERS or the USDA. Journal paper number J-19000 of the Iowa Agriculture and Home Economics Experiment Station, Ames, Iowa, Project 3077, supported by Hatch Act and State of Iowa funds.

References

Akerlof, G. (1970). The Market for Lemons: Quality Uncertainty and the Market Mechanism. *Quarterly Journal of Economics* 84, 488–500.

Alston, J.M., G.W. Norton and P.G. Pardey (1995). *Science Under Scarcity: Principles and Practice for Agricultural Research Evaluation and Priority Setting.* Ithaca, New York: Cornell University Press.

Avery, C., P. Resnick and R. Zeckhauser (2000). The Market for Evaluations. *American Economic Review* 89, 564–584.

Convention on Biological Diversity (1993). User's guide to the clearing house mechanism. www.biodiv.org/chm/userguide/CHM_User_Guide.html

Cornes, R.C. and T. Sandler (1996). *The Theory of Externalities, Public Goods and Club Goods,* 2nd Edn. New York: Cambridge University Press.

David, P.A. (2000). "The political economy of public science." In H.H. Smith (Ed.), *The Regulation of Science and Technology.* London: Macmillan.

European Commission (2000). *Guidelines to the Approximation of European Union Environmental Legislation.* Part 2G. *Chemicals and Genetically Modified Organisms.* Brussels: European Commission.

Falck-Zepeda, J.B., G. Traxler and R.G. Nelson (2000a). Surplus Distribution from the Introduction of a Biotechnology Innovation. *American Journal of Agricultural Economics* 82, 360–369.

Falck-Zepeda, J.B., G. Traxler and R.G. Nelson (2000b). Rent Creation and Distribution from Biotechnology Innovations: The Cause of Bt Cotton and Herbicide-Tolerant Soybeans in 1997. *Agribusiness.*

Frey, K.J. (2000). *National Plant Breeding Study-IV. Future Priorities for Plant Breeding.* Special Report 102, Iowa State University, Iowa Agri. Home Econ. Experiment Station.

Friends of the Earth (2000). Genetically engineered food: farm scale trial locations. Friends of the Earth, GM sites. www.foe.co.uk/camps/foodbio/sites/index.html.

Gaskell, G., M.W. Bauer, J. Durant and N.C. Allum (1999). Worlds Apart? The Reception of Genetically Modified Foods in Europe and the U.S. *Science* 285, 384–387.

Gibbons, R. (1998). Incentives in Organizations. *Journal of Economic Perspective* 12, 115–132.

Golan, E., F. Kuckler and L. Mitchell (2000). *Economics of Food Labeling.* US Department of Agriculture, Agricultural Economics Report No. 793.

Greenpeace (2000a). EPA responds to Greenpeace lawsuit defends rubber stamping of bud science?' Greenpeace USA Media Center, 20 April.

Greenpeace (2000b). Kellogg's tiger targeted in action condemning use of Franken Food. Greenpeace USA Media Center, 23 May.

Greenpeace (2000c). Illegal GE crops found in several European countries: Greenpeace calls for urgent action to prevent further imports. Greenpeace Genetic Engineering Highlights, 19 May. www.greenpeace.org/-geneng/highlights/gmo/00_5_19.html.

Greenpeace (2000d). Greenpeace and farmer decontaminate oil seed rape field in Germany. Greenpeace Genetic Engineering Press Release, 29 May. www.greenpeace.org/pressreleases/geneng/2000may29.html

Heijs, W.J.M. and C.J.H. Midden (1996). *Biotechnology. Attitudes and Influencing Factors: Fourth Survey.* Eindhoven University of Technology, Department of Philosophy and Social Science, Eindhoven, Netherlands.

Hoban, T.J. (1997). Consumer Acceptance of Biotechnology: An International Perspective. *Nature Biotechnology* 15, 232–234.

Hoban, T.J. (1999). Consumer Acceptance of Biotechnology in the U.S. and Japan. *Food Technology* 53, 50–53.

Holmstrom, B. (1979). Moral Hazard and Observability. *Bell Journal of Economics* 10, 74–91.

Holmstrom, B. (1989). Agency Costs and Innovation. *Journal of Economic Behaviour and Organization* 12, 305–327.

Holmstrom, B. and P. Milgrom (1987). Aggregation and Linearity in the Provision of Intertemporal Incentives. *Econometrica* 55, 303–328.

Huffman, W.E. (2002). "Human Capital: Education and Agriculture." In B.L. Gardner and G.E. Rausser (Eds.), *Handbook for Agricultural Economics*, Vol. 1. Amsterdam: Elsevier Science (in press). (Also, Iowa State University, Department of Economics Staff Paper No. 222, August 1999.)
Huffman, W.E. and R.E. Evenson (1993). *Science for Agriculture: a Long Term Perspective*. Ames, Iowa: Iowa State University Press.
Huffman, W.E. and R.E. Just (2000). Setting Efficient Incentives for Agricultural Research: Lessons from Principal-Agent Theory. *American Journal of Agricultural Economics* 82, 828–841.
Macilwain, C. (1999). U.S. Food-Safety Body Hears Protests Over Genetically Modified Food. *Nature* 402, 571.
Martin, S. and J. Tait (1992). *Release of Genetically Modified Organisms: Public Attitudes and Understanding*. London: The Open University, Centre for Technology Strategy.
Maurer, M.T. and N. Harl (2000). *Obtaining Approval to Market GMOs in the European Union*. Working paper, Iowa State University, Department of Economics.
Milgrom, P. and J. Roberts (1986). Relying on the Information of Interested Parties. *RAND Journal of Economics* 17, 18–32.
Molho, I. (1997). *The Economics of Information*. Malden, Massachusetts: Blackwell.
Moschini, G., H. Lapan and A. Sobolevsky (1999). *Roundup Ready Soybeans and Welfare Effects in the Soybean Complex*. Staff working paper No. 324, Iowa State University, Department of Economics.
National Research Council (2000a). *Transgenic Plants and World Agriculture*. Washington, DC: National Academy Press.
National Research Council (2000b). Committee on Genetically Modified Pest-Protected Plants. *Genetically Modified Pest-Protected Plants: Science and Regulation*. Washington, DC: National Academy Press.
Nettleton, J.A. (1999). The Frankenfood myth. *Food Technology* 53, 20.
OECD (2000). Rapporteurs' Summary: GM food safety: facts, uncertainties, and assessment. From *OECD Edinburgh Conference on the Scientific and Health Aspects of GM Foods*, 28 February–1 March.
Olson, M. (1965). *The Logic of Collective Action*. Cambridge, Massachusetts: Harvard University Press.
Sandler, T. (1992). *Collective Action*. Ann Arbor, Michigan: The University of Michigan Press.
Sandler, T. (1997). *Global Challenges: an Approach to Environmental, Political, and Economic Problems*. New York: Cambridge University Press.
Sandler, T. (1998). Global and Regional Public Goods: A Prognosis for Collective Action. *Fiscal Studies* 19, 221–247.
Scheermeier, Q. and D. Dickson (2000). Europe Lifts Patent Embargo on Transgenic Plants and Animals. *Nature* 403, 3.

Serageldin, I. (1999). Biotechnology and Food Security in the 21st Century. *Science* 285, 387–389.

Shapiro, C. and H.R. Varian (1999). *Information Rules: a Strategic Guide to the Network Economy.* Boston, Massachusetts: Harvard Business School Press.

Sobel, J. (1985). A Theory of Credibility. *Review of Economic Studies* 52, 557–573.

World Bank (1999). *Knowledge for Development.* World Development Report 1998/99. New York: Oxford University Press.

Chapter 5

Media Coverage, Public Perceptions, and Consumer Behavior: Insights from New Food Technologies*

Jill J. McCluskey[†,||], Nicholas Kalaitzandonakes[‡,**], and Johan Swinnen[§,¶,††]

[†]*School of Economic Sciences, Washington State University, Pullman, Washington 99164*

[‡]*Economics and Management of Agrobiotechnology Center, University of Missouri, Columbia, Missouri 65211*

[§]*Department of Economics and LICOS Centre for Institutions and Economic Performance, KU Leuven, Leuven, B-3000 Belgium*

[¶]*Center on Food Security and the Environment, Stanford University, Stanford, California 94305*

[||]*mccluskey@wsu.edu*

[**]*KalaitzandonakesN@missouri.edu*

[††]*jo.swinnen@kuleuven.be*

The media often play the role of translating new science to consumers. We discuss the recent literature that has examined the supply and demand factors that affect media coverage of new food technologies and the impact on public perceptions and consumer behavior toward food that utilizes these technologies. We start with a discussion of the ways in which the media influence public perceptions and consumer behavior related to foods made with new technologies. We then discuss the incentives of news media and the potential sources of biases in their reporting. We review empirical studies that have examined media reporting of new agricultural and food technologies, especially biotechnology, in terms of both their agenda setting and framing effects

*This article originally appeared in the **Annual Review of Resource Economics (2016), 8(1), 467–486**. This is published by permission of **Annual Reviews**.

and the social amplification of risk. We synthesize the findings of studies that have examined the influence of media coverage on public attitudes and consumer behavior. We conclude and discuss avenues for future research.

Keywords: agenda setting, biotechnology, framing, GM foods, media, media bias, new food technologies, political economy

Not that the media lie about the news they report; in fact, they have incentives not to lie. Instead, there [are] selection, slanting, decisions as to how much or how little prominence to give a particular news item (Posner 2005).

1. Introduction

Consumers are inundated with information about food through a myriad of media access points from cell phones to traditional television and newspapers. Still, the public is said to be poorly informed when its consumption declines in opposition to new food technologies, despite claims by scientists and officials that such new technologies are safe. Lusk (2015) points out the divergence in risk perceptions between the scientific community and the general public with results from his food demand survey. He found that a large majority (82%) support mandatory labels on genetically modified organisms (GMOs) but that curiously approximately the same proportion (80%) also support mandatory labels on foods containing deoxyribonucleic acid (DNA). Indeed, the media and the information they provide are often blamed for the sometimes irrational and disproportional consumer response to new technology. A concern is that information provided by the media may not be objective or may not foster the public good.

The media often play the role of translating new science to consumers. The information is provided by either privately or publicly owned media that may have incentives to slant the information they provide in particular ways. Who owns the news media appears to matter. Privately owned media have become increasingly important globally. Whereas in the United States news coverage has always been largely commercial, in Europe private companies have become the dominant source of information only over the last two decades.

Also, in other regions of the world, there is significant growth in private and commercial media, but state ownership of media is still more important in low-income countries. The structure of the private media market has also changed with the growth of 24-h news media, pay per view, satellite, cable, websites, and Internet blogs.

Commercial mass media have thus had an increasingly important influence on public opinion. Extensive media coverage affects consumer perceptions of products and risk and consequently consumer demand for goods and services. Yet the media are also criticized for being sensationalistic and biased in their reporting, focusing on sound bites and ignoring careful analyses. There is also concern that the increased commercialization of the media has produced a dumbing down of the news because this trend is leading to decreased quality and quantity of coverage of complex issues and because competitive pressures are associated with cutbacks in reporting and editorial quality (Alterman 2008). The emergence of the 24-h news cycle may have actually weakened journalism, and news reports include an increasing number of factual errors (Pew 2004).

These concerns have induced many European governments to continue to subsidize public news broadcasting to improve the overall news quality. However, it is not clear that public news is always unbiased or of higher quality. Furthermore, if subsidized public media offer higher-quality news, this may result in low-quality commercial news filling the leftover niche (Canoy & Nahuis 2005). This argument is similar to those in studies of the US media market, which show that the regional spread of so-called quality newspapers such as *The New York Times* and *The Washington Post* has induced a reduction in the quality of local and regional newspapers (George & Waldfogel 2006).

Although the literature on the (political) economics of the media is relatively recent, it is a burgeoning field, with applications in many areas, including political economy (impacts on electoral turnout and voting) and public goods provision (e.g., Della Vigna & Gentzkow 2010; Gentzkow & Shapiro 2008a,b; McCluskey & Swinnen 2011; and Prat & Strömberg 2011 for partial reviews). Similarly extensive is the literature on media coverage and its impacts on the public. In this article, we review the literature that has examined the supply and demand factors that shape media coverage of new food technologies and the impact of such coverage on public perceptions and consumer behavior toward these technologies. This is an important area of focus because Hoban & Kendall (1993), Marks *et al.* (2003), and Wildavsky (1995) find that the vast majority of consumers receive information about food technologies primarily through the popular press and television and first encounter claims of harm to health and the environment through the media.

More than 100 published studies have empirically analyzed how various news media have covered agricultural biotechnology and genetically modified (GM) foods, the use of synthetic pesticides and growth hormones in agriculture, the use of nanotechnology in food processing, and other new food technologies. These studies have generally focused on the agenda setting and framing roles of news media and their impacts on public perceptions and attitudes. Studies on media coverage of GM foods exemplify the wide-ranging hypotheses tested and the rich data and analytical methods used, and, thus, we review the primary findings of these studies here.

We structure the rest of the review as follows. In Section 2, we discuss the ways news media coverage of new technologies in food and agriculture may influence public perceptions and consumer behavior related to foods made with new technologies. In Section 3, we discuss the incentives of news media and the potential sources of biases in their reporting. In Section 4, we review empirical studies that examine media reporting of new agricultural and food technologies, especially biotechnology, in terms of both their agenda setting and framing effects and the social amplification of risk. In Section 5, we synthesize the findings of studies that examine the influence of media coverage on public attitudes and consumer behavior. Finally, in Section 6, we conclude and discuss avenues for future research.

2. Media and Controversies Over Food Technologies

Use of new technologies in agriculture and food production has consistently led to increased productivity and food supplies, reduced food prices, enhanced food attributes, and entirely new products (Huffman & McCluskey 2014). Such new technologies include synthetic pesticides, growth hormones, biotechnologies, nanotechnologies, and various new processing techniques. Despite assurances from scientists, governments, and industry that new food technologies are safe, consumers have often been skeptical and even polemic. These negative consumer attitudes are sometimes blamed on consumers being poorly informed. However, consumers from highly educated and affluent segments are often the most guarded (Curtis *et al.* 2008). Indeed, many of these consumers choose to opt out of consuming foods produced with new technologies by buying organic and other foods, which are produced without biotechnology seeds, growth hormones, and synthetic pesticides (Loureiro *et al.* 2001). Such consumers generally do not fit the profile of uninformed consumers. Rather, they seem willing to pay a small share of their typically high income to avoid perceived risks associated with foods made with the use of new technologies.

The use of new technologies in agriculture and food production therefore seems increasingly dependent on consumers' risk perceptions. Although consumers may like the lower prices and the potential product-enhancing attributes of foods produced with new technologies, they may also be suspicious of such products when the underlying production processes and technologies seem risky to them (Annunziata & Vecchio 2011). A typical example of such behavior is the frequently displayed negative attitudes toward GM foods (Lusk *et al.* 2005). The scientific consensus is that first-generation GM foods are equivalent to their conventional counterparts. However, on average, consumers say that they want a discount to choose GM foods over conventional ones (Lusk *et al.* 2005). The consumer response in the market determines product success because, in the end, consumers' perception of risk — rather than the perception of objective, technical risk — directly affects consumer choices and market demand.

What is behind the elevated perceived risks associated with new food technologies that consumers seem to display so readily? Lusk *et al.* (2014) primarily discuss how developments in the economic and sociopsychological literature shed light on consumer preferences for food technology and the determinants of food technology acceptance.

Heiman & Zilberman (2011) find that both positive framing and negative framing affected the likelihood of purchasing GM products at a given price but that negative framing had a stronger impact in deterring consumers. Other commentators point the finger at the news media. For instance, Bauer & Gaskell (2002) suggest that negative consumer perceptions and attitudes toward biotechnology may have been cultivated by plentiful media coverage and unbalanced reporting of its relative risks and benefits. Lusk *et al.* (2014) briefly discuss the impact of media on consumers' risk perceptions. They explain that media can frame food technologies by emotionalizing an issue and increasing salience with repetitive messaging.

Because in many countries the news media serve as primary sources of risk communication for the general public (Boholm 1998), they may also define the agenda of public concern. This process of mediated public attention is known as agenda setting or sizing and highlights the role of the media in focusing the public's attention on one issue and not on another (Cohen 1963, McCombs & Shaw 1972). The main theoretical claim of the agenda setting hypothesis is that the degree of emphasis placed on selected issues by the news media adds salience to these issues: The greater the volume and prominence of media coverage are, the more important the public will judge the issues to be (Entman 1991, McCombs & Ghanem 2001).

In addition to setting the agenda, news media may also frame the discussion around it. Tuchman (1976, 1978) argues that through reporting on an issue, news media select few events out of many and turn them into newsworthy stories. In telling these stories, journalists use frames that are shaped by their viewpoints, experiences and skills, deadlines, media space, and other practical parameters (Hornig 1990). When writing about a new food technology, journalists may choose to emphasize technical features, socioeconomic implications, environmental risks or benefits, human health concerns, and so on (Hornig 1990). Through frames, therefore, news media may highlight certain points of view and marginalize others (Hornig 1993). When risks are emphasized relative to the benefits of a new technology, the theory predicts that a more negative public sentiment can prevail and that the news media can become a means for the social amplification of risk.

The agricultural and food industries have inherent incentives to defend their ability to use new technologies by providing their view of risks and benefits through information campaigns and education activities and by influencing the reports of the news media (Annunziata & Vecchio 2011). However, consumer confidence in information about food products made with new technologies may vary depending on the source (Annunziata & Vecchio 2011, Siró *et al.* 2008). Several studies indicate that consumers are more confident in information about the potential benefits of foods made with new technologies when it comes from trusted and credible authorities and that consumers are less confident in information provided by industry (Cox *et al.* 2008, Urala & Lahteenmaki 2007). Yet the difficulties of industry in achieving more positive communication about new food technologies may go beyond a trust deficit. News media have their own objectives and incentives that may lead to biased reports. For instance, the underlying science of new food technologies can be complicated, and the media need to sell stories.

Another key constraint is that, because there are diminishing returns to time spent processing information from media stories and there are increasing amounts of information, consumers may still be rationally ignorant in a world with plenty of information and media (McCluskey & Swinnen 2004) — and are hence less likely to spend the time to become well informed about the benefits of such new technologies. Kiesel *et al.* (2011) discuss that consumers may not use nutritional labels because time and effort are required to process the information. The time required to process the information about the potential benefits and risks associated with new food technologies is likely greater than the time required to process information on nutrition labels, so we expect for most consumers to not be experts on new food technologies.

Furthermore, consumers give greater weight to negative information than to positive information. This is interesting because often-heard complaints are that news coverage is too negative and that there is not enough positive news in the stories that media produce. McCluskey *et al.* (2015) show that this generally recognized tendency of the popular news media is driven by demand. Thus, the news media may be more likely to highlight potential risks associated with new food technologies in their reporting. However, this tendency says little about the impact on perceptions or behavior of consumers. Consumers may realize the existence of bias and may thus discount the information provided by media. In the next sections, we first review potential sources of bias in the media and then review empirical studies that measure the impact of bias on consumers.

3. Media Bias and Framing

Is the general emphasis of news media on the potential risks over the benefits of new food technologies a systematic bias, and if so, are there specific structural factors that drive this bias? Media bias can take various forms, and there is no generally accepted definition. Anand *et al.* (2007, p. 637) write, "The phenomenon of bias in the media appears to be quite different than, say, a statistician's notion of bias — because bias lies in the eyes of the beholder (consumer)." Others define bias as the "absence of balance resulting in one side of a story receiving unwarranted attention" (Baron 2006, p. 4) or, in other words, as "sins of omission — cases where a journalist chose facts or stories that only one side of the ... spectrum is likely to mention" (Groseclose & Milyo 2005, p. 1205).

Studies have identified several possible theoretical explanations for the existence of bias. Bias can be induced by supply and/or demand factors. On the supply side, it can be due to ideology or partisan politics, in which owners, editors, or journalists present stories that support particular world views. Bias can also result from falsehoods or from information hidden or distorted by sources or journalists eager for a scoop or under pressure to attract attention. Finally, bias can be a response to consumer preferences.

The most obvious source of bias is preferences from owners, editors, or journalists who may affect the news coverage (Bovitz *et al.* 2002). This bias is most evident in mass media owned by the state (e.g., North Korea) and is present in many developing countries where the state continues to control mass media. In those countries, the media are used by governments to disseminate the political communication of the ruling parties and to control information that may threaten their legitimacy or their hold on power.

However, in less rigorously controlled media regimes, bias can still be significant. In many European countries, until recently, many of the nonstate-owned printed media and television stations were owned or were closely related to political parties, and the different media expressed the preferences of their parties. Similarly, public television organizations were often influenced by the parties in government. An interesting illustration is Italy, where the main leader of the right-wing political party, Silvio Berlusconi, owns many of the commercial TV stations, and their political news coverage switches when left- or right-wing parties take over government. Governments can also put strong pressure on media not to publish stories.

Owners of commercial media may wish to impose their personal preferences on their media reporting. In doing so, they may face a trade-off between political objectives (i.e., using the media to express the owners' ideological bias) and commercial objectives (Mullainathan & Shleifer 2005). First, commercial objectives may be affected by potential consumers' distaste for bias or by the negative utility they get from consuming media products that differ from their personal social and political preferences. Second, commercial media's profitability depends not only on consumers but also, to a large extent, on advertising revenues. Bagdikian (1992, p. 129) writes, "As mass advertising grew, the liberal and radical ideas — in editorials, in selection of news, and in investigative initiatives — became a problem. If a paper wished to attract maximum advertising, its explicit politics might create a disadvantage." Along the same lines, Gabszewicz *et al.* (2001) show that the media's incentives to appeal to a larger audience and to hence be more attractive to advertisers may induce editors to moderate the political messages they display to their readers.

Other supply-driven forms of media bias may be related to the availability of potential journalists who are willing to work for lower wages in positions that can advance their careers or who demonstrate influence by exercising the discretion granted by news organizations (Baron 2006). Dyck & Zingales (2002) focus on the relationship between journalists and their sources of information as the reason for media bias. Sources may release partial information that supports their preferences, or journalists may use partial information to reward sources for providing information. Baron (2005) models how the competition between information sources affects the news report. Private information may be held by two sources with opposing views. The sources have incentives to reveal only information that supports their own views, and it is costly for the media to obtain additional information from independent sources. Dyck & Zingales (2002) argue that to induce a source to reveal information, the journal puts a positive spin on stories to reward the source for providing the information.

On the demand side, McCluskey *et al.* (2015) explain that negative news coverage is likely to dominate positive news stories. Their argument is based on the premise that consumers use the information from positive media stories to take advantage of opportunities from positive shocks and use the information from negative stories to avoid negative shocks. If utility is concave, the marginal loss in utility from not consuming the first bad news story is greater than the marginal gain in utility from consuming the first positive news story. As a result, consumers will choose to consume more negative stories than positive ones.

Spatial models of firm location also provide a consumer-driven rationale for bias based on product differentiation. Mullainathan & Shleifer (2005) argue that readers or viewers have a preference for news that is consistent with their initial beliefs and that media organizations therefore have an incentive to bias their reporting toward confirming their readers' or viewers' initial beliefs. When readers are heterogeneous in their beliefs, accuracy increases due to cross-checking of facts across newspapers. This is a wisdom-of-crowds argument (Surowiecki 2004) that aggregation of signals reduces noise. Anand *et al.* (2007) assume that facts are not always verifiable and that consumers have heterogeneous ideologies. They find that when facts are verifiable, there is no bias. However, when a news item comprises information that is mostly nonverifiable, then consumers may care about both opinion and editorials, and the media firm's report will contain both of these aspects. The diversity of opinion and editorials results in a differentiated product market. A dynamic version of this type of argument can be made when media organizations attempt to obtain a reputation for accuracy, which induces bias in reporting. Gentzkow & Shapiro (2006) consider the Bayesian consumer who is uncertain about the quality of an information source. The consumer infers that the source is of higher quality when its report conforms to the consumer's prior expectations. Consequently, media may slant their reports toward the prior beliefs of their customers to build a reputation for quality.

The incentive to attract large numbers of readers or viewers may also lead to bias. One mechanism is explained by Strömberg (2004), who argues that media coverage is biased toward large groups, as the media are more likely to cover issues that are of interest to such groups. This bias can result from the need to attract as many readers as possible or from economies of scale in the media. Kuzyk & McCluskey (2006) provide empirical support for Strömberg's (2001) theoretical model with content analysis.

An interesting empirical study on these issues is by Gentzkow & Shapiro (2010). They analyze text from a large set of US media and come to the conclusion that "newspapers' actual slant is neither to the right nor to the left of the profit maximizing level, on average." They conclude that the slant (or bias) in newspapers is strongly related to the political distribution of their potential readers, much more so than to the political preferences of their owners or the journalists.

Competition may also play an important role in media organizations' trade-off between ideology and profits. Some researchers, such as Baron (2006), show that bias can persist in the face of competition. Gentzkow & Shapiro (2006, 2008b) argue that competition reduces supply side–induced bias because it increases the likelihood that erroneous reports will be exposed ex post, but these researchers argue that the impact of competition on demand side–induced bias is less clear. Mullainathan & Shleifer (2005) in their two-firm location model show that price competition results in greater product differentiation — e.g., more slanting of news. When advertising revenues are included in the product differentiation models of media, minimum differentiation can result (Barros *et al.* 2004, Gabszewicz *et al.* 2004, Gal-Or & Dukes 2003).

Hence, many supply and demand factors have been identified in the literature as potential sources of bias and framing in news media reporting. In the next sections, we review evidence on bias and framing in media reporting of new food technologies, especially agricultural biotechnology and GM foods, which have been widely studied.

4. Empirical Studies of Media Coverage of New Food Technologies

For more than 25 years, news media reporting on biotechnology and GM foods has been closely examined in many parts of the world, including various European countries, the United States, Japan, China, Australia, the Philippines, Kenya, and Turkey. Studies have compared media coverage across two or more countries (e.g., Bauer 2005b, Botelho & Kurtz 2008, Lewison 2007, Listerman 2010, Marks *et al.* 2003, Vilella-Vila & Costa-Font 2008); across different biotechnology applications, including medical as well as crop and livestock improvements (e.g., Hibino & Nagata 2006, Marks *et al.* 2007, Marques *et al.* 2015, Müller *et al.* 2010); across different types of news media, including elite press, tabloids, and television (e.g., Carver *et al.* 2013, Maeseele & Schuurman 2008, Vicsek 2013); and over time. These comparative analyses have demonstrated systematic differences in the amount of coverage and tone across geographies, in types of media, and in biotechnology applications and systematic differences over time and have teased out some of the underlying causes of such differences. Furthermore, they have often linked media coverage of biotechnology and GM foods to public attitudes and perceptions toward them.

4.1. *The agenda setting role of news media*

Mazur (1981) proposes that the quantity of coverage, independent of tone, would tend to lead to a conservative bias or increasingly negative attitudes toward a technology. As such, a large number of published studies have measured the frequency of news media reporting on biotechnology and GM foods, and many have described changes in the amount of coverage over time.

Building on Downs's (1972) idea of issue attention cycles, Nisbet & Huge (2006) describe cycles of increased and then decreased media attention to plant biotechnology and propose underlying social mechanisms that lead to cyclical media coverage. Several other studies document the cyclical nature of media coverage of biotechnology and GM foods. For instance, Listerman (2010) observes an increase, peak, and decrease in US, UK, and German media coverage of biotechnology over the 2000–2002 period. Similar media attention cycles are also visible over longer time periods when data from various studies across many countries and news media are pieced together. From early studies in European countries (e.g., Bauer & Gaskell 2002, Bonfadelli *et al.* 2002, Durant *et al.* 1998, Jesuino *et al.* 2001, Kohring & Matthes 2002, Wheeler *et al.* 2002), it can be inferred that in France, Austria, the United Kingdom, the Netherlands, Sweden, and other countries, news media attention to biotechnology was quite low in the early 1990s and increased between 1995 and 1997. Shanahan *et al.* (2001) show further growth in media coverage of agricultural biotechnology until 2000, with a dramatic increase in attention occurring between 1997 and 2000. McInerney *et al.* (2004) examine patterns of media coverage of biotechnology and GM foods between 1992 and 2002 and report a peak in the coverage in 2000. Marks *et al.* (2003, 2007) report similar results.

The dramatic increase in media attention to biotechnology in the late 1990s and the observed peak in the early 2000s were driven mostly by events (Botelho & Kurtz 2008; Cook *et al.* 2006; Flipse & Osseweijer 2013; Marks *et al.* 2002, 2003, 2007) that were triggers to jump-start media coverage and public attention. Abbott & Eichmeier (1998) term this phenomenon the hoopla effect. Trigger or flash point events were both global and local in nature.

Two scientific controversies in the late 1990s attracted global media attention. First, in 1998, Árpád Pusztai claimed, on national television in the United Kingdom, that biotech potatoes fed to laboratory rats had caused damage to their organs and overall development. Although his work was disputed by scientific societies and his institution, it was published in *The Lancet* and attracted media attention in a large number of countries. Then in 1999, Losey and his colleagues reported in the journal *Nature* that monarch butterflies could be harmed by biotech corn pollen. Although the study's findings proposed a potential hazard that was more real in the lab than in the field, it quickly attracted international media attention. On the heels of the monarch butterfly controversy, another one centered around a market disruption unfolded and maintained the attention of global media to biotechnology. The discovery of StarLink corn in the US food supply (a biotech corn approved only for animal consumption) led to recalls of hundreds of food products and to trade disruptions in international corn markets, leading to numerous media reports across the globe.

Local events also played a role in the increased media attention to biotechnology and GM foods in the late 1990s and early 2000s. For instance, in May 2000, the Prince of Wales publicly condemned GM foods while in the same month Scottish farmers discovered that they had mistakenly sowed unapproved GM seeds in their fields. These events were heavily reported in the United Kingdom, but not elsewhere. In contrast, the US president's statement on biotechnology and the first field trial destruction were widely reported in the Philippines, while other local events signaling controversies attracted similar media attention in other countries.

Overall, increased media attention has heightened public attention to biotechnology and GM foods and has had an agenda-setting role in fueling public debate. Within this context, many studies have examined what sort of information was provided to the public through the media and what kind of balance this information flow assumed over time.

4.2. *The framing role of news media*

News stories are constructed realities that pull together facts and information within a narrative structure, or frame, which communicates an event or story to the reader (Tuchman 1976). Fresh information about new food technologies, especially in fast-changing scientific fields such as biotechnology, is ongoing and continuously shapes the balance of risks and rewards of such technologies. There is extensive research on how reporters have interpreted and narrated such information, that is, how they have framed the public debate. The definition of a frame tends to vary from one study to another, depending on the research interest. Pan & Kosicki (1993) explain that frame analysis in news media text must be operationalized not only through examination of syntactic and script structures, but also through the media's themes and rhetoric. Empirically detecting specific frames and objectively measuring the messages that such frames communicate are not straightforward, however.

Story frames might include cognitive elements, which convey specific information about certain aspects of the technology, or affective elements, which carry a positive or negative tone in the overall representation of the technology. Entman (1991) suggests that story frames can be detected in practice by probing for particular words and visual images that consistently appear in the narrative. Researchers have therefore surveyed news media articles for the presence of both affective and cognitive elements through textual or content analysis — a systematic method for analyzing and quantifying message content and message handling in news stories (Budd *et al.* 1967). For instance, Gaskell & Bauer (2001), Marks *et al.* (2007), and others have looked for frames consistent with certain alleged risks or benefits by using content analysis of news media articles.

Researchers have also been interested in general frames as well as issue-specific frames used in the news media coverage of biotechnology and GM foods. For instance, Augoustinos *et al.* (2010) analyze newspaper articles published in the United Kingdom, with an interest on how media, in general, framed the position of major stakeholders in the national debate on GM foods: the British public, the British government, and biotechnology companies. The authors find that a pervasive representation of the British public as uniformly opposed to GM foods served rhetorically to position the British government as undemocratic and beholden to industry. In contrast, Veltri & Suerdem (2013) examine news media framing of biotechnology and GM foods in Turkey in relation to the introduction of the national biosafety law in 2010. Within this issue-specific context, the authors find that some media outlets framed their discussion on biotechnology by emphasizing political controversies, although this discussion was ideologically nuanced. For example, some Islamist and extreme nationalist outlets characterized the issue as a conspiracy by Jewish multinational companies for world domination, whereas left-leaning outlets characterized the law as a product of capitalist development with potential future hazards. Other news media framed biotechnology and the introduction of the biosafety law as a technical issue that should be placed under government supervision.

Despite the variety of the frames selected by researchers or the words and syntactic features of the articles examined, most published studies have ultimately assessed the overall tone and balance of news media coverage on biotechnology and GM foods. In this context, a number of researchers have concluded that reporters across the globe have consistently drawn parallels between agricultural biotechnology and other known food hazards (e.g., mad cow disease and dioxins) and have continued to raise the possibility of unknown long-term health effects — essentially framing biotechnology as a food safety risk. Similarly, a number of studies have established that reporters have emphasized potential environmental and social risks over benefits.

For instance, Hagedorn & Allender-Hagedorn (1997) examine media coverage of agricultural biotechnology (microbes, modified plants, animals, and foods) from 1987 to 1994. They find that the popular press focused mostly on social risks associated with biotechnology: health and ethical issues, the labeling of biotech foods, and public safety and regulatory input.

Bonfadelli *et al.* (2002) measure the balance of risks and benefits in Swiss media reporting from 1997 to 1999. They find that 25 per cent of articles on biotechnology and GM foods reported only risks, 25 per cent covered risks and benefits, 21 per cent focused only on benefits, and 29 per cent mentioned neither risks nor benefits. The authors then conclude that the overall tone of media coverage was slightly negative. Similarly, Marks *et al.* (2002, 2003) examine media frames of GM crops in US and UK newspapers, finding that environmental risks (e.g., irreversible transgenes) were emphasized over environmental benefits (e.g., reduced pesticide use and associated benefits from land savings).

Lewison (2007) examines the risk frames used by 14 news media in six countries (Canada, France, Germany, Spain, the United Kingdom, and the United States) between 2002 and 2004. The author finds that environmental, financial, health, and political risks (listed in order of magnitude) were broadly covered in the media narratives and concludes that UK news media tended to be the most "scary," whereas Spanish media were the most "robust."

Media coverage of biotechnology and GM foods has not been uniformly negative, however. For instance, Navarro *et al.* (2011) report that biotechnology media coverage in the Philippines over the 2000–2009 period was generally positive and that, although the coverage was marked by occasional peaks brought about by local controversial events that triggered attention, the volume of coverage remained low, and reporting was balanced. Still, such findings have become exceptional in recent years.

4.3. *Changes in the frames used over time*

Whereas recent studies describe a predominantly negative media view of agricultural biotechnology and GM foods across most countries, earlier analyses indicate that news media coverage was more positive in the 1980s and the early 1990s and became more negative during the late 1990s and 2000s (Abbott & Lucht 2000; Bauer & Gaskell 2002; Besley & Shanahan 2005; Bonfadelli *et al.* 2002; Gaskell & Bauer 2001; Marks *et al.* 2002, 2003; Ten Eyck & Williment 2003). Hence, as with the frequency of reporting, the narrative has been cyclical in nature. This result is not surprising, however. Because both the volume of articles and the frames used for biotechnology and GM foods have been driven by media attention to the same controversial events, they should be expected to coevolve and to demonstrate similar cyclical patters. More than anything, these cyclical effects emphasize what a large number of studies have demonstrated: News media have limited their communication on biotechnology and GM foods mostly to controversies and have frequently chosen words and images to emphasize potential risks over benefits. In deciding which issues to emphasize and which to ignore or which arguments to publish and how to frame them, news media in many parts of the world have shed an increasingly negative light on this new food technology.

5. Impacts of Media Coverage on Public Perceptions and Consumer Demand

5.1. *Heterogeneity and bidirectional impacts*

Citizens may anticipate that information from media may be biased or framed. They can take this possibility into account in evaluating the information. There is evidence from media impact studies on political behavior that voters adjust their media (information) consumption to the expected bias in the information provision and discount the information that is provided by different media sources (Chiang & Knight 2011, Durante & Knight 2012, Gentzkow & Kamenica 2011). Yet some behavioral studies suggest that, even when audiences know that the media sources are biased, they insufficiently discount the information and do not fully take into account the bias. Exposure to media can thus systematically alter beliefs and consumer behavior (Della Vigna & Gentzkow 2010).

However, the impact is likely heterogeneous. Both social and individual factors amplify or dampen perceptions of risk (Flynn *et al.* 1998, Kone & Mullet 1994). Gender and education are consistent demographic predictors of perceived food risk; the natures of the perceived threat, trust, communication sources and channels, and health and environmental concerns are the most relevant nondemographic predictors (Ellis *et al.* 2009).

The decision about how much information to process depends on consumers' ex ante risk perceptions. Van Ravenswaay & Hoehn (1991), in one of the first surveys on consumer perceptions of health risks in food, conclude that most consumers acknowledge the existence of food risks but that most consumers perceive the risk to be quite small. Slovic (1987) suggests that risk is influenced by two major factors: dread risk and unknown risk. Dread risks are those judged as uncontrollable and involuntary, potentially impacting large numbers of people with consequences that are potentially catastrophic. New food technologies typically rate highly on unknown risks. Thus, variations in levels of knowledge across consumers may influence their risk perceptions.

The nature of information matters as well. As we discuss above, consumers may give greater weight to negative information than to positive information (McCluskey & Swinnen 2004). Siegrist & Cvetkovich (2001) conducted experiments and find that people place greater trust in research results indicating health risks and that the confidence in the results increases with an increasing indication of health risks.[1]

Some criticize these studies, warning about possible confusion between pure negativity bias and confirmatory bias in exploring the impact of new information and media reports on citizens' perception. Prior beliefs may play an important role in selecting and processing the information provided by media. Poortinga & Pidgeon (2004) performed an experiment focusing on the perception of British people toward GM food and find that confirmatory bias was strong, as people with clear positive or negative beliefs interpreted the same events in line with their existing attitude position. Frewer *et al.* (1999) also determine that one's initial attitude toward genetic engineering is the most important determinant of attitudes after information provision. These attitudes remain fairly stable even after persuasive arguments regarding the technology are provided. In fact, initial attitudes also affect perceptions of informational quality: Respondents with a more negative view are likely to perceive positive information about the technology as less accurate and informative and are more biased than those who hold more positive views. However, after this confirmatory bias is controlled for, negativity

[1]The authors suggest three possible explanations: diagnosticity (i.e., negative information may be given greater weight because it is more diagnostic than positive information), loss aversion (i.e., for most people avoiding losses is more important), and credibility (i.e., negative information may be more credible than positive information because positive information can be self-serving, whereas negative information often lacks this quality).

bias still plays a role: Negative items still have greater impacts than do positive items.

The source of information may matter as well for risk perceptions, although there is some dispute about the importance of this factor. Some studies find that source credibility plays a key role in determining the impact of a message on public opinion (Renn 2005). However, others find that source credibility appears to have a limited impact and that it is far less important than initial attitude as a determinant of individual reactions to incoming information. Kumkale *et al.* (2010) show through a meta-analysis that the credibility of the source matters mostly for attitude-formation conditions, whereas its impact in attitude-change conditions is lower. In contrast, recent studies show that Internet users pay little or no attention to source credibility when seeking out health information on the web (Bates *et al.* 2006, Benotsch *et al.* 2004, Eysenbach & Kohler 2002).

Hence, the impact of bias in media reporting on consumer attitudes may be substantial under some conditions but is also bidirectional and complex. Consumer personal preferences and beliefs affect media reporting strategies to convince these consumers to buy informational products. In turn, consumer perceptions, attitudes, and behaviors may be affected by media reports. In the rest of this section, we review key insights from specific studies on food technologies: first studies on perceptions and then studies on behavior and regulation.

5.2. *Impact on perceptions*

Whatever the underlying causes or cognitive processes, there is some empirical evidence that broad and often negative media coverage of new food technologies has raised public awareness and has influenced public perceptions. This is especially the case for biotechnology and GM foods (e.g., Bauer 2005a,b; Bauer & Gaskell 2002; Durant *et al.* 1998; Marks *et al.* 2007).

In one of the most data-rich studies, Bauer (2005b) examines the impact of news media coverage on public perceptions of biotechnology across 12 European countries (Austria, Denmark, Finland,

France, Germany, Greece, Italy, the Netherlands, Portugal, Sweden, Switzerland, and the United Kingdom). Using a longitudinal design, the author examines changes in public attitudes toward medical and agricultural biotechnology recorded through large surveys carried out across all 12 countries from 1996 to 1999 in the presence of sharp increases in media coverage and a progressively negative narrative. Bauer (2005b) concludes that media coverage and public perceptions about biotechnology converged over time and that such convergence was apparent among readers of news media but was not present among nonreaders.

Cook *et al.* (2006) evaluate, in an even more direct way, the mechanism by which public attitudes could be shaped by news media stories about GM foods in the United Kingdom. Using focus groups, the authors show that participants, who were largely unaware of the ongoing public debate, formed views on biotechnology on the basis of the narratives of the articles they were given to read. References to entities such as Monsanto acted as framing devices. Participants also made explicit some of the obscured cultural associations in newspaper articles, linking GM foods to historical narratives about power relations and the trustworthiness of institutions.

Vilella-Vila & Costa-Font (2008) examine whether news media coverage and reporting influenced public attitudes toward GM foods in Spain and in the United Kingdom for the period 1999 to 2004. Their analysis indicates that press coverage in the United Kingdom focused on risks for and potential hazards to public health, essentially framing GM foods as a highly controversial issue. The authors also find that differences in media reporting along with attitudes toward journalism correlated with attitudes to and risk perception of GM food in the two countries. Marques *et al.* (2015) also examine how the amount and tone of media coverage might have influenced public attitudes in Australia over a 10-year period, and they report that Australians were less positive toward GM animals compared with GM plants for food, especially in years when media coverage of biotechnology was high.

Kasperson *et al.* (1988) describe a mechanism by which broad and negative media coverage of biotechnology and other new food technologies can create negative public perceptions and attitudes. The authors propose that news media can act as so-called amplification stations by focusing on potential risks. In turn, the amplified risk can lead to behavioral responses that cause secondary impacts and ripple effects. In this context, minor risks, as assessed by technical experts, may elicit strong public concerns and responses. Kasperson *et al.* term this mechanism risk amplification. Frewer *et al.* (2002) provide empirical support for risk amplification in the context of media coverage of GM foods in the United Kingdom. By collecting attitude data before, during, and after an increase in reporting of potential risks of GM foods in the United Kingdom in the spring of 1999, Frewer *et al.* (1999) demonstrate that people's risk perceptions increased and decreased in line with media coverage and the narrative of the time. Public perceptions of benefits from GM foods, however, appeared to be permanently depressed from negative news media reporting.

5.3. *Impact on behavior and regulation*

Although the impact of media coverage of biotechnology on public perceptions is well established, whether changes in perceptions translate into changes in consumer behavior is not clear. There is only limited evidence from analysis of media coverage of new food technologies. There is, however, more evidence from the media coverage of food scares. The evidence in this regard suggests that short-run consumer reactions are strong but that the long-run effects are minor. The most important long-run effects are indirect, e.g., by inducing changes in public regulation.

The short-term impacts of media coverage of food scares on consumer demand can be significant. A prime example is the coverage of bovine spongiform encephalopathy (BSE). The discovery of BSE was broadly covered in mass media, and such coverage had a significantly negative effect on consumer demand for beef. Verbeke & Ward (2001) find that there was considerable misperception by consumers, lack of knowledge, and bias between perception and scientific-indicator criteria related to the health and safety characteristics of meat.

Television coverage on meat safety had a negative effect on the demand for red meat after the BSE outbreak (Verbeke *et al.* 2000). Younger people were the most susceptible to such negative media coverage.

However, in the longer run, consumption and sales typically recover after a food scare (Piggott & Marsh 2004). Studies find that there is little or no long-term demand response to food safety concerns (Henneberry *et al.* 1999). An example is the 1993 *Escherichia coli* outbreak in the Jack in the Box restaurant chain, during which 144 people were hospitalized and 3 died. Jack in the Box almost went out of business in the wake of the event, but after 2 years sales recovered to their pre–food scare levels (Entine 1999).

The limited empirical evidence available on the impacts of media coverage of biotechnology and GM foods seems to be consistent with the above findings. Kalaitzandonakes *et al.* (2004) examine consumer response to two kinds of biotech media coverage environments. In the first environment, media coverage of biotechnology and GM foods was substantial and sustained over a 5-year period. The tone of generally negative reporting frequently raised the possibility of health risks. In this environment, Dutch consumers could choose between food products that were labeled as containing GM ingredients and unlabeled food products containing only conventional ingredients. Under these circumstances, the authors find that consumers did not respond to the media coverage and did not change their purchasing patterns for the duration of the analysis. In the second environment, media coverage was acute but was brief, negative, and focused almost exclusively on potential food safety risks related to the unauthorized presence of StarLink corn and food product recalls in the US market. In this environment, consumer demand for identified products was affected by the media coverage, but the overall consumer response was limited.

Indeed, the most significant long-term effect of mass media is likely through their impact on public policy. One mechanism is the agenda-setting effect of the media, which has sometimes been referred to as the CNN factor (Hawkins 2002). It refers to the process by which the media influence policy by invoking responses in their audiences through concentrated and emotionally based coverage, which in turn applies pressure to governments to react. Similarly, the absence of media coverage reduces priority in agenda setting. Robinson (2001) suggests that the media can be a powerful source in leading policy makers, especially when there is great uncertainty or limited information. For example, following the media frenzy surrounding the Jack in the Box *E. coli* outbreak in 1993, President Clinton called congressional hearings regarding the safety of the food supply. The US Food and Drug Administration raised the recommended internal temperature of cooked hamburgers to 155°F. Ordering a hamburger cooked less than "medium" at US restaurants is now difficult.

More closely related to the issue of food technologies is the change in the use of the precautionary principle in regulation in the European Union and the United States. The precautionary principle is now used as a major regulatory principle in food regulation, in particular for GM food, in the European Union. However, historically, the precautionary principle was used much more in the United States from the 1960s through the mid-1980s (Vogel 2003). Several European food scares in the 1990s, heavily publicized in the mass media, changed this trend. Such events pushed politicians to introduce a series of new regulations and caused consumers to be much more concerned about food safety issues. Although ex post studies showed that several of these food safety problems turned out to be exaggerated, the massive press coverage induced strong political reactions, leading to regulations and shifts in consumer preferences that are having long-lasting effects on European food risk perceptions and on the regulation of the entire food system. Yet the induced change in policy had major implications for consumer behavior in a dynamic perspective, as it influenced the use of food technologies in the long run in the United States versus the European Union (Swinnen & Vandemoortele 2010, Swinnen *et al.* 2012).

A fascinating case on the turnaround of public opinion in a highly mediatized public debate is Zilberman *et al.*'s (2014) analysis of California's Proposition 37, which would have instituted mandatory labeling on food containing GM ingredients. The debate was widely covered by many media outlets, including newspapers, television, social media, and editorial blogs. Celebrity food author Michael Pollan wrote an opposition piece for *The New York Times Magazine* (Pollan 2012). Proponents, who spent approximately $8.7 million, made the arguments that consumers had a right to know whether their food contained GM ingredients and that GMOs did not provide real benefits and may introduce unknown risks. Opponents, who spent $45.6 million, made the case that the debate should be between mandatory labeling and voluntary labeling and that mandatory GM labeling would increase food prices for all. With voluntary labeling, consumers who are indifferent about GM will not have to pay for information they do not value. Zilberman *et al.* (2014) document how, after high initial support, the measure was narrowly defeated as support declined with extensive media coverage and opponents' advertisements of new economic studies showing that mandatory GM labeling could cause economic losses for California.[2]

6. Conclusions and Future Research

A conclusion from our review is that the interactions between mass media, public perceptions of new food technologies, and consumer behavior are bidirectional and complex. Consumers' preferences and beliefs, and consumers' disproportionate interest in negative news, affect media reporting strategies to attract these consumers to buy certain products. Media market conditions, competition, and other supply-side factors may also augment the biased coverage of new food technologies by mass media. In turn, citizens who receive information from the media anticipate that this information from media sources may be biased. Yet citizens only partially discount this bias, so there is real influence.

News coverage tends to be concentrated around events that are often controversial in nature. The volume and the tone of the coverage tend to be affected by such events with significant agenda

[2]Opinion polls showed a rise in opposition from 22% to 51% and a decline in support from 65% to 39% in the final month of the campaign (Zilberman *et al.* 2014).

setting and framing effects. The impact of such agenda setting and framing effects on public perceptions may be durable, as the effects on consumption appear to be important in the short run, even if they are more limited in the long run. The same underlying demand and supply conditions that lead early on to an intensification of media attention also lead to a rapid decline in attention afterward. The most important long-term effects of media attention seem to be indirect via media influence on changes in public policy.

There are implications for risk communication, education, and management. First, the public's initial beliefs affect its overall risk perceptions and how consumers process new information provided by the mass media (or other sources). Therefore, it is important to enhance consumer understanding of the benefits and objective risks associated with new agricultural and food technologies through proactive education and early information to establish a strong framework for people to assess risk when certain events occur. Preemptive risk communication and development of institutions that are responsive to problems can mitigate the negative long-term consequences. Second, firms, scientists, and governments should be prepared to provide information when crises occur. The media will report on the issues either way. If firms, scientists, and government officials are not ready to put events and facts into perspective, the media will draw on whatever "expert" they can find.

Finally, this field is an exciting area for further research because progress can be made on various fronts. The emergence of improved research methodologies (such as improved understanding of laboratory and field experiments) and new media technologies (such as web-based media applications; social media; and, especially, much better, digitalized, and personalized media databases and consumer databases, such as those that include retail scanner data, and store loyalty cards) should allow for greater quality and variety of research in this field.

The availability of data on the use of content created on social media platforms creates new opportunities to understand media content demand and impact. For example, Smith & Wooten (2014) analyze a billion tweets and find that the popularity of a political pundit hinges more on whether she is confident than on whether she is correct. Blogs and social media represent a change in the way people receive information. McKenzie & Ozler (2014) find evidence that a blog can increase knowledge of the topics it covers for the average, but not the marginal, reader. Researchers should also consider the influence of charismatic authors and journalists and the impacts of blogs and other newer forms of media on acceptance of new technologies. This was a factor in the Proposition 37 policy debate over mandatory GM labeling.

There is also room for enhanced insights from more interdisciplinary research on consumers and media choice and the factors that affect such choice. Future directions could include insights from sensory sciences about the factors affecting taste and perceptions and how experience affects perceptions of technology and the product. Psychology could be integrated into studies to understand how values, social norms, identity, and personality affect choices of food made with new technology. We expect research to increasingly draw upon findings of behavioral economics to learn more about the interaction of media and consumer choice.

7. Disclosure Statement

The authors are not aware of any affiliations, memberships, funding, or financial holdings that might be perceived as affecting the objectivity of this review.

Literature Cited

Abbott, E. and A. Eichmeier (1998). *The hoopla effect: toward a theory of regular patterns of mass media coverage of innovations.* Presented at Annu. Meet. Assoc. Educ. J. Mass Commun. 81st, Aug. 5–8, Baltimore, MD.

Abbott E. and T. Lucht (2000). *How triggering events affect mass media coverage and source use concerning genetically modified organisms (GMOs) in Britain and the United States.* Presented at Agric. Commun. Educ. Congr., July 25, Washington, DC.

Alterman, E. (2008). *What Liberal Media? The Truth About Bias and the News.* New York: Basic.

Anand, B., R. Di Tella and A. Galetovic (2007). Information or Opinion? Media Bias as Product Differentiation. *Journal of Economics & Management Strategy* 163, 635–682.

Annunziata, A. and R. Vecchio (2011). Factors Affecting Italian Consumer Attitudes Toward Functional Foods. *AgBioForum* 14(1), 20–32.

Augoustinos, M., S. Crabb and R. Shepherd (2010). Genetically Modified Food in the News: Media Representations of the GM Debate in the UK. *Public Understanding of Science* 19(1), 98–114.
Bagdikian, B.H. (1992). *The Media Monopoly*, 4th Edn. Boston: Beacon.
Baron, D.P. (2005). Competing for the Public Through the News Media. *Journal of Economics & Management Strategy* 142, 339–376.
Baron, D.P. (2006). Persistent Media Bias. *Journal of Public Economics* 90, 1–36.
Barros, P.P., H.J. Kind, T. Nilssen and L. Sørgard (2004). Media Competition on the Internet. *Topics in Economic Analysis & Policy* 41, 32.
Bates, B.R., S. Romina, R. Ahmed and D. Hopson (2006). The Effect of Source Credibility on Consumers' Perceptions of the Quality of Health Information on the Internet. *Informatics for Health & Social Care* 31(1), 45–52.
Bauer, M.W. (2005a). Distinguishing Red and Green Biotechnology: Cultivation Effects of the Elite Press. *International Journal of Public Opinion Research* 17(1), 63–89.
Bauer, M.W. (2005b). Public Perceptions and Mass Media in the Biotechnology Controversy. *International Journal of Public Opinion Research* 17(1), 5–22.
Bauer, M.W. and G. Gaskell (Eds.) (2002). *Biotechnology: The Making of a Global Controversy.* Cambridge, UK: Cambridge University Press.
Benotsch, E.G., S. Kalichman and L.S. Weinhardt (2004). HIV-AIDS Patients' Evaluation of Health Information on the Internet: The Digital Divide and Vulnerability to Fraudulent Claims. *Journal of Consulting And Clinical Psychology* 72(6), 1004–1011.
Besley, J.C. and J. Shanahan (2005). Media Attention and Exposure in Relation to Support for Agricultural Biotechnology. *Science Communication* 26(4), 347–367.
Boholm, A. (1998). Comparative Studies of Risk Perception: A Review of Twenty Years of Research. *Journal of Risk Research* 1(2), 135–163.
Bonfadelli, H., U. Dahinden and M. Leonarz (2002). Biotechnology in Switzerland: High on the Public Agenda, But Only Moderate Support. *Public Understanding of Science* 11(2), 113–130.
Botelho, D. and H. Kurtz (2008). The Introduction of Genetically Modified Food in the United States and the United Kingdom: A News Analysis. *Social Science Journal* 45(1), 13–27.
Bovitz, J., N. Druckman and A. Lupia (2002). When Can a News Organization Lead Public Opinion? Ideology versus Market Forces in Decisions to Make News. *Public Choice* 113, 127–155.
Budd, R.W., R.K. Thorp and L. Donohew (1967). *Content Analysis of Communications.* New York: Macmillan.
Canoy, M. and R. Nahuis (2005). *Public service broadcasting and the quality of news on commercial channels: Substitutes or complements?* Work. Pap., CPB Neth. Bur. Econ. Policy Anal./Utrecht Sch. Econ., Utrecht, Neth.
Carver, R.B., E.A. Rødland and J. Breivik (2013). Quantitative Frame Analysis of How the Gene Concept is Presented in Tabloid and Elite Newspapers. *Science Communication* 35(4), 449–475.
Chiang, C.F. and B. Knight (2011). Media Bias and Influence: Evidence from Newspaper Endorsements. *Review of Economic Studies* 78(3), 795–820.

Cohen, B. (1963). *The Press and Foreign Policy*. Princeton, NJ: Princeton University Press.
Cook, G., P.T. Robbins and E. Pieri (2006). Words of Mass Destruction: British Newspaper Coverage of the Genetically Modified Food Debate, Expert and Non-Expert Reactions. *Public Understanding of Science* 15(1), 5–29.
Cox, D.N., G. Evans and H.J. Lease (2008). Australian Consumers' Preferences for Conventional and Novel Sources of Long Chain Omega-3 Fatty Acids: A Conjoint Study. *Food Quality and Preference* 19, 306–314.
Curtis, K.R., J.J. McCluskey and J.F.M. Swinnen (2008). Differences in Global Risk Perceptions of Biotechnology and the Political Economy of the Media. *International Journal of Global Environmental Issues* 8(1–2), 77–89.
DellaVigna, S. and M. Gentzkow (2010). Persuasion: Empirical Evidence. *Annual Review of Economics* 2, 643–669.
Downs, A. (1972). Up and Down with Ecology — the 'Issue-Attention Cycle.' *Public Interest* 28, 38–50.
Durant, J., M.W. Bauer and G. Gaskell (1998). *Biotechnology in the Public Sphere: A European Source Book*. London: Science Museum.
Durante, R. and B. Knight (2012). Partisan Control, Media Bias, and Viewer Responses: Evidence from Berlusconi's Italy. *Journal of the European Economic Association* 10(3), 451–481.
Dyck, A. and L. Zingales (2002). "The Corporate Governance Role of the Media." In R. Islam (Ed.), *The Right to Tell: The Role of Mass Media in Economic Development*, pp. 107–140. Washington, DC: World Bank.
Ellis, B.J., A.J. Figueredo, B.H. Brumbach and G.L. Schlomer (2009). Fundamental Dimensions of Environmental Risk. *Human Nature* 20(2), 204–268.
Entine, J. (1999). The ethical edge: How "Jack" turned crisis into opportunity. *Bus. Dig.*, Nov./Dec. http://www.jonentine.com/ethical_corporation/jack_crisis.htm.
Entman, R.M. (1991). Framing U.S. Coverage of International News: Contrasts in Narratives of the KAL and Iran Air Incidents. *Journal of Communication* 41(4), 6–27.
Eysenbach, G. and C. Köhler (2002). How Do Consumers Search for and Appraise Health Information on the World Wide Web? Qualitative Study Using Focus Groups, Usability Tests, and In-Depth Interviews. *BMJ* 324(7337), 573–577.
Flipse, S.M. and P. Osseweijer (2013). Media Attention to GM Food Cases: An Innovation Perspective. *Public Understanding of Science* 22(2), 185–202.
Flynn, J., E. Peters, C.K. Mertz and P. Slovic (1998). Risk Media and Stigma at Rocky Flats. *Risk Analysis* 18(6), 715–727.
Frewer, L.J., C. Howard, D. Hedderley and E. Shepherd (1999). Reactions to Information About Genetic Engineering: Impact of Source Characteristics, Perceived Personal Relevance and Persuasiveness. *Public Understanding of Science* 8, 35–50.
Frewer, L.J., S. Miles and R. Marsh (2002). The Media and Genetically Modified Foods: Evidence in Support of Social Amplification of Risk. *Risk Analysis* 22(4), 701–711.
Gabszewicz, J.J., D. Laussel and N. Sonnac (2001). Press Advertising and the Ascent of the Pensée Unique? *European Economic Review* 45, 645–651.

Gabszewicz, J.J., D. Laussel and N. Sonnac (2004). Programming and Advertising Competition in the Broadcasting Industry. *Journal of Economics & Management Strategy* 134, 657–669.

Gal-Or, E. and A. Dukes (2003). Minimum Differentiation in Commercial Media Markets. *Journal of Economics & Management Strategy* 123, 291–325.

Gaskell, G. and M.W. Bauer (Eds.) (2001). *Biotechnology 1996–2000: The Years of Controversy.* London: Science Museum.

Gentzkow, M. and E. Kamenica (2011). *Competition in persuasion.* NBER Working Paper 17436.

Gentzkow, M. and J. Shapiro (2006). Media Bias and Reputation. *Journal of Political Economy* 114(2), 280–316.

Gentzkow, M. and J. Shapiro (2008a). Competition and Truth in the Market for News. *Journal of Economic Perspectives* 22(2), 133–154.

Gentzkow, M. and J. Shapiro (2008b). Preschool Television Viewing and Adolescent Test Scores. Historical Evidence from the Coleman Study. *Quarterly Journal of Economics* 123(1), 279–323.

Gentzkow, M. and J. Shapiro (2010). What Drives Media Slant? Evidence from US Daily Newspapers. *Econometrica* 78(1), 35–71.

George, L.M. and J. Waldfogel (2006). The *New York Times* and the Market for Local Newspapers. *American Economic Review* 102, 435–447.

Groseclose, T. and J. Milyo (2005). A Measure of Media Bias. *Quarterly Journal of Economics* 120(4), 1191–1237.

Hagedorn, C. and S. Allender-Hagedorn (1997). Issues in Agricultural and Environmental Biotechnology: Identifying and Comparing Biotechnology Issues from Public Opinion Surveys, the Popular Press and Technical/Regulatory Sources. *Public Understanding of Science* 6(3), 233–245.

Hawkins, V. (2002). The Other Side of the CNN Factor: The Media and Conflict. *Journal of Studies* 3(2), 225–240.

Heiman, A. and D. Zilberman (2011). The Effects of Framing on Consumers' Choice of GM Foods. *AgBioForum* 13(3), 9.

Henneberry, S.R., K. Piewthongngam and H. Qiang (1999). Consumer Food Safety Concerns and Fresh Produce Consumption. *Journal of Agricultural and Resource Economics* 24(1), 98–113.

Hibino, A. and M. Nagata (2006). Biotechnology in the Japanese Media: Comparative Analysis of Newspaper Articles on Genetic Engineering in Japan and Europe. *Asian Journal of Social Psychology* 9(1), 12–23.

Hoban, T.J. and P.A. Kendall (1993). *Consumer Attitudes About Food Biotechnology.* Proj. Rep., NC Coop. Extens. Serv., Raleigh, NC.

Hornig, S. (1990). Science Stories: Risk, Power and Perceived Emphasis. *Journalism & Mass Communication Quarterly* 67(4), 767–776.

Hornig, S. (1993). Reading Risk: Public Response to Print Media Accounts of Technological Risk. *Public Understanding of Science* 2(2), 95–109.

Huffman, W. and J.J. McCluskey (2014). "Labeling of Genetically Modified Foods." In P.W.B. Philips, S. Smyth and D. Castle (Eds.), *The Handbook on Agriculture, Biotechnology and Development*, pp. 467–487. Cheltenham, UK: Edward Elgar.

Jesuino, J.C., J.A. Nunes, C. Diego, P. Alcantara, S. Costa and M. Matias (2001). Representation of Biotechnology in Portugal. See Gaskell & Bauer 2001, pp. 258–266.

Kalaitzandonakes, N., L.A. Marks and S.S. Vickner (2004). Media Coverage of Biotech Foods and Influence on Consumer Choice. *American Journal of Agricultural Economics* 86(5), 1238–1246.

Kasperson, R.E., O. Renn, P. Slovic, H.S. Brown, J. Emel *et al.* (1988). The Social Amplification of Risk: A Conceptual Framework. *Risk Analysis* 8(2), 177–187.

Kiesel, K., J.J. McCluskey and S.B. Villas Boas (2011). Nutritional Labeling and Consumer Choices. *Annual Review of Resource Economics* 3, 141–158.

Kohring, M. and J. Matthes (2002). The Faces of Biotech in the Nineties: How the German Press Framed Modern Biotechnology. *Public Understanding of Science* 11, 143–154.

Koné, D. and E. Mullet (1994). Societal Risk Perception and Media Coverage. *Risk Analysis* 14, 21–24.

Kumkale, G.T., D. Albarracín and P.J. Seignourel (2010). The Effects of Source Credibility in the Presence or Absence of Prior Attitudes: Implications for the Design of Persuasive Communication Campaigns. *Journal of Applied Social Psychology* 40(6), 1325–1356.

Kuzyk, P. and J.J. McCluskey (2006). The Political Economy of the Media: Coverage of the US–Canadian Lumber Trade Dispute. *World Economy* 29(5), 637–654.

Lewison, G. (2007). The Reporting of the Risks from Genetically Modified Organisms in the Mass Media, 2002–2004. *Scientometrics* 72(3), 439–458.

Listerman, T. 2010. Framing of Science Issues in Opinion-Leading News: International Comparison of Biotechnology Issue Coverage. *Public Understanding of Science* 19(1), 5–15.

Loureiro, M.L., J.J. McCluskey and R.C. Mittelhammer (2001). Assessing Consumers Preferences for Organic, Ecolabeled and Regular Apples. *Journal of Agricultural and Resource Economics* 26(2), 404–416.

Lusk, J.L. (2015). Food demand survey. *Jason Lusk Blog*, Jan. 15. http://jaysonlusk.com/blog/2015/1/15/food-demand-survey-foods-january-2015.

Lusk, J.L., M. Jamal, L. Kurlander, M. Roucan and L. Taulman (2005). A Meta-Analysis of Genetically Modified Food Valuation Studies. *Journal of Agricultural and Resource Economics* 30, 28–44.

Lusk, J.L., J. Roosen and A. Bieberstein (2014). Consumer Acceptance of New Food Technologies: Causes and Roots of Controversies. *Annual Review of Resource Economics* 6, 381–405.

Maeseele, P.A. and D. Schuurman (2008). Biotechnology and the Popular Press in Northern Belgium: A Case Study of Hegemonic Media Discourses and the Interpretive Struggle. *Science Communication* 29(4), 435–471.

Marks, L.A., N. Kalaitzandonakes, L. Wilkins and L. Zakharova (2007). Mass Media Framing of Biotechnology News. *Public Understanding of Science* 16(2), 183–203.

Marks, L.A., N. Kalaitzandonakes and L. Zakharova (2002). On the Media Roller-Coaster: Will GM Foods Finish the Ride? *Choices* 2002(Spring), 34–38.

Marks, L.A., N. Kalaitzandonakes and L. Zakharova (2003). Media Coverage of Agrobiotechnology: Did the Butterfly Have an Effect? *Journal of Agribusiness* 21(1), 1–20.

Marques, M.D., C.R. Critchley and J. Walshe (2015). Attitudes to Genetically Modified Food Over Time: How Trust in Organizations and the Media Cycle Predict Support. *Public Understanding of Science* 24(5), 601–618.

Mazur, A. (1981). Media Coverage and Public Opinion on Scientific Controversies. *Journal of Communication* 31(2), 106–115.

McCluskey, J.J. and J. Swinnen (2011). The Media and Food-Risk Perceptions. *EMBO Reports* 12(7), 624–629.

McCluskey, J.J., J. Swinnen and T. Vandemoortele (2015). You Get What You Want: A Note on the Economics of Bad News. *Information Economics and Policy* 30, 1–5.

McCluskey, J.J. and J.F.M. Swinnen (2004). Political Economy of the Media and Consumer Perceptions of Biotechnology. *American Journal of Agricultural Economics* 86, 1230–1237.

McCombs, M. and S.I. Ghanem (2001). "The Convergence of Agenda Setting and Framing." In S.D. Reese, O.H. Gandy Jr., and A.E. Grant (Eds.), *Framing Public Life: Perspectives on Media and Our Understanding of the Social World*, pp. 67–81. New York: Routledge.

McCombs, M.E. and D.L. Shaw (1972). The Agenda-Setting Function of Mass Media. *Public Opinion Quarterly* 36(2), 176–187.

McInerney, C., N. Bird and M. Nucci (2004). The Flow of Scientific Knowledge from Lab to the Lay Public: The Case of Genetically Modified Food. *Science Communication* 26(1), 44–74.

McKenzie, D. and B. Ozler (2014). Quantifying Some of the Impacts of Economics Blogs. *Economic Development and Cultural Change* 62(3), 567–597.

Mullainathan, S. and A. Shleifer (2005). The Market for News. *American Economic Review* 95, 1031–1053.

Müller, M., B. Freitag and F. Köder (2010). Plant Biotechnology in German Media: A Linguistic Analysis of the Public Image of Genetically Modified Organisms. *Biotechnology Journal* 5(6), 541–544.

Navarro, M.J., J.A. Panopio, D.B. Malayang and N. Amano Jr. (2011). Print Media Reportage of Agricultural Biotechnology in the Philippines: A Decade's (2000–2009) Analysis of News Coverage and Framing. *Journal of Science Communication* 10(3), A01.

Nisbet, M.C. and M. Huge (2006). Attention Cycles and Frames in the Plant Biotechnology Debate. *International Journal of Press-Politics* 11(2), 3–40.

Pan, Z. and G.M. Kosicki (1993). Framing Analysis: An Approach to News Discourse. *Political Communication* 10(1), 55–75.

Pew (Pew Res. Cent. People Press) (2004). *How journalists see journalists in 2004*. Washington, DC: Pew. http://www.people-press.org/files/legacy-pdf/214.pdf

Piggott, N.E. and T.L. Marsh (2004). Does Food Safety Information Impact US Meat Demand? *American Journal of Agricultural Economics* 86(1), 154–174.
Pollan, M. (2012). Vote for the dinner party. *New York Times Mag.*, Oct. 10. http://michaelpollan.com/articles-archive/vote-for-the-dinner-party/
Poortinga, W. and N.F. Pidgeon (2004). Trust, the Asymmetry Principle, and the Role of Prior Beliefs. *Risk Analysis* 24(6), 1475–1486.
Posner, R.A. (2005). Bad news. *New York Times*, July 31. http://www.nytimes.com/2005/07/31/books/review/31POSNER.html.
Prat, A. and D. Strömberg (2011). *The political economy of mass media.* Discuss. Pap. DP8246, Lond. Sch. Econ, London.
Renn, O. (2005). Risk Perception and Communication: Lessons for the Food and Food Packaging Industry. *Food Additives and Contaminants* 22(10), 1061–1071.
Robinson, P. (2001). Theorizing the Influence of Media on World Politics Models of Media Influence on Foreign Policy. *European Journal of Communication* 16(4), 523–544.
Shanahan, J., D. Scheufele and E. Lee (2001). Polls and Trends: Attitudes About Agricultural Biotechnology and Genetically Modified Organisms. *Public Opinion Quarterly* 65(2), 267–281.
Siegrist, M. and G. Cvetkovich (2001). Better Negative than Positive? Evidence of a Bias for Negative Information About Possible Health Dangers. *Risk Analysis* 21(1), 199–206.
Sirö, I., E. Kápolna, B. Kápolna and A. Lugasi (2008). Functional Food. Product Development, Marketing and Consumer Acceptance — A Review. *Appetite* 51(3), 456–467.
Slovic, P. (1987). Perception of Risk. *Science* 236, 280–285.
Smith, B. and J. Wooten (2014). *Estimating demand for information attributes using Twitter data.* Unpublished manuscript, Coll. Bus. Admin., Univ. Nebraska, Omaha, NE.
Strömberg, D. (2001). Mass Media and Public Policy. *European Economic Review* 45, 652–663.
Strömberg, D. (2004). Mass Media Competition, Political Competition, and Public Policy. *Review of Economic Studies* 71, 265–284.
Surowiecki, J. (2004). *The Wisdom of Crowds.* New York: Doubleday.
Swinnen, J., A. Olper and T. Vandemoortele (2012). Impact of the WTO on Agricultural and Food Policies. *World Economy* 35(9), 1089–1101.
Swinnen, J.F.M. and T. Vandemoortele (2010). Policy Gridlock or Future Change? The Political Economy Dynamics of EU Biotechnology Regulation. *AgBioForum* 13(4), 291–296.
Ten Eyck, T.A. and M. Williment (2003). The National Media and Things Genetic Coverage in the *New York Times* (1971–2001) and the *Washington Post* (1977–2001). *Science Communication* 25(2), 129–152.
Tuchman, G. (1976). The News' Manufacture of Sociological Data. *American Sociological Review* 41(6), 1065–1067.
Tuchman, G. (1978). *Making News.* New York: Free.

Urala, N. and L. Lahteenmaki (2007). Consumer's Changing Attitudes Toward Functional Foods. *Food Quality and Preference* 18, 1–12.

Van Ravenswaay, E.O. and J.P. Hoehn (1991). "The Impact of Health Risk Information on Food Demand: A Case Study of Alar and Apples." In J. Caswell (Ed.), *Economics of Food Safety*, pp. 155–74. Dordrecht, Neth.: Springer.

Veltri, G.A. and A.K. Suerdem (2013). Worldviews and Discursive Construction of GMO-Related Risk Perceptions in Turkey. *Public Understanding of Science* 22(2), 137–154.

Verbeke, W. and R.W. Ward (2001). A Fresh Meat Almost Ideal Demand System Incorporating Negative TV Press and Advertising Impact. *Agricultural Economics* 25, 359–374.

Verbeke, W., R.W. Ward and J. Viaene (2000). Probit Analysis of Fresh Meat Consumption in Belgium: Exploring BSE and Television Communication Impact. *Agribusiness* 16, 215–234.

Vicsek, L. (2013). Gene-Fouled or Gene-Improved? Media Framing of GM Crops and Food in Hungary. *New Genetics and Society* 32(1), 54–77.

Vilella-Vila, M. and J. Costa-Font (2008). Press Media Reporting Effects on Risk Perceptions and Attitudes Towards Genetically Modified (GM) Food. *Journal of Socio-Economics* 37(5), 2095–2106.

Vogel, D. (2003). The Hare and the Tortoise Revisited: The New Politics of Consumer and Environmental Regulation in Europe. *British Journal of Political Science* 33(4), 557–580.

Wheeler, D.L., D.M. Church, A.E. Lash, D.D. Leipe, T.L. Madden *et al.* (2002). Database Resources of the National Center for Biotechnology Information. *Nucleic Acids Research* 30(1), 13–16.

Wildavsky, A. (1995). *But Is It True? A Citizen's Guide to Environmental Health and Safety Issues.* Cambridge, MA: Harvard University Press.

Zilberman, D., S. Kaplan, E. Kim and G. Waterfield (2014). "Lessons from the California GM Labeling Proposition on the State of Crop Biotechnology." In P.W.B. Philips, S. Smyth and D. Castle (Eds.), *Handbook on Agriculture, Biotechnology and Development*, pp. 538–549. Cheltenham, UK: Edward Elgar.

Part II

New Traits without Added Nutrition: GM Input Traits, Organic, and All Natural Foods

Chapter 6

Consumer Willingness to Pay for Genetically Modified Food Labels in a Market with Diverse Information: Evidence from Experimental Auctions*

Wallace E. Huffman[†], Jason F. Shogren[‡], Matthew Rousu[§], and Abebayehu Tegene[¶]

[†]*Department of Economics, Iowa State University*
[‡]*Natural Resource Conservation and Management, Department of Economics and Finance, University of Wyoming*
[§]*RTI International*
[¶]*Food and Rural Economics Division, ERS, U.S. Department of Agriculture*

With the continuing controversy over genetically modified (GM) foods, some groups advocate mandatory labeling of these products, while other groups oppose labeling. An important issue is how GM labels affect consumers' willingness to pay for these food products in the market. Using a statistically based economics experiment with adult consumers as subjects, we examine how willingness to pay changes for three food products — vegetable oil, tortilla chips, and potatoes — when GM labels are introduced. Participants in the experiments discounted GM-labeled foods by approximately 14% relative to their standard-labeled counterparts. The evidence also showed that sequencing of food labels affects willingness to pay, and that randomizing treatments is an important methodological feature in experiments of willingness to pay.

Keywords: consumer demand, corn chips, experimental economics, food labels, genetic modification, GM foods, laboratory auctions, potatoes, vegetable oil, willingness to pay

*This article originally appeared in **Journal of Agricultural and Resource Economics (2003), 28(3), 481–502**. This is published by permission of **Western Agricultural Economics Association**.

1. Introduction

The growing controversy over genetically modified (GM) food products and consumers' attempts to make better food purchasing decisions have stimulated interest in new objective information, including food labeling. Labeling has become an important public policy issue worldwide. In the United States, truthful labeling has been used to provide consumers with information on calories, nutrients, and food ingredients. But the federal government requires explicit labeling only if a GM food has distinctive characteristics relative to the non-GM version. In contrast, the European Commission adopted genetically modified food labels in 1997. The Commission requires each member country to enact a law requiring labeling of all new products containing GM organisms. Japan, Australia, and many other countries have also passed laws requiring GM labels for major foods. Labeling involves real costs — fixed costs of testing, segregation or identity preservation, and risk premium for being out of contract — and variable costs of monitoring for truthfulness. A key question is: Does consumer behavior change with the presence of different labels?

Mendenhall and Evenson report on a telemarketing survey conducted in February 1999 of 55 adults in New Haven, Connecticut. Eight-two percent of these respondents strongly believe foods made with GM ingredients should be labeled. This finding is consistent with most surveys on GM foods, but it is strikingly different from the results of the referendum vote in Oregon on November 2002, in which approximately 70% of voters rejected a ballot referendum that would have mandated labels for GM foods.

This study uses the tools of survey design, statistical experimental design, experimental economics theory, and the random n^{th}-price auction to elicit consumers' willingness to pay (WTP) for both GM-labeled and standard-labeled foods using a random sample of adult consumers drawn from two Midwestern cities.[1] By gathering

---

[1]Currently, many unlabeled GM food products are available in U.S. grocery stores, and labeling is voluntary. If GM labels were present, testing consumers' willingness to pay for foods with GM food labels and foods with standard food labels would be an appropriate measure.

information on three goods — vegetable oil, tortilla chips, and potatoes — dislike for genetic modification can be distinguished from dislike for a particular food item. In an experimental auction with divergent information about risks and benefits, we examine whether consumers value information provided in GM labels.

In this study, tests of the following hypotheses are reported: (H1) GM food labels have no effect on WTP for food items; (H2) attributes of participants, including prior beliefs, do not affect WTP for food items; and (H3) no difference in willingness to pay occurs because of the sequencing of labels in laboratory trials (i.e., whether the consumer first bids on foods with or without GM food labels).

2. Background on GM Food and GM Labels

Few experimental auctions have been conducted to elicit information about genetic modification. Noussair, Robin, and Ruffieux (2001) used experimental auctions with a sample of 97 Europeans, and found that bids for biscuits decreased by 37% when participants were told a product was GM. In a subsequent experimental auction paper, Noussair, Robin, and Ruffieux (2002) found that French consumers ignored the information on GM labels when labels were placed on the back of a package. These consumers initially price-discounted GM chocolate bars by less than 2% relative to non-GM chocolate bars. When the monitors emphasized the presence of the GM food label on the back of a chocolate bar, the consumers discounted the GM chocolate by more than 25%. This latter study, however, focused solely on French consumers.

To our knowledge, only one other study has used experimental auctions to elicit U.S. consumer preferences for GM or non-GM foods. Lusk *et al.* conducted experimental auctions with 50 undergraduate agricultural business students at Kansas State University. All subjects were endowed with a bag of GM corn chips. Using first- and second-price auctions, Lusk *et al.* observed that a majority of the student participants would not pay to upgrade to a bag of non-GM corn chips. One potential problem in their experimental design was the inability to distinguish dislike for corn chips from dislike for GM chips. Because we use multiple products, employ a larger participant base, and work with adult, U.S. consumers, the present study can produce inferences extending beyond these earlier studies.

Caswell (1998, 2000) and Caswell and Padberg have shown that many policies are possible, including mandatory labeling of GM foods, voluntary labeling of GM foods, or bans on all labeling identifying whether a food is GM. The policies each country chooses are likely to be determined by the information demanded by domestic consumers. An informed decision on whether to implement a labeling policy on GM foods should only be made after a benefit-cost analysis is conducted. Then politicians can weigh the net economic benefit against other purely political factors when deciding whether to require GM labeling.

Two prominent environmental groups, Greenpeace and Friends of the Earth, believe GM labeling would benefit consumers, and these groups advocate labels on GM foods to give consumers the right to choose whether to consume GM products. Many environmental and consumer advocacy groups have demanded mandatory labeling, which they argue will benefit consumers. While there may be benefits, implementing a labeling policy would be costly. The USDA's Economic Research Service considered the range of costs associated with implementing a labeling policy (Golan *et al.*). If a mandatory labeling policy on GM foods is enacted, significant costs would be incurred. Potential fixed costs include testing, identity preservation, accidental contamination of non-GM crops by their GM counterparts, and reformulation costs. All items imply real supply-side costs when a labeling policy is implemented.

Relatively few estimates of the cost of labeling GM foods exist in the literature. A study by Wilson and Dahl, however, estimates wheat identity preservation would cost \$ 1.45 (\$ 3.36) per bushel for a 5% (1%) tolerance level (although labeling costs could be lower than identity preservation costs). Moreover, in estimates published in the *AGBiotech Reporter* (August 2001), the Philippine Chamber of Food Manufacturers warned that mandatory GM food labels would increase production costs by 15%, and these increased costs would be passed on to consumers. These added labeling, handling, and storage costs would lead to higher prices for consumers and possibly lower prices to producers. One conclusion seems apparent: Implementing a labeling policy on GM foods is costly and involves uncertain effects on firms, consumers, and the industry.

3. Experimental Design

The research project was designed to incorporate the private information-revealing feature of experimental auction markets, the rigorous randomized treatments of statistical experimental design (see, for example, Hoffman *et al.*; Fox *et al.*; and Shogren, List, and Hayes), and random sampling methods from survey design.[2] The primary purpose of the project was to identify the effects of GM food labels and different perspectives on genetic modification on consumers' WTP for food items that might be GM. In this analysis, we concentrate on the set of experimental auctions in which food items differed only by the presence or absence of GM food labels.[3]

There are two factors in this experiment: first, the sequence in which consumers bid on the food products with GM labels, and second, the type of information consumers could receive. For the first factor, consumers could bid on food products with GM labels before they bid on plain-labeled food products, or they could bid on food products with GM labels after they bid on plain-labeled food products. For the second factor, associated with the perspective on genetic modification, consumers could receive a combination of the biotech industry perspective, the environmental group perspective, and the third-party perspective, or verifiable biotech information (see Huffman and Tegene).

There are two levels for the labeling-order factor and six levels for the information factor (biotech industry perspective, environmental

[2]Professors Phil Dixon and Wayne Fuller, Department of Statistics, Iowa State University, provided assistance with the statistical design of the project.

[3]The conclusions from the statistical analysis of this paper are not changed significantly by including information treatment effects which were randomly assigned. In a major companion paper, we examine the analytics and measure the value of third-party or verifiable information (see Rousu *et al.*).

group perspective, biotech industry and environmental group perspectives, biotech industry and third-party perspectives, environmental group and third-party perspectives, and all three perspectives). The experimental design has two levels by six levels, yielding a total of 12 treatments. Each of the 12 treatments is randomly assigned to one of 12 experimental units, comprised of 13 to 16 consumers. This design was anticipated to achieve a desired sample size of 165 to 190 participants, a size considered necessary for finding statistically significant results without being prohibitively costly.

Using statistical sampling methods, a random digital dialing procedure was employed to select consumers who were adults living in two major Midwestern metropolitan areas, Des Moines, Iowa, and St. Paul, Minnesota. These cities were chosen because they represent two large metropolitan areas of the Midwest (in contrast to small university cities like Ames), and yet were easily accessible by researchers centered in Ames, Iowa. Conducting experiments in two urban areas rather than one is also seen as enhancing credibility by demonstrating that the experiments can be replicated across urban areas. Nevertheless, caution must be exercised in generalizing to the broader population of the Midwest or to the entire U.S. population (also see relevant comments in Lusk *et al.*).

We now describe the four elements in this study's GM labeling experiments — the GM food, the auction mechanism, the experimental units, and the specific steps in the experiment (which includes the detailed information labels).

3.1. *The GM food*

We anticipated that consumers might react differently to GM content for foods of different types. Because one food item was unlikely to reveal enough information, three items were used: vegetable oil made from soybeans, tortilla chips made from yellow corn, and russet potatoes. In the distilling and refining process for vegetable oils, essentially all of the proteins (which are the components of DNA and the source of genetic modification) are removed, leaving pure lipids. Minimal human health concerns should arise from consumption of the oil, but people might still fear that GM soybeans could harm the natural environment. Tortilla chips are highly processed foods which may be made from GM or non-GM corn, and consumers might have human health or environmental concerns or both. Russet potatoes are purchased as a fresh product and generally fried or baked before eating. Consumers might reasonably see the potential concentration of genetic modification as being different in potatoes than in processed corn chips. Consequently, consumers might have concerns about both human health and environmental risks from eating russet potatoes.

3.2. *The random n^{th}-price auction*

Over the past decade, scientists conducting valuation experiments have used auction mechanisms to induce individuals to reveal their preferences for new goods and services (e.g., see Shogren *et al.* 1994; Fox *et al.*; Shogren, List, and Hayes). Vickrey's sealed bid, second-price auction has been a popular mechanism. The popularity of the second-price auction mechanism is largely because it is demand revealing in theory, is relatively simple to explain, and has an endogenous market-clearing price. Also, evidence from induced-value experiments suggests the auction mechanism can produce efficient outcomes in the aggregate (Kagel).

Although a second-price auction is better than a first-price auction, a problem arises in second-price auctions in that they tend not to engage bidders who anticipate being off the margin (i.e., bidders whose value for a good is far below or above the market-clearing price). These bidders have a low opportunity cost from an insincere bid, making it difficult to measure accurately the entire demand curve for a new good like GM food (see Miller and Plott; Franciosi *et al.*). Insincere bidding can be sustained if the behavior is undetected and unpunished by the institutional structure of the auction mechanism (see Cherry, Crocker, and Shogren).

We chose the random n^{th}-price auction for our GM food experiments over the second-price auction because it engages both the on- and off-the-margin bidders and helps ensure that consumers reveal their demand truthfully (see Shogren *et al.* 2001). The auction combines elements of two classic demand-revealing mechanisms: the Vickrey auction and the Becker-DeGroot-Marschak random-pricing mechanism. The key characteristic of the random n^{th}-price auction is a *random but endogenously determined* market-clearing price. Randomness is used to give all participants a positive probability of being a purchaser of the auctioned good; the endogenous price guarantees the market-clearing price is related to the bidders' private values.

The random n^{th}-price auction works as follows. Each of k bidders submits a bid for one unit of a good; then each of the bids is rank ordered from highest to lowest. The auction monitor then selects a random number (the n in the nth-price auction) which is drawn from a uniform distribution between 2 and k, and the auction monitor sells one unit of the good to each of the $n - 1$ highest bidder at the nth-price. For instance, if the monitor randomly selects $n = 4$, the three highest bidders each purchase one unit of the good priced at the fourth-highest bid. Ex ante, bidders who have low or moderate valuations now have a nontrivial chance to buy the good because the price is determined randomly. This auction increases the probability that insincere bidding will be costly (Shogren *et al.* 2001).

3.3. *The experimental units*

Auctions were planned and conducted at two Midwestern U.S. cities — Des Moines, Iowa, and St. Paul, Minnesota. An independent institution (the Iowa State University Statistics Laboratory) called telephone numbers chosen by a random digital dialing method, and each adult was asked if he or she was willing to participate in a group session in Des Moines (St. Paul) relating "to how people select food and household products."[4] They were informed the session would last about 90 minutes, and were also told that at the end of the session each participant would receive \$40 in cash for his or her time.

[4]Because common grocery store products were auctioned, participants were not required to be the main grocery shopper in the household.

Three different times were available each auction day — 9 am, 11:30 am, and 2 pm. Willing participants were asked to choose a time most convenient to their schedule. The 12 experimental units were split equally between Des Moines and St. Paul and were held on April 7 and April 21, 2001. After adjusting for unusable phone numbers, the ISU Statistics Laboratory contacted approximately 950 people; 200 individuals accepted the invitation to participate, although 20 people failed to show, leaving 180 people who came to the lab to participate. The response rate, after adjusting for unusable telephone numbers, was 19%. In addition, eight individuals had to be turned away because the quota for their time slot was full by the time they arrived (a maximum of 16 people per group was set), leaving 172 total participants.[5]

3.4. *Steps in the experiment*

The 10 specific steps in each experimental unit are detailed below.[6]

- In Step 1, when participants arrived at the experiment, they signed a consent form agreeing to participate in the auction. After signing this form, each individual was given $40 for participating and an ID number to preserve the participant's anonymity. The participants then read brief instructions and filled out a pre-auction questionnaire. The questionnaire was designed to collect demographic information and to determine consumers' perceptions

[5]We went to considerable expense to obtain participants who were selected by a random digital dialing process. Potential participants were told they were being invited to participate in an Iowa State University project dealing with household and food products, but they were not told the purpose of the project was to test consumers' willingness to pay for genetically modified foods. This information was purposely withheld to ensure the participation decision was unrelated to an individual's attitudes toward agricultural biotechnology. Of course, some individuals chose not to participate. It is unknown whether these unwilling subjects are statistically different from the participants. Based on the summary data reported in Table 1, however, the demographic characteristics of our sample match closely those reported by the 2000 U.S. Census for the two areas where participants were chosen. Caution is warranted in making generalizations beyond the Midwest.

[6]The complete set of instructions given to participants is available from the authors upon request.

of the safety of vaccinations for diseases, eating meat from animals fed growth hormones, and of irradiated and GM foods.[7]

- In Step 2, participants were given detailed instructions about how the random n^{th}-price auction works, including an example written on the blackboard. After the participants learned about the auction, they were given a short quiz to ensure everyone understood how the auction worked.
- Step 3 was the first practice round of bidding, in which participants bid on a brand-name candy bar. The participants were all asked to examine the product and then place a bid on the candy bar. The bids were collected and the first round of practice bidding was over. Throughout the auctions, when the participants were bidding on items in a round, they had no indication of what other items they might be bidding on in future rounds.
- Step 4 was the second practice round of bidding, and in this round the participants placed separate bids on three different items. The products were the same brand-name candy bar, a deck of playing cards, and a box of pens. Participants knew that only one of the two rounds would be chosen at random to be binding, which prevented anyone from taking home more than one unit of any product. Following Melton *et al.*, this random binding round eliminates the threat of a person reducing his or her bids because the individual potentially buys more than one unit.[8]

---

[7]It seems unlikely that the pre-auction questions about perceived GM risks significantly affected the results. The question dealing with GM food products was at the very beginning of each auction session and placed near questions about the frequency at which the population was made ill during the past year from eating irradiated meat or poultry and consuming meat from animals fed growth hormones. Also, after the participants went through several steps in the auction, they were given at least one full page of information from an interested party (either the biotech industry or the environmental group perspective) before bidding on food products. Any potential effect from the GM information question was likely to have been overwhelmed by other information presented to consumers in step 6.

[8]If one assumes there is little or no income effect from the deck of cards and box of pens, the two bids on the candy bar should be the same — i.e., because the deck of cards and box of pens are neither complements nor substitutes to the candy bar, they should not affect the bids on the candy bar. A Wilcoxon signed-rank test confirmed the bids for

The consumers first examined the three products and then submitted their bids.

- After the two practice auction rounds were completed, the binding round and the binding n^{th}-prices were revealed in Step 5. All bid prices were written on the blackboard, and the nth-price was circled for each of the three products. Participants could see immediately what items they won, and the price they would pay. The participants were told that the exchange of money for goods was in another room nearby and would take place after the entire experiment was completed.
- In Step 6, participants received information about biotechnology. Three info-packets were produced as follows: *(a) the environmental group perspective* — a collection of statements and information on genetic modification from Greenpeace, a leading environmental group; *(b)* the *industry perspective* — a collection of statements and information on genetic modification provided by a group of leading biotechnology companies, including Monsanto and Syngenta; and *(c)* the *independent, third-party perspective* — a statement on genetic modification approved by a third-party group, consisting of a variety of people knowledgeable about GM goods, including scientists, professionals, religious leaders, and academics, who do not have a financial stake in GM foods. These information sheets are provided in the appendix.

These three info-packets were used to create six information treatments: *(a)* only the biotech industry perspective, *(b)* only the environmental group perspective, *(c)* biotech industry and environmental perspectives, *(d)* biotech industry and third-party perspectives, *(e)* environmental group and third-party perspectives, and *(f)* all three perspectives. In the context of this study, focus is exclusively on the important issue of consumers' preferences for

the candy bars are not significantly different in the two rounds, with a test statistic of 0.03. This result does not contradict the notion that the subjects' bidding behavior was reasonable.

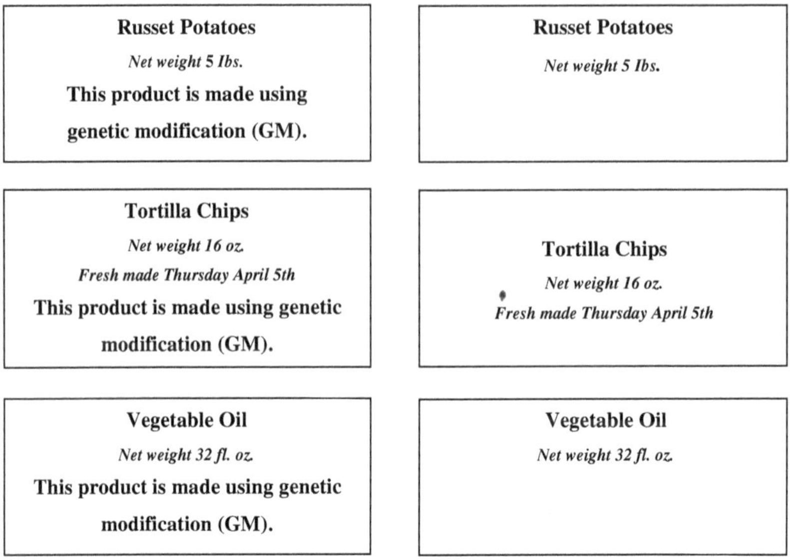

Fig. 1. GM and standard labels used for the three food items.

GM- versus plain-labeled food items. The impact of information on consumer behavior is not addressed.[9]

Once the appropriate info-packet was distributed to the participants in a given unit, two auction rounds were then conducted. The rounds were differentiated by the food label — either the food had a standard food label or a GM label (see Figure 1). In one round, which could be round 1 or round 2 depending on the experimental unit, participants bid on the three food products each with the standard food label. These labels were made as plain as possible to avoid any influence on the bids from the label design. In the other round, participants bid on the same three food products, but with a GM label, which differed from the standard label by the inclusion of only one extra sentence: "This product is made using genetic

[9]To more intricately examine the issue of consumer acceptance of GM labeling, this analysis is devoted entirely to the consumer reaction to labeling, without examining impact of information treatments. The interested reader is referred to a companion paper that presents a detailed evaluation of the information effects (Rousu *et al.*).

modification (GM)." For each experimental unit, participants knew that only one round would be chosen as the binding round which determined auction winners.[10] The sequencing of the standard food and GM labels was randomized across experimental units. One unit had the standard label in round 1, and the GM label in round 2; the second unit had the GM label in round 1, and the standard label in round 2.

- In Step 7, participants bid on three different food products: a bag of potatoes, a bottle of vegetable oil, and a bag of tortilla chips, either with the standard or GM label. The participants were instructed to examine the three products and then to write down their sealed bid for each of the three goods. Participants bid on each good separately. The monitor then collected the bids from the participants and told them they were next going to look at another group of food items.[11]
- Step 8 had participants examine the same three food products, each with a different label, from round 1. Again the participants examined the products and bid on the three products separately. The bids were then collected from all of the individuals.
- Step 9 selected the binding round, and the binding random n^{th}-prices for the three goods. Winners were notified. Each individual was asked to complete a brief post-auction questionnaire to see if the participant's risk attitudes changed, and to inquire about whom he or she would trust for information on genetic modification (see Huffman *et al.*).

[10]By bidding on foods with GM labels in one round and standard labels in the other round, participants are bidding on labels with statements which are similar to those found on labels in major markets with GM labels, like those in the European Union and Australia (see CNN, Australia-New Zealand Food Authority 2000, 2001).

[11]Unlike some food safety experiments (e.g., Shogren *et al.* 1994), we do not require immediate consumption of the products for people who win the auction. At least two of our three products are ones that are *not* usually consumed at the point of sale — vegetable oil and raw potatoes. Furthermore, adults regularly make family food-purchase decisions in grocery stores and supermarkets (e.g., see Carlson, Kinsey, and Nadav), which suggests that having an adult place bids in our experiment seemed entirely appropriate for most households.

- In Step 10, the monitors dismissed the participants who did not win. The monitors and the winners exchanged money for goods, and the winners were then dismissed.

Although standard experimental-auction valuation procedures were followed (e.g., Shogren *et al.* 1994), several refinements were made to our experimental design to better reflect consumer purchases. First, our subjects submitted only one bid per product. The protocol of using multiple repeated trials and posting market-clearing prices was not followed, to avoid any question of creating affiliated values that can affect the demand-revealing nature of a laboratory auction (see, for example, List and Shogren). Second, subjects were not endowed with any food item and then asked to "upgrade" to another food item; rather, participants were paid $40, and then they bid on different foods in only two trials. This approach avoids the risk of an in-kind endowment effect distorting the participant's bidding behavior (e.g., see Lusk and Schroeder, or Corrigan and Rousu) and the risk of any credit constraint. Third, each consumer bid on three unrelated food items. Thus, if he or she did not have positive demand for one or two products, we could still obtain information on the participant's taste for genetic modification based on the second and (or) third products. Fourth, treatments to the experimental units were randomly assigned; now estimation of treatment effect is simply the difference in means across treatments (see Wooldridge, pp. 603–607).

Fifth, we used adult consumers over 18 years of age from two different Midwestern metropolitan areas, who were chosen using a random digital dialing method.[12] Because we use common food items available to shoppers in grocery stores and supermarkets, we wanted adult rather than student subjects, to better reflect a typical household of consumers. Although several studies have used college undergraduates in laboratory auctions of food items (e.g., Hayes *et al.*; Lusk *et al.*), students are not the best choice for participants

[12]Information on other characteristics, like the age of children in the household and membership in environmental groups, was collected but is not reported to simplify the exposition. All non-reported information is available from the authors upon request.

when the items being auctioned are ones sold in grocery stores or supermarkets. Using a national random sample of grocery store shoppers, Katsaras *et al.* found that the share of college-age (18 to 24 years) shoppers falls far below their share in the population — 8.5% of shoppers versus 12.8% in the U.S. Census of Population. College students obtain a large share of their food from school cafeterias and only a small share from grocery stores and supermarkets compared to older shoppers (Carlson, Kinsey, and Nadav). Although our participants are slightly skewed toward women, Katsaras *et al.* note that women make up a disproportional share of grocery shoppers — 83% of shoppers versus 52% in the U.S. Census of Population. A sample primarily of grocery store shoppers also weakens the sometimes-stated need for having students participate in several rounds of bidding to stabilize bids for food items. The Iowa State University Statistics Laboratory selected potential participants from a randomly generated list of telephone numbers in the Des Moines and St. Paul areas by using a random digital dialing method. A summary of the demographic characteristics of the participants is displayed in Table 1.

Finally, this experimental design and implementation allow for results going beyond those of the two earlier experimental auction studies of GM foods (Lusk *et al.*; and Noussair, Robin, and Ruffieux 2002). The Noussair, Robin, and Ruffieux study was conducted in France; our experiments were conducted in two cities in the Midwestern United States. Compared to the one published experimental economics study measuring tastes for genetic modification using U.S. consumers (Lusk *et al.*), our experimental design extends the earlier work in three areas. First, our 172 participants are adult consumers, unknown to us, and randomly chosen from a list of telephone numbers in two Midwestern cities. Our sample is more than three times larger than the 50 students used in Lusk *et al.*, and consists of adults responsible for household food shopping rather than university students. Second, the incentive-compatible random n^{th}-price auction is used, rather than splitting the sample into a second-price auction and a nondemand-revealing first-price auction, and, to avoid learning effects, consumers are randomly assigned to

Table 1. Characteristics of the auction participants and characteristics of the populations of Polk and Ramsey counties (which contain Des Moines, IA, and St. Paul, MN).

Variable	Definition	Sample		County Population Mean[a]	
		Mean	Standard Deviation	Polk	Ramsey
Gender	1 if female	0.62	0.49	0.52	0.52
Age	Participant's age (years)	49.5	17.5	45.7	45.7
Married[b]	1 if the individual is married	0.67	0.47	0.59	0.51
Education[c]	Years of schooling	14.54	2.25	13.52	13.76
Household	Number of people in participant's household	2.78	1.65		
Income	Household's income level ($000s)	57.0	32.6	46.1	45.7
White	1 if participant is white	0.90	0.30	0.90	0.80
Read_L	1 if never reads labels before a new food purchase	0.01	0.11		
	1 if rarely reads labels before a new food purchase	0.11	0.31		
	1 if sometimes reads labels before a new food purchase	0.31	0.46		
	1 if often reads labels before a new food purchase	0.37	0.48		
	1 if always reads labels before a new food purchase	0.20	0.40		
Informed	1 if individual is self-reported as at least somewhat informed regarding genetically modified (GM) foods	0.42	0.49		
Labels 1	1 if the treatment bid on foods with GM labels occurs in round 1	0.52	0.50		

Note: For the Polk and Ramsey county figures, all variables are for individuals of all ages, except for *Married,* which is for individuals 18 or older; *Education,* which is for individuals 25 or older; and *Age,* which is for individuals 20 or older.

[a] *Source*: U.S. Bureau of the Census.

[b] For the Polk and Ramsey county estimates, the number of married people who are 18 or older was obtained by taking the number of people married over 15 and assuming that the number of people who were married at ages 15, 16, and 17 was zero. This gives the percentage of married people who are 18 or older.

[c] For Polk and Ramsey counties, the years of schooling were estimated by placing a value of 8 for those who have not completed 9th grade, 10.5 for those who have not completed high school, 12 for those who have completed high school but have had no college, 13.5 for those with some college but no degree, 14 for those with an associate's degree, 16 for those with a bachelor's degree, and 18 for those with a graduate or professional degree.

treatments that did not have a multiple sequence of trials. Third, in the Lusk *et al.* study, it is impossible to correctly interpret a zero bid — i.e., whether it meant a student disliked the food for sale or disliked genetic modification. In contrast, we asked consumers to bid on three different food items rather than on one product. This approach permits identification of a consumer's preferences for food versus his or her preferences for genetic modification. If a participant did not like one kind of food, reliable information on the other two food products can be obtained.

4. Experimental Results and Statistical Analysis

Table 2 reports the number of participants who bid less for the GM-labeled food than for the standard-labeled food. For each good, 35% to 41% of the participants bid less for the GM-labeled variety. Most participants showed a general dislike for genetic modification — about 60% of the participants bid less for the GM-labeled version for at least one of the three products. About 26% of the subjects bid less for all three GM-labeled products than for standard-labeled products, which indicates a strong preference against GM foods.

Table 3 gives the mean bids for each of the three products: a 32-ounce bottle of vegetable oil, a 1-pound bag of tortilla chips, and a 5-pound bag of potatoes. Panel A of Table 3 shows the mean bids

Table 2. Number and percentage of individuals who bid less for the GM-labeled food than for the standard-labeled food ($N = 172$).

	Individuals Bidding Less for GM-Labeled Version of the Product	
Description of Product	Number	Percent (%)
Tortilla Chips, 16 oz. bag	70	41
Vegetable Oil, 32 fl. oz. bottle	63	37
Russet Potatoes, 5 lb. bag	60	35
All three items	44	26
At least one of the three items	100	58

Table 3. Mean bids for each of the three products.

Product Description	No. Of Observations	Mean Bid	Standard Deviation	Maximum	No. of Zero bids
PANEL A. Mean Bids, All Participants					
Vegetable oil (GM)	172	0.91	0.84	3.99	39
Vegetable oil (standard)	172	1.05	0.85	3.79	28
Tortilla chips (GM)	172	0.93	0.86	3.99	37
Tortilla chips (standard)	172	1.08	0.85	4.99	17
Russet potatoes (GM)	172	0.78	0.67	3.00	33
Russet potatoes (standard)	172	0.91	0.67	3.89	14
PANEL B. Mean Bids When Participants Bid on Food with GM Food Labels in Round 1					
Vegetable oil (GM)	88	0.98	0.91	3.99	23
Vegetable oil (standard)	88	1.04	0.89	3.79	19
Tortilla chips (GM)	88	0.95	0.87	3.25	22
Tortilla chips (standard)	88	1.05	0.81	4.99	12
Russet potatoes (GM)	88	0.90	0.69	2.50	15
Russet potatoes (standard)	88	0.94	0.63	2.51	7
PANEL C. Mean Bids When Participants Bid on Food with GM Food Labels in Round 2					
Vegetable oil (GM)	84	0.83	0.77	3.25	16
Vegetable oil (standard)	84	1.06	0.80	3.00	9
Tortilla chips (GM)	84	0.90	0.86	3.99	15
Tortilla chips (standard)	84	1.11	0.89	4.99	5
Russet potatoes (GM)	84	0.65	0.63	3.00	18
Russet potatoes (standard)	84	0.88	0.72	3.89	7

for all participants. For each of the three food products, participants were willing to pay less when the product had the GM label than when it displayed the standard food label. Both one-tailed t-tests and Wilcoxon signed-rank tests show that the difference is statistically significant at the 5% significance level.[13]

Both Table 2 and panel A of Table 3 provide univariate tests of the first hypothesis: (H1) GM labels have no effect on consumers' WTP for food items. This hypothesis is rejected; GM labels are found to

[13]These tests are univariate tests. Like all univariate tests, they suggest a simple relationship, indicating here that consumers paid less for food products with GM labels. However, they provide no information about why consumers paid less for food products with GM labels.

affect consumer WTP for food products.[14] The percentage discount for each product was approximately 14% , indicating consumers displayed a general dislike for genetic modification, but they did not seem to dislike one type of genetically modified product more or less than another. Consumers, for instance, did not appear to view genetic modification of the soybeans used to make vegetable oil better than genetic modification in potatoes. This result suggests respondents concerns about GM technology are largely environmental and not human health.

In most economics experiments of food products (e.g., Melton *et al.*; Hayes *et al.*; Lusk *et al.* 2001; Shogren *et al.* 2000), all participants have been subjected to the same sequence of treatments. It is impossible to disentangle treatment effects from sequencing effects. In Table 3, panel B, bids are shown for the subset of auction participants who bid on food products having GM food labels in round 1, and panel C shows the bids for participants who bid on food products having the GM labels in round 2. The bids for the GM-labeled and standard-labeled food are much closer when participants bid first on GM food (round 1) compared to those who bid on GM-labeled food second (round 2). For two of the three products (potatoes and vegetable oil), the difference due to the order is statistically significant at the 5% level (both *t*-tests and Wilcoxon rank-sum tests show the same results). We therefore reject hypothesis 3: (H3) no difference exists in consumers' WTP for food items due to the GM food label treatment sequence. In our experiment, the order in which consumers bid on food products influences their bid. This is an important finding because many experimental economics studies use repeated, non-randomized trials to gain information on consumer preferences. The results of this

[14]The mean bid prices are slightly lower than a person might see in a grocery store. This is logical because a rational bidder should not bid more for a product than the price he or she can find in a store (although some participants could add travel costs to a bid price). This upper bound suggests that mean bids in most auctions should be less than the average price for a product one finds in a store, although this remains to be established in the lab.

analysis indicate randomization is important to ensure results are not an artifact of the sequence used in the experiment.

To graphically explore the distribution of the difference in bid prices between the GM-labeled and plain-labeled products, histograms are presented in Figure 2. For all three products, the majority of participants have no difference in bid prices between GM-labeled and plain-labeled food products. However, for all three products, there are some participants who bid more for the plain-labeled products, with several participants placing premiums of $1 or more. For all three products, very few people bid less for the plain-labeled product than for the GM-labeled product.

To further explore the bids for GM-labeled and plain-labeled food products of participants, consider the following equation:

$$P_j^K = \beta_j + \beta_2^k \mathbf{X}_{j2} + \mu_j^k, \quad k = \text{plain-labeled, GM-labeled}, \tag{1}$$

where the price of the kth labeled commodity by the jth participant, P_j^k, is a function of a linear unmeasured effect (β_j) that is constant across commodity labels for a given individual, regressors $\mathbf{X}_{j2}$ that vary across individuals but not across labels for a given individual (i.e., a participant's age, education, household income, gender), and a zero mean random disturbance term across labels and individuals (μ_j^k). Now taking differences across label types gives:

$$\begin{aligned} P_j^{\textit{plain-labeled}} - P_j^{\textit{labeled}} &= (\beta_2^{\textit{plain-labeled}} - \beta_2^{\textit{labeled}})\mathbf{X}_j \\ &\quad + \mu_j^{\textit{plain-labeled}} - \mu_j^{\textit{labeled}}. \end{aligned} \tag{2}$$

Note that the intercept term, the unmeasured effect across labels for a given individual, vanishes and the vector of coefficient differences is a measure of the difference in impact of a factor on a plain-labeled versus a GM-labeled food item, and the differenced disturbance term across individuals and labels. The coefficients and error terms in (2) can be condensed and rewritten as:

$$P_j^{\textit{plain-labeled}} - P_j^{\textit{labeled}} = \beta_1^* + \beta_2^* X_j + \mu_j^*. \tag{3}$$

The coefficient of β_1^* is expected to be small and not significantly different from zero; the vector of parameters β_2^* is expected to be

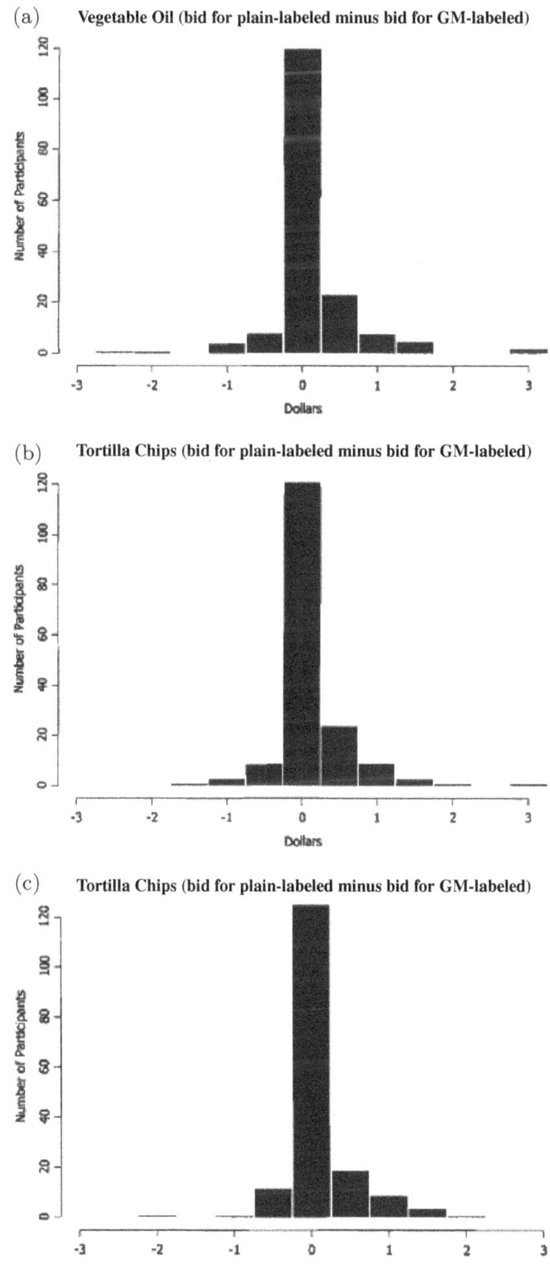

Fig. 2. Histogram of the difference in bid prices ($s) for plain-labeled and GM-labeled versions of each of the three commodities ($N = 172$).

significantly different from zero *only* if a regressor has a different *marginal* impact on the price of plain-labeled than on GM-labeled food items. Otherwise these coefficients should be approximately zero. The disturbance term μ_j^* is expected to have a zero mean after the censoring in equation (1).

All participants placed a bid on GM-labeled and plain-labeled potatoes, tortilla chips, and vegetable oil. Equation (3) is then estimated using a fixed-effects model in which each product is a fixed effect. All bids are pooled over the three products, yielding $172 \times 3 = 516$ observations. Although the minimum bid for any product is zero, some participants may have viewed a product as a "bad" and preferred a negative valuation.

In equation (1), for {k= plain-labeled, GM-labeled}, the price could be censored at zero, so we take account of this censoring in fitting equation (3). The censored regression model examines the two components of the dependent variable, the bid for the plain-labeled product and the bid for the GM-labeled product. The dependent variable is considered censored if *either* (*a*) the bid for the GM-labeled product is zero, (*b*) the bid for the plain-labeled product is zero, or (*c*) the bids for the plain-labeled product and the GM-labeled product are both zero. Note that in our model, the dependent variable itself can be zero and still not be censored (e.g., if a participant bids $2 for both the GM-labeled and plain-labeled version of a product). Our findings show this result arises quite frequently. This model is estimated using a maximum-likelihood procedure in the LIFEREG PROCEDURE in SAS, version 8 (the data and code used are available from the authors upon request).

The censoring has four cases:[15]

- Case 1. Consumer *j* bids a positive amount for both the GM-labeled and the plain-labeled product. The measured difference in bid prices is the difference between the two bid prices (non-censored case).

[15]The dependent variable in equation (2) has two parts, each of which could be censored. This is not a typical Tobit model, but rather a more general censored regression model. For a good renew of censored regression models, see Wooldridge (2002, pp. 517–550).

- Case 2. Consumer j bids zero for the GM-labeled product and a positive amount for the plain-labeled product. The "true" difference in bid prices with the censored regression will be greater than the difference between the two observed bid prices. This outcome arises because the bids on the GM-labeled product are censored at zero.
- Case 3. Consumer j bids a positive amount for the GM-labeled product and zero for the plain-labeled product. This is similar to case 2 in which the true difference in bid prices for the censored regression is absolutely larger than the measured difference between the two bid prices.
- Case 4. Consumer j bids zero for both products. This case does not give any information about the consumer's true demand for GM products.[16]

Because our model treats zero bids for both the GM-labeled and plain-labeled food products as censored bids, all four cases are correctly accounted for.

Censored regression results from fitting equation (3) with commodity fixed effects to explain the difference in bid price (plain-labeled minus GM-labeled) for the sample are reported in Table 4. In regressions (1) and (2), the estimate of the intercept is small, as expected, and it is not significantly different from zero at the 5% level in regression (2). The coefficients of the sociodemographic variables in regression (2) are not significantly different from zero, i.e., a participant's gender, education, age, or prior beliefs about GM technology individually do not have a significant effect on bid price differences.

Also, we performed a joint test that all four of the coefficients for the sociodemographic variables are zero. Using the information in regressions (1) and (2), the sample value of the chi-squared

[16]A participant might bid zero for each version of a product for several reasons. For example, the participant may dislike a particular food product. Or the participant may be placing a "protest bid" because he or she does not like the procedures. (See Shogren, List, and Hayes for a more detailed explanation on why consumers bid as they do in experimental auctions.)

Table 4. Censored regression, fixed-effects model explaining difference in bid price for plain-labeled and GM-labeled foods (N = 516).

(Dependent Variable: *Bid Price Plain-Labeled Food minus Bid Price GM-Labeled Food*

	Regression(1)		Regression(2)	
Regressors	Coefficient	Std. Error	Coefficient	Std. Error
Intercept	0.27***	0.05	0.14	0.22
GM Labels in Round 1	−0.17***	0.05	−0.15***	0.06
Female			−0.05	0.06
Education			0.01	0.01
Age			−0.00	0.00
Informed			0.06	0.06
Vegetable Oil (fixed effect)	0.00	0.07	0.00	0.07
Tortilla Chips (fixed effect)	0.02	0.07	0.02	0.07
Log Likelihood	−414.02		−412.45	

***Denotes statistical significance at the 1% level.

statistic for this test is 3.04, and the critical value is 9.5 with four degrees of freedom at the 5% significance level. Hypothesis H2 cannot be rejected: Sociodemographic attributes of participants, including prior beliefs about genetic modification, individually and jointly do not have a significant effect on bid price differences.[17] However, hypothesis H3 is rejected: The estimated coefficient for the dummy variable denoting a respondent was presented with GM labels in round 1 of bidding is significantly different from zero. Randomizing the labeling treatment is an important methodological feature of our study. The coefficients of the two dummy variables for commodity

[17]The method used is a linear labeling-variant fixed effect, censored regression model to carry out the tests of H2. The fixed effect being removed is any common unmeasured linear, labeling-invariant fixed effect that is product specific (see Wooldridge, pp. 247–249), and it is not a Tobit model. The differencing method safeguards against potential coefficient bias which would be caused by an unmeasured individual effect being correlated with demographic attributes.

fixed effects, however, are not significantly different from zero.[18] No commodity-specific fixed effects are observed in the results in Table 4.[19]

4.1. *Discussion*

The results presented here go beyond earlier studies in three key areas. First, it was observed that 58% of the adult participants bid less for at least one of the three GM-labeled products, while 26% of the participants disliked GM foods intensely enough that they bid less for every GM-labeled product. Second, the method allows us to estimate the price discount U.S. consumers might register in grocery stores if GM foods were labeled. In our lab experiments, consumers discounted the GM-labeled food by an average of 14% relative to the standard-labeled food.

The GM food industry seems likely to continue to oppose GM food labels, but the non-GM food industry may see new opportunities to supply non-GM foods at a premium. Although the results of this analysis differ from the experimental auction results reported by Lusk *et al.*, they acknowledged the limitations placed on their results by using only undergraduate students: "If experiments were conducted with a larger, potentially more representative sample, we would expect a larger percentage of participants to bid [for non-GM food], and bid at higher levels" (Lusk *et al.*, p. 55). Our results confirm the expectation expressed by Lusk *et al.* specifically, the average adult in two major Midwestern cities discounted GM-labeled foods

[18]Our treatments include two labels and three information packets for a total of six labeling and information treatments. The econometric model for equation (3) was also refitted including dummy variables for information treatments, and results were not materially different from those reported in Table 4. (See Rousu *et al.* for an analysis of the impacts of diffuse information on participants' willingness to pay.) Some of the information effects are statistically significant.

[19]Other censored regression models were fitted for equation (3), including regressors for a participant's pre-auction risk perceptions associated with irradiated foods, meat from animals fed growth hormones, and vaccines to prevent diseases. We also fitted models including a participant's household income, marital status, race, and other variables. None of these variables had a coefficient that was statistically different from zero for explaining bid differences for the GM-labeled and plain labeled food products.

by about 14% relative to standard-labeled foods, and almost 60% of our sample showed a dislike for genetic modification, as revealed by a lower bid for a GM-labeled food relative to the standard-labeled counterpart. Third, participants in our lab experiments placed about the same percentage discount for each of the three GM-labeled food products, suggesting consumers are less concerned about *how* genetic modification enters the food product than the fact it *has* entered the food product, and that their concerns are primarily environmental.

Two interpretations are possible from our results. One interpretation is that without the GM labels, a respondent is bidding more for GM food than he/she otherwise would. Under this interpretation, GM food labels would therefore benefit consumers by informing them of GM content, which would lower their bids. The alternative interpretation is that consumers do not understand genetic modification. Given that GM food products are deemed to be "substantially equivalent" to their non-GM counterparts, this interpretation implies consumers would be better off by not being exposed to GM labels. If this is the case, more information on the risks or benefits of GM foods could help consumers make better informed decisions. Whatever the interpretation, our results provide evidence that consumers' willingness to pay for GM-labeled foods is significantly lower than their willingness to pay for the plain-labeled counterpart.

5. Conclusion

This study has demonstrated that consumers' willingness to pay for a food product decreases when the food label indicates the food product is genetically modified. The evidence shows consumers were willing to pay a 14% premium for food items they perceived as not genetically modified. Sociodemographic attributes of the participants — gender, education, household income, and prior beliefs about GM — do not alter significantly a consumer's WTP for GM foods. Our results, however, provide evidence that the order in which consumers bid on foods (i.e., whether they bid on the food with GM labels in round 1 or round 2) affects WTP for GM foods. Participants who bid on food with GM labels in round 1 had a higher

WTP for GM food than the participants who bid on food with GM food labels in round 2. Because sequencing matters, we emphasize that future economics experiments randomize all treatments.

The implications of this study are notable given the ongoing global controversy over the issue of labels on GM foods. This debate has forced many countries around the world to consider or to implement new food-labeling policies. Given that the average adult consumer in a major Midwestern city revealed a significant premium for foods perceived as being non-GM, a mandatory GM labeling policy seems *unlikely* to be in the best interest of the biotechnology industry — but may be in the interest of consumers. From the perspective of the biotech industry, the use of GM labels would likely reduce consumers' willingness to pay for products which otherwise would be plain labeled.

Future research might examine the robustness of these results by replicating experiments in other areas of the country. Another avenue for future research would be to examine consumers' reaction to GM foods having specific consumer benefits, e.g., enhanced protein, vitamins, micro-nutrient content, shelf life. Two such products are Flavor-Saver tomatoes which were genetically engineered to have a longer shelf life, and "golden rice" which was genetically modified to provide more vitamin A, an attribute having potential benefits for people in third-world countries.

Acknowledgments

The authors gratefully acknowledge comments from two anonymous referees. Thanks to Wayne Fuller and Phil Dixon for developing the statistical experimental design and to Daniel Monchuk and Terrance Hurley for their generous help in conducting the auctions, and to Monsanto for providing some of the products used in the experiment. This work was supported through a grant from the Cooperative State Research, Education, and Extension Service, U.S. Department of Agriculture, under Agreement No. 00-52100-9617, and from the U.S. Department of Agriculture, Economic Research Service, under Agreement No. 43-3AEL-8-80125 and by the Iowa Agriculture and

Home Economics Experiment Station. Views expressed are those of the authors, and not necessarily those of the funding agencies.

Review coordinated by Gary D. Thompson.

References

Australia-New Zealand Food Authority (2000). "Labelling Genetically Modified Foods: Compliance Guide to Standard A18." Online. Available at http://www.anzfa.gov.au/whatsinfood/gmfoods/complianceguidea18gm.cfm.

Australia-New Zealand Food Authority (2001). "Genetically Modified Foods." Online. Available at http:/www.anzfa.gov.au/GMO/.

Becker, G., M. DeGroot and J. Marschak (1964). Measuring Utility by a Single Response Sequential Method. *Behavioral Science* 9, 226–236.

Carlson, A., J. Kinsey and C. Nadav (1998). "Who Eats What, When, and From Where?" Working Paper No. 98-05, Retail Food Industry Center, University of Minnesota, St. Paul.

Caswell, J.A. (1998). Should Use of Genetically Modified Organisms Be Labeled? *AgBioForum* 1, 22–24. Online. Available at http://www.agbioforum.missouri.edu.

Caswell, J.A. (2000). Labeling Policy for GMOs: To Each His Own? *AgBioForum* 3, 53–57. Online. Available at http://www.agbioforum.missouri.edu.

Caswell, J.A. and D.I. Padberg (1992). Toward a More Comprehensive Theory of Food Labels. *American Journal of Agricultural Economics* 74, 460–467.

Cherry, T., T. Crocker and J. Shogren (2003). Rationality Spillovers. *Journal of Environmental Economics and Management* 45, 63–84.

CNN (2001). "Europe Approves Tough GM Food Rules." News Release. Online. Available at http://www.cnn.com/2001/WORLD/europe/02/14/eu.gm/index.html.

Corrigan, J.R. and M. Rousu (2003). "The Endowment Effect in the Absence of Loss Aversion." Unnumb. Work. Pap., RTI International, Research Triangle Park NC.

Fox, J.A., J.F. Shogren, D.J. Hayes and J. Kliebenstein (1998). CVM-X: Calibrating Contingent Values with Experimental Auction Markets. *American Journal of Agricultural Economics* 80, 455–465.

Franciosi, R., R.M. Isaac, D. Pingry and S. Reynolds (1993). An Experimental Investigation of the Hahn-Noll Revenue Neutral Auction for Emissions Licenses. *Journal of Environmental Economics and Management* 24, 1–24.

Friends of the Earth (2001). "The Need for Labeling Genetically Engineered Foods." Online. Available at http://www.foe.org/safefood/factshtgelabel.htm.

Golan, E., F. Kucher, L. Mitchell, C. Green and A. Jessup (2000). "Economics of Food Labeling." Agr. Econ. Rep. No. 793, USDA/Economic Research Service, Washington DC.

Greenpeace International (2001). "We Want Natural Food!" Online. Available at http://www.greenpeace.org/~geneng/.

Hayes, D.J., J.F. Shogren, S.Y. Shin and J.B. Kliebenstein (1995). Valuing Food Safety in Experimental Auction Markets. *American Journal of Agricultural Economics* 77, 40–53.

Hoffman, E., D. Menkhaus, D. Chakravarti, R. Field and G. Whipple (1993). Using Laboratory Experimental Auctions in Marketing Research: A Case Study of New Packaging for Fresh Beef. *Marketing Science* 12, 318–338.

Huffman, W.E., M. Rousu, J.F. Shogren and A. Tegene (2003). "Who Do Consumers Trust for Information? The Case of Genetically Modified Foods." Department of Economics, Iowa State University, Ames.

Huffman, W.E. and A. Tegene (2002). "Public Acceptance of and Benefits from Agricultural Biotechnology: A Key Role for Verifiable Information." In V. Santaniello, R. E. Evenson and D. Zilberman (Eds.), *Market Development of Genetically Modified Foods*, pp. 179–90. New York: CABI Publishing.

Kagel, J.H. (1995). "Auctions: A Survey of Experimental Research." In J. Kagel and A. Roth (Eds.), *Handbook of Experimental Economics*, pp. 501–85. Princeton, NJ: Princeton University Press.

Katsaras, N., P. Wolfson, J. Kinsey and B. Senauer (2001). "Data Mining: A Segmentation Analysis of U.S. Grocery Shoppers." Working Paper No. 01-01, the Retail Food Industry Center, University of Minnesota, St. Paul.

List, J.A. and J.F. Shogren (1999). Price Information and Bidding Behavior in Repeated Second-Price Auctions. *American Journal of Agricultural Economics* 81, 942–949.

Lusk, J., M.S. Daniel, D. Mark and C. Lusk (2001). Alternative Calibration and Auction Institutions for Predicting Consumer Willingness to Pay for Nongenetically Modified Corn Chips. *Journal of Agricultural and Resource Economics* 26, 40–57.

Lusk, J.L. and T.C. Schroeder (2002). "An Alternative Test for the Endowment Effect." Working Paper, Department of Agricultural and Economics, Mississippi State University, Starkville.

Melton, B.E., W.E. Huffman, J.F. Shogren and J.A. Fox (1996). Consumer Preferences for Fresh Food Items with Multiple Quality Attributes: Evidence from an Experimental Auction of Pork Chops. *American Journal of Agricultural Economics* 78, 916–923.

Mendenhall, C.A. and R.E. Evenson (2002). "Estimates of Willingness to Pay a Premium for Non-GM Foods: A Survey." In V. Santaniello, R.E. Evenson and D. Zilberman (Eds.), *Market Development of Genetically Modified Foods*, pp. 55–62. New York: CABI Publishing.

Miller, G. and C. Plott (1985). "Revenue Generating Properties of Sealed-Bid Auctions: An Experimental Analysis of One-Price and Discriminative Actions." In V. Smith (Ed.), *Research in Experimental Economics*, Vol. 3, pp. 159–82. Greenwich CT: JAI Press, Inc.

Noussair, C., S. Robin and B. Ruffieux (2001). "Genetically Modified Organisms in the Food Supply: Public Opinion vs. Consumer Behavior." Paper No. 1139, Institute for Research in the Behavioral, Economic, and Management Sciences, Krannert Graduate School of Management, Purdue University.

Noussair, C., S. Robin and B. Ruffieux (2002). Do Consumers Not Care About Biotech Foods or Do They Just Not Read the Labels? *Economics Letters* 75, 47–53.

Rousu, M., W.E. Huffman, J.F. Shogren and A. Tegene (2002). "The Value of Verifiable Information in a Controversial Market." Department of Economics, Iowa State University, Ames.

Shogren, J.F., J.A. List and D.J. Hayes (2000). Preference Learning in Consecutive Experimental Auctions. *American Journal of Agricultural Economics* 82, 1016–1021.

Shogren, J.F., M. Margolis, C. Koo and J. List (2001). A Random n^{th}-Price Auction. *Journal of Economic Behavior and Organization* 46, 409–421.

Shogren, J.F., S.Y. Shin, D.J. Hayes and J.B. Kliebenstein (1994). Resolving Differences in Willingness to Pay and Willingness to Accept. *American Economic Review* 84, 255–270.

U.S. Bureau of the Census (2000). *U.S. Census of the Population, 2000.* Online. Available at http://www.census.gov/. [Retrieved June 21, 2002.]

U.S. Food and Drug Administration (2001). "Guidance for Industry: Voluntary Labeling Indicating Whether Foods Have or Have Not Been Developed Using Bioengineering." FDA, Washington DC. Online. Available at http://vm.cfsan.fda.gov/~dms/biolabgu.html.

U.S. Food and Drug Administration (1992). Department of Health and Human Services. Statement of Policy: Foods Derived from New Plant Varieties: Notice. *Federal Register* 57, 22,984.

Vickrey, W. (1961). Counterspeculation, Auctions, and Competitive Sealed Tenders. *Journal of Finance* 16, 8–37.

Wilson, W.W. and B.L. Dahl (2002). "Costs and Risks of Testing and Segregating GM Wheat." Rep. No. 501, Dept. of Agribus. and Appl. Econ., North Dakota State University.

Wooldridge, J.M. (2002). *Econometric Analysis of Cross-Section and Panel Data.* Cambridge, MA: The MIT Press.

Appendix: Biotechnology Information Released to Participants During the Experiment

A. *International environmental group information*

The following is a collection of statements and information on genetic modification from Greenpeace, a leading environmental group.

General information

Genetic modification is one of the most dangerous things being done to your food sources today. There are many reasons that genetically modified foods should be banned, mainly because unknown adverse effects could be catastrophic! Inadequate safety testing of GM plants, animals, and food products has occurred, so humans are the ones testing whether or not GM foods are safe. Consumers should not have to test new food products to ensure that they are safe.

Scientific impact

The process of genetic modification takes genes from one organism and puts them into another. This process is very risky. The biggest potential hazard of genetically modified (GM) foods is the unknown. This is a relatively new technique, and no one can guarantee that consumers will not be harmed. Recently, many governments in Europe assured consumers that there would be no harm to consumers over mad-cow disease, but unfortunately, their claims were wrong. We do not want consumers to be harmed by GM food.

Human impact

Genetically modified foods could pose major health problems. The potential exists for allergens to be transferred to a GM food product that no one would suspect. For example, if genes from a peanut were transferred into a tomato, and someone who is allergic to peanuts eats this new tomato, they could display a peanut allergy. Another problem with genetically modified foods is a moral issue. These foods are taking genes from one living organism and transplanting them into another. Many people think it is morally wrong to mess around with life forms on such a fundamental level.

Financial impact

GM foods are being pushed onto consumers by big businesses, which care only about their own profits and ignore possible negative side effects. These groups are actually patenting different life forms that they genetically modify, with plans to sell them in the future. Studies have also shown that GM crops may get lower yields than conventional crops.

Environmental impact

Genetically modified foods could pose major environmental hazards. Sparse testing of GM plants for environmental impacts has occurred.

One potential hazard could be the impact of GM crops on wildlife. One study showed that one type of GM plant killed Monarch butterflies. Another potential environmental hazard could come from pests that begin to resist GM plants that were engineered to reduce chemical pesticide application. The harmful insects and other pests that get exposed to these crops could quickly develop tolerance and wipe out many of the potential advantages of GM pest resistance.

B. *Biotech industry information*

The following is a collection of statements and information on genetic modification provided by a group of leading biotechnology companies, including Monsanto and Syngenta.

General information

Genetically modified plants and animals have the potential to be one of the greatest discoveries in the history of farming. Improvements in crops so far relate to improved insect and disease resistance and weed control. These improvements using bioengineering/GM technology lead to reduced cost of food production. Future GM food products may have health benefits.

Scientific impact

Genetic modification is a technique that has been used to produce food products that are approved by the Food and Drug Administration (FDA). Genetic engineering has brought new opportunities to farmers for pest control and in the future will provide consumers with nutrient enhanced foods. GM plants and animals have the potential to be the single greatest discovery in the history of agriculture. We have seen just the tip of the iceberg of future potential.

Human impact

The health benefits from genetic modification can be enormous. A special type of rice called "golden rice" has already been created which has higher levels of vitamin A. This could be very helpful because the disease Vitamin A Deficiency (VAD) is devastating in third-world countries. VAD causes irreversible blindness in over 500,000 children and is also responsible for over one million deaths annually. Since rice is the staple food in the diets of millions of people in the third world, golden rice has the potential of improving millions of lives a year by reducing the cases of VAD. The FDA has approved GM food for human consumption, and Americans have been consuming GM foods for years. While every food product may pose risks, there has never been a documented case of a person getting sick from GM food.

Financial impact

Genetically modified plants have reduced the cost of food production, which means lower food prices, and that can help feed the world. In America, lower food prices help decrease the number of hungry people and also let consumers save a little more money on food. Worldwide the number of hungry people has been declining, but increased crop production using GM technology can also help further reduce world hunger.

Environmental impact

GM technology has produced new methods of insect control that reduce chemical insecticide application by 50% or more. This means less environmental damage. GM weed control is providing new methods to control weeds, which are a special problem in no-till farming. Genetic modification of plants has the potential to be one of the most environmentally helpful discoveries ever.

C. *Independent, third-party information*

The following is a statement on genetic modification approved by a third-party group consisting of a variety of individuals knowledgeable about genetically modified foods, including scientists, professionals, religious leaders, and academics. These parties have no financial stake in genetically modified foods.

General information

Bioengineering is a type of genetic modification where genes are transferred across plants or animals, a process that would not otherwise occur. (In common usage, genetic modification means bioengineering.) With bioengineered pest resistance in plants, the process is somewhat similar to the process of how a flu shot works in the human body. Flu shots work by injecting a virus into the body to help make a human body more resistant to the flu. Bioengineered plant-pest resistance causes a plant to enhance its own pest resistance.

Scientific impact

The Food and Drug Administration standards for GM food products (chips, cereals, potatoes, etc.) are based on the principle that they have essentially the same ingredients, although they have been modified slightly from the original plant materials. Oils made from bioengineered oil crops have been refined, and this process removes essentially all the GM proteins, making them like non-GM oils. So even if GM crops were deemed to be harmful for human consumption, it is doubtful that vegetable oils would cause harm.

Human impact

While many genetically modified foods are in the process of being put on your grocer's shelf, there are currently no foods available in the U.S. where genetic modification has increased nutrient content. All foods present a small risk of an allergic reaction to some people. No FDA-approved GM food poses any known unique human health risks.

Financial impact

Genetically modified seeds and other organisms are produced by businesses that seek profits. For farmers to switch to GM crops, they must see benefits from the switch. However, genetic modification technology may lead to changes in the organization of the agribusiness industry and farming. The introduction of GM foods has the potential to decrease the prices to consumers for groceries.

Environmental impact

The effects of genetic modification on the environment are largely unknown. Bioengineered insect resistance has reduced farmers' applications of environmentally hazardous insecticides. More studies are occurring to help assess the impact of bioengineered plants and organisms on the environment. A couple of studies reported harm to Monarch butterflies from GM crops, but other scientists were not able to recreate the results. The possibility of insects growing resistant to GM crops is a legitimate concern.

Chapter 7

Who Do Consumers Trust for Information: The Case of Genetically Modified Foods?*

Wallace E. Huffman†, Matthew Rousu‡,
Jason F. Shogren§, and Abebayehu Tegene¶

†*Curtiss Distinguished Professor of Agriculture and Professor of Economics, Iowa State University, Ames*
‡*Assistant Professor, Susquehanna University*
§*Stroock Distinguished Professor of Natural Resource Conservation and Management, Department of Economics and Finance, University of Wyoming*
¶*Senior Economist, Food and Rural Economics Division, ERS, USDA*

During the twentieth century, research and development (R&D) has produced a steady stream of inventions and new consumer goods, many of which have been adopted and proven to be the source of a rising standard of living (Boskin *et al.*, 1998). The introduction of new goods, however, creates a disequilibrium (Hausman, 1996), which in turn creates a demand by economic agents for objective information to assist in making decisions on adoption and use (Schultz, 1975).[1] The consumer's challenge is to sort

*This article originally appeared in the **American Journal of Agricultural Economics (2004), 86(5), 1222–1229**. This is published by permission of **Agricultural & Applied Economics Association**.

This article was presented in a principal paper session at the AAEA annual meeting (Denver, Colorado, August 2004). The articles in these sessions are not subjected to the journal's standard refereeing process.

[1]The appearance of new goods (or new attributes) has the same effect that the appearance of a new means of production has on a firm (Bianchi, 2002; Becker, 1976, pp. 131–149).

through the various, competing and sometimes conflicting, sources of information.

Consider, for example, the controversial products — genetically modified (GM) foods. The agricultural biotechnology firms (e.g., Monsanto, Syngenta, and the industry's Council on Biotechnology Information) have hailed the use of biotechnology to create new products as a major revolution in product innovation (Hoban, 1997, 2001). They have disseminated information claiming that GM crops will lower food costs worldwide and improve environmental quality. The council has even created and distributed children's coloring books that promote the positive aspects of GM foods.

In contrast, two international environmental NGOs, Greenpeace and Friends of the Earth, have distributed negative information through web sites, press releases, and demonstrated claiming risks to human health, environment, and biodiversity. They also claim that consumers have the right to know with respect to GM labeling, and that new technology benefits only large multinationals and not consumers (Friends of the Earth, Greenpeace International, 1997). The federal government also supplies information on genetic modification. In 1992, the Food and Drug Administration (FDA) issued a statement saying that GM foods do not have to be labeled if the new product has the same characteristics as its non-GM counterpart (FDA, 1992). If, however, a new vitamin was introduced into a food crop using genetic modification, the food product would have to be labeled (FDA, 2001). In contrast, Australia, China, European Union (EU), Mexico, New Zealand, and Russia have mandatory labeling policies for GM foods. In 2004, the EU removed a four-year moratorium on approvals of GM crops, but in April it implemented traceability requirement for food and feed.

In principle, society can recapture some of the losses from diverse information disseminated by interested parties if decision makers have access to independent, third party information. For example, see the discussion of verifiable information in Milgrom and Roberts and Huffman and Tegene. In recent research, Huffman *et al.* have defined verifiable information for genetic modification as information that is from a knowledgeable independent, third-party source that has no significant financial ties to the agricultural biotechnology industry. Although government agencies provide information, we do not claim that federal government agencies present an independent, third-party perspective. Federal regulatory agencies do not have the resources required to engage in significant independent research, and they must rely heavily upon the firms that they regulate to provide them the information they use to make recommendations. Although biotech firms are not always pleased with decisions made by the FDA, USDA, or EPA, the international NGOs believe that these agencies have a decidedly probiotech perspective (Greenpeace International, 1997, 2002; Gates, 2001). Also, elected government officials may receive campaign contributions from interested parties who are hoping to obtain future favorable treatment. Consequently, Huffman and Tegene have proposed that a quasi-governmental agency, that is funded, perhaps by the federal government, but not controlled by government officials, would be the most credible source of independent, third-party information on genetic modification. Rousu *et al.* have concluded that verifiable information on GM food products can have an annual value to U.S. consumers of over $2 billion annually.

Interested and disinterested parties disseminate information with the goal of affecting consumers' (and producers') decisions on GM technology and other products. However, for these groups to be influential, they must garner "trust". Recent evidence by Glaeser *et al.* shows that individuals who are closer in social status or who have similar personal capital (PC) are more likely to trust one another. For example, individuals who were raised with a particular religious tradition place more trust in others who were raised within the same religious tradition, *ceteris paribus.* More generally, Becker (1996) argues that consumer's social capital (SC) and PC are important determinants of his/her tastes or preferences. Social capital is defined as the capital the individual acquires through his or her surroundings, upbringing, and social network. Personal/human capital is defined as capital that the individual personally acquires, such as, schooling, habits, or experience. Becker shows when PC and SC are incorporated into economic models, economic theory can explain many previously puzzling outcomes such as the effect of advertising on consumers' purchasing behavior and human addictions.

Understanding the formation of trust in information sources is an important step in understanding consumers' preferences for information on new products. With the aid of a model, we formulate hypotheses about the role of measurable attributes of a consumer, which are related to his or her household income, PC and SC, and prior beliefs in the formation of trust. For this study, unique data were collected by an independent agency from a random sample of adults chosen from two major Midwestern cities.[2] These individuals were paid $40 to come to a central location, to provide sociodemographic information and information on prior beliefs about technologies, and to participate in a set of experiments (as explained in detail in a companion paper, Huffman *et al.*, 2003). At the end of the experiments, they were asked to complete a short questionnaire including the following question: "If a source of information were to give you verifiable information on GM foods, who would you trust most?" This information was coded into six different categories: third party, government, environmental or consumer group, private industry or organization, none or nobody, and "other" (including no response). A multinomial logit model is fitted to the sample of postexperiment participants to explain their relative trust in sources to provide verifiable information on GM foods.

We find an individual's household income has no significant effect on relative trust, but an increase in his or her schooling lowers the probability of trusting information from government, private

[2]A telephone interview of a random sample of adults is an alternative route to obtaining observations and data. See for example, Johnston *et al.* who undertake a contingent-choice telephone survey of a national sample of households. Contingent valuation has the well-known shortcoming of asking only "hypothetical questions" rather than budget-constrained questions. Our laboratory auction market required participants to "pay" what they "say," that is, bids are real. Because of major costs of getting participants to a laboratory location and setting up experiments and paying participants, a laboratory auction can be conducted in only few locations and with a modest number of participants. We have a total of 318 individuals participant, which large by experimental economics standards.

industry or organizations, environmental or consumer groups, or "other" sources relative to information from an independent third-party source. Older individuals have significantly lower odds of trusting "nobody" for GM information relative to an independent, third-party source. People who claimed to be informed about genetic modification before the experiments were more likely to trust the government than the third-party sources. People who had a conservative religious upbringing had a lower odds of trusting private industry or organizations and higher odds of trusting "nobody" relative to an independent, third-party source. This article has five sections.

1. Model

Following Becker (1996), consider the strictly quasi-concave utility function shown as:

$$U = U(X_l, X_{pl}; T_1, \ldots, T_j). \tag{1}$$

Utility is based on the consumption of two choice variables: foods labeled as GM (X_l) and foods that have a plain label (X_{pl}). The utility of these two goods is hypothesized to be affected by information from j sources. This information differs in quality for each type (i.e., level of trust). Assume information quality or trust in the jth type is a function of the consumer's PC and SC:

$$T_j = f_j(\mathrm{SC}, \mathrm{PC}). \tag{2}$$

The market price for foods labeled as GM is p_l and the price of plain-labeled foods is p_{pl}. At time t, the consumer maximizes his or her utility, subject to the budget constraint M, and stock of PC and SC:

$$\begin{aligned} &\max U(X_l, X_{pl}, T_1, \ldots, T_J), \\ &T_j = f_j(\mathrm{SC}, \mathrm{PC}) \quad \text{s.t. } p_l X_l + p_{pl} X_{pl} \leq M. \end{aligned} \tag{3}$$

The first-order conditions are as follows:

$$MU_l(X_l, X_{pl}; T_1, \ldots, T_J) - \lambda p_l = 0 \tag{4}$$

$$MU_{nl}(X_l, X_{pl}; T_1, \ldots, T_J) - \lambda p_{pl} = 0 \tag{5}$$

$$p_l X_l + p_{pl} X_{pl} - M = 0. \tag{6}$$

Equations (5) and (6) can be rearranged to show the marginal rate of substitution between GM- and plain-labeled foods, as shown in equation (7):

$$\frac{MU_l(X_l, X_{pl}; T_1, \ldots, T_J)}{MU_{pl}(X_l, X_{pl}; T_1, \ldots, T_J)} = \frac{p_l}{p_{pl}}. \tag{7}$$

A consumer's marginal rate of substitution between GM- and plain-labeled foods is a function of the relative prices of the goods and PC and SC, which influences the trust for the j providers of information. By moving the ratio of prices to the left-hand side, we can differentiate with respect to PC or SC. Consider the following equation, which examines the impact of a change in a consumer's PC for the two goods:

$$\frac{MU_{pl}(\cdot)\left(\sum_{j=1}^{J} \frac{\partial MU_l(\cdot)}{\partial f_j}\frac{\partial f_j}{\partial \mathrm{PC}}\right) - MU_l(\cdot)\left(\sum_{j=1}^{J} \frac{\partial MU_{pl}(\cdot)}{\partial f_j}\frac{\partial f_j}{\partial \mathrm{PC}}\right)}{[MU_{pl}(\cdot)]^2}. \tag{8}$$

A change in PC seems likely to have differential impacts across the j information quality types and is not neutral on the marginal rate of substitution between GM- and plain-labeled foods. To simplify the analysis and without loss of generality, assume a change in T_j, $j = 1, \ldots, J$, does not impact the marginal utility for plain-labeled (non-GM) foods. Then, equation (8) becomes

$$\frac{MU_{pl}(\cdot)\left(\sum_{j=1}^{J} \frac{\partial MU_l(\cdot)}{\partial f_j}\frac{\partial f_j}{\partial \mathrm{PC}}\right)}{[MU_{pl}(\cdot)]^2}. \tag{9}$$

Now if an increase in an individual's PC reduces his or her trust in agri-business information which then lowers his or her marginal utility of GM-labeled food, this causes the consumer's marginal rate of substitution between GM- and plain-labeled foods to decrease. Hence, the consumer will purchase more plain-labeled foods at given relative prices. This example illustrates that, when an individual's PC (or SC) changes his or her trust in an interested party (by changing the perceived quality of the information), it can affect consumer's demand for GM- and plain-labeled foods.

2. The Survey

The participants in our postexperiment survey were adult consumers over eighteen years of age chosen by a random digit-dialing method from two major Midwestern metropolitan areas (Tegene *et al.*, 2003). Three hundred eighteen individuals participated in our project out of 1,673 contacted, which was a response rate of approximately 19%. When individuals arrived at the lab site, they were asked to complete a preauction questionnaire giving Sociodemographic information and information about prior knowledge on GM technology. After the laboratory experiments were completed, they then completed a postauction questionnaire. This questionnaire included the question: "If a source were to give you verifiable information on GM foods, who would you trust most?"[3] This was an open-ended question, and participants wrote their answer down on the questionnaire. "We then arbitrarily coded the responses into six categories": government; university, scientists/researchers, or third-party groups; environmental or consumer group; private industry or organization; none or nobody; and "other," including media.

The first category is independent third-party sources. It contains responses from individuals who would most trust universities, scientists, or an independent third-party group that does not have financial ties to genetic modification. The second category "government" contains responses from individuals who named a government (national, state, or local) or a governmental entity (e.g., FDA). The third category "environmental or consumer group" is for participants who indicated they would most trust an environmental or a consumer group to provide verifiable information on GM foods. The fourth category is "private industry or organizations," which contains the response for any individual who listed a private entity or business as the group they would trust most. Most of these responses were for agri-business firms or grocery stores. The fifth category is "none or no body" and it is for individuals who said they would not trust any source. The last category is classified as "other," and it

[3]To see the format of these experiments, see Huffman *et al.* (2003) or Tegene *et al.* (2003).

contains responses by individuals who would trust the media, and some responses that were too sparse for their own category (e.g., one person said he or she most trusted God to provide verifiable information on GM foods).

Excluding the "other" category, the most frequently reported trusted source for information on genetic modification the first group, "third-party including university, scientists/researches," accounting for 30% of the responses (see Table 1, Panel A). The "government" was listed by 20% of the respondents. The "environmental or consumer group," "private industry or organizations," and "none" each received less than 6% of the responses. The media is included in a residual category that includes "other" preferences.

Although the demographics of our sample do not perfectly match the U.S. Census demographic characteristics for these regions, they are similar and provide a sufficient representation to examine which consumers trust for information on GM foods (see Tegene *et al.*, 2003). Our participants were slightly skewed toward women, but

Table 1. Attributes of the sample ($N = 318$).

	Number	Percentage
Panel A. "Who Individuals Trust for Information on Genetic Modification"		
All	318	100
Third party including university, scientists/researchers	94	29.6
Government	62	19.5
Environmental or consumer group	12	3.8
Private industry or organization	16	5
None or nobody	19	6
Other, media, or no answer	115	36.1

Variables	Definition	Mean	SD
Panel B. Attributes of Participants			
Income	household's annual income ($1,000)	54.70	34.10
Education	years of form schooling completed	14.50	4.50
Age	the participant's age	50.10	17.50
Gender	1 if female, and 0 for male	0.62	0.49
Informed	1 if participant considered him/herself at least somewhat about GM preexperiment	0.48	0.44
Religion	1 if conservative religious upbringing	0.59	0.25

Katsaras *et al.* (2001) show that women make up a disproportional share of grocery shoppers — 83% of shoppers versus 52% in the U.S. Census of Population.

Personal capital is proxied by an individual's education, age, and prior knowledge about GM technology. An individual's education not only affects his or her opportunity cost of his or her time, but also his or her ability to acquire and process information, to make decisions (Schultz, 1975; Huffman, 1977), and formulate trust. Twenty-five percent of the participants had not gone beyond high school. About 39% of the participants had completed four or more years of college.

An individual's age is a proxy for years of experience as a decision maker, which is expected to affect the formation of trust, and also an indicator of length of expected remaining length of life. As an individual becomes older, he or she has fewer expected years over which to obtain benefits from acquired information.

If a participant is "informed" about genetic modification before the experiment, this may affect his or her preferences for information sources. Being "Informed" is represented as dummy variable taking a value of 1 if the respondent perceives him or herself as being "at least somewhat informed regarding GM foods."

Religious upbringing is a form of SC, and, hence, could play a role in trust formation. Our survey asked people to indicate their religious affiliation when they were young. Fifteen percent of the participants were raised as Baptists, slightly more than 26% of participants were raised as Catholics, and about 17% were raised as Lutherans. Forty-seven percent had other affiliations. Individuals were designated as having a "conservative" religious upbringing relative to "trust" in information sources for GM food if they were raised as Baptists, Catholic, or Lutheran. These three religions have among the strictest religious upbringings on the origin of life. See Panel B of Table 1 for sample means of these variables.

3. Econometric Model

Consider a random indirect utility model in which the utility of a consumer's choice j is determined by x_j, consumer's household income, goods' prices, and attributes:

$$U_{ij} = \beta' x_{ij} + \varepsilon_{ij}. \tag{10}$$

The utility of consumer i is based on information choice $j \in J$. If he or she chooses j, it must be the one yielding the highest utility. When disturbance terms are independently and identically distributed Weibull, the probability of consumer i choosing source j is

$$\text{Prob}(Y_i = j) = e^{\beta_j' x_i} \Big/ \sum_{k=1}^{J} e^{\beta_k' x_{ii}} \quad \text{for } j = 0, 1, \ldots, J. \tag{11}$$

Equation (11) is the multinomial logit model (see Greene, 2003, pp. 720–722). To solve the model, however, one must first define $\beta_j^* = \beta_j + q$ for a vector $\mathbf{q}$, and then normalize $\beta_0 = 0$.[4] The probability of choosing source j is then

$$\text{Prob}(Y = j) = e^{\beta_j' x_i} \Big/ \left(1 + \sum_{k=1}^{J} e^{\beta_k' x_{ii}}\right) \quad \text{for } j = 1, 2, \ldots, J \tag{12}$$

$$\text{Prob}(Y = 0) = 1 \Big/ \left(1 + \sum_{k=1}^{J} e^{\beta_k' x_{ii}}\right). \tag{13}$$

We can represent the probability a consumer prefers source $j (j = 1, \ldots, J)$ as the log-odds ratios:

$$\ln\left(\frac{P_{ij}}{P_{i0}}\right) = \beta_j' x_i. \tag{14}$$

Equation (14) shows the probability that a consumer prefers (trust) source j over choice 0, the reference choice.[5] If β_j' is positive, then a marginal increase in x_i increases the odds that the consumer prefers

[4]This arises because the alternatives are mutually exclusive and exhaustive and the associated probabilities sum to 1. Only J parameter vectors are needed to determine the $(J + 1)$ probabilities.

[5]From the point of view of estimation, it is a major advantage that the odds ratio does not depend on the other choices, which follows from the independence of disturbances in the original model. From a behavioral viewpoint, however, this fact is not so attractive.

source j over the reference source 0, which is the "independent third-party source." The regressors are variables proxying an individual's PC and SC, his or her beliefs, and his or her household income.

4. Econometric Results

The fitted model provides empirical evidence for the odds that a consumer trusts one of the five sources of information more or less than he or she trusts an "independent third-party source" to provide verifiable information on GM foods. Five regressors were included in this multinomial logit model: a participant's household income, education, age (which can be thought of as a proxy for experience), a dummy variable for prior beliefs, and a dummy for conservative religious affiliation.

The maximum likelihood estimates of the coefficients for the multinomial logit model for an individual's relative trust in information sources are reported in Table 2. The independent variables are listed on the far-left column in Table 2, while the information sources are listed along the top. Increasing a participant's household income did not change the odds of choosing any of the five sources relative to third-party information, and hence, was excluded in the final fitting. Income may increase the demand for information of all types equally (or not at all). An individual who is well educated is shown to more likely to trust an "independent third-party" source relative to the other five sources. Also, increasing a participant's schooling lowers significantly the odds that he or she trusts government, private industry or organization sources, or no "body", and "other" relative to a third-party source. Education has no significant effect on the choice of environmental or consumer sources relative to a third-party source.

As an individual ages, the odds he or she trusts an environmental or a consumer group or "nobody" falls significantly relative to trusting a third-party source. His or her age, however, has positive effect on the odds of trusting government, private industry or organization, or other relative to a third-party source. These effects, however, are statistically weak. If a participant reported in

Table 2. Estimates of multinomial logit model: Who would you trust to provide verifiable information on genetically modified foods? ($N = 318$).

Variable	Government/ Third Party	Env. or Con. Group/ Third Party	Private Ind. or Org./ Third Party	Nobody/ Third Party	Other/ Third Party
Intercept	1.169	−1.658	2.369	4.043*	5.568***
	(1.286)	(2.420)	(2.172)	(1.810)	(1.141)
Household income	−0.0010	−0.0099	0.0076	−0.0100	0.0017
	(0.0052)	(0.0110)	(0.0085)	(0.0096)	(0.0046)
Education	−0.138*	0.081	−0.347**	−0.242*	−0.389***
	(0.080)	(0.149)	(0.147)	(0.124)	(0.074)
Age	0.010	−0.025	0.001	−0.035**	0.004
	(0.010)	(0.019)	(0.017)	(0.016)	(0.009)
Informed	0.344**	0.442	0.003	0.050	0.158
	(0.170)	(0.328)	(0.286)	(0.269)	(0.149)
Religious upbringing	−0.042	0.055	−0.840**	0.556**	−0.074
	(0.171)	(0.316)	(0.399)	(0.280)	(0.151)

*Indicates that an estimated coefficient is significantly different from zero at the 10% level.
**At the 5% significance level.
***At the 1% level.
Note: The reference group is an independent, third-party source, and standard errors are in parentheses.

the preauction questionnaire that he or she was "informed about genetically modified foods," he or she is significantly more likely to trust the government relative to a third-party source. If a participant had a strict religious upbringing, he or she has significantly lower odds of trusting private industry or organization and higher odds of trusting "nobody" relative to a third-party source. The person's religious upbringing, however, did not have a significant effect on the odds of any of the other choices relative to third-party information.[6]

[6]Multinomial logit models including a participant's gender and marital status were also fitted. These variables did not have estimated coefficients that were statistically different from zero at the 5% significance level.

5. Conclusion and Implications

Although many organizations disseminate information on a wide range of topics, they must gain the trust of a constituent group before they can affect decisions. In the case of GM foods, the international environmental NGOs and agricultural biotech industry disseminate conflicting information. International environmental NGOs disseminate negative information; agricultural biotech industry disseminates positive information. In fact, the international environmental NGOs, agricultural biotechnology industry, and U.S. government all have different interpretations of the role of GM foods should play in our society. Hence, the federal government is not a valid third-party source, for example, some groups are not in favor of the FDA policies on voluntary GM food labels, but a quasigovernmental entity funded by the government and staffed with informed but financially disinterested scientists not answering to the government may be the best possible source to provide information on foods labeled as GM.

Although the literature on the economics of trust is growing rapidly, few studies have examined the contribution of an individual's PC and SC to his or her trust. This article has provided new econometric evidence that PC and SC of adults who are consumers affect significantly their trust in five different sources of information on genetic modification relative to an independent, third-party source. We have shown that an individual's PC — schooling, age, religion, and self-reporting status as being informed about GM foods — and SC in religious affiliation contributed significantly to explaining the odds of particular information source preferences.

Acknowledgments

The authors gratefully acknowledge assistance from Daniel Monchuk and Terrance Hurley in conducting the surveys and assistance from Monsanto in providing some of the products used in the experiment. The views expressed here are those of the authors, and may not be attributed to the Economic Research Service or the U.S. Department of Agriculture. This work was supported through a grant from the U.S. Department of Agriculture, Cooperative State Research,

Education, and Extension Service, under Agreement 00-52100-9617 and from the U.S. Department of Agriculture, Economic Research Service, under Agreement 43-3AEL-8-80125.

References

Becker, G.S. (1996). *Accounting for Tastes.* Cambridge, MA: Harvard University Press.

Becker, G.S. (1976). *The Economic Approach to Human Behavior.* Chicago, IL: University of Chicago Press.

Bianchi, M. (2002). Novelty, Preferences, and Fashion: When Goods Are Unsettling. *Journal of Economics Behaviour and Organization* 47, 1–18.

Boskin, M.J., E.R. Dulberger, R.J. Gordon, Z. Griliches and D.W. Jorgenson (1998). Consumer Prices, the Consumer Price Index, and the Cost of Living. *Journal of Economic Perspectives* 12, 3–26.

Friends of the Earth (2001). "The Need for Labeling Genetically Engineered Foods." Available at http://www.foe.org/safefood/factshtgelabel.htm.

Gates, B. (2001). "Will Frankenfood Feed the World?" *Time.* Available at http://www.time.com/time/reports/v21/tech/mag_food.html.

Glaeser, E.L., D.I. Laibson, J.A. Scheinkman and C.L. Soutter (2000). Measuring Trust. *Quarterly Journal of Economics* 115(2000), 811–46.

Greene, W.H. (2003). *Econometric Analysis*, 5th Edn. Upper Saddle, NJ: Prentice Hall, 2003.

Greenpeace (2000). "EPA Responds to Greenpeace Lawsuit Defends Rubber Stamping of Bud Science?" Greenpeace USA Media Center.

Greenpeace International (2002). "Getting It On for the Good of the Planet: The Greenpeace Guide to Environmentally Friendly Sex." Available at http://www.greenpeace.org/features/details?features_id=25885, Accessed September 17, 2002.

Greenpeace International (1997). "Greenpeace Launches Genetech Labeling Policy as European Commission Fails to Do So." Available at http://www.greenpeace.org/pressreleases/geneng/1997nov3.html.

Hausman, J. (1996). "Valuation of New Goods under Perfect and Imperfect Competition." In T. Bresnahan and R.J. Gordon (Eds.), *The Economics of New Goods.* Chicago, IL: The University of Chicago Press.

Hoban, T.J. (2001). "American Consumers' Awareness and Acceptance of Biotechnology." In A. Eaglesham, S.G. Pueppke and R.W.F. Hardy (Eds.), *Genetically Modified Food and the Consumer*, pp. 51–68, NABC Report No. 13, Ithaca, NY: National Agricultural Biotechnology Council.

Hoban, T.J. (1997). Consumer Acceptance of Biotechnology: An International Perspective. *Nature Biotechnology* 15, 232–234.

Huffman, W.E. (1977). Allocative Efficiency: The Role of Human Capital. *Quarterly Journal of Economics* 91, 59–80.

Huffman, W.E., J.F. Shogren, M. Rousu and A. Tegene (2003). Consumer Willingness to Pay for Genetically Modified Food Labels in a Market

with Diverse Information: Evidence from Experimental Auctions. *Journal of Agricultural and Resource Economics* 28, 1309–1318.
Huffman, W.E. and A. Tegene (2002). "Public Acceptance of and Benefits from Agricultural Biotechnology: A Key Role for Verifiable Information." In V. Santaniello, R.E. Evenson and D. Zilberman (Eds.), *Market Development for Genetically Modified Food.* New York, NY: CAB International.
Johnston, R.J., C.R. Wessells, H. Donath and F. Asche (2001). Measuring Consumer Preferences for Ecolabeled Seafood: An International Comparison. *Journal of Agricultural and Resource Economics* 26, 200–239.
Katsaras, N., P. Wolfson, J. Kinsey and B. Senauer (2001). *Data Mining: A Segmentation Analysis of U.S. Grocery Shoppers.* St. Paul, MN: The University of Minnesota, The Retail Food Industry Center, Working paper 01-01.
Milgrom, P. and J. Roberts (1986). Relying on the Information of Interested Parties. *RAND Journal of Economics* 17, 18–32.
Rousu, M., W.E. Huffman, J.F. Shogren and A. Tegene (2004). "The Value of Verifiable Information in a Controversial Market: Evidence from Lab Auctions of Genetically Modified Foods." Ames, IA: Iowa State University, Department of Economics, Working Paper No. 334.
Schultz, T.W. (1975). The Value of the Ability to Deal with Disequilibria. *Journal of Economic Literature* 13, 827–846.
Tegene, A., W.E. Huffman, M. Rousu and J. Shogren (2003). "The Effects of Information on Consumer Demand for Biotech Foods: Evidence from Experimental Auctions." U.S. Department of Agriculture, Economic Research Service, Technical Bulletin No. 1903.
U.S. Food and Drug Administration (2001). *Guidance for Industry: Voluntary Labeling Indicating Whether Foods Have or Have Not Been Developed Using Bioengineering.* Available at http://vm.cfsan.fda.gov/~dms/biolabgu.html.
U.S. Food and Drug Administration (1992). Statement of Policy: Foods Derived from New Plant Varieties: Notice. *Federal Register* 57, 22984.

Chapter 8

Effects and Value of Verifiable Information in a Controversial Market: Evidence from Lab Auctions of Genetically Modified Food*

Matthew Rousu[†,||], Wallace E. Huffman[‡,**], Jason F. Shogren[§,††], and Abebayehu Tegene[¶,‡‡]

[†]*Susquehanna University, 514 University Avenue, Selinsgrove, PA 17870*

[‡]*Department of Economics, Iowa State University, Ames, IA 50011*

[§]*Natural Resource Conservation and Management, Department of Economics and Finance, University of Wyoming, Laramie, WY 82070*

[¶]*Food Economics Division, ERS, U.S. Department of Agriculture, Washington DC 20036*

[||]*rousu@susqu.edu*

[**]*whuffman@iastate.edu*

[††]*Jramses@uwyo.edu*

[‡‡]*ategene@ers.usda.gov*

Food products containing genetically modified (GM) ingredients have entered the market over the past decade. The biotech industry and environmental groups have disseminated conflicting private information about GM foods. This paper develops a unique methodology for valuing independent third-party information in such a setting and applies this method to consumers willingness to pay for food products that might be GM. Data are collected from real consumers in an auction market setting

*This article originally appeared in **Economic Inquiry (2007), 45(3), 409–432**. This is published by permission of **John Wiley & Sons, Inc.**

with randomized information and labeling treatments. The average value of third-party information per lab participant is small, but the public good value across U.S. consumers is shown to be quite large. (JEL C91, D12, D82).

1. Introduction

If all information regarding new products is public, consumers can search out this information and make informed decisions (Hirshleifer and Riley 1992; Stigler 1961), but the available information is typically incomplete and asymmetric. That is, some effects of the new commodities/technologies on consumers and the environment are either unknown or only privately known (private information) by nonconsumers. With private information, the informed and interested parties frequently have an incentive to use their private information strategically (Akerlof 1970; Molho 1997), for example, to beneficially affect market conditions.

Foods made using bioengineering, often referred to as genetically modified (GM) foods, are examples of new goods with largely the same set of attributes as conventional foods, but that are produced using new agricultural technologies about which consumers might have positive or negative preferences. These new raw materials have been produced with crop varieties that contain one or more transgenic genes — genes that have been moved across species using the gene splicing technology of Cohen and Boyer and that could not be moved using conventional sexual reproduction (Cohen *et al.*, 1973; Huffman and Evenson 2006, 167–75). The transgenic traits of these innovations relate to the insertion of herbicide-tolerant and insect-resistant genes from soil bacteria into crop varieties, which can be expected to reduce farmers' production costs for raw food and thereby possibly to reduce food prices to consumers. However, this technology has not created *commercially available* foods that consumers' value directly, for example, reduced bruising or enhanced protein or vitamin content. Genetically engineered crop varieties for enhanced input traits burst upon the frontier of agricultural technologies in the mid-1990s with herbicide-tolerant soybeans and insect-resistant cotton varieties.[1] Since then, herbicide-tolerant cotton, corn, and canola varieties and insect-resistant corn varieties have been successfully developed and marketed to American farmers (Fernandez-Cornejo and Caswell 2006). Soybeans and corn and canola seeds are important U.S. sources of vegetable oil, corn is a principal source of sweetener, and corn and soybeans are significant sources of other ingredients for processed foods. Tomatoes bioengineered for extended shelf life (Flavr-Savr) and potatoes for insect resistance were also approved by the Food and Drug Administration (FDA) in the early 1990s, but these products failed commercially after a few years on the market and were withdrawn.

Abbreviations

CSSM: Center for Survey Statistics and Methods
DNA: Deoxyribonucleic Acid
EU: European Union
FDA: Food and Drug Administration
GM: Genetically Modified
NGOs: Nongovernmental Organizations
WTP: Willingness To Pay

The early GM crop varieties made available to U.S. farmers were developed and marketed principally by two large chemical companies: Monsanto and Syngenta. As has been the case in U.S. agriculture for at least six decades, these companies have extensively advertised their perceived positive benefits of these new crop varieties, sharing a type of private information. Given that financial interests in biotechnology, these probiotech parties have used their private information

[1]A herbicide-tolerant crop variety is created by transferring a soil bacteria into the plant that is resistant to a chemical herbicide, for example, Round-Up, and then when this variety is planted, farmers can apply Round-Up and kill all the plants in the field except for the Round-Up tolerant crop variety. An insect-resistant crop variety is created by transferring soil bacteria that is toxic to common plant insects, for example, the cotton boll and bud worm, European Corn borer. In this case, biological control rather than a toxic chemical pesticide is used to control the target insect.

selectively or strategically to expand the size of the market for GM crops by emphasizing that GM crops would enhance environmental quality, increase food availability, and reduce world hunger (Council for Biotechnology Information 2001).

In contrast, environmental nongovernmental organizations (NGOs), including Greenpeace and Friends of the Earth, have countered by asserting that foods made from GM plant materials might harm human health by introducing new allergens (e.g., see "Frankenfoods," Greenpeace International 2003), harm the environment by possibly out crossing with native species, reduce genetic diversity, prevent farmers from saving their own seed for future plantings, or give excessive market power to multinational companies (Greenpeace International 2001a; 2000b; 2000c; Friends of the Earth 2003). Such diverse and conflicting information leads to potential confusion among consumers (and possibly agricultural producers). Similarly, public officials are also faced with a situation of information asymmetry and may not be able to discern socially optimal policies for GM crop varieties and GM foods, or their short- and long-run social welfare effects (Hausman 2003).

In principle, society can avoid losses due to strategic behaviors of interested parties toward new technologies and products if decision makers can access independent third-party or *verifiable information* (Huffman and Tegene 2002; Milgrom and Roberts 1986). Verifiable information provides an objective assessment of the benefits and costs, including environmental and health risks, of GM crop varieties and foods made from these raw materials.

The overall objective of this paper is to develop a unique methodology for valuing independent third-party information in such a setting and to apply this method to consumers' willingness to pay (WTP) for food products that might be GM. Data are collected from real consumers participating in an auction market setting with randomized information and labeling treatments. Our methodology improves upon the standard protocol of experimental economics, where undergraduate students are the auction participants and they are put through a long set of nonrandomized treatments (e.g., Fox *et al.*, 2002; Shogren *et al.*, 1994).[2] Applying our method for valuing third-party information, we show that verifiable information in the GM food market has potentially large and statistically significant social value.

2. Background

For millennia, individuals involved in cultivation of land have engaged in a rudimentary form of genetic modification in an attempt to augment the desirable attributes of their crops. They have selected the most productive and hardiest seeds from each year's harvest for use in subsequent plantings. However, since the early 1990s, genetic modification has been associated with a much narrower set of techniques that use recombinant deoxyribonucleic acid (DNA) or gene splicing technology to facilitate the transfer of genes across species.[3] Foods made using this type of GM material are known commonly as GM foods.

The on-going controversy over GM food products and the perceived importance of facilitating informed food purchases have stimulated interest in food labeling, identity preservation, and new sources of information (Caswell 2000). For example, Greenpeace and Friends of the Earth advocate mandatory labeling for GM content, asserting that labeling would benefit consumers by enabling them to easily chose whether or not to consume products containing GM content (Friends of the Earth 2001; Greenpeace International 2006). Some international NGOs have made GM foods their number one issue (Nestle 2003).

[2]In designing the experiments, we combine the best attributes of survey design, statistical experimental design, and experimental economics to obtain a superior overall experimental design. Our methodology differs significantly from the telephone survey used by Mendenhall and Evenson (2002) to solicit information about consumers risk perceptions of GM foods and hypothetical WTP a premium for non-GM foods.

[3]In 1973, Cohen and Boyer discovered the basic technique for recombinant DNA, which launched a new field of genetic engineering. The Cohen-Boyer patent on gene splicing technology was awarded in 1980 to Stanford University and the University of California (Office of Technology Assessment 1989). They built on the 1953 discovery by Watson and Crick of the structure of DNA and of the suggestion about how it replicates.

Before 1990, food labels were largely unregulated. But the 1990 Nutrition Labeling and Education Act dramatically changed nutrition labels on packages of food sold in U.S. supermarkets (Balasubramanian and Cole 2002; Caswell 2000). This law requires packaged foods to display nutritional information prominently in a new label format, namely the Nutrition Facts panel. It also regulates health claims and descriptor terms on food packages. If the proposal by environmental groups for GM food labels were adopted, the label would be a *negative* label denoting the "use of GM ingredients." However, GM content has not been proven scientifically to have human health consequences, except for the transport of some known allergens to new locations. Hence, today, GM food labels would not meet the nutrition labeling requirement of nutrient intake, leading to proven health outcomes.

In 1992, the U.S. Department of Health and Human Services issued a regulation that GM food did not need to be labeled if the food product had the same food characteristics as its non-GM counterparts, that is, contained the same basic attributes. In January 2001, the FDA issued a "Guidance for Industry" statement reaffirming this policy stating that the only GM foods requiring labels are foods with characteristics different from those of their non-GM counterpart. Currently, labeling food for GM content is not required for any other food. Under the current rules and regulations governing the food industry, a GM labeling initiative could be implemented either by manufacturers voluntarily or the FDA would need to change its policy. However, if a voluntary label is affixed to a food product, the FDA has mandated that it cannot use the phrase "genetically modified." The FDA preferred the phrase "genetically engineered" or "made through biotechnology." In contrast, the European Union (EU) and most other Western European countries have strong labeling and tracking requirements for GM content.

In Europe, international environmental groups were effective in lobbying governments with their platform that "consumers have the right to know whether their food is GM or not" (Greenpeace International 2001a). In 1997, the European Commission enacted a mandatory GM food labeling policy. The Commission required each member country to enact a law requiring labeling of all new products containing substances derived from GM organisms. In October 2003, the European Parliament passed more stringent labeling legislation requiring that food and feed that contain at least 0.9% of 1% of GM content must be labeled as GM and that traceability of the product from the farm to the consumer must be maintained.

In the EU, mandatory labeling has led to the curious result that no (or very little) GM foods are being sold in grocery stores and supermarkets. Once the label requirement went into effect, the environmental groups mounted a major negative information campaign against GM products. The result was reduced demand by EU consumers and then the gradual disappearance of GM products from store shelves. In effect, the implementation of mandatory labeling of GM food resulted in a reduction in the variety of foods available for consumers (Carter and Gruere 2003).

Japan, Australia, New Zealand, Norway, Switzerland, and at least nine other countries have also passed mandatory GM labeling requirements (Carter and Gruere 2003). However, labeling involves additional real costs, such as testing for the presence of GM, segregating crops, monitoring for truthfulness in labeling, enforcement of existing regulations, and risk premiums for being out of contract (Wilson and Dahl 2005).

An effective labeling policy requires the use of scarce resources to maintain an effective "segregation" system or an "identity preservation system."[4] To the extent that there is a market for non-GM crops, buyers would be expected to specify in their purchase contracts some limit on GM content and/or precise prescriptions regarding production/marketing/handling processes (Wilson and Dahl 2005). One can envision a marketplace of buyers who have differentiated demand according to their risk tolerance for GM content. To make this differentiation meaningful, new costs and risks are incurred. Additional testing for GM content involves the costs of deciding which test to use, among those with varying accuracy, and then

[4]Wilson and Dahl (2005) estimate that the cost of segregation with a varietal declaration system is much less expensive than for an identity preservation system.

conducting the test. If non-GM is the "superior commodity," then there is the risk that GM and non-GM varieties will be commingled, and then with some positive probability, the GM content will be detected in shipments to customers, and the shipment will be rejected. This is a serious economic problem as agents seek to determine the optimal strategy for testing and mitigating risk.

Limited evidence from related consumer acceptance studies suggests that consumers behave as if they have an asymmetric value function in that they seem to act on "bad" news in a controversial market while heavily discounting "good" news (Kahneman and Tversky, 1979). Viscusi (1997) and Fox *et al.* (2002) concluded that individuals amplify the risks of a neoteric product and discount benefits. Using a survey in which consumers received divergent information on environmental risks, Viscusi showed that consumers place greater weight on information from the expert who provided a high-risk assessment. They did so regardless of whether the low-risk assessment came from a government or an industry source. A similar "alarmist" reaction to a new product was observed by Fox *et al.* (2002) in laboratory auctions of irradiated meat. In agreement with intuition, they found that a favorable description of irradiation increased demand and an unfavorable description decreased demand. However, when consumers were presented with both favorable and unfavorable descriptions, they bid as if they had used only the negative information, resulting in a dramatic decline in demand for-irradiated meat despite the fact that the negative information was presented in a nonscientific manner by a consumer advocacy group.

Consistent with models of choice under risk (e.g., loss aversion, status quo bias, Bayesian updating), this result illustrates the incentive for partisan groups to promote unscientific claims for private gain. The important question that studies of Viscusi and Fox *et al.* do not address is the social value of introducing verifiable information into a market where a controversial product is for sale. From this, we might hypothesize that verifiable information is most valuable under a voluntary labeling policy, as in the United States, Canada, and Argentina, but not a mandatory policy, as in the EU and Japan.

Stigler (1961), Hirshleifer and Riley (1992), Molho (1997), and Morris and Shin (2002) provide general frameworks for placing an economic value on information. Foster and Just (1989) devised an applied methodology for assessing the value of government information about insecticide contamination (heptachlor) of milk for human consumption in Hawaii. They used market-level data to calculate the value of government information as the difference between rational consumers' choices under incomplete and more complete information. In our study, we use information obtained from experimental auctions where participants/consumers are provided with randomly assigned information and labeling treatments. By using an experimental market, we are able to obtain precise information on consumers' reactions to the injection of verifiable information when information from interested parties is already available. Further, our methodology can be applied to assess consumer responses to other new goods when interested parties might have propagated scientifically unfounded conflicting information.

3. Experimental Design

With two interested parties disseminating conflicting information into a market for a new good, what are rational consumers likely to do? Consumers should be able to make informed decisions provided (a) they are sophisticated enough to understand the technical processes at work and to recognize that interested parties provide information reflecting their own private interests and (b) they can verify all the information provided (Milgrom and Roberts 1986). Unfortunately, a full and verifiable information environment does not characterize the market for GM food products [also see Mendenhall and Evenson (2002)]. Genetic modification through bioengineering is a complex process, and not all GM information is currently verifiable. Additionally, the search costs for most consumers to find neutral information is very high, as there are contradictory messages describing GM food as "food to feed the planet," as well as "Frankenfood" [see, e.g., Gates (2001)]. Because of these high search costs, an independent third-party source providing verifiable information about genetic modification has potentially large public good value (Huffman and Tegene 2002).

We design a set of experiments to incorporate the private information revealing features of experimental auction markets similar to the experiments described by Smith (1976) and Fox *et al.* (2002), and then introduce rigorous randomized treatment effects from the field of statistical experimental design. Although we conduct two practice rounds of bidding on common goods to learn our auction process, our main results follow from information collected from two rounds of bidding on the experimental food products where the experimental design consists of six biotech information labeling treatments randomly assigned to sessions or trials with two replications. Each session consists of 13–16 adult consumers drawn randomly from the households of two major Midwestern cities.[5] Iowa State University's Center for Survey Statistics and Methods (CSSM) contacted prospective participants and offered them the opportunity to participate in an Iowa State University project dealing with consumers' decisions on food and household products. Individuals were told that they would need to come to a common location, a lab or classroom, that the project would take about 90 min of their time, and that they would be paid $40 for participating.

We anticipated that consumers would have heterogeneous tastes for food ingredients and methods of production. Given the sizeable fixed cost of conducting a laboratory experiment even with one commodity, we conjectured that using three somewhat different experimental food items in an auction would be only marginally more expensive. We chose to use in our auctions vegetable oil (made from soybeans), tortilla chips (made from yellow corn), and Russet potatoes. At the time of the planned experiments, these food products could have been made using GM ingredients or technology. Furthermore, in the distilling and refining processes for vegetable oils, essentially all the proteins (which are the components of DNA and the source of genetic modification) are removed, leaving pure lipids.

[5]A "session" in experimental economics is equivalent to an "experimental unit" in statistical experimental design literature.

Hence, for GM oils, minimal human health concerns should arise, but consumers might be concerned about GM soybeans having adverse effects on the environment. Tortilla chips are highly processed foods that may be made from GM or non-GM corn, and consumers might have human health and environmental concerns. Russet potatoes are purchased as a fresh product and are generally baked or fried before eating. As with tortilla chips, consumers might perceive human health and environmental risks from eating GM Russet potatoes.

The CSSM obtained randomly selected residence telephone numbers from each of the metropolitan areas and contacted these numbers to find eligible participants. Participants were asked to come on a particular Saturday in April, 2001 and to choose one of three different starting times: 9 a.m., 11:30 a.m. or 2 p.m. Participation per household was limited to two adults, and they were assigned to different sessions. The CSSM followed up by sending willing participants a letter containing more information, including the date and location, a map, and a telephone number to contact for more information. The response rate was about 19%, and the total number of participants or sample size for this study is 172, which is large compared to most experimental auctions.

The protocol for an auction day and sessions is critical to the success of the experiments. Each auction session is conducted by one of two session monitors, and it consists of the same exact ten steps, which are summarized in Figure 1. In Step 1, when participants arrive, the host assigns them to a particular session and monitor. The monitor asks them to sign a consent form agreeing to participate in the session. After participants sign this form, the monitor gives them $40 for participating and an ID number to preserve their anonymity. The participants then are asked to read a brief set of instructions and to complete a short questionnaire asking for socio-demo-economic information and attitudes toward several types of biological technologies.

In Step 2, participants learn the auction mechanism of a random nth price auction. In the random nth price auction (Shogren *et al.*, 2001), an object is auctioned to n bidders, for example $n = 13$, and auction rules require each player to submit a private

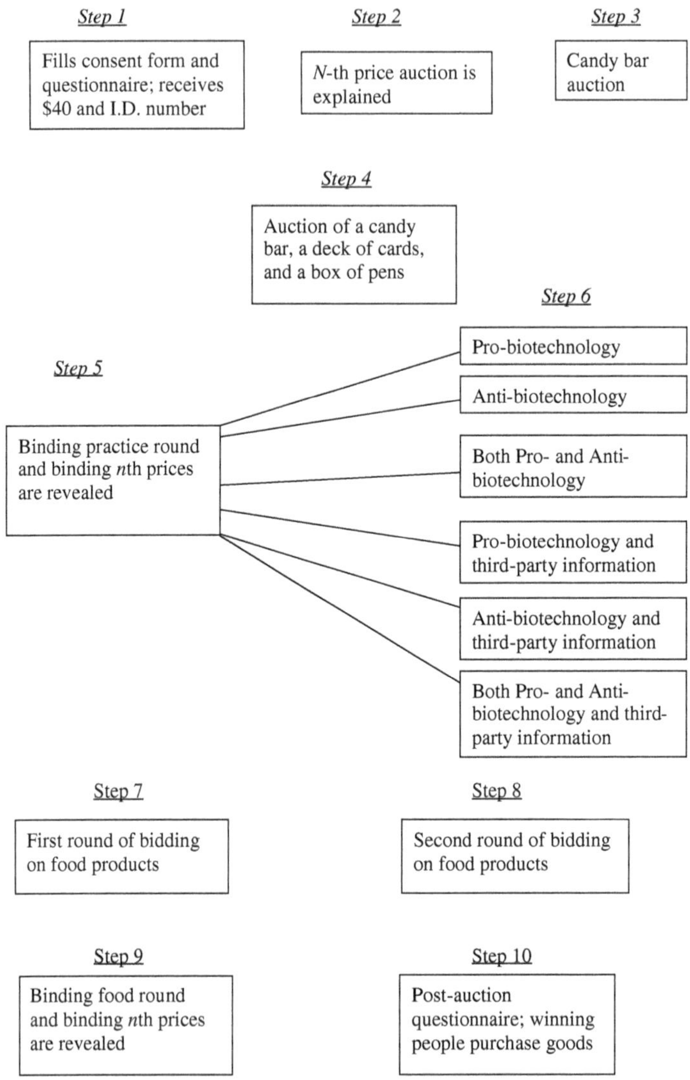

Fig. 1. Steps in the experiments.

bid (a nonnegative number). The envelopes are then collected by the auctioneer or session monitor, and the n bids are ranked from highest to lowest. Next, the session monitor draws a single number from a discrete uniform distribution over 1 to the subset of positive integers from 1 to n, for example, 5. In this case, the four highest bidders are the winners, and they each pay the fifth highest price.[6] This auction mechanism is incentive compatible, as it gives every bidder an incentive to bid his or her own true value. However, unlike the Vickery or second price auction, $n - 1$ of the bidders have a chance to be winners. Thus, the random nth price auction does a better job of keeping all bidders engaged in the auction relative to other demand-revealing auctions like the Vickery auction and the Becker-DeGroot-Marschak (Lusk and Rousu forthcoming; Shogren *et al.*, 2001). We do not conduct a repeated-round auction with posted prices, where bidders place bids on identical items across a number of rounds and adjust their bids based on information revealed in other bidders' prices. For a summary of how repeated rounds in auctions affect bidding behavior, see Corrigan and Rousu (2006).[7]

The session monitor then writes an example of the nth price auction on the blackboard. After participants learn the general attributes of the auction, a short quiz is administered to the participants to test their understanding of the auction mechanism. Note that throughout the auction, when a participant is bidding on an item in a particular round, he or she does not know if additional rounds will occur. Participants are assured, however, that at most they will purchase only one unit of any auctioned item. This has two advantages. First, it eliminates price reduction because a bidder has a negatively sloped demand curve for any commodity. Second, it significantly reduces concerns that participants might have about financial liquidity, given that they were not told in advance that they would be participating in a real auction.

Step 3 is the first "practice round of bidding" in which participants bid on a brand-name candy bar. The participants are asked to examine the product and then place a (sealed) bid. The bids are

---

[6]If there are ties, this can increase the number of winners to greater than $n - 1$.

[7]The view of our advisors from the Department of Statistics, Iowa State University, was that putting an individual through many rounds of bidding would change his/ her behavior because of the experience, and so we were encouraged to keep the total number of bidding rounds per participant to a very small number, two. This is in contrast to most economics experiments where individuals are put through a long nonrandomized sequence of rounds of bidding.

collected by the session monitor and ordered from 1 to n. The first round of practice bidding is then completed.

Step 4 is a second round of practice bidding. In this round, each participant in a session bids separately on a set of three items. This is to provide experience in bidding on more than one item at a time. The items are a brand-name candy bar, a deck of playing cards, and a box of pens. The participants are asked to examine the three items and place bids on each product separately. Then the bids are collected by the session monitor and ordered from highest to lowest. Of the two practice rounds, only one is chosen as binding (valid).

In Step 5, the binding round (randomly chosen between practice sessions 1 and 2) and the binding nth prices are revealed to bidders. All the bids from the two practice rounds are written on the blackboard, and the nth price is circled for each of the products. Bidders are then told whether they are a "winner" and the market-clearing price that they are to pay on binding bids. The bidders are told by the session monitor that all payment of money for goods will take place after the experimental session.

Step 6 is the first round of bidding on the experimental food products — vegetable oil, tortilla chips, and potatoes — that might be GM. A randomly assigned information treatment is injected into each session at this point for these food items. The information treatments consist of packages of one or more of the following perspectives on agricultural biotechnology: (1) *the industry (pro-biotech) perspective* — a collection of mainly positive or optimistic statements and information on genetic modification provided by a group of leading biotechnology companies, including Monsanto and Syngenta; (2) *the environmental group (antibiotech) perspective* — a collection of mainly negative statements and information on genetic modification from Greenpeace, a leading environmental group; and (3) *the third-party (verifiable information) perspective* — an objective statement on genetic modification, as of the date of the experiments, approved by a third-party group consisting of a variety of individuals knowledgeable about GM goods, including scientists, professionals, religious leaders, and academics, none of whom have a financial stake in GM foods. To assist the participants in processing these different

The following is a collection of statements and information on genetic modification from Greenpeace, a leading environmental group.

General Information

Genetic modification is one of the most dangerous things being done to your food sources today. There are many reasons that genetically modified foods should be banned, mainly because unknown adverse effects could be catastrophic! Inadequate safety testing of GM plants, animals, and food products has occurred, so humans are the ones testing whether or not GM foods are safe. Consumers should not have to test new food products to ensure that they are safe.

Scientific Impact

The process of genetic modification takes genes from one organism and puts them into another. This process is very risky. The biggest potential hazard of genetically modified (GM) foods is the unknown. This is a relatively new technique, and no one can guarantee that consumers will not be harmed. Recently, many governments in Europe assured consumers that there would be no harm to consumers over mad-cow disease, but unfortunately, their claims were wrong. We do not want consumers to be harmed by GM food.

Human Impact

Genetically modified foods could pose major health problems. The potential exists for allergens to be transferred to a GM food product that no one would suspect. For example, if genes from a peanut were transferred into a tomato, and someone who is allergic to peanuts eats this new tomato, they could display a peanut allergy.

Another problem with genetically modified foods is a moral issue. These foods are taking genes from one living organism and transplanting them into another. Many people think it is morally wrong to mess around with life forms on such a fundamental level.

Financial Impact

GM foods are being pushed onto consumers by big businesses, which care only about their own profits and ignore possible negative side effects. These groups are actually patenting different life forms that they genetically modify, with plans to sell them in the future. Studies have also shown that GM crops may get lower yields than conventional crops.

Environmental Impact

Genetically modified foods could pose major environmental hazards. Sparse testing of GM plants for environmental impacts has occurred. One potential hazard could be the impact of GM crops on wildlife. One study showed that one type of GM plant killed Monarch butterflies.

Another potential environmental hazard could come from pests that begin to resist GM plants that were engineered to reduce chemical pesticide application. The harmful insects and other pests that get exposed to these crops could quickly develop tolerance and wipe out many of the potential advantages of GM pest resistance.

Fig. 2. Environmental group perspective on GM.

perspectives, the volume of information is limited to one $8.5'' \times 11''$ page that is organized into five categories — general information, scientific impact, human impact, financial impact, and environmental impact — in order to ease the information processing load on participants. Figures 2–4 display the exact layout and wording of the three perspectives on genetic modification.

The following is collection of statements and information on genetic modification provided by a group of leading biotechnology companies, including Monsanto and Syngenta.

General Information

Genetically modified plants and animals have the potential to be one of the greatest discoveries in the history of farming. Improvements in crops so far relate to improved insect and disease resistance and weed control. These improvements using bioengineering/GM technology lead to reduced cost of food production. Future GM food products may have health benefits.

Scientific Impact

Genetic modification is a technique that has been used to produce food products that are approved by the Food and Drug Administration (FDA). Genetic engineering has brought new opportunities to farmers for pest control and in the future will provide consumers with nutrient enhanced foods. GM plants and animals have the potential to be the single greatest discovery in the history of agriculture. We have just seen the tip of the iceberg of future potential.

Human Impact

The health benefits from genetic modification can be enormous. A special type of rice called "golden rice" has already been created which has higher levels of vitamin A. This could be very helpful because the disease Vitamin A Deficiency (VAD) is devastating in third-world countries. VAD causes irreversible blindness in over 500,000 children, and is also responsible for over one million deaths annually. Since rice is the staple food in the diets of millions of people in the third world, Golden Rice has the potential of improving millions of lives a year by reducing the cases of VAD.

The FDA has approved GM food for human consumption, and Americans have been consuming GM foods for years. While every food product may pose risks, there has never been a documented case of a person getting sick from GM food.

Financial Impact

Genetically modified plants have reduced the cost of food production, which means lower food prices, and that can help feed the world. In America, lower food prices help decrease the number of hungry people and also let consumers save a little more money on food. Worldwide the number of hungry people has been declining, but increased crop production using GM technology can also help further reduce world hunger.

Environmental Impact

GM technology has produced new methods of insect control that reduce chemical insecticide application by 50 percent or more. This means less environmental damage. GM weed control is providing new methods to control weeds, which are a special problem in no-till farming. Genetic modification of plants has the potential to be one of the most environmentally helpful discoveries ever.

Fig. 3. Agricultural biotech industry perspective on GM.

The three types of information are organized into six information treatments: (i) probiotechnology information, (ii) antibiotechnology information, (iii) pro- and antibiotechnology information, (iv) probiotechnology and third-party verifiable information, (v) antibiotechnology and third-party verifiable information, and (vi) probiotechnology, antibiotechnology, and third-party verifiable information. These six combinations are randomly assigned to

The following is a statement on genetic modification approved by a third-party group, consisting of a variety of individuals knowledgeable about genetically modified foods, including scientists, professionals, religious leaders, and academics. These parties have no financial stake in genetically modified foods.

General Information

Bioengineering is a type of genetic modification where genes are transferred across plants or animals, a process that would not otherwise occur (in common usage, genetic modification means bioengineering). With bioengineered pest resistance in plants, the process is somewhat similar to the process of how a flu shot works in the human body. Flu shots work by injecting a virus into the body to help make a human body more resistant to the flu. Bioengineered plant-pest resistance causes a plant to enhance its own pest resistance.

Scientific Impact

The Food and Drug Administration standards for GM food products (chips, cereals, potatoes, etc.) are based on the principle that they have essentially the same ingredients, although they have been modified slightly from the original plant materials.

Oils made from bioengineered oil crops have been refined, and this process removes essentially all the GM proteins, making them like non-GM oils. So even if GM crops were deemed to be harmful for human consumption, it is doubtful that vegetable oils would cause harm.

Human Impact

While many genetically modified foods are in the process of being put on your grocers' shelves, there are currently no foods available in the U.S. where genetic modification has increased nutrient content.

All foods present a small risk of an allergic reaction to some people. No FDA approved GM food poses any known unique human health risks.

Financial Impact

Genetically modified seeds and other organisms are produced by businesses that seek profits. For farmers to switch to GM crops, they must see benefits from the switch. However, genetic modification technology may lead to changes in the organization of the agri-business industry and farming. The introduction of GM foods has the potential to decrease the prices to consumers for groceries.

Environmental Impact

The effects of genetic modification on the environment are largely unknown. Bioengineered insect resistance has reduced farmers' applications of environmentally hazardous insecticides. More studies are occurring to help assess the impact of bioengineered plants and organisms on the environment. A couple of studies reported harm to Monarch butterflies from GM crops, but other scientists were not able to recreate the results. The possibility of insects growing resistant to GM crops is a legitimate concern.

Fig. 4. Independent, third-party, verifiable perspective on GM.

12 experimental sessions with two replications (Figure 5).[8] The sequence of information in treatments containing more than one

[8]When participants in a session received both probiotech and antibiotech information, the order was randomized, so that some participants received the probiotech information first and others received the antibiotech information first. When third-party information was distributed in a session, it always was distributed after the other information types.

Session	Pro-biotech and Anti-biotech Information Treatments Disseminated	Third-party Information Disseminated	Round with GM Labels
1.	Pro-biotech	Yes	1
2.	Anti-biotech	Yes	1
3.	Pro-biotech, anti-biotech	Yes	1
4.	Pro-biotech	Yes	2
5.	Anti-biotech	Yes	2
6	Pro-biotech, anti-biotech	Yes	2
7	Pro-biotech	No	1
8	Anti-biotech	No	1
9.	Pro-biotech, anti-biotech	No	1
10.	Pro-biotech	No	2
11.	Anti-biotech	No	2
12.	Pro-biotech, anti-biotech	No	2

Fig. 5. Treatments: Information and labeling combinations assigned.

type of information is as follows. If the treatment contained both probiotech and antibiotech information, the sequencing of this information to bidders in a session is randomized. If the treatment contains verifiable information, this information is always distributed last. This was done to most closely simulate the environment that exists in the United States and to be the one from which the value of verifiable information should be assessed.

We also use two types of food labels, a "plain label" identifying only the contents of a food package and a "GM label" indicating both the content of the package and whether the product is produced using GM ingredients or GM technology. These labels are plainly designed and clearly displayed on the front of the food containers (Figure 6). For each session or trial, the sequencing of food labels is also randomized; some sessions receive the plain label first and the GM label second, and the others receive the GM labels first and plain labels second. The participants in each auction session then

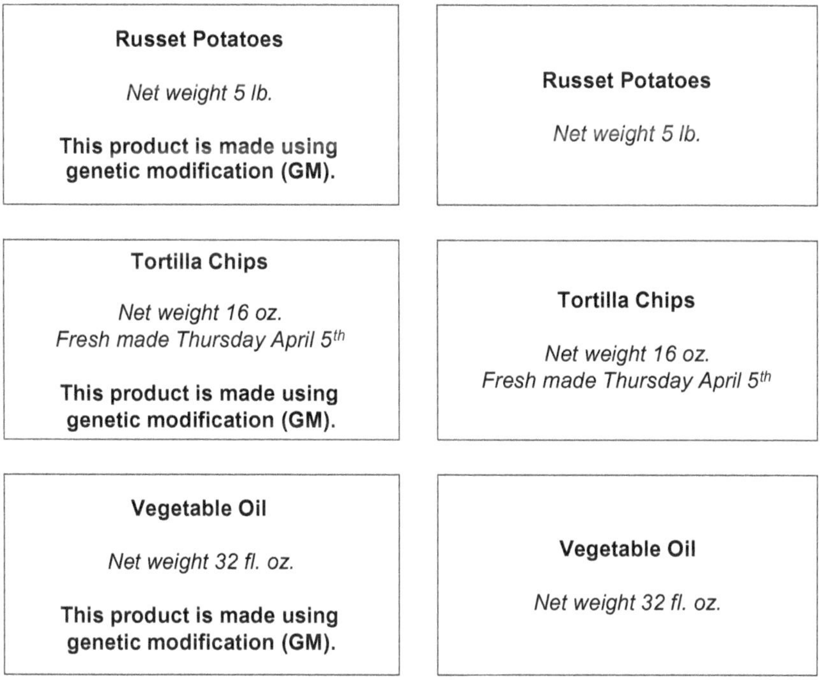

Fig. 6. GM- and plain-food labels used: Potatoes, chips, and oil.

bid in two rounds on vegetable oil, tortilla chips, and potatoes that are identical except for food labels. Hence, each bidder in a session places bids on a total of six experimental food items: GM-labeled vegetable oil, plain-labeled vegetable oil, GM-labeled tortilla chips, plain labeled tortilla chips, GM-labeled Russet potatoes, and plain-labeled Russet potatoes.[9]

[9]Because all consumers bid on GM-labeled and plain-labeled food products, our experimental market emulates a market where mandatory GM labeling is required (similar to markets in Europe, Japan, and many other parts of the world). However, in another set of experiments Huffman *et al.* (2004) showed that consumers interpreted labeling in a market with mandatory GM labeling the same as they interpreted labeling in a market with voluntary labeling, so our results should not be constrained to markets that require mandatory labeling of GM foods.

In Step 7, bidders are told that only one round of bidding on vegetable oil, tortilla chips, and potatoes will be binding and then instructed to examine the three food products and write down their (sealed) bid for each of them separately. The bids are then collected by the session monitor and ordered from high to low. The bidders are now told by the monitor that they are about to look at another group of food items.

In Step 8, bidders are informed by the session monitor again that only one round of bidding on food products will be binding, and then they are asked to examine another set of experimental commodities — vegetable oil, tortilla chips, and potatoes. These food products now have a different label. After the participants examine these products, they are instructed by the monitor to place bids on each of them separately. The bids are then collected from all the bidders or participants and ordered from high to low by the session monitor. In Step 9, the session monitor randomly selects among the two rounds of bidding on vegetable oil, tortilla chips, and potatoes for the binding round and selects the random n. After both the binding round and the binding nth prices are revealed to bidders by the session monitor, the participants who are winners are notified and all participants/bidders are asked to complete a brief post-auction questionnaire.

In Step 10, the participants/bidders who are winners, either in one of the two practice rounds or one of two rounds of bidding on experimental food items, are told to go to the stockroom next door to exchange money for their winning goods, and then they are free to leave. Bidders who did not win any product are informed that they are free to leave immediately.

4. Effects of Verifiable Information on WTP

With the information obtained from the experimental auction sessions, it is possible to assess the impact of diverse information and the social value of verifiable information on consumers' WTP for food products that might be GM at the time of the experiments. We first examine bidding behavior for GM labeled food items with and without verifiable information on genetic modification. Following Viscusi and Evans (1990), we expect that bidders will react rationally to biotech industry and environmental group perspectives. Without verifiable information, we expect bidder's demand for GM foods to shift rightward when the biotech industry perspective is injected into the experiment and leftward when the environmental group perspective is provided. We perform both unconditional and conditional analyses of the bid prices submitted by the auction participants, denoted "bidders."

We first present summary statistics for our sample and then an unconditional analysis of our data on bidders' behavior. The attributes of the auction participants or bidders are summarized in Table 1, and they are representative of the populations from which they are drawn. In Table 2, participants' bids in the laboratory auction experiments are summarized. In Part A, mean bid prices across all information and labeling treatments are displayed. On average, bidders discounted GM-labeled food items by 14% relative to exactly the same food items having a plain-food label. Part B shows that bidders who received only the biotech industry perspective are willing, on average, to pay a small premium for GM-labeled food items for two of the three products. This occurred despite the fact that genetic modification is only used to enhance attributes that might be expected to lower farmers' production costs and not otherwise to directly benefit consumers, for example, by enhancing nutrient content or improving shelf life. Part C shows that when bidders receive only the environmental group perspective, they discount the GM-labeled food items by an average of 35% relative to the plain-labeled food item. Part D shows that bidders who receive both the biotech industry and environmental group perspectives discount GM-labeled foods by an average of 17%–29% across the three experimental food products.

An important issue is the size of the impact of verifiable information. Table 2, Part E, shows that bidders who receive both the biotech industry and environmental group perspectives discount GM-labeled food items only slightly relative to the plain-labeled food items. These differences, however, are not statistically significant

Table 1. Variables for auction participants-bidders and sample summary statistics ($N = 172$).

Variable	Definitions	Mean	Standard Deviation
Gender	1 if female	0.62	0.49
Age	The Participant's age	49.5	17.5
Married	1 if the individual is married	0.67	0.47
Education	Years of schooling	14.54	2.25
Household	Number of people in participant's household	2.78	1.65
Income	The households income level (in thousands)	57.0	32.6
White	1 if participant is white	0.90	0.30
Read_L[a]	1 if never reads labels before a new food purchase	0.01	0.11
	1 if rarely reads labels before a new food purchase	0.11	0.31
	1 if sometimes reads labels before a new food purchase	0.31	0.46
	1 if often reads labels before a new food purchase	0.37	0.48
	1 if always reads labels before a new food purchase	0.20	0.40
Informed[a]	1 if individual considered themselves at least somewhat informed regarding GM foods	0.42	0.49
Labels1	1 if the treatment bid on foods with GM labels in round 1	0.52	0.50

[a]Preexperiment information.

relative to the bids from those who only receive the probiotech (biotech industry) perspective. In contrast, bidders who receive only the biotech industry perspective value GM-labeled foods higher than the plain-labeled counterpart.

Table 2, Part F, shows that bidders who receive both antibiotech and verifiable information discount the GM-labeled foods by an average of 17%–22% across the three experimental food items. Bidders who receive antibiotech, probiotech, and verifiable information are more accepting of the GM-labeled foods than those who receive probiotech (biotech industry) and antibiotech (environmental group) perspectives. The bidders who receive the sixth information

Table 2. Mean bids for participants-bidders, excluding double-zero bids.

Commodity/ Type	N	Mean Bid	Standard Deviation	Median	Minimum	Maximum
Mean bids — all bidders and treatments						
GM oil	146	1.07	0.81	0.99	0	3.99
Oil	146	1.24	0.78	1.00	0	3.79
GM chips	155	1.03	0.85	0.99	0	3.99
Chips	155	1.20	0.81	1.00	0.05	4.99
GM potatoes	159	0.84	0.66	0.75	0	3
Potatoes	159	0.98	0.65	0.89	0	3.89
A. Mean bids when bidders received only probiotech information						
GM oil	26	1.56	0.73	1.50	0	2.99
Oil	26	1.54	0.79	1.55	0	3.50
GM chips	30	1.31	0.72	1.13	0	2.99
Chips	30	1.36	0.72	1.18	0.05	2.99
GM potatoes	27	1.30	0.71	1.25	0	2.50
Potatoes	27	1.26	0.67	1.25	0	2.00
B. Mean bids when bidders received only antibiotech information						
GM oil	26	0.79	0.82	0.50	0	3.25
Oil	26	1.22	0.65	1.00	0.25	2.49
GM chips	29	0.81	0.94	0.50	0	3.99
Chips	29	1.25	1.02	1.00	0.05	4.99
GM potatoes	29	0.61	0.68	0.50	0	2.75
Potatoes	29	0.98	0.88	0.75	0.05	3.89
C. Mean bids when bidders received both probiotech and antibiotech information						
GM oil	24	0.68	0.55	0.50	0	1.79
Oil	24	0.90	0.72	0.85	0	3.00
GM chips	23	0.68	0.74	0.35	0	2.25
Chips	23	0.81	0.79	0.49	0.05	2.75
GM potatoes	26	0.50	0.39	0.50	0	1.50
Potatoes	26	0.70	0.43	0.50	0.05	1.60

(*Continued*)

Table 2. (*Continued*)

Commodity/ Type	N	Mean Bid	Standard Deviation	Median	Minimum	Maximum
D. Mean bids when bidders received probiotech and verifiable information						
GM oil	26	1.12	0.62	1.00	0	2.39
Oil	26	1.14	0.57	1.00	0.10	2.39
GM chips	25	1.24	0.77	1.19	0	2.79
Chips	25	1.33	0.73	1.16	0.20	2.89
GM potatoes	26	0.92	0.45	0.99	0	1.85
Potatoes	26	0.93	0.39	0.99	0.25	1.90
E. Mean bids when bidders received antibiotech and verifiable information						
GM oil	21	1.33	1.05	1.25	0	3.99
Oil	21	1.60	0.97	1.50	0.49	3.79
GM chips	25	1.12	0.97	0.99	0	3.50
Chips	25	1.38	0.77	1.01	0.49	3.00
GM potatoes	27	0.89	0.77	0.89	0	3.00
Potatoes	27	1.14	0.67	0.99	0.50	3.00
F. Mean bids when bidders received probiotech, antibiotech, and verifiable information						
GM oil	23	0.94	0.77	0.95	0	2.75
Oil	23	1.06	0.82	1.00	0.05	3.29
GM chips	23	0.95	0.81	0.85	0	3.25
Chips	23	0.95	0.66	0.99	0.1	2.89
GM potatoes	24	0.82	0.61	1.00	0	1.99
Potatoes	24	0.84	0.55	0.84	0.01	2.00

treatment — probiotech, antibiotech, and verifiable information — discount the GM-labeled food by an average of 0%–11% across the three food items.

Given the interest in the effects of verifiable information on bidders' behavior, we perform a test of a null hypothesis of no effect versus significant effects using a Wilcoxon rank sum test. This testing is conducted as part of a more extensive analysis of the unconditional analysis of bidders' behavior. The effects of verifiable information on bidders who receive the antibiotech perspective are reported in Table 3, and they show a statistically significant effect of verifiable information on bidding behavior at the 5% level for vegetable oil,

Table 3. Results from test of null hypothesis of no significant difference in bid prices between pairs of information treatments.

Commodity/Information Type	Reduction in the Premium for Plain-Labeled Version of Food Products When Third-Party, Verifiable Information Is Introduced	One-Sided ρ Value from Wilcoxon Rank Sum Test Statistic
Potatoes, pro-GM info	\$–0.05	0.34
Vegetable oil, pro-GM info	\$0.00	0.30
Tortilla chips, pro-GM info	\$–0.04	0.38
Potatoes, anti-GM info	\$0.14	0.07
Vegetable oil, anti-GM info	\$0.16**	0.05
Tortilla chips, anti-GM info	\$0.18	0.13
Potatoes, both pro- and anti-GM info	\$0.18**	0.03
Vegetable oil, both pro- and anti-GM info	\$0.06	0.28
Tortilla chips, both pro- and anti-GM info	\$0.13	0.28

**Statistical significance at 5% level.

10% level for tortilla chips, and 5% level for potatoes. The differences in bids when verifiable information is introduced for those who previously received both the biotech industry and environmental group perspectives are statistically significant at the 10% level using the Wilcoxon rank sum test for potatoes, but are not significantly different from zero for tortilla chips or vegetable oil (Table 3).

While information from unconditional analysis of the bid prices is suggestive, multiple regression analysis of bid prices allows us to control for possible confounding factors, for example, a bidder's gender, age, and income, and for the censoring of bids. The regression model holds bidders' tastes constant for each of the experimental food products by defining the dependent variable to be the difference in a bidder's bid prices for plain-labeled (nonlabeled) and GM-labeled (labeled) food items. The bidder's inverse demand equation for the plain-labeled food item is

$$P_j^{\text{nonlabeled}} = \beta_1^{\text{nonlabcled}} + \beta_2^{\text{nonlabeled}} \mathbf{X}_{j2} + \mu_j^{\text{nonlabeled}} \qquad (1)$$

and for the GM-labeled food item is

$$P_j^{\text{labeled}} = \beta_1^{\text{labeled}} + \beta_2^{\text{labeled}} X_{j2} + \mu_j^{\text{labeled}}, \tag{2}$$

where P_j represents the price bid for a food item by bidder j, β_1 is an intercept term, X_{j2} is a vector of exogenous variables associated with the bidder and other information, β_2 is the associated vector of regression coefficients, and μ_j is a random disturbance representing other unmeasured effects. It is assumed to have a zero mean and to be normally distributed.

Subtracting Equation (2) from Equation (1), we obtain an equation in which the dependent variable is the bidder's difference in bid prices for the two products:

$$\begin{aligned} P_j^{\text{nonlabeled}} - P_j^{\text{labeled}} &= \beta_1^{\text{nonlabeled}} - \beta_1^{\text{labeled}} \\ &\quad + \left(\beta_2^{\text{nonlabeled}} - \beta_2^{\text{labeled}}\right) X_{j2} + \mu_j^{\text{nonlabeled}} \\ &\quad - \mu_j^{\text{labeled}}. \end{aligned} \tag{3}$$

The coefficients and error terms can be condensed and rewritten as

$$P_j^{\text{nonlabeled}} - P_j^{\text{labeled}} = \beta_1^* - \beta_2^* X_{j2} + \mu_j^*. \tag{4}$$

Hence, the difference in bid prices is then explained by an intercept term β_1^*, a vector of slope terms β_2^* that is multiplied by a vector of exogenous characteristics X_{j2}, and a random disturbance term $\mu_j^* = \mu_j^{\text{nonlabeled}} - \mu_j^{\text{labeled}}$.[10]

If a bidder submits a zero bid for nonlabeled food product or for a labeled food product, disturbance term μ_j^* in Equation (4) is censored (Greene 2003, 761–66). The censoring scenarios are as follows. Case 1: bidder j bids a positive amount for both the GM-labeled and the plain-labeled food item; the measured difference in bid prices for bidders is the difference between the two bid prices. Case 2: bidder j bids zero for the GM-labeled product and a positive amount for

[10]Because no bid price is revealed until all bids are placed and participants in a session were restricted from talking with each other, we anticipate no contemporaneous correlation of random disturbance terms across bidders in a session.

the plain-labeled product. The "true difference" in a bidder's bid price with the censored regression will be greater than the difference between the two observed bid prices. This arises because the bids on the GM-labeled food items are censored at zero, and this outcome is transmitted directly to the disturbance in Equation (4). Case 3: bidder j bids a positive amount for the GM-labeled product and zero for the plain-labeled product. As in Case 2, the true difference in a bidder's bid prices for the censored regression is absolutely larger than the measured difference between his or her two bid prices. Case 4: bidder j bids zero for both the plain- and GM-labeled food items. This latter outcome does not give any information about a bidder's true demand for GM products.

A positive aspect of using the censored regression model for a bidder's bid price differences[11] and associated random disturbances is that zero bid prices are accounted for correctly and bias effects from zero bids are minimized. The disadvantage is that the model is not linear in its unknown parameters and compound disturbance. Least squares estimation is no longer feasible, but maximum likelihood estimation of Equation (4), including censoring, is feasible and reasonable (Greene 2003). Tests of joint null hypotheses on the regression coefficients in Equation (4) against an alternative of negation can be performed using the likelihood ratio test statistic, $\lambda = L_r/L_u$. L_u and L_r are the values of the maximum of the "unrestricted" likelihood function [Equation (4), including censoring] and the maximum of the likelihood function after restrictions contained in the null hypothesis are imposed on the unrestricted model (Greene 2003, 484–86). λ is the sample value of the likelihood ratio for a particular unrestricted model and null hypothesis. In large samples, $-2 \ln (L_r/L_u)$ is distributed approximately as chi-squared, with degrees of freedom equal to the number of restrictions imposed

[11]Our censored regression model is a more general form of the common tobit model (Greene 2003). While a tobit model, the dependent variable or associated disturbance is censored, but given Equations (1) and (2) and zero bids, we have a disturbance term $\mu_j^* = \mu_j^{\text{nonlabeled}} - \mu_j^{\text{labeled}}$ which can have censoring on one or both of its components — the bid for the plain- and/or GM-labeled products.

by the null hypothesis on the parameters of the unrestricted model (Greene 2003, 484–86).

Now we turn to maximum likelihood estimates of the regression coefficients in Equation (4), including censoring. To increase the power of the statistical tests with the econometric model, we pool the observations on bid price differences across the three commodities. Hence, the sample size for this fitted model is 516 (= 172 × 3). Dummy variables are defined for each of the three types of information (probiotech, antibiotech, and verifiable), and indicators for probiotech and antibiotech separately are included, plus interactions between probiotech and antibiotech, probiotech and verifiable, antibiotech and verifiable, and all three information types. Other regressors include a bidder's gender and household income, a dummy variable taking a value of 1 if a bidder saw the GM label in the first trail (and zero otherwise), and a dummy variable indicating a value of 1 if the bidder perceives herself/himself to be informed about GM foods before the auction (and zero otherwise).[12] These latter regressors allow us to control for selective demographic attributes of bidders and to examine how a bidder's prior knowledge about GM foods or GM technologies affects WTP and demand.

Table 4 displays the basic maximum likelihood regression results and associated standard errors.[13] Commodity fixed effects are included, but no intercept term is included. The estimated coefficients for information treatment effects show that in auction sessions where bidders receive only antibiotech information the bid price differences are larger (between GM and plain labeled) and statistically significant at the 5% level. When bidders receive only probiotech information, bid price differences are reduced, but the difference is not significantly different from zero at the 5% (or 10%) level. When bidders receive both probiotech and antibiotech

[12]Several other models that included the bidder's age, marital status, religious upbringing, and educational attainment as regressors were fitted. None of these variables, however, impacted the difference in bid prices in a statistically significant way (at the 10% level).

[13]Ordinary least squares regressions were also fitted to 172 observations and to the observations remaining after "double-zero" bids were excluded. The results for these regressions are similar and are available from the authors upon request.

Table 4. Estimates of pooled censored regressions explaining bid price differences in GM-labeled and plain-labeled foods using a commodity fixed effects model ($n = 516$, standard errors are in parentheses).[a]

Regressors	Coefficients
Dependent variable: bid price plain-labeled food less bid price GM-labeled food	
Probiotech	−0.073 (0.097)
Antibiotech	0.432** (0.093)
Probiotech × antibiotech	−0.199 (0.131)
Probiotech × verifiable	0.007 (0.088)
Antibiotech × verifiable	−0.205** (0.090)
Probiotech × antibiotech × verifiable	0.065 (0.157)
Other variables	
Labels-Round 1	−0.115** (0.053)
Gender	−0.045 (0.054)
Income	0.00001** (0.0000)
Informed	0.093* (0.054)
Chi-squared excluding three regressors containing verifiable information	7.177*
Chi-squared for no explanatory power of regressors	76.44**

[a]The estimated coefficients for the commodity fixed effects are not reported.
*Statistical significance at the 10% level.
**Statistical significance at the 5% level.

information, the bid price difference is reduced, but it also is not statistically significant. From these results, we conclude that in those sessions where bidders receive only antibiotech or both probiotech and antibiotech information, they bid differently than where they receive only probiotech information.

When bidders are in sessions that receive probiotech and verifiable information, the impact of this combination is not statistically significant. When bidders are in sessions that receive antibiotech and verifiable information, bid price differences are reduced, and the difference is significantly different from zero. Hence, those who receive antibiotech information along with verifiable information discount GM foods less than those who receive only antibiotech information. When bidders are in sessions that receive all three types of information (probiotech, antibiotech, and verifiable), the impact of this treatment on bid price differences is small and not statistically significant. Hence, in this complex setting, verifiable information does not have a distinguishable effect.

To focus more specifically on the potential significance of verifiable information on bidders' behavior, we perform a joint test that verifiable information has no effect on bidder's behavior. We do this by deleting the three information treatments from Equation (4) that include verifiable information as a component of a treatment. The sample value of the chi-squared statistic for this test is 7.17 with 3 degrees of freedom, and at a 10% significance level, the tabled chi-squared is 6.97. Hence, verifiable information has a statistically significant effect on bidders' WTP for commodities that might be GM.

Other results include the following: bidders who have larger household incomes discount GM by a larger amount than those with less household income. This result is statistically significant at the 5% level and is consistent with non-GM products being viewed on average by bidders as a superior product. Those bidders who consider themselves to be at least "somewhat informed about GM foods" (as recorded in the preauction survey) discount GM-labeled foods by more than other bidders. This effect is statistically significant (at the 10% level). Moreover, this result suggests that bidders in our experiments who are "GM informed" have on average acquired/received negative information about GM foods prior to the experiment.[14] Bids also are affected by the labeling sequence. Bidders who are in sessions that bid on the GM-labeled food products in Round 1 (and the plain-labeled food products in Round 2) discount GM-labeled foods by less than those who are in sessions that bid on the products in the opposite order. The regression coefficient for this regressor is statistically significant at the 5% level, and the result reinforces the importance of randomized assignment of treatments to sessions in experimental auctions, which is an innovation in our methodology. Finally, we perform a joint test of no explanatory power of Equation (4) when fitted to the pooled data with commodity

[14]See Huffman *et al.* (2006) for an analysis of the impact of bidders' prior beliefs about GM technology on their WTP for food items that might be GM.

fixed effects and censoring. The sample value of the chi-squared statistic is 18.6 with 10 degrees of freedom. The tabled value of the chi-squared statistic at the 5% significance level is 18.3. Hence, this null hypothesis of no explanatory power of Equation (4) is rejected.

5. Methodology for Valuing Verifiable Information

In the previous section, we showed that independent verifiable information affects bidders' bids on food products that may be GM, given the presence of information from interested parties. We now summarize the methodology used to estimate the public good value of verifiable information. First, consider the empirical specification of the model leading to the public good value of verifiable information. Our approach is similar to the approach taken by Foster and Just (1989) and Teisl *et al.* (2001). Information has value if an agent's observable behavior changes. For our case, information has social value if a participant/consumer changes his/her behavior as a result of receiving the information, that is, they "switched products that they purchased" — from GM-labeled to plain-labeled foods, or vice versa.[15]

Consider the two types of bidders that benefit from verifiable information. One type purchases GM-labeled food before receiving verifiable information and then switches to plain-labeled food after receiving verifiable information. The second type purchases plain-labeled food before receiving verifiable information and switches to GM-labeled food after receiving the same information.

The economist's task is to approximate the net welfare change for bidders who change their observed behavior after receiving verifiable information. Because we are trying to assess the average value of information for each product, we assume all bidders purchase either the GM-labeled version or the plain-labeled version of a product.

[15]Note that our model does not assume an auction market but a conventional market. But auctions are essential for this analysis because our auction market elicits the nonhypothetical WTP under different information treatments that is not obtainable in a conventional market.

The bidder's surplus is approximated by the difference between his/her WTP and the "market price" (i.e., the price consumers would pay for a product in a store) for the product he/she purchases. Bidder j's consumer surplus from purchasing plain-labeled food or GM-labeled food is defined to be:

$$\text{surplus}^j_{\text{PL}} = \text{WTP}^j_{\text{PL}} - \text{MP}^j_{\text{PL}} \tag{5}$$

$$\text{surplus}^j_{\text{GML}} = \text{WTP}^j_{\text{GML}} - \text{MP}^j_{\text{GML}}. \tag{6}$$

In Equations (5) and (6), the bidder's WTP is revealed in the experimental auctions, MP is the price the bidder faces for the product in the marketplace, the superscript j refers to bidder j, and the subscripts PL and GML refer to the plain-labeled and GM-labeled products in this section.[16]

A consumer or bidder purchases either the GM-labeled or plain-labeled food product. The product that bidder j purchases is assumed to be the one that gives him/her the higher surplus. Formally, if $\text{surplus}^j_{\text{PL}} > \text{surplus}^j_{\text{GML}}$ then $\text{buy\_PL}^j_I = 1$ and $\text{buy\_GML}^j_I = 0$ and if $\text{surplus}^j_{\text{PL}} < \text{surplus}^j_{\text{GML}}$ then $\text{buy\_PL}^j_I = 0$ and $\text{buy\_GML}^j_I = 1$, where the subscript I refers to the information setting (whether or not the consumer has received verifiable information). Those who purchase the plain-labeled product after receiving verifiable information obtain

$$\text{PREMGAIN}^j_{\text{PL}} = \text{surplus}^j_{\text{PL}} - \text{surplus}^j_{\text{GML}}. \tag{7}$$

Similarly, those who purchase the GM-labeled product after receiving verifiable information gain:

$$\text{PREMGAIN}^j_{\text{GML}} = \text{surplus}^j_{\text{GML}} - \text{surplus}^j_{\text{PL}}. \tag{8}$$

[16]In computing this value of verifiable information, we used "market" prices that we paid for plain-labeled vegetable oil, yellow tortilla chips, and Russet potatoes as the market price for plain-labeled product: $1.65 for the 32 ounce bottle of vegetable oil, $2.99 for the 16 ounce bag of tortilla chips, and $1.79 for the 5 pound bag of potatoes as the prices for the plain-labeled products. For market price of GM-labeled product, we adjusted the market price of plain-labeled price for the average GM discount in our experiment. Therefore, the estimated market prices for the GM-labeled products were $1.48 for the oil, $2.82 for the tortilla chips, and $1.65 for the potatoes.

Although all bidders enjoy the premium gained by consuming one product instead of another, as shown in Expressions (7) and (8), the premium gained represents the increase in welfare (i.e., the value of information) only for those *who switch products.*

We next discuss the method used to estimate the percentage of bidders who change purchases when verifiable information is introduced. First, the percentage of bidders who purchase GM-labeled products is denoted:

$$\text{percentbuyGM}_I = \left(\sum_j \text{buy\_GML}_K^j \right) / \text{ N.} \tag{9}$$

Equation (9) shows that this number can be represented as the summation across bidders who purchase the GM-labeled version of the commodity divided by the total number of bidders. Therefore, the percentage of bidders who purchase the plain-labeled version is $1 - \text{percentbuyGM}_I$.

Verifiable information causes a bidder to switch purchases if his or her surplus is higher for one product version (e.g., the plain-labeled product) prior to receiving verifiable information, but higher for the other product version (e.g., the GM-labeled product) after receiving verifiable information. The net change in the percentage who purchase the GM-labeled product due to the introduction of verifiable information is the (absolute) difference between the "percentage who purchase GM-labeled foods when treated to verifiable information" and the "percentage who purchase GM-labeled foods but do not receive the verifiable information" given the other information they have received:

$$\begin{aligned} \text{Percentswitch}^K = |&\text{percentbuyGM}_{\text{verifiable}} \\ &- \text{percentbuyGM}_{\text{no verifiable}}|. \end{aligned} \tag{10}$$

In Equation (10), the percentage of bidders who switched purchases is estimated as the absolute value of the difference in the percentage who purchase the GM-labeled food with and without verifiable information. The superscript K represents either GM-labeled or plain labeled foods, depending on which product bidders are switching to.

Who switches purchases once verifiable information is introduced? Because bidders who receive given information treatments are in distinct experimental sessions, we do not know the specific persons who switch, but we can compute the percentage of the sample that switched after the introduction of verifiable information. To do this, we assume that the bidders who switch have relative preferences for the food products that are uniformly distributed across the population that consumes the good that was abandoned. For example, we assume that bidders who switched to GM-labeled foods after receiving verifiable information had relative valuations of plain-labeled foods that were evenly distributed throughout the population of consumers who purchased the plain-labeled foods before information was introduced. Thus, without verifiable information, treated and untreated participants have the same behavior.

We now compute the probability of a participant being a "switcher" — one who changes his or her behavior after verifiable information is introduced:

$$\text{prob\_switch}_{\text{GML}} = \text{Percentswitch}^{\text{GML}}/\text{percentbuyGM}_{\text{verifiable}}. \tag{11}$$

$$\text{prob\_switch}_{\text{PL}} = \text{percentswitch}^{\text{PL}}/\text{percentbuynonGM}_{\text{verifiable}}. \tag{12}$$

To determine the expected value of verifiable information to a participant, we multiply his or her premium surplus (PREMGAIN) by the probability that he or she switched products:

$$\begin{aligned}\text{EVperson}^j = {} & \text{PREMGAIN}^j_{\text{GML}} \times \text{prob\_switch}^j_{\text{GML}} + \text{PREMGAIN}^j_{\text{PL}} \\ & \times \text{prob\_switch}^j_{\text{PL}}.\end{aligned} \tag{13}$$

In Equation (13), $EVperson^j$ is the expected value of information to bidder j.[17] One can also think of this as the average value of verifiable information across all bidders or participants. What is

[17]Note that because it is assumed that auction participants consume either GM- or plain-labeled food products, that only one of the two PREMGAIN coefficients will be positive, while the other is zero. The PREMGAIN coefficients will also differ across participants.

important is that we can compute this value for each of the three products and over different initial information treatments.

Next, we need the expected value of information to a bidder who switches purchases. This is computed by dividing the expected value of verifiable information per person by the percentage of bidders who switched purchases:

$$\text{EVswitcher} = \text{EVperson}^j/\text{percentswitch}. \quad (14)$$

In Equation (14), EVswitcher is the average value of verifiable information *to a bidder who switches* his or her purchase of a food product, either to GM-labeled food from plain-labeled, or vice versa.[18]

In summary, the experimental auction data collected for this study allow us to calculate the percentage of bidders who switch in each of the information settings: receiving the probiotech perspective, the antibiotech perspective, and both pro- and antibiotech perspectives. We then estimate an expected value of verifiable information per experiment participant or bidder and suggest that it might be appropriate to apply this number to the whole U.S. population.

Now consider the estimated value of verifiable information. For each commodity, Table 5 presents the marginal percentage of bidders who would switch, the value of verifiable information to a person who would switch, and the expected value of verifiable information to a person in society. When a bidder receives only the biotech industry perspective, one might expect the introduction of verifiable information to cause some individuals to switch to plain-labeled foods. For example, among those bidders who receive both biotech industry and verifiable information, some are more likely to purchase GM-labeled potatoes but are less likely to purchase the GM-labeled tortilla chips than bidders who receive only the biotech industry perspective. The share of bidders who would switch to either of these goods, however, is small. Among those bidders who first received biotech industry information, *the expected value of the verifiable*

[18]The SAS code used to estimate the value of information is available from the authors upon request.

Table 5. Value of verifiable information about genetic modification and GM foods to participants-bidders.

	Participants Switching to GM		
	Percentage	Value per Person	Average Overall Value
A. Value to bidders who originally received pro-biotech information[a]			
Tortilla chips (N = 25)	3.3[a]	\$0.112/bag	\$0.004/bag, t value = 1.54
Vegetable oil (N = 26)	15.4	\$0.209/bottle	\$0.032/bottle**, t value = 4.25
Potatoes (N = 26)	3.3	\$0.184/bag	\$0.006/bag**, t value = 6.01
B. Value to bidders who originally received only antibiotech information[b]			
Tortilla chips (N = 25)	18.6	\$0.250/bag	\$0.051/bag**, t value = 2.91
Vegetable oil (N = 21)	28.2	\$0.190/bottle	\$0.054/bottle*, t value = 2.75
Potatoes (N = 27)	25.0	\$0.179/bag	\$0.045/bag**, t value = 5.89
C. Value to bidders who originally received both pro-biotech and antibiotech information[c]			
Tortilla chips (N = 23)	8.7	\$0.233/bag	\$0.020/bag**, t value = 3.77
Vegetable oil (N = 23)	15.9	\$0.280/bottle	\$0.044/bottle**, t value = 3.29
Potatoes (N = 24)	21.5	\$0.298/bag	\$0.064/bag**, t value = 3.48

[a]On average, more individuals purchased the GM-labeled potatoes and GM-labeled vegetable oil when they received probiotech and verifiable information as opposed to just getting probiotech information, but fewer individuals purchased the GM-labeled tortilla chips than their plain-labeled counterpart when they received probiotech and verifiable information.

[b]Consumers who received antibiotech and verifiable information were more accepting of GM foods than individuals who only received antibiotech information.

[c]Consumers who received probiotech, antibiotech, and verifiable information were more accepting of GM foods than individuals who only received probiotech and antibiotech information.

*Statistically significant at the 5% level using a t-test.

**Statistically significant at the 1% level using a t-test.

information per person was about one-half cent per product but was only statistically significant for the potatoes.

While verifiable information brought about virtually no change in bidding behavior for tortilla chips and potatoes, bidders who received biotech industry and verifiable information were much more likely to purchase GM-labeled vegetable oil than bidders who receive only the industry perspective. Approximately 15% of the bidders who received the biotech industry perspective would switch from plain-labeled vegetable oil to GM labeled vegetable oil after the introduction of the verifiable information. This outcome is consistent with verifiable information revealing that vegetable oils do not contain DNA because all the protein-carrying DNA is boiled out in refining vegetable oils for human consumption. Vegetable oils made from GM and non-GM soybeans are chemically and nutritionally indistinguishable. Bidders worried about their own health now become more likely to purchase GM-labeled vegetable oil, even if they do not change their attitude toward other GM-labeled products. Among those bidders who receive biotech industry information, *the average value per person of verifiable information to those who switch to the GM-labeled vegetable oil is almost 21 cents per switcher, and the expected value per person is just over 3 cents per bottle.* These value estimates are statistically significant at the 1% level using a t-test (Table 5).[19] This is interesting because among bidders who received biotech industry information, virtually all the surplus gain arose from vegetable oil and little from the other two products.

In another scenario, some bidders received either the environmental group perspective only or both the biotech industry and environmental group perspectives. We now consider the value of verifiable information in these cases. We expect that bidders who initially received only the environmental group perspective on GM foods and later were given the verifiable information will be more likely to purchase GM-labeled foods. Our results confirm this hypothesis. For all three experimental food products, a significant

[19]For all value of information test statistics, we also ran a nonparametric Wilcoxon signed rank test and found similar results.

share of the sample would switch from plain labeled to the GM-labeled food under these circumstances: between 18.6% and 28.2% of bidders would switch to the GM-labeled food, across the three commodities. The value of verifiable information for each bidder who switches ranges from 18 to 25 cents per food item, and *the expected value of information per person in the experiments is 5.1 cents per bag of tortilla chips, 5.4 cents per bottle of vegetable oil, and 4.5 cents per bag of potatoes.* These values of information estimates are all statistically significant at either the 1% or 5% level (Table 5).

Bidders who receive biotech industry and environmental group perspectives first and then receive verifiable information are more likely to purchase GM-labeled foods than bidders who do not receive verifiable information. The share of bidders who switch from plain labeled foods to GM-labeled foods is smaller for each of the three goods in this auction when compared to the bidders who receive the environmental group perspective, but larger for each food item when compared to bidders who receive only the biotech industry perspective. Only 8.7% of bidders would switch to the GM-labeled tortilla chips, while 15.9% and 21.5% would switch to the GM-labeled vegetable oil and GM-labeled potatoes. Moreover, the average value of verifiable information per person who would switch from the plain-labeled to GM-labeled food ranged from 23 to 30 cents across the three food products. In this scenario, the *implied expected value of verifiable information was 2 cents per bag of tortilla chips, 4.4 cents per bottle of vegetable oil, and 6.4 cents for each bag of potatoes* (Table 5).[20]

Suppose we take a bold step and prepare a "ballpark" estimate of the value of verifiable information on genetic modification for the

[20]Recall that we had to use estimated prices for the GM-labeled products to compute the value of information, and we are using the mean discount for GM-labeled foods to estimate these prices. We consider the impact of used alternative prices. One alternative set of prices yielded a greater premium for plain-labeled foods (prices for GM-labeled products estimated to be $1.38 for the oil, $2.72 for the tortilla chips, and $1.58 for the potatoes), and one that yielded a smaller premium for plain-labeled foods (prices for GM-labeled products estimated to be $1.57 for the oil, $2.91 for the tortilla chips, and $1.72 for the potatoes). The average value of information for the three food products was 5 cents and 3 cents per product for the greater and smaller plain-labeled premiums, respectively. Thus, our results are relatively robust to the estimated prices.

whole U.S. population. Generalizing is not without some risks, but the generalization is also illustrative. Our estimate is that the value of verifiable information is about 4 cents per food product for participants who have heard either the environmental group perspective or both environmental group and biotech industry perspectives on GM foods. Because the prices for these three food products in units that we used in our experiment were typically between $1.50 and $2.50, verifiable information has a value of approximately 2% of the purchase price for products that could be GM.

Estimates of the share of grocery store foods that are GM vary. Some observers suggest that two-thirds of *all processed foods* in the United States contain some GM material (Davis 2001). Others argue that one-third of *all products* in a grocery store contain GM material (Friends of the Earth 2001). To provide a conservative estimate, we approximate the aggregate value of verifiable information by assuming that only one-third of all products on a grocers shelf were GM at the time of our experiments in April 2001.

In 1997, U.S. citizens spent $390 billion for food at home (Putnum and Allshouse 1999). Applying the one-third rule, we suggest that Americans spent roughly $130 billion on foods that might be GM. If verifiable information has a value of about 2% of the price for these foods and if one generalizes these results, our estimate of the potential public good value to U.S. consumers of verifiable information is about $2.6 billion annually. While large, the aggregate value does not seem totally unrealistic. Two and six-tenths billion dollars is only an average value of approximately $9.00 per year for every U.S. man, woman, and child. As a reference point, Foster and Just (1989) reported a value of government information about pesticide-contaminated milk in Hawaii of approximately $10.00 per person per month ($120.00 per year), using similar techniques.

6. Concluding Comments

We have presented a new methodology for testing and valuing the effects of objective third-party or verifiable information in a market with conflicting information. Our results show that the perspectives of interested parties are consequential in an auction market setting; probiotech information distributed by the biotech industry has significantly negative effects on bid price differences between plain- and GM-labeled product, and antibiotech information distributed by environmental NGOs has significantly positive effects on bid price differences for food that might be GM. However, no single type of information is dominant, although the empirical impacts of probiotech and verifiable information have similarities. In an environment that is roughly similar to the situation that exists in the United States today, which is one of voluntary GM labeling, we show that verifiable information has a significant effect on bid price differences when participants have not received probiotech information, and more generally, when our data on bidding behavior are pooled across all three commodities. We expect similar results in other developed countries that have voluntary labeling policies. Furthermore, we argue that verifiable information has its greatest impact and hence greatest potential social value, when interested parties are disseminating information that is quite negative about the new technologies. However, if the resistance were to disappear, the value of verifiable information would be greatly reduced.

Verifiable information has value when it changes the behavior of consumers. We make several simplifying assumptions and produce a calculation suggesting that verifiable information on genetic modification might have an annual social value to U.S. consumers through food purchases of $2.6 billion annually. This might seem high, but it is expected to decrease over time if reputable sources of verifiable information for GM foods and technologies were established in the United States. The exact details of how such a system might function or be financed are summarized in Huffman and Tegene (2002).

Our new methodology for assessing the impacts and value of conflicting and third-party information on consumers' WTP for new goods items can be adapted to a range of new consumer goods, for example, fresh fruit with reduced bruising, food fortified with protein and vitamins, and new drugs. In particular, policy makers could use our methodology to assess the public good benefit of new information for controversial commodities and compare these benefits to their expected social cost.

Acknowledgment

We gratefully acknowledge financial assistance from the USDA under Cooperative Agreement No. 43-3AEL-8-80125; the Cooperative State Research, Education, and Extension Service, USDA, under Grant No. 00-52100-9617; and the Iowa Agriculture and Home Economics Experiment Station. Also, we thank Wayne Fuller and Phil Dixon in the Department of Statistics, Iowa State University, for assistance with the statistical experimental design; Dianne Anderson of the Iowa State University Statistics Laboratory for supervising the identification of the sample; Dan Monchuk and Terry Hurley for assistance in conducting the experiments while Ph.D. candidates at Iowa State University; an anonymous reviewer for extensive comments; Walt Fehr, John Miranowski, and Greg Colson of Iowa State University for helpful suggestions and comments; and Monsanto for providing food products. All errors or shortcomings of the paper remain the fault of the authors. The views expressed here are the authors' and not necessarily those of the USDA.

References

Akerlof, G. (1970). The Market for Lemons: Uncertainty and the Market Mechanism. *Quarterly Journal of Economics* 84, 488–500.

Balasubramanian, S.K. and C. Cole (2002). Consumers' Search and Use of Nutrition Information: The Challenge and Promise of the Nutrition Labeling and Education Act. *Journal of Marketing Research* 66, 113–127.

Carter, C.A. and G.P. Gruere (2003). Mandatory Labeling of Genetically Modified Foods: Does It Really Provide Consumer Choice? *AgBioForum* 6, 68–70.

Caswell, J.A. (2000). Labeling Policy for GMOs: To Each His Own? *AgBioForum* 3, 53–57. http://www.agbioforum.missouri.edu.

Cohen, S.N., A.C.Y. Chang, H. Boyer and R.B. Helling (1973). Construction of Biologically Functional Bacterial Plasmids In Vitro. *Proceedings of the National Academy of Sciences U.S.A.*, 70, 3240–3244.

Corrigan, J. and M. Rousu (2006). Posted Prices and Affiliated Values: Evidence from Experimental Auctions. *American Journal of Agricultural Economics* 88, 1078–1091.

Council for Biotechnology Information (2001). "Frequently Asked Questions". http://www.whybiotech.com/en/faq/default.asp?MID=10 (accessed October 2001).

Davis, J. (2001). "'Frankenfoods' — Friend of Foe?" *WebMD* http://my.webmd.com/content/article/1728.80187 (accessed May 22, 2001).

Fernandez-Cornejo, J. and M. Caswell (2006). "The First Decade of Genetically Engineered Crops in the United States". USDA, ERS EIB-11. http://www.ers.usda.gov.

Foster, W. and R.E. Just (1989). Measuring Welfare Effects of Product Contamination with Consumer Uncertainty. *Journal of Environmental Economics and Management* 17, 266–283.

Fox, J.A., D.J. Hayes and J.F. Shogren (2002). Consumer Preferences for Food Irradiation: How Favorable and Unfavorable Descriptions Affect Preferences for Irradiated Pork in Experimental Auctions. *Journal of Risk and Uncertainty* 24, 75–95.

Friends of the Earth (2001). "The Need for Labeling Genetically Engineered Foods". http://www.foe.org/safefood/factshtgelabel.htm.

Friends of the Earth (2003). "Environmental Hazards of Genetically Engineered Plants". http://www.foe.org/camps/comm/safefood/gefood/factsheets/envirohazfacts.html.

Gates, B. (2001). "Will Frankenfood Feed the World?" Visions of the 21st Century. *Time.* http://www.time.com/time/reports/v21/tech/mag_food.html.

Greene, W.H. (2003). Econometric Analysis. 5th ed. Upper Saddle River, NJ: Prentice Hall.

Greenpeace International (2001a). "We Want Natural Food!" http://www.greenpeace.org/~geneng/.

Greenpeace International (2001b). "7. Public Concern." http://www.greenpeace.org/%7Egeneng/reports/food/intrfo07.htm.

Greenpeace International (2001c). "11. The Potential for Allergic Reactions". http://www.greenpeace.org/%7Egeneng/reports/food/intrfo11.htm.

Greenpeace International (2003). "GeneticEngineering." http://www.greenpeace.org/international_en/campaigns/intro?campaign%5fid=3942.

Greenpeace International (2006). "Say No to Genetic Engineering". http://www.greenpeace.org/international/campaigns/geneticengineering.

Hausman, J. (1997). "Valuation of New Goods under Perfect and Imperfect Competition." In T.F. Bresnahan and R.J. Gordon (Eds.), *The Economics of New Goods*, pp. 209–237. NBER Studies in Income and Wealth Vol. 58. Chicago: The University of Chicago Press.

Hirshleifer, J. and J.R. Riley (1992). *The Analytics of Uncertainty and Information.* New York, NY: Cambridge University Press.

Huffman, W.E. and R.E. Evenson (2006). *Science for Agriculture: A Long-Term Perspective.* Ames, IA: Blackwell Publishing.

Huffman, W.E., M. Rousu, J.F. Shogren and A. Tegene (2004). "The Welfare Effects of Implementing Mandatory GM labelling in the USA." In R.E. Evenson and V. Santaniello (Eds.), *Consumer Acceptance of Genetically Modified Food*, pp. 41–53. Oxfordshire, UK: CAB International.

Huffman, W.E., M. Rousu, J.F. Shogren and A. Tegene (forthcoming). The Effects of Prior Beliefs and Learning on Consumers' Acceptance of Genetically Modified Foods. *Journal of Economic Behavior and Organization* 59, 2006.

Huffman, W. E. and A. Tegene (2002). "Public Acceptance of and Benefits from Agricultural Biotechnology: A Key Role for Verifiable Information."

In V. Santaniello, R. E. Evenson and D. Zilberman (Eds.), *Market Development for Genetically Modified Food*, pp. 179–190. Oxfordshire, UK: CAB International.

Kahneman, D. and A. Tversky (1979). Prospect Theory: An Analysis of Decision under Risk. *Econometrica* 47, 263–291.

Lusk, J.L. and M. Rousu (forthcoming). "Market Price Endogeneity and Accuracy of Value Elicitation Mechanisms." In J. List (Ed.), *Using Experimental Methods in Environmental and Resource Economics*. North Hampton, MA: Edward Elgar Publishing.

Mendenhall, C.A. and R. E. Evenson (2002). "Estimates of Willingness to Pay a Premium for Non-GM Foods: A Survey." In V. Santaniello, R.E. Evenson and D. Zilberman (Eds.), *Market Development for Genetically Modified Foods*, pp. 55–62. New York, NY: CABI Publishing.

Milgrom, P. and J. Roberts (1986). Relying on the Information of Interested Parties. *Rand Journal of Economics* 17, 18–32.

Molho, I. (1997). *The Economics of Information: Lying and Cheating in Markets and Organization*. Malden, MA: Blackwell Publishing.

Morris, S. and H.S. Shin (2002). Social Value of Public Information. *The American Economic Review* 92, 1521–1534.

Nestle, M. (2003). *Safe Food: Bacteria, Biotechnology, and Bioterrorism*. Berkeley, CA: The University of California Press.

Putnum, J.J. and J.E. Allshouse (1999). "Food Consumption, Prices, and Expenditures, 1970–1997." Statistical Bulletin No. 965, Economic Research Service, U.S. Department of Agriculture.

Shogren, J.F., M. Margolis, C. Koo and J.A. List (2001). A Random *n*th-Price Auction. *Journal of Economic Behavior and Organization* 46, 409–421.

Shogren, J.F., S.Y. Shin, D.J. Hayes and J.B. Kliebenstein (1994). Resolving Differences in Willingness to Pay and Willingness to Accept. *The American Economic Review* 84, 255–270.

Smith, V.L. (1976). Experimental Economics: Induced Value Theory. *American Economic Review, Papers and Proceedings* 66, 274–279.

Stigler, G.J. (1961). The Economics of Information. *Journal of Political Economy* 69, 213–225.

Teisl, M.F., N.E. Bockstael and A. Levy (2001). Measuring the Welfare Effects of Nutrition Information. *American Journal of Agricultural Economics* 83, 133–149.

Viscusi, W.K. (1997). Alarmist Decisions with Divergent Risk Information. *The Economic Journal* 107, 1657–1670.

Viscusi, W.K. and W.N. Evans (1990). Utility Functions that Depend on Health Status: Estimates and Economic Implications. *American Economic Review* 80(3), 353–374.

Wilson, W.W. and B. Dahl (2005). Costs and Risks of Testing and Segregating GM Wheat. *Review of Agricultural Economics* 27, 212–228.

Chapter 9

The Effects of Prior Beliefs and Learning on Consumers' Acceptance of Genetically Modified Foods*

Wallace E. Huffman[†,‡,**], Matthew Rousu[§], Jason F. Shogren[¶], and Abebayehu Tegene[||]

[‡]*Department of Economics, Iowa State University, Ames, IA 50011, United States*

[§]*Susquehanna University*

[¶]*Natural Resource Conservation and Management, Department of Economics and Finance, University of Wyoming*

[||]*Food and Rural Economics Division, ERS, U.S. Department of Agriculture*

[**]*whuffman@iastate.edu*

New food products using genetically modified crops appeared in U.S. supermarkets starting in 1996, and consumers' perceived some risks. This paper examines the role of consumers' prior beliefs about genetic modification and of diverse, new information on their willingness to pay for foods that might be genetically modified. We use data from economics experiments and show that participants who had informed prior beliefs discounted GM-labeled food products more highly than those who had uninformed prior beliefs. Uninformed participants were especially susceptible to information from interested and third parties. In contrast, informed participants were generally not affected significantly by new information.

JEL classification: C91; D12; D82

Keywords: Laboratory experiments, nth-Price auction, Food labels, Diverse information, Learning, Prior beliefs, Consumers, Willingness to pay

---

*This article originally appeared in the **Journal of Economic Behavior & Organization (2007), 63(1), 193–206**. This is published by permission of **Elsevier**.

[†]Corresponding author.

New food products made from genetically modified crops appeared in U.S. supermarkets starting in 1996. The genetic modification consisted of herbicide tolerance and insect resistance that have been introduced into field crops through the use of techniques in modern biotechnology. Herbicide tolerance and insect resistance are so-called input traits that reduce the expected cost of production to farmers (Fernandez-Cornejo and McBride, 2002; FAO, 2004) but have no direct benefit to consumers and pose some risks (Chern and Rickertsen, 2004; FAO, 2004). Consequently, GM products have been subject to much controversy (claims by environmental groups of reducing biodiversity, new food safety concerns due to allergens, and ethical concerns regarding the movement of genes across species), and they have raised important new issues in trade talks especially between the U.S. and the European Union.

Huffman *et al.* (2003) examined the effects of GM-food labels on bidding behavior of participants in an experimental auction, and Rousu *et al.* (2004) developed a methodology to value the contribution of third-party information in a setting with conflicted information. These papers, however, have not focused on the contributions of subjective prior beliefs about genetic modification in an experimental auction-market setting. Consumers have subjective prior beliefs about attributes of goods (Akerlof, 1970; Hirshleifer and Riley, 1992; Molho, 1997; Stigler, 1961) and frequently obtain new information to update these prior beliefs, for example, Bayesian learning (DeGroot, 1970; Molho, 1997, pp. 248–249, Tirole, 2003, p. 373).[1] This paper examines in depth the role of consumer's prior beliefs about genetic modification and diverse, new information about genetic modification on their "willingness to pay" for foods that might be genetically modified. The information is of two broad types: subjective prior beliefs that arise from prior investments in information and new information from interested and disinterested parties. In the GM-food debate, the interested parties are the agricultural biotech industry and environmental groups. The agricultural biotech industry consists of the private companies that market crop

[1]These beliefs could, however, be uninformative or diffuse (DeGroot, 1970).

input traits and distribute pro-biotech information such as Monsanto, Syngenta, and Pioneer Hibred, and the Council for Biotechnology Information, a private trade association. The environmental groups are largely Greenpeace, Friends of the Earth, and Action Aid, which disseminate anti-biotech information. Furthermore, Huffman and Tegene (2002) have speculated about the potential value of independent, third-party information in such a conflicted market, and Rousu *et al.* have injected independent, third-party information into a set of economic experiments to assess its value.[2] In these experiments, food-label types and information treatments were randomly assigned to sessions or trials. In this setting, participants who perceived themselves to be at least somewhat informed about genetic modification bid significantly less for GM-labeled foods than those who considered themselves to be uninformed. This raises an important issue of how prior information affects the interpretation of new information, which this study emphasizes.

If consumers place heavy weight on information from interested parties, including cheap talk, their welfare will be lower than if they use objective information (Akerlof, 1970; Molho, 1997; Morris and Shin, 2002). One hypothesis is that consumers who have uninformed priors have their bidding behavior affected by information from one or more interested parties, but consumers who have informed priors are relatively unaffected (Schultz, 1975; Huffman, 2001; Tversky and Kahneman, 1981; Kahneman, 2003). Another hypothesis is that the presence of third-party information affects the way that consumers use information from interested parties in placing bids.

This research builds on the strength of a unique data set that we collected from laboratory auctions that combined the methods of economic experiments, statistical experimental design, and survey design (Rousu *et al.*). Our auctions were "willingness to pay" experiments constructed such that it was in each participant's best interest to state her true preferences, and furthermore, the winners were expected to "pay" what they "say" or "bid." Because of the

---

[2]Independent, third-party information is sometime referred to as verifiable information (Milgrom and Roberts, 1986).

cost of running a large number of experiments at different locations, we chose two locations (major cities) and a modest number of participants.[3] The model, results, and conclusions follow.

1. The Conceptual Framework

We review a few pieces of relevant literature and then present our model. Other studies have shown that people place much less weight on prior beliefs than on new information when bidding on lotteries (e.g., see Grether, 1980 or Tversky and Kahneman, 1974).We extend this work in two important ways. First, we adopt the Bayesian concept of subjective beliefs instead of assuming that prior beliefs are always "fact" or objective beliefs (DeGroot). Second, we examine how prior beliefs and new information from interested and disinterested parties affect a consumer's willingness to pay for a new product that might cause environmental or human harm (i.e., they pose some potential risks).

A consumer's utility from consuming genetically modified food products is modeled as depending on her prior beliefs about genetic modification. Following Kivi and Shogren (2005), state-dependent indirect utility functions are defined for a "good outcome" $U(Y)$ and a "bad outcome" $V(Y - L)$ that are independent of an individual's prior beliefs or the amount of new information that she has acquired/received. The indirect utility functions give maximum utility for a given level of income, Y and $Y - L$, and they are weighted by the individual's subjective probabilities conditioned on her prior beliefs: p^{I}, probability of a bad outcome, given the participant has informed prior beliefs, and $p^{\mathrm{no-I}}$, probability of bad outcome given the participant has uninformed prior beliefs, to obtain a consumer's

[3]An alternative methodology is a state preference or contingent value survey. In these surveys a random sample of households are contacted and asked to respond to hypothetical product preference or willingness to pay options. This approach has the advantage of being able to be administered relatively inexpensively to a large sample, which can give precision (for examples see Johnston *et al.*, 2001). However, because responses are hypothetical, they are not bound by a budget constraint. This can lead to hypothetical biases or credibility problems that do not vanish as the sample size increases. For examples, see Diamond and Hausman (1994) and Fox *et al.* (1998).

expected utility:

$$U(Y - \mathrm{WTP}^{\mathrm{I}}) = \mathrm{EU}^{\mathrm{I}} = p^{\mathrm{I}}(I, \mathrm{inf})V(Y - L) \\ +(1 - p^{\mathrm{I}}(I, \mathrm{inf}))U(Y), \quad (1)$$

$$U(Y - \mathrm{WTP}^{\mathrm{no-I}}) = \mathrm{EU}^{\mathrm{no-I}} = p^{\mathrm{no-I}}(\mathrm{inf})V(Y - L) \\ +(1 - p^{\mathrm{no-I}}(\mathrm{inf}))U(Y). \quad (2)$$

A consumer's indirect utility is a function of household income, Y, minus her willingness to pay for the food products. A consumer is assumed to perceive that the "bad state" will occur with probability p, which differs between those who have informed and uninformed prior beliefs. The posterior probability of a bad (good) outcome is determined by prior beliefs and new information on GM technologies and food, denoted by "inf" .

After normalizing utility such that $U = 1$ and $V = 0$, we simplify the representation of a consumer's expected utility:

$$\mathrm{EU}^{\mathrm{I}} = (1 - p^{\mathrm{I}}(I, \mathrm{inf})), \quad (3)$$

$$\mathrm{EU}^{\mathrm{no-I}} = (1 - p^{\mathrm{no-I}}(\mathrm{inf})). \quad (4)$$

Next, consider a consumer's determination of the subjective probability of a bad outcome on GM foods, given the following parameterization:

$$p^{\mathrm{I}} = \alpha_0 + \alpha_{\mathrm{inf}}, \quad (5)$$

$$p^{\mathrm{no-I}} = \alpha_{\mathrm{inf}}. \quad (6)$$

Now α_0 is the effect of an informed consumer's prior beliefs about GM on her posterior beliefs about GM, and α_{inf} is the effect of new information on her posterior beliefs. If prior beliefs are informative, that is, they are not diffuse priors (DeGroot, 1970), then we assume $0 < \alpha_0 < p$. We hypothesize that consumers who have uninformed prior beliefs place more weight on information from interested parties than participants who have informed priors. Also we hypothesize that the injection of third-party information affects the way that consumers use information from interested parties in placing bids.

Rearranging Eqs. (5) and (6), we obtain the following relationship between posterior probabilities:

$$p^{\mathrm{I}} = p^{\mathrm{no}-p^{\mathrm{I}}} + \alpha_0. \tag{7}$$

Substitute for p^{I} in Eq. (3), we obtain

$$\mathrm{EU}^{\mathrm{I}} = (1 - p^{\mathrm{no-I}} - \alpha_0), \tag{8}$$

and taking the difference in expected utility between Eqs. (4) and (8), we obtain

$$U(Y - \mathrm{WTP}^{\mathrm{I}}) - U(Y - \mathrm{WTP}^{\mathrm{no-I}}) = -\alpha_0. \tag{9}$$

We have shown formally that consumers who have the same indirect utility values and receive the same information treatment have differences in expected utility that are due only to *prior beliefs*. More generally, differences in expected utility are due to differences in prior beliefs and information treatments.

Willingness to pay for GM-labeled foods could be higher or lower, depending on the prior beliefs and the content of the new information they obtain/receive. Our experimental design and data allow us to test for the effects of prior beliefs under different information treatments.

2. Data

The observations and data for this paper are from a unique set of economic experiments reported in Rousu *et al.* and in Huffman *et al.* A brief overview of the design is presented here; the full design can be found in Huffman *et al.* or obtained from the authors upon request. We were interested in auctioning three rather dissimilar food products: vegetable oil, tortilla chips, and potatoes. In the refining process for vegetable oils, say from soybean oil, all of the protein is removed leaving pure lipid or fat, and the resulting oil will be chemically indistinguishable irrespective of the use of GM or non-GM soybeans to make the oil. The tortilla chips are a processed product made from yellow corn that might be contaminated by GM-traits, and consumers might have different concerns for GM content here than in russet potatoes that were genetically modified for virus resistance.

Because we use common food items available to shoppers in grocery stores and supermarkets, we wanted adults who were not primarily students to reflect better a typical household's decisions on grocery store food purchases.[4] Our participants are adults from the population of individuals 18 years of age or older and were chosen randomly from two major Midwestern metropolitan areas by a random digital dialing method. They were contacted by an independent agency to obtain their agreement to participate, given instructions of how to get to the project site, and told that they would be paid $40 for their participation.[5]

We ran two concurrent sessions for each time slot on auction days, and participants were alternately assigned to each group/session as they arrived, each group/session consisting of 13–16 individuals. Upon arrival at the lab site, participants were given an ID number, asked to complete a questionnaire containing questions about their social-economic characteristics and prior beliefs about new technologies, and paid $40. Each session or trial/session followed exactly the same protocol. Participants were first asked to engage in a round of bidding on a candy bar to learn the mechanism of the random nth-price auction. The random nth-price auction mechanism has been shown to be superior to a 2nd-price Vickery auction for eliciting consumers' entire demand curve for new goods (Shogren *et al.*, 2001). After winners were chosen, they were asked to engage in a second practice round containing a candy bar, box of pens,

[4]Although several studies have used only college undergraduates in laboratory auctions of food items (including Lusk *et al.*, 2001; Hayes *et al.*, 1995), they are not the best choice for participants when the items being auctioned are ones sold in grocery stores or supermarkets. For example, Katsaras *et al.* (2001), using a national random sample of grocery store shoppers, showed that the share of college-age (18–24 years) shoppers falls far below their share in the population, 8.5 percent of shoppers versus 12.8 percent in the U.S. Census of Population. College students obtain a large share of their food from school cafeterias and a small share from grocery stores and supermarkets compared to older shoppers (Carlson *et al.*, 1998).

[5]They were told that the study was being conducted by Iowa State University and that it involved consumer decision making on food and household products. No mention was made of experiments or GM-products.

and a deck of cards to get them familiar with placing three bids simultaneously. Winning bids were announced, and participants were given a short quiz on the auction mechanism that was followed by discussion and clarification. This was done to check up on participants' understanding of the auction mechanism.

Next, one of the six information treatments was randomly assigned and released in each session or trial. These treatments were constructed from the three basic information types defined for these experiments. They were the: (1) *industry perspective* — provided by a group of leading biotechnology companies, including Monsanto and Syngenta (Council for Biotechnology Information, 2001); (2) *environmental group perspective* — from Greenpeace, a leading environmental group or biotech antagonist (Greenpeace International, 2001a,b; Friends of the Earth, 2001, 2003); (3) *third-party perspective* — from an independent group of scientists, professionals, religious leaders, and academics, none with a financial stake in GM foods. This third type of information is to be viewed as an informed objective assessment, given the state of science, and without significant direct financial interest in genetic modification. The three information types were packaged into *six information treatments*: (1) the biotech industry perspective; (2) the environmental group perspective; (3) agricultural biotech industry and environmental perspectives; (4) agricultural biotech industry and third-party perspectives; (5) environmental group and third-party perspectives; or (6) all three perspectives. When a trial/session received pro- and anti-biotech information, the order was randomized among the participants. Third-party information, however, was always distributed last.

Next, they were asked to view the three experimental foods (vegetable oil made from soybeans, tortilla chips made from yellow corn, and russet potatoes) and to place their bids. They were then asked to view another set of these three goods, having different GM-food-label information, and place a new set of bids.[6] In one

[6]Participants were never told how many total rounds of bidding they would engage in. All participants engaged in a total of four rounds of bidding including the two practice rounds. At the start of the auction of the food products that might be GM, the participants were told that they would be expected to purchase a maximum of one

Table 1. Characteristics of auction participants ($N = 172$).

Variable	Definition	Mean	S.D.
Gender	1 if female	0.62	0.49
Age	The Participant's age	49.5	17.5
Married	1 if the individual is married	0.67	0.47
Education	Years of schooling	14.54	2.25
Household	Number of people in participant's household	2.78	1.65
Income	The household's income level (in thousands)	57.0	32.6
White	1 if participant is white	0.90	0.30

round participants were bidding on food products that were labeled as genetically modified (GM) and in the other round the food products had a plain-label.[7] Among the two rounds of bidding on experimental goods, one was chosen randomly by the auction monitor to be a binding round in which winners were expected to complete their exchange of money for experimental goods. The actual exchange took place in an adjacent room.[8]

Table 1 summarizes the demographic characteristics of participants. Although our participants are slightly skewed toward women, Katsaras *et al.* showed that women make up a disproportional share of grocery shoppers: 83 percent of shoppers versus 52 percent in the U.S. Census of Population. Although the demographics of the sample do not perfectly match the population reported by the U.S. census demographic characteristics for these regions, they are similar and provide a sufficient representation for our initial probe into labeling and information for GM products (see Appendix A, available on the JEBO website, for the demographic characteristics of the areas).

The pre-auction questionnaire contained the question: "Regarding genetically modified foods: How informed do you consider yourself?" The participants were offered the following six options: extremely

unit of each auctioned commodity. After the second round of bidding on food products that might be GM, the auction part of the experiment ended.

[7]The sequence was determined randomly.

[8]Contrary to most economics experiments our participants engage in very few rounds of bidding, which reduces behavior modification associated with the experience of participating in our experiment.

Table 2. Response of participants to the pre-experiment question of "How informed are you about genetic modification?" ($N = 172$).

Category	Relative Frequency (percent)
Extremely well informed	3.5
Well informed	5.8
Somewhat informed	32.6
Not very informed	40.1
Not informed at all	15.7
I do not know	2.3

Source: Pre-auction questionnaires.

well informed, well informed, somewhat informed, not very informed, not informed at all, and I do not know. In particular, this information was collected before the start of the lab auction of GM-foods and release of new information.[9] Table 2 summarizes these responses and shows that 41.9 percent of the 172 participants were at least somewhat informed about genetic modification before our auction experiments. The other 58.1 percent were classified as "uninformed."

3. Results

Sample mean information for average bid prices for GM- and plain-labeled foods for all participants and for participants who had informed and uninformed priors are reported in Table 3.[10] Among all participants and commodities, average bids were 14 percent lower for GM than for plain-labeled commodities (part A). Among participants who had informed prior beliefs about GM foods, the average difference for the bid price of a plain-labeled product less the

[9]We made no effort to test participant's beliefs for objectiveness; they are subjective beliefs.

[10]We tested whether there were differences in the other demographic characteristics of those who perceived themselves as informed and uninformed, such as age and gender, and there were no differences between these groups that were statistically significant at the 10-percent level.

Table 3. Mean bids for participants by commodity and type.

	Mean Bid	S.D.	Median	Minimum	Maximum
(A) Mean bids–all participants and treatments ($N = 172$)					
GM OIL	0.91	0.84	0.75	0	3.99
OIL	1.05	0.85	1.00	0	3.79
GM CHIPS	0.93	0.86	0.70	0	3.99
CHIPS	1.08	0.85	0.99	0	4.99
GM POTATOES	0.78	0.67	0.69	0	3
POTATOES	0.91	0.67	0.80	0	3.89
(B) Mean bids for participants who had informed prior beliefs about genetic modification ($N = 72$)					
GM OIL	0.93	0.88	0.77	0	3.99
OIL	1.10	0.89	1.00	0	3.79
GM CHIPS	0.86	0.81	0.75	0	3.50
CHIPS	1.05	0.74	1.00	0	2.99
GM POTATOES	0.73	0.61	0.75	0	2.30
POTATOES	0.92	0.59	0.88	0	2.00
(C) Mean bids for participants who had uninformed prior beliefs about genetic modification ($N = 100$)					
GM OIL	0.90	0.82	0.68	0	3.25
OIL	1.01	0.82	0.99	0	3.29
GM CHIPS	0.97	0.90	0.69	0	3.99
CHIPS	1.10	0.92	0.99	0	4.99
GM POTATOES	0.81	0.72	0.60	0	3.00
POTATOES	0.90	0.73	0.75	0	3.89

bid price for GM-labeled counterpart was 18 percent (part B). Among participants who had uninformed prior beliefs, the average difference for the bid price of a plain-labeled product less the bid price for the GM-labeled counterpart was 11 percent (part C). Across the two groups, informed and uninformed, the difference is 49 percent. The difference between these two percentages is not significantly different from zero at the 5-percent level.

What is the impact of subjective prior beliefs about genetic modification in a market with new information from interested parties? Table 4 summarizes the mean differences in bid prices from participants for GM- and plain-labeled food products due to the

Table 4. Mean difference in bid prices of participants for plain-labeled less bid price for GM-labeled food products due to information from interested parties, conditional on prior beliefs.

Information Treatments	Vegetable Oil	Tortilla Chips	Potatoes
(A) Participants who did not receive third-party GM-information			
Received only pro-biotechnology information, given informed priors ($N = 13$)	−\$0.10, $p = 0.63$	\$0.11, $p = 0.32$	\$0.02, $p = 0.93$
Received only pro-biotechnology GM-information, given uninformed priors ($N = 17$)	\$0.03, $p = 0.68$	\$0.00, $p = 0.94$	−\$0.08, $p = 0.13$
Received only anti-biotechnology GM-information, given informed priors ($N = 8$)	\$0.50, $p = 0.12$	\$0.61*, $p = 0.09$	\$0.52**, $p = 0.05$
Received only anti-biotechnology GM-information, given uninformed priors ($N = 21$)	\$0.34***, $p < 0.01$	\$0.37***, $p < 0.01$	\$0.32***, $p < 0.01$
Received pro-biotechnology and anti-biotechnology GM-information, given informed priors ($N = 10$)	\$0.51, $p = 0.12$	\$0.29*, $p = 0.08$	\$0.40**, $p = 0.05$
Received pro-biotechnology and anti-biotechnology GM-information, given uninformed priors ($N = 18$)	\$0.00, $p = 0.97$	\$0.01, $p = 0.90$	\$0.07, $p = 0.24$

Table 4. (*Continued*)

Information Treatments	Vegetable Oil	Tortilla Chips	Potatoes
(B) Participants who received third-party GM-information			
Received pro-biotechnology and third-party GM-information, given informed priors ($N = 14$)	\$0.13, $p = 0.17$	\$0.09, $p = 0.52$	\$0.07, $p = 0.28$
Received pro-biotechnology and third-party GM-information, given uninformed priors ($N = 14$)	−\$0.09, $p = 0.26$	\$0.07, $p = 0.43$	−\$0.05, $p = 0.38$
Received anti-biotechnology and third-party GM-information, given informed priors ($N = 13$)	\$0.07, $p = 0.17$	\$0.10, $p = 0.55$	\$0.20*, $p = 0.08$
Received anti-biotechnology and third-party GM-information, given uninformed priors ($N = 16$)	\$0.29, $p = 0.21$	\$0.33*, $p = 0.10$	\$0.26*, $p = 0.08$
Received pro-biotechnology, anti-biotechnology and third-party GM-information, given informed priors ($N = 14$)	\$0.16*, $p = 0.08$	\$0.10*, $p = 0.07$	\$0.10, $p = 0.15$
Received pro-biotechnology, anti-biotechnology and third-party GM-information, given uninformed priors ($N = 14$)	\$0.03, $p = 0.77$	−\$0.08, $p = 0.28$	−\$0.06, $p = 0.42$

*Statistically significant at the 10-percent level using a two-sided t-test.
**Statistically significant at the 5-percent level using a two-sided t-test.
***Statistically significant at the 1-percent level using a two-sided t-test.

release of information from interested parties, the agricultural biotech industry or environmental NGOs, given prior beliefs.

Part A examines the bid prices in a market without third-party information. When participants — both those who had informed and uninformed prior beliefs — received only the industry perspective, the mean bid price differences between GM- and plain-labeled products were very small, less than plus or minus 11 cents per product. For participants who had informed prior beliefs about genetic modification and received only the environmental group perspective, the mean bid price differences were 50–60 cents per product, but only the 52 cent difference for potatoes was significantly different from zero at the 5-percent level. For participants who had uninformed prior beliefs, mean bid price differences were less, 32–37 cents per product, and all are significantly different from zero at the 5-percent level. For participants who had informed priors and received both the agricultural biotech and environmental group perspectives, the discount for GM products is an average of 29–51 cents per product, but the 40 cent differential for potatoes is the only one of these three differences that is significantly different from zero at the 5-percent level. In contrast, those participants who had uninformed prior beliefs did not discount GM-products.

Part B of Table 4 examines bids from consumers who received third-party information. When participants, both with informed and uninformed prior beliefs, received only pro-biotech information followed by third-party information, the mean bid-price difference between GM- and plain-labeled food items was small and similar to those reported in part A. When participants had informed priors and received anti-biotech information followed by third-party information, mean bid price differences were 7–20 cents per food item, which is much smaller than the outcome without third-party information (as reported in part A). The most significant of these three outcomes is again for potatoes, but now the bid price difference is only significant at the 10-percent level. When those who had uninformed priors receive this same set of information treatments, the bid price differences are similar to those in part A in which third-party information was excluded. With third-party information, the differences are at best significant at the 10-percent level. When participants who had informed priors received all three types of new information, their mean bid price differences were only 10–16 cents per product, which is much lower than the outcome when third-party information was not injected (as reported in part A). Two of these differences are different from zero at only the 10-percent level. When participants who had uninformed priors received all three types of information, however, their mean price differences were similar to those reported when third-party information was not injected. Bidding behavior was affected by prior beliefs, new information from interested parties, and by injection of third-party information.

In Table 5, we report the results for statistical tests of bid price differences for plain and GM-labeled products, conditional on participant's prior beliefs about genetic modification. Part A reports the results for statistical tests for participants who did not receive third-party information. This is a "difference-in-differences" test (see Wooldridge, 2002, pp. 283–291, 128–131). We report results from tests of the null hypothesis that the difference-in-mean bid-price difference for plain labeled and GM-labeled food items is zero across participants who had informed and uninformed prior beliefs. When the participants received only the agricultural biotech industry perspective or the environmental group perspective, no significant difference-in-differences existed at the 5-percent significance level.[11] If they received the pro-biotech and anti-biotech perspectives, the estimate of the difference-in-differences estimator was significantly different from zero for potatoes (5-percent level) and tortilla chips (6-percent level). The estimate of the difference-in-differences estimator, however, was not significantly different for vegetable oil (p-value of 8 percent).[12]

Part B of Table 5 reports additional difference-in-differences results for a situation in which third-party information was injected.

[11]We use a consistent method for labeling the construction of differences-in-differences, but we do not specifically examine the signs of these differences-in-differences estimates, but focus instead on their statistical significance.

[12]Wilcoxon Rank-Sum tests were also conducted and provided similar results.

Table 5. Do the informed and uninformed bid differently?

Information Treatments	Vegetable Oil	Tortilla Chips	Potatoes
(A) Participants who did not receive third-party information			
Received pro-biotechnology GM-information only ($N = 30$)	$t = 0.67$, $p = 0.51$	$t = -0.92$, $p = 0.37$	$t = -0.55$, $p = 0.59$
Received anti-biotechnology GM-information only ($N = 29$)	$t = -0.61$, $p = 0.54$	$t = -0.92$, $p = 0.36$	$t = -0.99$, $p = 0.33$
Received both pro-biotechnology and anti-biotechnology GM-information ($N = 28$)	$t = -1.83$, $p = 0.08$	$t = -1.94$, $p = 0.06$	$t = -2.22$, $p = 0.04$
(B) Participants who received third-party information			
Received pro-biotechnology and third-party GM-information ($N = 28$)	$t = -1.85$, $p = 0.08$	$t = -0.15$, $p = 0.89$	$t = -1.45$, $p = 0.16$
Received anti-biotechnology and third-party GM-information ($N = 29$)	$t = 0.87$, $p = 0.39$	$t = 0.92$, $p = 0.36$	$t = 0.36$, $p = 0.71$
Received pro-biotechnology, anti-biotechnology, and third-party GM-information ($N = 28$)	$t = -0.93$, $p = 0.36$	$t = -2.00$, $p = 0.06$	$t = -1.66$, $p = 0.11$

Test of null hypothesis that difference in mean bid price differences for plain-labeled less bid price for GM-labeled food items is zero for participants who had informed and uninformed prior beliefs.

The information treatments are: (i) pro-biotech and third-party, (ii) anti-biotech and third-party, and (iii) all three types of information. None of the estimates of the difference-in-difference estimator for these tests is significantly different from zero at the 5-percent level. The third-party information lowers the relative weight placed on information received from interested parties when bids are placed on food items that might be genetically modified. Furthermore, when third-party information is injected into the experiment, prior beliefs about genetic modification have relatively little impact on bidding behavior of participants. Both participants who had informed and uninformed prior beliefs behaved in our experiments as if they took the third-party information seriously.

Table 6. Results for difference-in-differences estimator.

Information Treatments	Vegetable Oil	Tortilla Chips	Potatoes
(A) Participants whose prior beliefs were uninformed about genetic modification			
Received only pro-biotechnology GM-information vs. received only anti-biotechnology GM information	$t = 2.11$, $p = 0.04$	$t = 2.71$, $p = 0.01$	$t = 3.53$, $p = 0.00$
Received only pro-biotechnology GM-information vs. received pro-biotechnology and anti-biotechnology GM-information	$t = 0.20$, $p = 0.84$	$t = -0.04$, $p = 0.97$	$t = -1.96$, $p = 0.06$
Received only anti-biotechnology GM-information vs. received pro-biotechnology and anti-biotechnology GM-information	$t = 2.03$, $p = 0.05$	$t = 2.68$, $p = 0.01$	$t = 2.26$, $p = 0.03$
(B) Participants who had informed prior beliefs about genetic modification			
Received only pro-biotechnology GM-information vs. received only anti-biotechnology GM-information	$t = 1.78$, $p = 0.09$	$t = 1.76$, $p = 0.09$	$t = 1.72$, $p = 0.10$
Received only pro-biotechnology GM-information vs. received pro-biotechnology and anti-biotechnology GM-information	$t = -1.76$, $p = 0.09$	$t = -0.96$, $p = 0.35$	$t = -1.45$, $p = 0.16$
Received only anti-biotechnology GM-information vs. received pro-biotechnology and anti-biotechnology GM-information	$t = -0.02$, $p = 0.98$	$t = 0.98$, $p = 0.34$	$t = 0.43$, $p = 0.68$

Null hypothesis that difference in bid prices for plain-labeled less bid price for GM-labeled food items under different information treatments is zero, given participants' prior beliefs (no third-party information).

Next we test for differences in bid prices due to information treatment effects for people who received third-party information, conditional on prior beliefs. Table 6 reports this new round of difference-in-differences tests. The null hypothesis is that the difference in mean bid prices for plain-labeled and GM-labeled food items under two different information treatments is zero, holding prior beliefs of participants constant. For participants who held uninformed prior beliefs, the estimate of difference-in-differences estimator is significantly different from zero (5-percent level) for all three food items when participants receive pro- versus anti-biotech information. They are also significantly different from zero when participants receive anti-biotech information versus pro-biotech and anti-biotech information. In fact these two sets of hypotheses produced t-values that look very similar (part A). When participants received pro- versus pro- and anti-biotech information the estimate of the difference-in-difference estimator was not significantly different from zero. When participants had informed priors, none of the estimates of the differences-in-differences estimator was significantly different from zero at the 5-percent level (part B). Clearly, prior beliefs affect the relative weight placed on new information from informed parties when third-party information is unavailable.

Now we examine the effects of different combinations of information from interested parties when third-party information is injected, conditional on prior beliefs. This is another test employing the difference-in-differences estimator. For participants who had uninformed priors, three of nine tests in Table 7, part A, for the estimate of the difference-in-differences estimator are rejected at the 6-percent level: for potatoes when participants received only pro-biotech information versus received only anti-biotech, and for tortilla chips and potatoes when participants received only anti-biotech information versus pro- and anti-biotech information. For participants having informed prior beliefs (part B), none of the coefficients of the difference-in-differences estimator is significantly different from zero at the 6-percent level. With the injection of third-party information, we continue to find that participants who have uninformed prior beliefs bid differently on plain-labeled versus GM-labeled food products than for those who had informed prior beliefs. Those who are uninformed behave as if they place greater weight on new information than the informed even when it comes from interested parties.

Individuals who had informed prior beliefs about genetic modification coming into our experiments discounted GM-labeled food

Table 7. Results for differences-in-differences estimator when participants received third-party GM-information.

Information Treatments	Vegetable Oil	Tortilla Chips	Potatoes
(A) Participants who had uninformed prior beliefs about genetic modification			
Bid prices when received pro-biotechnology and verifiable GM-information vs. bids when received anti-biotechnology and third-party GM-information	$t = 1.53$, $p = 0.14$	$t = 1.22$, $p = 0.23$	$t = 1.99$, $p = 0.06$
Bids when received pro-biotechnology and verifiable GM-information vs. bids when received pro-biotechnology, anti-biotechnology, and third-party GM-information	$t = -0.93$, $p = 0.36$	$t = 1.37$, $p = 0.18$	$t = 0.16$, $p = 0.87$
Bids when received anti-biotechnology and verifiable GM-information vs. bids when received pro-biotechnology, anti-biotechnology, and third-party GM-information	$t = 0.99$, $p = 0.33$	$t = 1.95$, $p = 0.06$	$t = 1.98$, $p = 0.06$
(B) Participants who had informed prior beliefs about genetic modification			
Bids when received pro-biotechnology and verifiable GM-information vs. bids when received anti-biotechnology and third-party GM-information	$t = -0.55$, $p = 0.58$	$t = 0.03$, $p = 0.97$	$t = 1.13$, $p = 0.27$
Bids when received pro-biotechnology and verifiable GM-information vs. bids when received pro-biotechnology, anti-biotechnology, and third-party GM-information	$t = -0.25$, $p = 0.80$	$t = -0.03$, $p = 0.98$	$t = -0.42$, $p = 0.68$
Bids when received anti-biotechnology and verifiable GM-information vs. bids when received pro-biotechnology, anti-biotechnology, and third-party GM-information	$t = -0.89$, $p = 0.38$	$t = 0.02$, $p = 0.98$	$t = 0.78$, $p = 0.44$

Null hypothesis that difference in bid price for plain-labeled less bid price for GM-labeled food items under different information treatments is zero, conditional on participants' prior beliefs.

products more highly than those participants having uninformed prior beliefs. This behavior suggested that their prior information was somewhat negative. Participants who had uninformed prior beliefs about genetic modification before our experiments exhibited the greatest change in bidding behavior due to the injection of new information. Information from interested parties caused a significant change in their bidding behavior, and when third-party information was injected these outcomes were further modified. Those who have informed prior beliefs were relatively unaffected by all types of new information. They may already have made a significant investment in information about genetic modification, and the marginal impact of new information including third-party information was small and not statistically significant.

4. Discussion and Conclusions

This paper has shown that prior beliefs and new information (both from interested parties and third-party sources) affect bidding behavior of people who participated in our auction market experiments for food items that might be genetically modified. This contradicts the earlier findings of Viscusi (1997) and Tversky and Kahneman (1974, 1992) who argued that people frequently ignore base rates. One potential explanation for this difference of outcome is that instead of measuring prior beliefs as objective knowledge (e.g., monetary lotteries), we asked lab participants to give us information about their prior knowledge about genetic modification. They were asked the following question: "How informed are you about genetic modification?" and they were given five options: extremely well informed, well informed, somewhat informed, not very informed, not informed at all. This information is subjective, and we use this as their subjective prior belief. Furthermore, we examined their use of prior beliefs and new information to inform decision on willingness to pay for common food items available in grocery stores and supermarkets and not in a lottery.

These results have implications for information policies by showing how both skeptics and proponents of new technologies (i.e., interested parties) might try to manage information to achieve private objectives. This is most likely to occur when much is unknown scientifically about the impacts of new technologies or when third-party information is limited or unavailable (Milgrom and Roberts, 1986; Huffman and Tegene, 2002). Opponents to a new technology may try to target people who are relatively uninformed about the technology. Proponents of the technology may try to target people who have informative prior beliefs for maximum effectiveness. This reasoning might explain why the Council for Biotechnology Education (a pro-GM organization) funds TV commercials during family and sports programming but the anti-GM groups invest in carrying out dramatic and spectacular shows of opposition for new technologies (e.g., with colorful, vocal demonstrations) that may be carried by a variety of media outlets. People who are uninformed seem less likely to use common media sources regularly.

In this study we have used a relatively blunt measure of prior beliefs about genetic modification; future research might examine the objectiveness of these beliefs and attempt to illicit strengths of pro-biotechnology and anti-biotechnology prior beliefs and the way that they affect willingness to pay.

Acknowledgements

The authors acknowledge comments from participants at a "Big Problems; Big Topics" lecture at the University of Chicago, Economics Department at Rensselaer Polytechnic Institute, and to David Grether, former co-editor JEBO. Thanks to Wayne Fuller and Phil Dixon for developing the statistical experimental design, to Daniel Monchuk and Terrance Hurley for their generous help in conducting the auctions, and to Monsanto for providing some of the products used in the experiment. This work was supported through a grant from the Cooperative State Research, Education, and Extension Service, U.S. Department of Agriculture, under Agreement 00-52100-9617 and from U.S. Department of Agriculture, Economic Research Service, under Agreement 43-3AEL-8-80125. Views expressed are those of the authors, and not necessarily those of ERS or the U.S. Department of Agriculture.

Appendix A. Supplementary Data

Supplementary data associated with this article can be found, in the online version, at doi:10.1016/j.jebo.2005.04.019.

References

Akerlof, G.A. (1970). The Market for 'Lemons': Quality Uncertainty and the Market Mechanism. *The Quarterly Journal of Economics* 84, 488–500.

Carlson, A., J. Kinsey and C. Nadav (1998). Who eats what, when, and from where? University of Minnesota, St. Paul, Retail Food Industry Center, Working Paper 98-05.

Chern, W.D. and K. Rickertsen (2004). "A Comparative Analysis of Consumer Acceptance of GM Foods in Norway and in the USA." In R. Evenson and V. Santaniello (Eds.), *Consumer Acceptance of Genetically Modified Foods*, pp. 95–110. Cambridge, MA: CABI.

Council for Biotechnology Information (2001). *Frequently Asked Questions.* http://my.webmd.com/en/faq/default.asp?MID=10.

DeGroot, M.H. (1970). *Optimal Statistical Decision.* New York, NY: McGraw-Hill.

Diamond, P.A. and J.A. Hausman (1994). Contingent Valuation: is Some Number Better Than no Number? *Journal of Economic Perspectives* 8, 45–64.

FAO (2004). *The State of Food and Agriculture: Agricultural Biotechnology, 2003–2004.* Rome, Italy: FAO.

Fernandez-Cornejo, J. and W.D. McBride (2002). Adoption of bioengineered crops. USDA, ERS, Agricultural Economics Report no. 810.

Fox, J.A., J.F. Shogren, D.J. Hayes and J.B. Kliebenstein (1998). CVM-X: Calibrating Contingent Values with Experimental Auction Markets. *American Journal of Agricultural Economics* 80, 455–465.

Friends of the Earth (2001). The need for labeling genetically engineered foods. http://www.foe.org/safefood/factshtgelabel.htm.

Friends of the Earth (2003). Environmental hazards of genetically engineered plants. http://www.foe.org/camps/comm/safefood/gefood/factsheets/envir ohazfacts.html.

Greenpeace International (2001a). Public concern. http://www.greenpeace.org/ %7Egeneng/reports/food/intrfo07.htm.

Greenpeace International (2001b). The potential for allergic reactions. http:// www.greenpeace.org/%7Egeneng/reports/food/intrfo11.htm.

Grether, D.M. (1980). Bayes Rule as a descriptive model: the representativeness Heuristic. *The Quarterly Journal of Economics* 95, 537–557.

Hayes, D.J., J.F. Shogren, S.Y. Shin and J.B. Kliebenstein (1995). Valuing Food Safety in Experimental Auction Markets. *American Journal of Agricultural Economics* 77, 40–53.

Hirshleifer, J. and J.R. Riley (1992). *The Analytics of Uncertainty and Information.* New York, NY: Cambridge University Press.

Huffman, W.E. (2001). "Human capital: education and agriculture." In B.L. Gardner and G.C. Rousser (Eds.), *Handbook of Agricultural Economics*, vol. A, pp. 334–381. Amsterdam, The Netherlands: Elsevier Science/North-Holland.

Huffman, W.E., J.F. Shogren, M. Rousu and A. Tegene (2003). Consumer Willingness to Pay for Genetically Modified Food Labels in a Market with Diverse Information: Evidence from Experimental Auctions. *Journal of Agricultural and Resource Economics* 28, 481–502.

Huffman, W.E. and A. Tegene (2002). "Public acceptance of and benefits from agricultural biotechnology: a key role for verifiable information." In V. Santaniello, R.E. Evenson and D. Zilberman (Eds.), *Market Development for Genetically Modified Food*, pp. 179–190. CAB International, City.

Johnston, R.J., C.R. Wessells, H. Donath and F. Asche (2001). Measuring Consumer Preferences for Ecolabeled Seafood: An International Comparison. *Journal of Agricultural and Resource Economics* 26, 20–39.

Kahneman, D. (2003). Maps of Bounded Rationality: Psychology for Behavioral Economics. *American Economic Review* 93, 1449–1475.

Katsaras, N., P. Wolfson, J. Kinsey and B. Senauer (2001). Data mining: a segmentation analysis of U.S. grocery shoppers. University of Minnesota, St. Paul, The Retail Food Industry Center, Working Paper 01-01.

Kivi, P.A. and J.F. Shogren (2005). Ambiguity in food safety valuation. University of Wyoming, Working Paper.

Lusk, J.L., M.S. Daniel, D. Mark and C.L. Lusk (2001). Alternative Calibration and Auction Institutions for Predicting Consumer Willingness to Pay for Nongenetically Modified Corn Chips. *Journal of Agricultural and Resource Economics* 26, 40–57.

Milgrom, P. and J. Roberts (1986). Relying on the Information of Interested Parties. *RAND Journal of Economics* 17, 18–32.

Molho, I. (1997). *The Economics of Information.* Malden, MA: Blackwell Publishing Inc.

Morris, S. and H.S. Shin (2002). Social Value of Public Information. *The American Economic Review* 92, 1521–1534.

Rousu, M., W.E. Huffman, J.F. Shogren and A. Tegene (2004). The value of verifiable information in a controversial market: evidence from lab auctions of GM foods. Iowa State University, Department of Economics, Staff Working Paper Series #344.

Schultz, T.W. (1975). The Value of the Ability to Deal with Disequilibria. *Journal of Economic Literature* 13, 827–846.

Shogren, J.F., M. Margolis, C. Koo and J.A. List (2001). A Random nth-Price Auction. *Journal of Economic Behavior and Organization* 46, 409–421.

Stigler, G.J. (1961). The Economics of Information. *Journal of Political Economy* 69, 213–225.

Tversky, A. and D. Kahneman (1974). Judgment Under Uncertainty: Heuristics and Biases. *Science* 185, 1124–1131.

Tversky, A. and D. Kahneman (1981). The Framing of Decisions and the Psychology of Choice. *Science* 211, 453–458.

Tversky, A. and D. Kahneman (1992). Advances in Prospect Theory: Cumulative Representation of Uncertainty. *Journal of Risk and Uncertainty* 5, 297–323.

Tirole, J. (2003). *The Theory of Industrial Organization.* Cambridge, MA: The MIT Press.

Viscusi, W.K. (1997). Alarmist Decisions with Divergent Risk Information. *The Economic Journal* 107, 1657–1670.

Wooldridge, J.M. (2002). *Econometric Analysis of Cross Section and Panel Data.* Cambridge, MA: MIT Press.

Chapter 10

Willingness-to-pay for Natural, Organic, and Conventional Foods: The Effects of Information and Meaningful Labels*,†

Jonathan R. McFadden[‡,**] and Wallace E. Huffman[§,¶,||]

[‡]*Economic Research Service, U.S. Department of Agriculture, 1400 Independence Ave., SW Mail Stop 1800, Washington, DC 20250-1800, United States*

[§]*Department of Economics, Iowa State University, 460E Heady Hall, Ames, IA 50011-1054, United States*

[¶]*whuffman@iastate.edu*

Vague food labels and distorted product claims have persisted in the "natural" food industry, while organic claims can be certified by the U.S. Department of Agriculture (USDA). Using experimental methods and a sample of randomly selected subjects, we test food label and information treatment effects on subjects' willingness-to-pay (WTP) for organic, "natural," and conventional foods. Random nth-price auctions are used to elicit WTP after subjects received one of five randomly assigned information treatments. We find large information effects, including asymmetric cross-market effects for natural and organic foods. Perhaps surprising is that organic premiums increase in response to subjects seeing the "natural" foods industry's perspective on its products. Demographics effects are also important. The results have practical implications for natural and organic food marketing and valuing products where there are vague claims about their attributes.

---

*This article originally appeared in **Food Policy (2017), 68, 214–232**. This is published by permission of **Elsevier**.

†The views expressed are those of the authors and should not be attributed to the Economic Research Service or USDA.

||Corresponding author.

**The experiment was completed while at Iowa State University.

Keywords: Experimental auctions, Organic foods, "Natural" foods, Information effects, Food labels, Random nth-price auction

1. Introduction

Organic farming and sales of organic foods and food products in the U.S. have grown steadily since the Organic Food Production Act (OFPA) of 1990 and implementation of the National Organic Program (NOP) in 2002. U.S. organic food sales have increased more than 160% from $13.3 billion in 2005 to an estimated $35 billion in 2014 (USDA-ERS, 2016). In recent years, roughly 9,000 to 14,000 U.S. certified organic operations have farmed 3.6 to 4.1 million acres (USDA-NASS, 2010; USDA-NASS, 2012). Domestic supply has responded to broad-based demand increases from consumers (e.g., farmers' markets, stands, and community-supported agriculture), wholesalers, and retailers (e.g., grocery stores and restaurants).

The growth of organic agriculture can be partially attributed to NOP's clear guidelines and certification procedures. Organic products cannot be made with conventional pesticides and fertilizers, herbicides, biotechnologies, irradiation, and sewage sludge, though there is a list of approved synthetic substances. Animals must have access to the outdoors and cannot receive growth hormones or antibiotics. Foods that have been organically produced and contain at least 95% organic ingredients may display the USDA organic seal (National Organic Program, 2002).

This contrasts greatly with "all natural" or "100% natural" food products (hereafter termed "natural" foods), which lack a set of guidelines or legally-binding regulations. The USDA's Food Safety and Inspection Service (USDA-FSIS), which regulates meat and poultry, implemented an informal policy on "natural" foods labeling starting in 1982. Its current "natural" informal policy for meat and poultry disallows artificial flavoring, coloring ingredients, chemical preservatives, and processing above minimal levels (USDA-FSIS, 2005). The Food and Drug Administration (FDA) regulates the majority of U.S. food products and has maintained a similar informal policy since 1991. Although it does not have a stringent definition for "natural" foods, FDA has considered "natural" to mean the exclusion of artificial or synthetic substances that would not normally be expected to be there (US-HHS-FDA, 1991).

Recent requests for legal "natural" standards from various interests, including Hormel Foods, Sara Lee Corporation, and The Sugar Association, have highlighted the adverse market consequences in the absence of a uniform and unambiguous policy. Differing interpretations of "natural" have contributed to increases in costly lawsuits against producers (Negowetti, 2013). There continues to be substantial misunderstanding among consumers about the differences (real or perceived) between organic and "natural" foods (Anstine, 2007). Indeed, FDA has recently filed a formal request for information and comments on use of the term "natural" in food product labeling (US-HHS-FDA, 2015). There is interest, inter alia, about the extent to which consumers are confusing or conflating "natural" foods with "organic" or "healthy" foods. The comment period has ended, and FDA is currently determining the appropriateness of a "natural" definition and appropriate use in food labeling.

Some of consumers' misunderstanding regarding the various attributes of "natural" and organic foods may result from differing or opposed perspectives contained in information sources. For example, profit-maximizing firms in the "natural" foods and organic foods industries provide information that promotes their products, while certain scientists, regulatory agencies, and public health officials may take a more independent and impartial view. Importantly, the provision of independent, third-party, verifiable information can have substantial effects on buyers' decisions (Milgrom and Roberts, 1986; Rousu *et al.*, 2007). Efficient decision making may hinge on a high degree of competition among information suppliers and skepticism on the part of the consumer. Given expanding consumer interest in "natural" and organic foods, an interesting economic question arises: how do individuals' willingness-to-pay (WTP) for these types of foods respond to industry information or both industry and independent, third party information?

Eliminating these sources of misunderstanding and asymmetric information would lead to better-informed choices, reduce suboptimal consumption, and decrease legal risks of production. This would likely increase welfare in organic, "natural," and other food markets

(Hamilton and Zilberman, 2006). The best methods for introducing higher quality information and improving access to such information is unknown, in part because the industry is much smaller than conventional agriculture. Though the majority of U.S. consumers have some experience with these foods, organic sales comprised roughly five percent of total U.S. food sales in 2015, with producers bearing relatively high production costs (Organic Trade Association, 2016; Greene *et al.*, 2009; Klonsky, 2012; McBride *et al.*, 2015). Expansion of the "natural" and organic sector hinges on inducing more consumers to pay price premiums over conventional products, e.g., influencing purchase decisions by increasing WTP.

The objective of this study is to test for food label and information treatment effects reflected in real consumers' WTP for organic, "natural," and conventional foods. To be able to do this, we collect unique data from lab experiments performed with a random sample of 102 adults aged 18–65 years in a Midwestern city during April 2013. Random nth-price auctions were used to elicit each subject's WTP because of its superior performance in practice and its incentive compatibility properties, i.e., the bidder's dominant strategy is to truthfully reveal her valuation. The experimental, labeled commodities used to assess WTP are Red Delicious apples, fresh broccoli, and large brown eggs. We chose these commodities because they are common and regularly-consumed fresh products available across the three types and are sufficiently different to capture subjects' heterogeneity in tastes. Each subject is randomly assigned to one of five information treatments: (i) no information baseline, (ii) "natural" food industry perspective, (iii) "natural" food industry perspective plus independent perspective on "natural" foods, (iv) organic food industry perspective, and (v) organic food industry perspective plus independent perspective on organic foods.

Our study is unique in that it focuses on a range of food products (fruits, vegetables, and eggs) across the conventional, "natural," and organic types. We expand the literature on information, labeling, and valuation of "natural" and organic foods with several contributions. First, our value of information estimates show that subjects' WTP is significantly affected by information treatments. Relative to those who are uninformed, subjects receiving the organic industry perspective plus independent perspective on organics have lower organic premiums, while those receiving the "natural" food industry perspective have higher premiums. This suggests there could be sizeable economic benefits for clearing up misconceptions about these foods with more accurate information and labeling. Second, we find that subjects' demographics have broader and more important effects than previously thought, especially regarding children in the household. Our results reflect the fact that the "natural" and organics industries are now reaching a broader customer base, with customers who cannot be adequately characterized by a handful of demographics. Third, we confirm that subjects are willing to pay premiums for "natural" foods over conventional, as well as for organic foods over "natural." Valuations are increasing in perceived wholesomeness and have remained generally high despite recent growth in the industry.

The remainder of the paper is organized as follows. Section 2 surveys the literature on valuation of "natural" and organic foods, with an emphasis on economics experiments. Section 3 describes the experimental design, while section 4 previews the econometric model and reviews the sample data. Section 5 presents the regression results and value of information estimates, and discusses economic implications for the food industry and organic agriculture. Section 6 concludes. Appendix A contains a table providing counts and proportions of zero bids, in addition to the four information perspectives used in the experiment.

2. Related Research

Early research on the modern "natural" and organics sector focused on growth in response to concerns about pesticide-risk reductions, and sought to identify setbacks related to market expansion, primarily on wholesalers, distribution channels, and consumer demand determinants (Jolly *et al.*, 1989; Misra *et al.*, 1991; Huang, 1996).[1] Park and Lohr (1996) document several sources of supply uncertainty

[1]See Govindasamy and Italia (1999) for a useful discussion about the early conflicting results surrounding the effects of consumer demographics on purchase intentions and WTP for pesticide-free and organic food products.

as standard explanations for the small share of organics in total agricultural output. However, their system of supply and demand estimates for organic produce indicate that demand factors are more important for market growth and recommend certain measures to increase demand, e.g., advertising targeted to high-income shoppers. Identifying the demographics of "natural" and organics grocery shoppers was particularly challenging before more recent industry expansion. Much of this research is analysis of survey data from private groups or academic researchers for foods in various locations. The primary shopper's age, education, income, marital status, and number and age of children are significant variables, but there have been conflicting findings (Thompson, 1998).

Additional research has centered on consumer demographics, implicitly recognizing the importance of demand considerations. Many studies in the literature analyze demand determinants and price premiums for organic foods using Nielsen Homescan datasets for years 2003–2006.[2] This line of inquiry shows that organic purchases and premiums vary by food product (e.g., fresh fruits and vegetables, meat and dairy, packaged foods), region, seasonality, store type (e.g., traditional grocery store, specialty food store, supercenter, or warehouse club), household attributes, and particular year in which the scanner data are used for analysis.

Using 2003 Nielsen data, Zhang *et al.* (2008) show that urban households with higher incomes and a household head with higher education have relatively more purchases of organic produce, though effects of young children and household size are negligible. In 2004, shoppers paid premiums of 7.4–17.3% for organic tomatoes over conventional tomatoes, with marginal WTP for packaged tomatoes ranging in \$0.26–1.02/lb (Huang and Lin, 2007). In the same year, Dettmann and Dimitri (2009) find that the main consumers of

[2]The Nielsen Homescan data comprise a stratified, random sample of several thousand representative U.S. households. After these households buy groceries, they use their home scanner to enter their purchases into the Nielsen database using the products' Universal Purchasing Codes (UPC). Random-weight produce without UPCs use a separate Nielsen coding system. The database also contains extensive information about the demographics of its sample.

organic vegetables tend to be white and well educated, with relatively high incomes. Using 2005 Nielsen data, the price premium for popular organic fruits, relative to conventional fruits, is found to be 20–42%, while the premium for organic vegetables ranges in 15–60% (Lin *et al.*, 2008). In 2006, the demand for conventional fruits was found to be price-elastic (–1.06 to –3.54), while the demand for organic fruits were found to be price-inelastic, –0.49 to –0.85 (Lin *et al.*, 2009). Similarly, in 2006, the price premiums for organic milk ranged in 60–109% above conventional milk, with little effect of shopper demographics (Smith *et al.*, 2009).

In a more recent examination of consumer demographics, Dimitri and Dettmann (2012) analyze the 2006 Nielsen Homescan dataset of 44,000 U.S. households. Higher education and income, access to organic food, and being married increase the likelihood of buying organics, but ethnicity is unimportant. However, these studies and similar research have well-known methodological weaknesses. Contingent choice studies using on-site interception can have self-selection and hypothetical biases. Analysis of scanner data can incur self-selection, non-reporting, and attrition biases. Our study eliminates these biases via an experiment with several layers of randomization and an incentive-compatible elicitation mechanism.

Batte *et al.* (2007) analyze WTP premiums for foods that are 100% organic, at least 95% organic, at least 70% organic, and foods containing some organic ingredients (less than 70% organic) in an intercepted sample of 199 Ohio grocery store shoppers. They find that prior awareness of the NOP increases the probability of willingness to pay a premium for foods with organic ingredients, whereas health concerns are insignificant. Families with children are willing to pay higher premiums for foods that are organic or made with organic ingredients, conditional on willingness to pay a premium.

Assessing consumer WTP with demand-revealing auctions in economics experiments has brought substantial improvements to the food valuation literature. This line of inquiry often uses labeling treatments and/or information treatments for inference on novel food products, production processes, or food consumption trends.

Menkhaus *et al.* (1992) is an early experiment implementing fifth-price auctions to value vacuum-sealed packaging for beef steaks. Information about the benefits of vacuum-sealing and demonstrations increase subjects' premiums for vacuum-sealed beef steaks over conventional packaging (overwrapped Styrofoam trays) by \$0.41–0.48, and steaks with an "all natural" label have a \$0.12 premium. In an experiment valuing tortilla chips, vegetable oil, and potatoes, Huffman *et al.* (2003) find a 14% discount of genetically modified (GM)-labeled foods relative to those with conventional labels, with no role for demographic effects. Subsequent analysis demonstrates that information effects on bidding behavior depend on subjects' prior GM knowledge (Huffman *et al.*, 2007).

Other experimental methods have been applied to examine underlying components of "natural" and organic foods. A series of recent papers tests whether WTP for organic foods, as a whole, equals the sum of WTP for individual components of organic production. Bernard and Bernard (2009) use Vickrey (1961) auctions to elicit subjects' valuations of conventional, rBST-free, antibiotics-free, and organic milk. Their study finds that the sum of WTP for rBST-free and antibiotics-free milk exceeds WTP for the remaining attributes of organic production. Subjects who are more educated, older, male, primary household shoppers, and have higher incomes have higher WTP for milk with differing organic components.

These results are similar to those in Bernard and Bernard (2010), a WTP experiment on conventional, pesticides-free, GM-free, and organic sweet corn and potatoes. Participants who believe that conventional foods are made with GM technologies or pesticides have higher WTP for GM-free or pesticides-free foods. This echoes another experimental finding of increased WTP for apples produced without pesticides, after controlling for cosmetic damage (Roosen *et al.*, 1998). However, some evidence suggests that valuations for organic foods decrease after learning about non-GM requirements (Gifford and Bernard, 2011). Analyzing bids for "natural," free-range, non-GM-fed, and organic boneless/skinless chicken, the authors conclude that premiums for organic over "natural" decrease after subjects receive information on legal requirements for both food types, likely due to lower criteria for "natural" foods. The experimental information does not contain perspectives from interested groups or statements about health, environmental, and animal welfare impacts.

In an extensive mail survey sent to households in Mid-Atlantic states during fall 2009, Onken *et al.* (2011) analyze consumer preferences and WTP for strawberry preserves with attributes for organic, "natural," locally-grown, and state marketing program promotions. Information was provided to households about organic and "natural" foods based on language from the NOP and FSIS' meat and poultry labeling terms, respectively. Among other results, they find evidence of higher marginal WTP for the "natural" attribute than the organic attribute. This could arise from consumer confusion of the two attributes. In a similar study, He and Bernard (2011) use fourth- and fifth-price auctions to analyze demand elasticities and premiums for conventional, non-GM, and organic sweet corn, potatoes, and tortilla/potato chips. The effects of subjects' demographics are mixed. The conventional foods tend to be most price-inelastic, while non-GM foods are most price-elastic, with subjects valuing organic over the non-GM attribute more in sweet corn than tortilla chips.

Most recently, Bazoche *et al.* (2014) and Meas *et al.* (2015) examine WTP and preferences for organic and related attributes in apples and blackberry jam. The former study uses an incentive-compatible mechanism to elicit WTP information in Portugal, France, Greece, and the Netherlands for apples produced conventionally, organically, and with reduced pesticides. They find that organic premiums increase after subjects receive information about organic production methods, though information has no impact on mean WTP for apples produced with integrated pest management practices. Meas *et al.* (2015) find a sizeable degree of overlapping values in WTP for blackberry jam with an organic and "local" attribute using stated-preference data.

3. Experimental Design

The use of random nth-price auctions is a central component of our experimental design. We use this elicitation device, rather than the

Becker *et al.* (1964) mechanism (BDM) or Vickrey (1961) auction, because of its frequent use in food values research and desirable properties in practice. The random nth-price auction builds on the theoretical framework of the BDM mechanism and Vickrey auction. After participants bid on the experimental product, a random number, n, is drawn uniformly on $[2, k]$, where k is the number of session participants. The bids are then ranked from highest to lowest. Those who bid higher than the nth bid are winners but pay the nth bid, not their submitted bid. As in other incentive compatible mechanisms, the bidders' dominant strategy is to truthfully reveal their valuations.

Shogren *et al.* (2001) argue that random nth-price auctions are desirable because they engage "off-margin" bidders, i.e., low-value subjects still have a positive probability of winning. Using an experiment with induced values, the authors find more sincere bidding from low-value individuals in the random nth-price auction than the Vickrey auction. A careful and thorough explanation of auction steps, combined with carefully answering subjects' questions, helps ensure a straightforward implementation with minimal potential for subject misconceptions.

3.1. *Background*

We performed the experiment on April 13, 2013 (Saturday) in a Midwestern city. The sessions began at 9:00 a.m., 11:30 a.m., and 2:00 p.m. There are two concurrent sessions at each starting time, so there are a total of six (3×2) sessions. The sample consists of 102 subjects.

The Survey and Behavioral Research Services (SBRS) at Iowa State University (ISU) developed the protocol for recruiting participants. A random sample of households was contacted to determine if there was at least one individual eligible to participate in group sessions in an ISU project on food preferences. Recruiters were seeking adults aged 18–65 years who could follow directions, write in English, and participate on the experiment date. Contacted households were told that the sessions involve a food preference experiment set up like an auction or bidding activity, and they would be paid $65 for about an hour of their time.[3] Willing subjects were asked to pick among the three available starting times. They were told that there would be follow-up communication confirming time, location, and directions to the lab site. They started with a set of random telephone numbers. Willing subjects are allowed to suggest other family members and friends that would also be willing to participate.

We use the following commodities: fresh apples (Red Delicious, 3 lbs.), fresh eggs (Grade A large, one dozen), and fresh broccoli (1.25 lbs.).[4] We use a diverse set of commodities because fresh produce comprises a large share of organic food sales, in addition to the experiment's high fixed costs and subjects' heterogeneous preferences. Each product has three types: conventional, "natural," and organic. "Natural" apples and broccoli were unavailable at local grocery stores, so we substituted organic products.[5] This is acceptable since organic apples and broccolis meet any reasonable criterion for "natural" produce. The apples are displayed on a paper plate, the broccoli is displayed over a bed of ice on a paper plate, and the eggs are displayed in a standard carton with the top removed and without commercial logos.

The experiments offer two types of information: food labels and prepared information statements. Conventional products have a plain label that describes the product and its weight. "Natural" products have a label that describes the product and its weight, plus the words "All-Natural" printed in green ink. Organic products carry a similar

[3]We pay participants at the beginning of the sessions to separate the act of paying them for the work to be done in the experiment from the act of deciding how much to pay in the auctions. It also eliminates concerns about cash constraints (Huffman *et al.*, 2003; Lusk and Shogren, 2007; Huffman *et al.*, 2007; Bazoche *et al.*, 2014).

[4]We calculate that organic apples, eggs, and broccoli account for 4.3%, 2.3%, and 5.5% of 2014 U.S. production and 8.6%, 5.4%, and 9.6% of 2014 U.S. sales, respectively (USDA-NASS, 2016a; USDA-NASS, 2016b). Using Nielsen Homescan Data, Carlson and Jaenicke (2016) find that apples and eggs comprised 6.6% and 3.4% of 2010 U.S. sales. Data on market or production shares of "natural" foods are unavailable since "natural" does not have a widely-accepted, legally-binding definition.

[5]For the purposes of our experiment, we consider "natural" foods to be those with a claim on the foods' packaging indicating "all natural" or "100% natural."

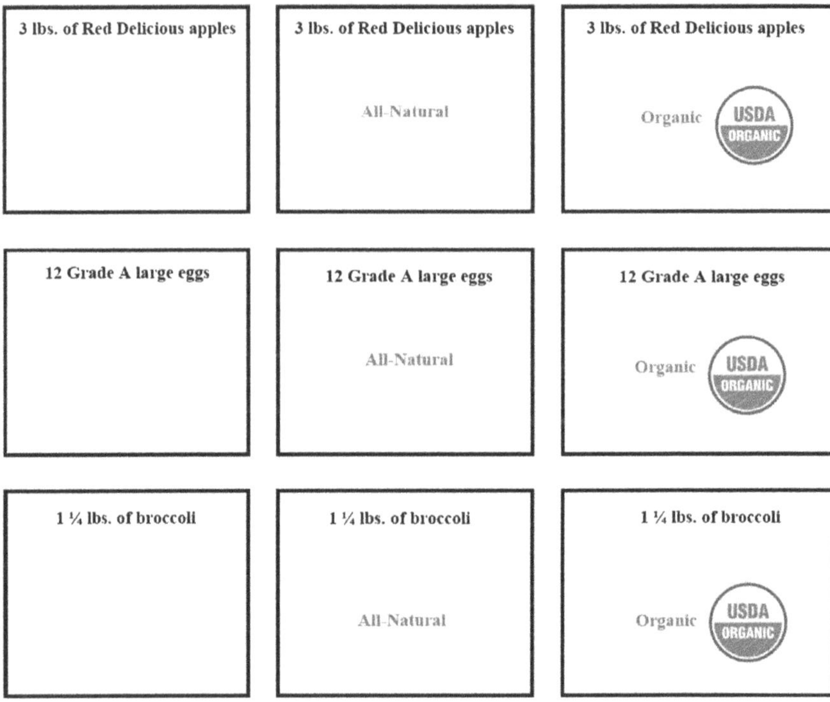

Fig. 1. Commodity labels for apples, eggs, and broccoli.

product and weight description, plus the USDA's official organic seal in color. Fig. 1 depicts the labels.

Information treatments are constructed from four perspectives, each limited to one 8.5″ × 11″ sheet of paper. These perspectives are (i) an organic food industry perspective on organic foods, (ii) independent, third-party, verifiable perspectives on organic foods, (iii) a "natural" food industry perspective on "natural" and "all natural" foods, and (iv) independent, third-party, verifiable perspective on "natural" foods. To assist subjects' reading and processing of information, each of these perspectives is organized under the same three headings (General Statement, Production Standards, and Nutrition, Taste, and Appearance; see Appendix A).

The information perspectives rely on information from nutrition research, government reports, non-governmental organization (NGO) reports, and websites. The "Production Standards" headings and organic and natural definitions draw on the NOP regulations and FSIS documentation (National Organic Program, 2002; USDA-FSIS, 2005). Much of the "natural" and organic food industry perspectives are adapted from statements on Whole Foods Market's website and The Organic Center's website. The information perspectives were pretested and subsequently revised prior to implementation.

There are a total of five information treatments: (i) no-information baseline, (ii) "natural" food industry perspective only, (iii) "natural" food industry perspective plus independent, third-party, verifiable perspective on "natural" foods, (iv) organic food industry perspective, and (v) organic food industry perspective plus independent, third-party, verifiable perspective on organic foods. Each subject is randomly assigned one of the five information treatments in their packet with the protocol for the experiment.[6]

The design of the experiments introduces randomness at several levels. First, subjects are recruited randomly.[7] Second, subjects are randomly assigned to a packet containing the protocol and information treatment. Third, they are alternately assigned to one of two concurrent sessions. Fourth, the order of bidding on products is randomized (e.g., we randomly select the order of conventional, "natural," and organic products). Fifth, we randomly select one of the three rounds of bidding as the binding round. This multilevel randomized design permits simple econometric identification by removing selection effects (Lusk and Shogren, 2007; Glennerster and Takavarasha, 2013).

3.2. *Subject activities in a session*

We now briefly review the steps of a typical auction session. Fig. 2 is a flowchart of subject tasks. Participants read and sign an informed

[6]No participant received a treatment consisting of three or four information perspectives or a combination of perspectives on both natural and organic foods. This reduces subject burden and session lengths.

[7]We acknowledge that — as with most economics experiments — if a different sample were to be randomly drawn, then the empirical results could slightly differ.

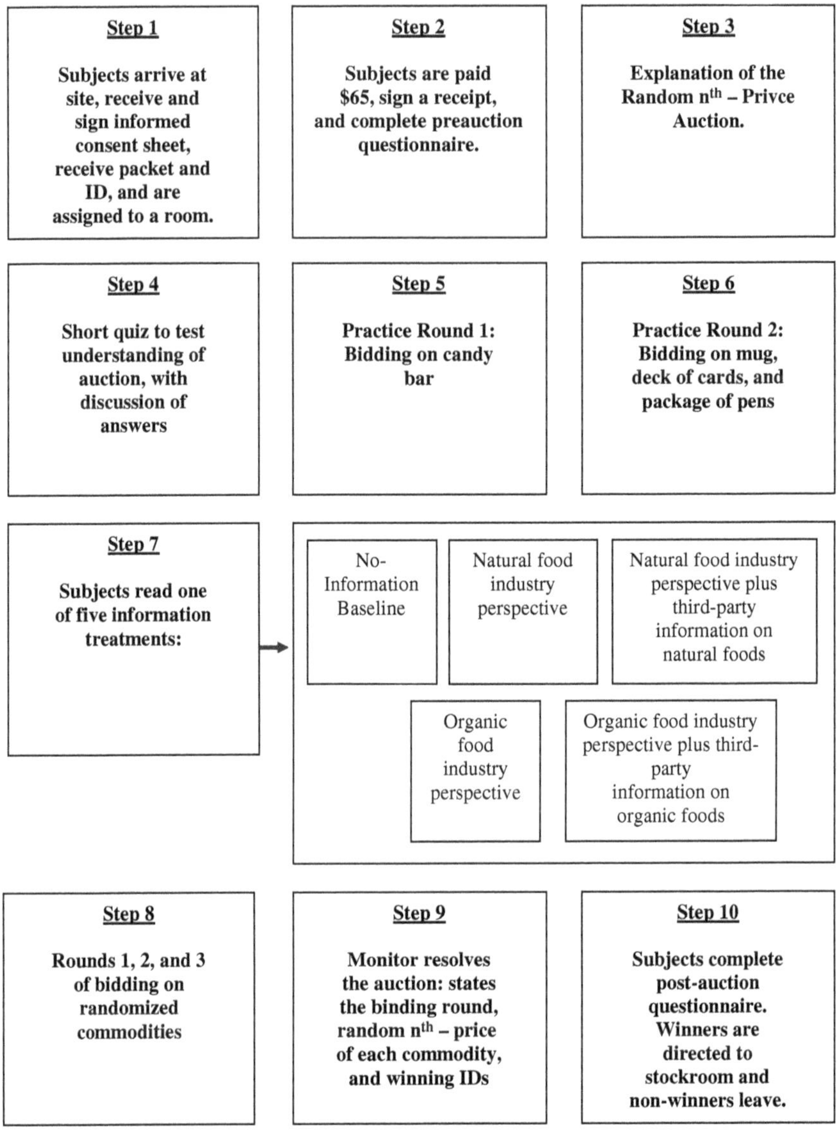

Fig. 2. Subject activities in a session.

consent sheet upon arrival at the experimental site. After agreeing to participate, subjects are given an ID number and are randomly assigned to one of two concurrent sessions and a packet containing all materials necessary for participation.[8] They are then instructed to begin completing the pre-auction questionnaire. Questions in the pre-auction questionnaire ask subjects about prior knowledge on "natural" and organic foods, frequency of looking at food labels when buying a new food product, and health. The demographics questions concern living in an urban/suburban/rural area, marital status, gender, age, household size, religion, number of children in certain age groups living in the household, level of schooling, ethnicity, household income, and occupation. The health questions assess overall health status and smoking status (at least once per day, less than once per day, or not at all). We distribute a $65 participation fee and have subjects sign a receipt while they begin completing the pre-auction questionnaire.

Subjects are told to turn off electronic devices and not talk with others. After all subjects have completed the pre-auction questionnaire, session monitors tell subjects that they will be participating in an experimental auction and explain how the auction works. Subjects are notified that they can win at most one unit of each commodity. At this stage of the experiment, subjects are allowed and encouraged to ask clarifying questions. A brief quiz is administered to check understanding of the auction rules. We allow subjects four to five minutes to finish the quiz and then review the answers with additional explanation.

We then proceed to one of two practice rounds once any questions have been addressed. The first practice round trains subjects on the experimental auction methods, while the second practice round is designed to familiarize subjects with bidding on three commodities simultaneously. Following previous economics experiments, the commodity in the first practice round is a candy bar. Subjects are asked

[8]Individuals who arrive together (e.g., friends and family members) are placed in separate sessions, unless special circumstances require assignment to the same session. In this case, we ask that they do not sit next to each other, unless this is also necessary for extenuating reasons.

to come to the front of the room, view the candy bar, return to their seats, and write down bids. The session monitor and assistant then collect and rank the bids (in a spreadsheet), randomly draw the binding round, and determine the winner(s) via the auction rules. In the second practice round, subjects gain experience placing three bids at one time on a mug, deck of cards, and package of pens. Bids for the three products are sorted separately and displayed, and "winner" IDs are revealed. Unlike for winners in the binding round of the actual auctions, winners of the practice rounds are not required to purchase the practice goods.

At the conclusion of the practice rounds, subjects are given approximately ten minutes to read their assigned information treatment.[9] Subjects then bid on the three commodities, with the order (conventional, "natural," or organic) randomized across three distinct rounds of bidding. Bid sheets are collected, and the only posted bids are those for commodities in the binding round at the end of the auctions.[10] Note that subjects are bidding on a separate set of products for each of the three rounds. However, the commodities across rounds are very similar and differ only with respect to labels and types (e.g., conventional, "natural," or organic).

In the final stage of a session, the binding round, random n, winning prices, and winning IDs are revealed. For product control purposes, we randomly choose the binding round and random n prior to all sessions. After inputting bids for the binding round into a spreadsheet, the bids for all products are projected at the front of each session's room, and subjects watch the session monitor rank the bids and reveal winning IDs.

Subjects are then asked to complete a post-auction questionnaire. We ask participants about monthly purchases, weekly consumption,

[9]To ensure that outside information does not contaminate any specific treatment, session monitors and assistants do not answer questions related to informational content. To avoid confusion, we state that some subjects will have information to read, while others will not have information to read.

[10]Reference prices are not given for each of the products prior to bidding. Providing reference prices could bias subjects' WTP if they anchor on these prices as reference points, rather than truthfully revealing their maximal WTP. Bids are not posted because this could induce changes in bids in subsequent rounds (Corrigan and Rousu, 2006).

and household inventories of apples, eggs, and broccoli. Additional questions relate to prior awareness of organic food labels, sources of food product information, and the experiment's contribution to how informed subjects feel about organic and "natural" foods. Upon completion of the post-auction questionnaire, winners are directed to a stockroom to purchase the won item(s), and non-winners leave the experimental site.

4. Econometric Model and Data

The nature of the experiment and detailed questionnaire data permit latitude for empirical methods. Regression analysis of WTP levels, however, could be problematic. First, some zero bids are truncated at zero, i.e., subjects would have preferred to bid a negative value but felt constrained to bid zero instead.[11] Other zero bids are legitimate bids with no truncation, i.e., zero was the preferred bid (Maddala, 1983). Second, valuation levels, although sincere, may be arbitrary (Ariely *et al.*, 2003). Our approach uses WTP differences across subjects for parameter estimation. A major benefit of this method is that unobserved heterogeneity (e.g., subject fixed effects) are differenced out in the estimation. Bids are pooled across commodities, with apples as the reference commodity.

4.1. *Regression specification*

We propose the following pooled linear regression model of WTP differences:

$$\begin{aligned} Bid_i^{Org} - Bid_i^{Con} &= \mathbf{x}_i'\boldsymbol{\beta}_1 + \mathbf{T}_i'\boldsymbol{\gamma}_1 + \mathbf{c}_j + \varepsilon_{1,i} \\ Bid_i^{Org} - Bid_i^{Nat} &= \mathbf{x}_i'\boldsymbol{\beta}_2 + \mathbf{T}_i'\boldsymbol{\gamma}_2 + \mathbf{c}_j + \varepsilon_{2,i} \\ Bid_i^{Nat} - Bid_i^{Con} &= \mathbf{x}_i'\boldsymbol{\beta}_3 + \mathbf{T}_i'\boldsymbol{\gamma}_3 + \mathbf{c}_j + \varepsilon_{3,i}. \end{aligned} \tag{1}$$

[11] Few of the zero bids would be in cases where an individual would have preferred to bid less than zero but felt constrained to bid zero. Thus, the number of problematic bids is very small, and a model taking into account censoring of WTP differences would not differ significantly from the reported results. Table A.1 in Appendix A provides counts of the zero bids.

In the above equations, $Bid_i^{Org} - Bid_i^{Con}$ is the premium for organic products relative to conventional for the ith bid. The premium for organic over "natural" is denoted $Bid_i^{Org} - Bid_i^{Nat}$, and the premium for "natural" over conventional is $Bid_i^{Nat} - Bid_i^{Con}$. Note that $\mathbf{x_i'}$ is a vector of controls on subjects' demographics, household food inventories, and previous knowledge about various labels and claims on food packaging. The indicator variables for information treatments are $\mathbf{T_i'}$, and $\mathbf{c_j}$ are fixed effects for $j \in \{eggs, broccoli\}$. Estimates of the γ are treatment effects. The error terms, $\varepsilon_{.,i}$, are assumed independent and identically distributed (i.i.d) normal.

4.2. *Data*

Table 1 contains descriptive statistics for the set of nine bids across subjects in the sample. Several interesting trends are apparent. We find that subjects' valuations are increasing across product types (conventional, "natural," and organic) for all commodities. The difference in subjects' mean WTP for organic and conventional commodities is \$0.54, \$0.57, and \$0.52 for apples, eggs, and broccoli. As a percentage of the conventional price, subjects organic premiums are 22%, 34%, and 30%, which are within the range in the literature (Park and Lohr, 1996; Bernard *et al.*, 2006; He and Bernard, 2011; Bazoche *et al.*, 2014). Differences in mean WTP for organic and "natural" types are \$0.21 (7%), \$0.30 (15%), and \$0.27 (14%) for apples, eggs, and broccoli. The "natural"-conventional differences in subjects' mean bids are \$0.33 (13%), \$0.27 (16%), and \$0.25 (14%). The raw data suggest that participants view "natural" products as intermediates (with respect to valuation) between conventional and organics, similar to the progression of bids in Bernard *et al.* (2006), He and Bernard (2011), and Bazoche *et al.* (2014). "Natural" foods are imperfect substitutes for organic foods, but they are deemed more valuable than conventional on average.

Variability in subjects' mean apple bids is consistent and relatively large across all commodity types. This is perhaps because of affinity for certain varieties of apples, either Red Delicious or another

Table 1. Consumer WTP levels: Commodities by type ($N = 102$).

Bid ($)	Mean	Std. Dev.	Min	Max
Conventional commodities				
Apples	2.51	1.21	0.00	6.75
Eggs	1.67	0.75	0.00	4.00
Broccoli	1.74	0.87	0.00	4.00
Natural commodities				
Apples	2.84	1.58	0.00	9.00
Eggs	1.94	1.00	0.00	6.00
Broccoli	1.99	1.07	0.00	6.00
Organic commodities				
Apples	3.05	1.58	0.00	7.80
Eggs	2.24	1.00	0.25	6.00
Broccoli	2.26	1.07	0.00	5.25

variety not represented in the experiment. Standard deviations for "natural" commodities are larger than conventional commodities, although standard deviations are quite similar for the set of "natural" and organic commodity bids. Given the lack of clear legal standards, participants may generally have varying interpretations of "natural" foods, which seem to be reflected in dispersed bidding. The opposite holds for conventional foods: an implicit designation as a reference or "status quo" food type may be contributing to subjects' greater uniformity of bids. Note that there is a low proportion of zero bids, with all bids strictly positive for organic eggs (see Table A.1 in Appendix A).

Descriptive statistics for the variables used in regression analysis are displayed in Table 2. There are 273 useable bid observations after removing bids for which the subject's demographics data are missing.[12] Statistics for subjects' food type premiums are broadly consistent with WTP level trends observed by commodity in Table 1. Mean premiums generally decrease from the $0.52 organic-conventional

[12] A total of 33 bid observations are removed. Removal of these bids does not substantially affect the sign, magnitude, and statistical significance of the information treatment effects in the regression analysis.

Table 2. Summary statistics: Bid differences and regressors ($N = 273$).

Variable	Mean	Std. dev.	Min	Max
Dependent variable ($)				
Org Commodity Bid — Con Commodity Bid	0.52	0.65	–1.19	2.80
Org Commodity Bid — Nat Commodity Bid	0.29	0.60	–2.15	2.61
Nat Commodity Bid — Con Commodity Bid	0.23	0.62	–1.25	4.15
General demographics				
Female (0–1)	0.73	0.45	0	1
Age (integer)	43.04	11.48	22	65
Age^2 (integer)	1,983.99	1,001.00	484	4225
Years of schooling (integer)	14.22	2.20	10	22
Per capita income ($ 1000 s)	23.16	15.75	1	68.75
Per capita income^2 ($ 1000 s)	783.50	993.15	1	4,726.56
Children aged 0–3 years in household (integer)	0.13	0.34	0	1
Children aged 4–7 years in household (integer)	0.21	0.48	0	2
Children aged 8–12 years in household (integer)	0.27	0.58	0	2
Children aged 13–18 years in household (integer)	0.33	0.67	0	3
Children older than 18 years in household (integer)	0.36	0.78	0	4
Previously informed about organic foods (0–1)	0.59	0.49	0	1
Previously informed about natural foods (0–1)	0.55	0.50	0	1
Household inventories (0–1)				
Conventional or natural apples at home	0.77	0.42	0	1
Conventional eggs at home	0.82	0.38	0	1
Conventional or natural broccoli at home	0.48	0.50	0	1
Labeling and food product claims (0–1)				
Looks at labels when buying new foods	0.89	0.31	0	1
Previously seen USDA organic seal	0.66	0.47	0	1
Previously seen natural claim	0.85	0.36	0	1
Previously seen 70% organic claim	0.14	0.35	0	1

Table 2. (*Continued*)

Variable	Mean	Std. dev.	Min	Max
Previously seen "Made with Natural Ingredients"	0.80	0.40	0	1
Commodity bid indicator (0–1)				
Eggs	0.33	0.47	0	1
Broccoli	0.33	0.47	0	1
Information treatments (0–1)				
Organic Industry	0.21	0.41	0	1
Organic Industry and Independent Organic	0.16	0.37	0	1
Natural Industry	0.19	0.39	0	1
Natural Industry and Independent Natural	0.20	0.40	0	1

difference to the $0.23 "natural"-conventional difference. There is a small non-linearity in mean premiums with respect to "natural" foods. Subjects are willing to pay $0.29 more for organic relative to "natural" commodities, but conventional products are worth $0.23 less than "naturals," on average. Standard deviations are $0.60–$0.65, though there is some asymmetry in magnitudes between premiums and discounts. The largest organic conventional premium ($2.80) for a subject is roughly twice the largest discount ($1.19). Such heterogeneity could have important marketing and segmentation implications.

The subjects have an average age of 43 years and have completed roughly 14 years of schooling (between a high school diploma and four-year college degree), and their households have an annual per-person income of $23,160.[13] Women comprise 73% of the sample, approximating the trend that 57% of U.S. primary grocery store shoppers are women (FMI, 2014). Importantly, our pre-auction questionnaire elicits information that captures the nonlinear impacts of number of children by age in a subject's household. Subjects

[13]Annual per-capita incomes are calculated as the midpoint of the income bracket chosen by the respondent in the pre-auction questionnaire, divided by the number of people in the household.

on average have more teenagers living at home than infants and toddlers, consistent with a higher frequency of older individuals in the sample. Subjects' households also have experience consuming and purchasing the experimental commodities. Roughly 77%, 82%, and 48% of subjects had the three experimental products (apples, eggs, and broccoli) in their home food inventories at the time of the experiments.

There is also ample variation in participants' previous knowledge of "natural" and organic foods and past awareness of seals and labels. A little over half of the sample is at least somewhat informed about "natural" foods (55%) and organic foods (59%). Nearly 90% of subjects look at food labels at least some of the time when buying a new food product. Yet there is disparity in subjects having previously seen seals for "naturals" and organics. The USDA organic seal had been noticed by 66% of subjects in their shopping trips prior to the experiment, while only 14% had previously seen food product packaging with a claim of being 70% organic. This contrasts with high visibility for "natural"-labeled foods (85%) and those made with "natural" ingredients (80%).[14] This might be indicative of the industrys relative advertising success to date.

Last, it should be pointed out that information treatments are randomly assigned in roughly uniform proportion. The organic industry perspective is assigned to 21% of the sample, while 16% of the sample receive both organic perspectives. The "natural" industry perspective is assigned to 19% of the sample, with both "natural" perspectives going to 20%. The remaining 24% receive the no-information baseline treatment.

5. Empirical Results and Discussion

5.1. *Regression analysis*

Table 3 presents regression results for the subjects' organic conventional premiums. Of crucial interest are the information impacts.

[14]Nearly all of the major supermarkets near the experiment site carry a broad assortment of organic foods. Further, a Whole Foods Market store is located 20 miles from the site. Thus, the summary statistics and regression results are not being influenced by inadequate access to organic foods.

Table 3. Organic-conventional WTP differences.

	(1)	(2)
Organic Industry information	0.039	0.043
Organic Industry and Independent Organic information	−0.335***	−0.347***
Natural Industry information	0.483***	0.430***
Natural Industry and Independent Organic information	0.138	0.131
Female	0.236***	0.243***
Age	−0.076***	−0.072***
Age^2	0.001***	0.001***
Years of schooling	−0.020	
Per capita income	0.029***	0.027***
Per capita income^2	−0.0004***	−0.0004***
Children aged 0–3 years in household	0.680***	0.688***
Children aged 4–7 years in household	−0.316***	−0.350***
Children aged 8–12 years in household	0.190***	0.185**
Children aged 13–18 years in household	0.091	0.072
Children older than 18 years in household	0.098	0.070
Conventional or natural apples at home	−0.160*	−0.201**
Conventional eggs at home	−0.180*	−0.119
Conventional or natural broccoli at home	0.366***	0.370***
Looks at labels when buying new foods	0.338***	0.350***
Previously informed about organic foods	0.164*	0.119
Previously informed about natural foods	−0.035	
Previously seen USDA organic seal	0.172*	0.192**
Previously seen natural claim	−0.029	−0.016
Previously seen 70% organic claim	0.175	
Previously seen "Made with Natural Ingredients"	−0.072	−0.064
Broccoli	−0.073	−0.073
Eggs	−0.009	−0.009
Constant	1.276***	0.922*
Observations	273	273
R^2	0.385	0.376
F-statistic	5.670***	6.224***

Note: Significance is denoted as *($p < 0.1$), **($p < 0.05$), ***($p < 0.01$). Dependent variable in columns (1) and (2) is the WTP for conventional commodities subtracted from the WTP for organic commodities. The F-statistic for testing the null hypothesis that the coefficients on the regressors deleted in column (2) are jointly zero is 1.15, with a p-value of 0.33.

Somewhat surprisingly, we find that the organic food industry perspective does not significantly affect subjects' premiums over conventional foods. This could be because individuals who are very familiar with conventionally-produced foods may not think organic foods are necessary or bring many benefits, thus discounting positive information about organic foods (Gifford and Bernard, 2006). However, when combined with independent, third-party, verifiable information on organic foods, there is a \$0.34 decrease in subjects' premiums relative to those without information. This is consistent with recent studies that find significant effects of information or message framing on WTP or purchase intentions for organic foods (Gifford and Bernard, 2004; Gifford and Bernard, 2006; Rousseau and Vranken, 2013; Bazoche *et al.*, 2014).

Information about "natural" foods have the opposite effect. The "natural" food industry perspective increases subjects' relative organic-conventional premium by \$0.48, whereas adding the independent information on "natural" foods is insignificant. The exact causes of these asymmetric effects cannot be identified in our study, but it is interesting that the results reflect current market characteristics. In recent years, consumers have had expanding access to both objective and subjective information about organics. Our results suggest that the net effect has reduced subjects' willingness to pay premiums, perhaps contributing to the organic industry's persistently small retail volumes. In the "natural" foods sector, there is much less access to independent information because of the lack of legal standards. It appears that greater WTP premiums for organic foods have increased as a result of a subject receiving and reading the "natural" food industry perspective.

We find that subject demographics have broad effects, in contrast to prior studies that find few or no demographic impacts on WTP premiums, preferences, or purchase intentions of organic or GM foods (Huang, 1996; Huffman *et al.*, 2003; Gifford and Bernard, 2004; Gifford and Bernard, 2006; Bazoche *et al.*, 2014; Stzrok and Huffman, 2015). The premium is \$0.24 higher among women than for men, in line with most prior findings of significant gender effects (Govindasamy and Italia, 1999; He and Bernard, 2011; Bazoche *et al.*, 2014).[15] Older participants also have higher premiums, with an additional year increasing their premiums by one cent (evaluated at the sample mean), comparable to Misra *et al.* (1991) and Govindasamy and Italia (1999). Similarly, a marginal increase in subjects' household per-capita income ($1,000) increases the premium by one cent. Subjects' years of schooling are insignificant, which counters past research (Misra *et al.*, 1991; Huang, 1996; Thompson and Kidwell, 1998; Bernard *et al.*, 2006; Stzrok and Huffman, 2015), but supports the claim that the demographic landscape has become much broader.

Last, subjects' household inventories have some influence. Those with non-organic apples in home inventories have reduced premiums by $0.16, and those with non-organic broccoli in home inventories have increased premiums of $0.37.[16] The former finding is analogous to Govindasamy and Italia (1999), who find that regular purchasers of organic foods are 45% more likely to pay a 10% premium for organic foods relative to those who are not regular purchasers. Those with broccoli in home inventories are perhaps more nutritionally-conscious or concerned about eating fresh produce, for whom organics may be especially appealing (Huang, 1996). It is also possible that frequency of organic food purchases could interact with information provision in determining WTP for organic foods (Rousseau and Vranken, 2013).

The regressions are also able to explain a likely source of conflicting evidence regarding the effects of household size and children on organic food values (Huang, 1996; He and Bernard, 2011; Bazoche *et al.*, 2014). Our results suggest that subjects' households with

[15]Govindasamy and Italia (1999) find that men are 12% less likely to pay a 10% premium for organic foods. He and Bernard (2011) find that the WTP ratio of organic to conventional potato and corn food products is 8–10% lower for men. Bazoche *et al.* (2014) find that women are willing to pay 0.195 EUR/kg (roughly $0.21/three pounds at the time of their study) more than men for organic apples relative to conventional apples. However, Thompson and Kidwell (1998) and Bernard *et al.* (2006) generally find that gender is not statistically significant.

[16]Yue *et al.* (2009) find that the share of subjects willing to pay premiums for organic apples is reduced if the apples have cosmetic imperfections. Our results are robust to inclusion of dummy variables indicating subjects' perceptions of blemishes or discolorations in any of the commodities.

more children in important developmental growth stages have higher valuations for organic over conventional foods. Subjects in households with an additional infant or toddler (0–3 years of age) are willing to pay \$0.68 more for organic relative to conventional commodities. This is a critical growth stage in very young children, and parents may feel encouraged to buy organic foods that are perceived to be healthier. Indeed, the organic baby food market has grown substantially in recent years, fueled by a willingness to pay premiums of \$0.03–0.04 per ounce (Maguire *et al.*, 2004). Subjects' households with an additional child aged 8–12 years, the beginning of puberty for many children, have increased premiums of \$0.19. An extra child aged 4–7 years decreases subjects' premiums by \$0.32. This shows that a simple indicator of "young" or "teen" children in the household is inadequate for explaining the organic premium. Effects over differing age ranges must be considered for a more complete understanding of the relationship between children and parents' food preferences (Thompson and Kidwell, 1998).

Labels and previous food type information remain significant predictors of subjects' organic-conventional premiums (Aertsens *et al.*, 2011). Subjects who regularly read labels have premiums that are \$0.34 higher than those who do not regularly read them. Those who read labels when buying new food products for the first time could be more nutritionally-conscious, which can translate into greater likelihood of purchasing organic foods and higher WTP (Huang, 1996; Nielsen Company, 2015). Relative to subjects who were previously uninformed or who had not seen the official seal, organic premiums are \$0.16–0.19 higher. This is similar to Gifford and Bernard's (2004) finding that increased prior knowledge of organic methods increases the likelihood of purchasing organic foods in the future. The 70% organic and "natural" foods packaging claims do not significantly affect subjects' WTP.

The second column of Table 3 presents a reduced model that excludes several of the individually insignificant regressors. The F-statistic associated with testing that the coefficients on these regressors are jointly zero is 1.15, less than the critical value of 1.57 at a 5% significance level. The estimated coefficients for the remaining variables are highly robust. It should also be emphasized that neither of the commodity fixed effects are significant, suggesting appropriateness of pooling the set of bids across commodities. The model fits well, with 38% of the premium variation explained by variation in treatment, demographics, and other predictors.

Regression results for subjects' organic-"natural" premiums are provided in Table 4. The focus is on coefficients in the reduced model of column (4), our preferred set of estimates. We cannot reject the null hypothesis that the set of deleted regressors are jointly equal to zero (F-statistic of 1.25 compared to the critical value of 1.57 with a 5% significance level). The information effects are comparable to those from the analysis of subjects' organic conventional premiums. It seems that the effect on subjects' WTP differences of independent organic information outweighs the effect of the organic food industry perspective, with a net negative effect of $0.27 relative to those without information. The effect of the "natural" foods industry perspective on subjects' WTP differences is again positive and increases subjects' premiums by $0.41 over those uninformed. This is similar in size and magnitude to the effect of information about organic methods on subjects' bids for organic apples (Bazoche *et al.*, 2014). The addition of independent information on "natural" foods, however, decreases subjects' WTP differences by $0.20. The general conclusion for subjects' organic-"natural" premiums is that adding independent information to industry food perspectives reduces excess valuations of organics over "naturals."

The impacts of subject demographics remain intuitive and comparable to those in Table 3. Women have $0.19 higher organic-"natural" premiums than men. However, the marginal effect of a subject being older decreases the premium by one-half cent (evaluated at the sample mean), but the marginal effect of an extra year of age for a subject is positive beyond 46 years.[17] An additional $1,000 of per-capita income in a subject's household increases the premium by

---

[17]The nonlinear relationship between subjects' age and WTP is not straightforward because older individuals could have dietary restrictions or be less inclined to change eating habits (Govindasamy and Italia, 1999).

Table 4. Organic-natural WTP differences.

	(3)	(4)
Organic Industry information	−0.021	−0.030
Organic Industry and Independent Organic information	−0.264**	−0.266**
Natural Industry information	0.403***	0.409***
Natural Industry and Independent Organic information	−0.179*	−0.195*
Female	0.164**	0.186**
Age	−0.096***	−0.092***
Age^2	0.001***	0.001***
Years of schooling	−0.011	
Per capita income	0.039***	0.039***
Per capita $income^2$	−0.0005***	−0.0005***
Children aged 0–3 years in household	0.241**	0.254**
Children aged 4–7 years in household	−0.075	−0.076
Children aged 8–12 years in household	0.179**	0.162**
Children aged 13–18 years in household	0.137**	0.119**
Children older than 18 years in household	0.266***	0.265***
Conventional or natural apples at home	−0.077	−0.057
Conventional eggs at home	−0.405***	−0.379***
Conventional or natural broccoli at home	0.048	0.046
Looks at labels when buying new foods	−0.147	−0.158
Previously informed about organic foods	0.340***	0.279***
Previously informed about natural foods	−0.108	
Previously seen USDA organic seal	0.118	
Previously seen natural claim	0.040	0.096
Previously seen 70% organic claim	0.360***	0.366***
Previously seen "Made with Natural Ingredients"	0.092	0.074
Broccoli	−0.024	−0.024
Eggs	0.037	0.037
Constant	1.914***	1.672***
Observations	273	273
R^2	0.336	0.325
F-statistic	4.583***	4.985***

Note: Significance is denoted as *($p < 0.1$), **($p < 0.05$), ***($p < 0.01$). Dependent variable in columns (3) and (4) is the WTP for natural commodities subtracted from the WTP for organic commodities. The F-statistic for testing the null hypothesis that the coefficients on the regressors deleted in column (4) are jointly zero is 1.25, with a p-value of 0.29.

1.6 cents at the sample mean, but additional per-capita income has negative effects beyond per-capita incomes of $39,000, comparable to Govindasamy and Italia's (1999) finding that willingness to pay an organic premium is greater among households with annual incomes over $70,000. Inventory effects are not as strong. Subjects that have conventional eggs in home inventories have $0.38 lower organic-"natural" premiums. Subjects that have the most experience with conventional eggs could be unwilling to switch to "natural" or organic types, perhaps perceiving them as close substitutes, thus driving down the premium.

The advantages of estimating disaggregated family size effects on the organic-"natural" premiums are also clear. Additional children in a subject's household for all age groups except the post-toddler years (4–7 years of age) have positive effects on subjects' organic-"natural" premiums. The largest effects occur in the youngest and oldest age groups. It is not surprising that parents of very young children are willing to pay $0.25 more for organic than conventional foods, given common perceptions that organic foods are healthier and contain lower concentrations of pesticides (Maguire *et al.*, 2006; Aertsens *et al.*, 2011). However, increased premiums in subject households with more adult children are less straightforward to interpret.[18] One potential explanation is that adult children in the household may have food preferences that are strongly influenced by peers or group membership (e.g., participation in student organic clubs or environmental organizations).

Last, we briefly mention that subjects' prior information and label awareness still help explain WTP differences, but the organic seal is no longer significant. Participants' prior information about organic foods raises the premium by $0.28, and his or her awareness of the 70% organic claim has a similar $0.37 impact. This could arise if those with greater prior awareness have more subjective knowledge

[18]As pointed out by Maguire *et al.* (2006), perceived pesticide risks to children are made by third parties (e.g., parents or other caregivers), whose perceptions about health risks and safety for their children are heterogeneous and generally unclear to researchers.

or enhanced attitudes about organic foods (Aertsens *et al.*, 2011), better opinions of organic foods (He and Bernard, 2011), or believe them to be "fairer" (Chang and Lusk, 2009), which could increase WTP. That the 70% organic claim has more consequence than the USDA organic seal (at least 95% organic) is puzzling, but Stzrok and Huffman (2015) report a similar outcome with respect to labels on food products of varying organic purity.

The final set of estimates in Table 5 reveal the difficult nature of making inference in markets with vague and distorted product claims. The reduced model in column (6) performs well (the coefficient of determination is 18%), but there are fewer important indicators that explain participants' valuations for "natural" over conventional foods. If subjects conflate "natural" with "pesticide-free" or "produced with Integrated Pest Management (IPM) methods," then these results are comparable to those of other studies. Specifically, Misra *et al.* (1991) find that 54% of their samples are not willing to pay a higher price for certified pesticide-free produce, and Bazoche *et al.* (2014) find no evidence of information effects on mean WTP for IPM-produced apples. If subjects already have high trust in conventional foods (Gifford and Bernard, 2004), there would be less incentive to pay a premium for poorly-understood "natural" foods.

Of the information treatments, only the combined "natural" industry perspectives plus third-party perspective on "natural" foods has a significant non-zero effect. In this case, subjects have a premium of $0.32 relative to those who do not receive information. Since biased information about "natural" foods is already prevalent in the marketplace, the $0.32 represents a benefit to greater clarification. Subjects with a more balanced and objective understanding of "natural" foods may be increasingly willing to substitute conventional for "natural" foods.

Some of the key demographic effects on subjects' organic premiums are also present in their "natural" premiums. Subjects in households with an additional very young child will pay $0.40 more for "natural" than conventional foods, though an opposite effect holds for those with 4–7 year-old children. There are similar impacts of a subject's household holding food inventories and of a subject

Table 5. Natural-conventional WTP differences.

	(5)	(6)
Organic Industry information	−0.059	0.096
Organic Industry and Independent Organic information	−0.071	−0.033
Natural Industry information	0.080	0.190
Natural Industry and Independent Organic information	0.317***	
Female	0.072	
Age	0.020	
Age^2	−0.0001	
Years of schooling	−0.009	
Per capita income	−0.010	
Per capita income^2	0.0001	
Children aged 0–3 years in household	0.439***	0.400***
Children aged 4–7 years in household	−0.242***	−0.248***
Children aged 8–12 years in household	0.011	0.049
Children aged 13–18 years in household	−0.046	−0.029
Children older than 18 years in household	−0.168***	−0.074
Conventional or natural apples at home	−0.083	0.020
Conventional eggs at home	0.226**	0.132
Conventional or natural broccoli at home	0.317***	0.304***
Looks at labels when buying new foods	0.485***	0.490***
Previously informed about organic foods	−0.175*	−0.097
Previously informed about natural foods	0.073	0.029
Previously seen USDA organic seal	0.053	
Previously seen natural claim	−0.069	−0.060
Previously seen 70% organic claim	−0.184	
Previously seen "Made with Natural Ingredients"	−0.163	−0.157
Broccoli	−0.050	−0.050
Eggs	−0.045	−0.045
Constant	−0.638	−0.322*
Observations	273	273
R^2	0.220	0.184
F-statistic	2.556***	2.994***

Note: Significance is denoted as *($p < 0.1$), **($p < 0.05$), ***($p < 0.01$). Dependent variable in columns (5) and (6) is the WTP for conventional commodities subtracted from the WTP for natural commodities. The F-statistic for testing the null hypothesis that the coefficients on the regressors deleted in column (6) are jointly zero is 1.42, with a p-value of 0.19.

reporting that she frequently reads food labels when purchasing new foods. Although the health or nutritional benefits accruing to "natural" foods are questionable, subjects who report having fresh broccoli in home food inventories and those who read food labels when purchasing new foods for the first time pay a premium of \$0.30–0.49 for "natural" foods. The remaining regressors have coefficients that are not significantly different from zero.

For ease of comparison, information effects on the three types of premiums are summarized graphically in Fig. 3. These bar charts depict the point estimate of coefficients on the information treatment indicators for the reduced models (second columns) from Tables 3–5. Both organic food perspectives decrease subjects' organic-conventional premiums by \$0.35, and the "natural" foods industry perspective increases their premiums by \$0.43. Similarly, both organic foods perspectives decrease subjects' organic-"natural" premiums by \$0.27, while the "natural" foods industry perspective has a positive \$0.41 effect. Last, both "natural" foods perspectives have a combined \$0.32 effect on participants' "natural"-conventional premiums.

5.2. *The value of independent, third-party information across commodities*

We now calculate the expected value of independent, third-party information on the three experimental commodities using the methodology outlined in Rousu *et al.* (2007). The first step is to calculate the consumer surplus by subtracting the market price (e.g., grocery story price) of each commodity from each subject's commodity bids.[19] Next, we assume that each subject purchases the particular version of each commodity (e.g., conventional, "natural,"

[19]The market prices used in the calculations are the prices we paid for the foods in the experiment. Specifically, we paid \$0.98/lb for conventional apples and \$2.99/lb for organic apples. The conventional and organic broccoli prices are \$2.49 and \$2.99. The per-dozen conventional, "natural," and organic eggs' prices are \$2.49, \$2.68, and \$4.18, respectively. Since "natural" apples and broccoli were unavailable, we use the organic prices for these foods in the calculations in Table 6.

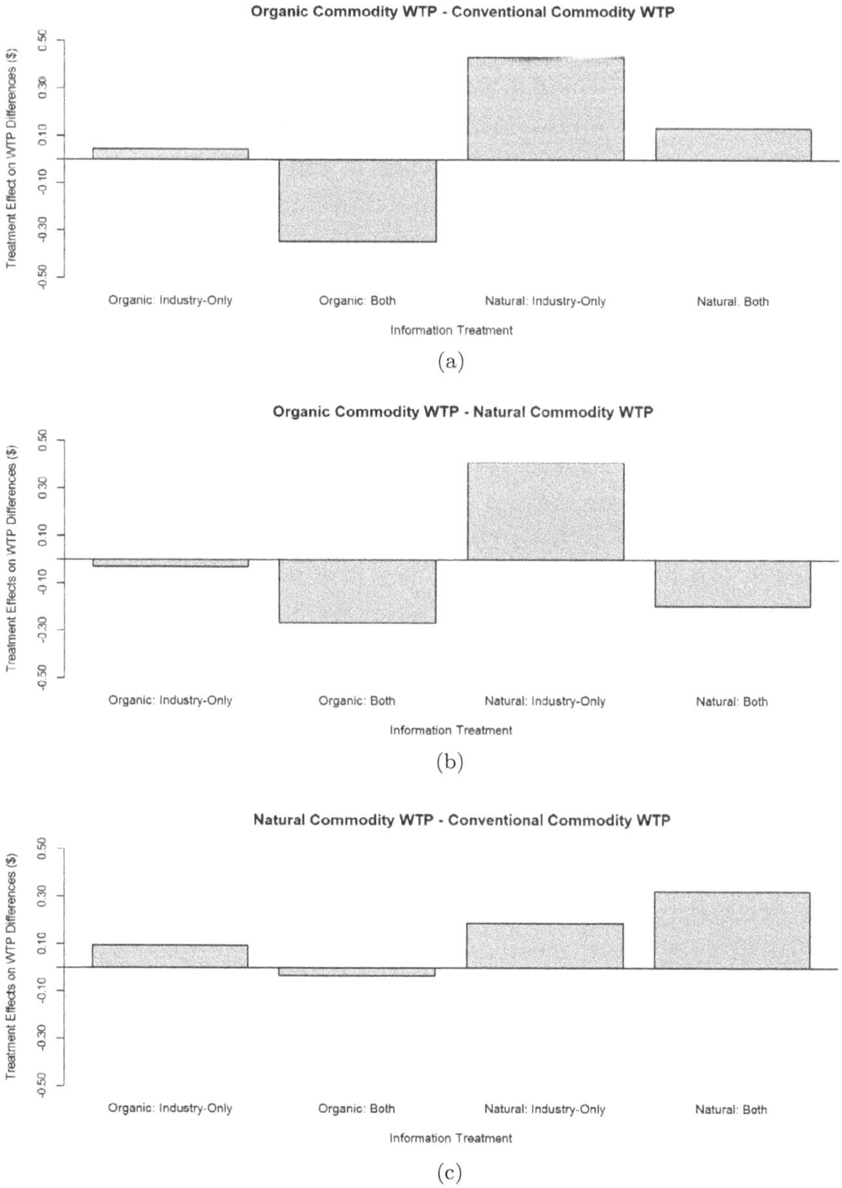

Fig. 3. Information treatment effects on WTP differences.

or organic) that yields the highest surplus.[20] For subjects who received independent information, premium gains are then calculated as the (positive) difference in surpluses between organic and conventional commodities or organic and "natural" commodities.

We then calculate the net change in the percentage of subjects who purchase conventional commodities as a result of reading independent, third-party information. This is the absolute value of the difference between the percentage who purchase conventional when treated to independent information and the percentage who purchase conventional but not treated to independent information. We are interested in two cases: (i) subjects whose surpluses are largest for conventional and second-largest for "natural" commodities ("switchers from "natural" to conventional foods") and (ii) subjects whose surpluses are largest for conventional and second largest for organic commodities ("switchers from organic to conventional foods"). For each of these cases, the first term in the difference is the fraction of subjects reading/receiving independent, third-party information, and the second term is the fraction of subjects that did not receive independent, third-party information.

For each commodity, switching probabilities are then calculated as the percentage that switch divided by the percentage that buy (e.g., those whose surpluses are largest for conventional and second-largest for "natural" commodities or those whose surpluses are largest for conventional and second-largest for organic commodities). Last, the expected value of information is the product of the switching probability and the mean premium gain. Interested readers are referred to Eqs. (5)–(13) of Rousu *et al.* (2007) for more information.

Table 6 provides estimates of the expected value of independent, third-party information for each of three commodities. In the first column of results, premium gains are calculated from the subsample

[20]Several subjects had surpluses for "natural" commodities equal to their surpluses for organic commodities. These ties are resolved by assuming that the highest surplus is from conventional commodities and the lowest surplus from either "naturals" or organics, depending on the context.

Table 6. Expected value of third-party information.

	Switching	
	Natural to Conventional	Organic to Conventional
Apples (per 3 lbs.)	\$2.90	\$2.97
Eggs (per dz.)	\$0.06	\$0.00
Broccoli (per 1.25 lbs.)	\$0.26	\$0.10

Note: Premium gains in the first column are calculated from sample sizes of 55, 36, and 50 subjects for apples, eggs, and broccoli (respectively). Premium gains in the second column are calculated from sample sizes of 49, 36, and 50 subjects for apples, eggs, and broccoli (respectively). The zero valuation of third-party information in switching from organic to conventional eggs arises because the percentage switching is zero, i.e., the number of subjects buying conventional who received third-party organic information equals the number of subjects buying conventional who received the organic industry information treatment.

of subjects who received either of the two "natural" information treatments, and whose surpluses were largest for conventional and then "natural" foods. In the second column of results, premium gains are calculated from the subsample of subjects who received either of the two organic information treatments, and whose surpluses were largest for conventional and then organic foods.

We find that the expected value of information for apples is largest at \$2.90–2.97 per three pounds. The expected value of information for broccoli, at \$0.10–0.26 per 1.25 lb, is also moderately large and similar to value of information estimates for tortilla chips, vegetable oil, and potatoes (Rousu *et al.*, 2007). Estimates for the dozen eggs are relatively small, a consequence of a smaller subsample size and fewer subjects switching. This suggests that independent, third-party information may have somewhat smaller impacts in the market for fresh eggs. This market has substantial variability in price premiums, and shoppers may be less willing to pay relatively high prices for certain kinds of eggs (Chang *et al.*, 2010). In general, our results relate to those of Onken *et al.* (2011) who find that subjects in Mid-Atlantic States would need a price subsidy between 1% and 13.7% to switch from "natural" to organic strawberry preserves.

5.3. *Policy implications for natural and organic markets*

Real production costs of conventional U.S. agriculture have decreased over the past several decades. U.S. organic and "natural" food producers do not seem to have experienced similar cost reductions (Kuminoff and Wossink, 2010). This hinders the growth in U.S. production for these products, though imports can help meet increasing demand (Jaenicke *et al.*, 2011). However, the current market structure continues to hinge on consumers willing to pay higher retail prices for organic foods (McBride *et al.*, 2015) and "natural" foods.

Past industry studies and academic research have focused on identifying demographics of those willing to pay premiums in these niche markets. This has presumably motivated market segmentation strategies: retailers appeal to observable characteristics of shoppers who are likely to be purchasers. Such strategies have helped to increase the presence and market share of "natural" and organic foods. Yet our experimental results reflect the fundamental shift in national market structure towards a much broader customer base. We find a large and diverse set of subject demographic effects, consistent with the observation that the national market is in transition away from niche status. This is important since major U.S. food processors and national supermarkets have been recently attempting major development and/or expansions of organic product lines (Howard, 2009; Harris and Strom, 2014; Nassauer, 2016). Long-term contracts and agreements with producers may help capture economies of scale, lowering premiums and thus reinforcing expansion.

Our finding of a cross-market asymmetric information effect, a kind of externality, has equally important implications. Subjects exposed to the "natural" food industry perspective are willing to pay $0.41–0.43 more for *organic* foods, whereas information about organic foods has no impact on their premiums for "natural" over conventional foods. It seems that the "natural" food industry could benefit from more objective information and meaningful labels. Organic markets have benefited from low-quality "natural" food information and vague labels, perhaps prompting some consumers to switch from "natural" to organic products. Some consumers have brought about class action lawsuits alleging deceptive business practices by "natural" food retailers (Benny, 2012; Negowetti, 2013; Thompson, 2014). Several of these suits have been filed by the Center for Science in the Public Interest in courts in California, a state with strong consumer protection laws, general support of organic agriculture, and aversion to GM foods in some counties. One potentially effective but costly method of reducing misinformation and deception is the development, implementation, and enforcement of a binding legal standard. However, the general welfare effects of such a standard are currently unclear.

Participants who regularly read organic labels pay relatively more for organic foods than those who do not. Although we find no effects from subjects' prior awareness of "natural" food labels, those who regularly read food labels when purchasing a new food item are willing to pay more for both organic and "natural" foods relative to conventional ones. Meaningful food labels, such as the USDA organic seal, can contribute to more effective information flow by summarizing relevant attributes of certified production processes. In effect, labels help ensure trust in credence goods without exhaustively listing all production practices.

In well-functioning markets, voluntary labeling might be expected to occur if demanded by a majority of market participants. For example, some private organizations have enacted independent certification procedures permitting foods to carry a "non-GM" seal if they meet voluntary standards and pass verification tests (Non-GMO Project, 2016). In contrast to these voluntary labeling efforts, a national bioengineered food disclosure standard was enacted in July 2016 (Public Law 114-216, 2016). It supersedes all state GM labeling legislation. If the FDA establishes a legal definition or standard for the term "natural" in food labeling (US-HHS-FDA, 2015) and the USDA establishes a verification system, this could have the potential to clear up the currently confusing "natural" designations.

6. Conclusion

U.S. demand for "natural" and organic foods have increased substantially in recent decades. This stems from renewed and increasing interest in nutrition, public health, animal welfare, and the environmental impacts of conventional agriculture. A general understanding of these diverse topics is complicated by a mix of labels and distorted claims. Conclusions from independent academic research about these foods are often at odds with arguments employed by private interest groups and trade associations. Food labels offer a similar dichotomy: certain seals like the USDA organic seal are a credible signal that the food product meets the NOP's standards. Foods with "natural" labels do not have an underlying binding legal standard. These foods are not made with uniform and certified production practices across farms and processors, and can contain ingredients that most consumers consider to be unnatural.

Our experimental design and results offer opportunities for future research. Although we can estimate the impact and value of third-party information by comparing treatment effects among subjects who receive both perspectives to those who receive only industry perspectives, another design approach would have been to randomly assign third-party perspectives only, in addition to the other randomly-assigned information treatments. However, a larger sample size would be required for identification of these two additional treatment effects. Moreover, a small set of food items display both the USDA organic seal and "natural" claims (e.g., eggs and processed snack foods). An experimental design that includes bidding on a fourth set of food items with both packaging claims would be an interesting and useful contribution, despite the potential increase in session length and experiment complexity.

Nevertheless, the results have broad policy implications for "natural" and organic retail markets and contribute to the ongoing discussion of the role of information and voluntary food labeling. First, subjects' organic food premiums (relative to both "natural" and conventional foods) tend to increase as a result of information favoring "natural" foods, but industry and independent, third-party perspectives on organic foods have no effect on subjects' "natural" food premiums. Pairing an independent, third-party perspective about organic foods with an organic industry perspective reduces both types of organic premiums. These findings imply greater scope for objective information to help dispel persistent misconceptions about the nutrition, taste, and environmental benefits of organic foods, but especially those surrounding "natural" foods. Second, our results suggest a demographic widening of demand. Retailers and other suppliers could potentially increase sales with more general advertising. However, subjects' demand (and likely price elasticities) varies by age of children in the household, implying a potential for additional segmentation strategies. Third, subjects' prior knowledge of organic foods and labeling influence their premiums, notably among those who actively read food labels when purchasing new foods. Higher-quality information, especially about "natural" foods, would promote greater effectiveness of labels. Misinformation about "natural" foods could be substantially eliminated through objective information campaigns (e.g., public service announcements), public-sector guidance for industry, and/or the development and enforcement of a legally binding standard.

Acknowledgments

The authors would like to thank Alicia Rosburg, Tim Lenell, Nicole Caviris, and Jennifer West for valuable assistance. We thank Rodolfo Nayga, Matthew Rousu, and Abebayehu Tegene for helpful comments. We acknowledge funding from USDA through the Economic Research Service (ERS) and the Iowa Agricultural and Home Economics Experiment Station. This research was approved by the ISU Institutional Review Board in ISU IRB #13-108.

Appendix A

Figs. A.1–A.4 and Table A.1.

Please take a few minutes to read the following information.

Organic Food Industry Perspective on Organic Foods

The following is a statement on organic foods expressing the views of leaders in the organic food industry, including the Organic Trade Association, organic food cooperatives, and Whole Foods Market.

General Statement:

- In the United States, the sales of organic food have grown very rapidly. This growth in demand is driven by many consumers seeking out foods that are perceived to be more healthful, tastier and more environmentally friendly than conventional foods.
- The term "organic" refers to foods produced in compliance with a comprehensive list of regulations from the USDA.
- The production of organic fruits, vegetables, grains, legumes, and other agricultural commodities does not use synthetic fertilizers or pesticides, and animals that produce organic milk and eggs eat organic feed and do not receive growth hormones or antibiotics.
- Certain livestock have access to pasture or outdoor areas for specified periods of time. Organic farming procedures also recycle resources and promote biodiversity. In the US, foods that contain at least 95% organic ingredients can be labeled as "organic" and carry the USDA organic seal USDA ORGANIC. Many consumers feel a strong affinity toward organic, integrated farming systems relative to more specialized farming systems.

Production Standards:

- Organic production and processing methods do not use synthetic fertilizers and pesticides, antibiotics, growth hormones, genetically modified organisms, irradiation, or sewage sludge. Organic producers frequently advertise their products as being pesticide and GMO free and environmentally friendly. These producers frequently integrate crop and livestock operations.
- Livestock are especially valued on organic farms because their manure, along with legume hay and beans, are the main sources of fertilizer nutrients and organic matter needed for growing organic vegetables, grains and fruits. Organic farmers use crop rotation and intensive seedbed preparation to prepare fields for planting and to give organic plants a head start on weeds. Post-emergent weed control is largely done by mechanical cultivation, mulching, and manually pulling weeds.
- Organic milk is produced from cows that are comfortably housed and obtain a significant amount of their feed from pasture grazing, with other feed coming from organic hay, grain and plant protein. Organic eggs are produced by hens that eat organic feed and are not kept in cages.

Nutrition, Taste and Appearance:

- The organic food industry sees organic foods as being more nutritious and tastier than conventional foods. In addition, The Organic Center has concluded that organic plant-based foods are on average nutritionally superior to conventional foods, especially when phytonutrients such as antioxidants are considered.
- Consumers purchase organic produce and eggs at local farmers markets because they believe that these foods are fresher, tastier and free of synthetic chemicals. For example, broccoli is considered to be one of the healthiest and most beneficial vegetables, containing vitamins and fiber, but organic broccoli is even better because of the rigorous production standards. One study found that organic apple puree contained more phenols, flavonoids and vitamin C than conventional apples. Apple puree is a popular food choice for babies, and if organic apples are used to make the puree, it is even better for children.

Fig. A.1. Organic food industry perspective on organic foods.

Please take a few minutes to read the following information.

Independent, Third Party, Verifiable Perspective on Organic Foods

The following is a statement on organic foods by a third-party group consisting of individuals knowledgeable about organic production systems and foods, including scientists at several leading land-grant universities.

General Statement:

- In 2002, the Agricultural Marketing Service implemented national uniform standards for the US production and processing of organic foods. It includes a list of permitted (prohibited) methods and practices or substances that should (should not) be used in organic production, processing and handling by domestic or foreign producers. These standards are strictly process based and do not make any claims about the nutrient value or food safety of organic foods.
- A USDA-certified organic producer must prepare an organic production plan that follows the National Organic Program (NOP) and have it certified by a USDA accredited certifier, must maintain records of practices and material used, and have an annual inspection. The NOP prohibits the use of genetic modification, synthetic fertilizers, synthetic pesticides, sewage sludge, and irradiation in production or processing.
- Only farms and processors that are USDA-certified organic, where foods contain at least 95 percent organic ingredients, can sell "organic" foods in the US and display the USDA's organic seal. In addition to natural foods grocery stores, organic foods are now available in almost all supermarkets and big box stores, such as Super Wal-Mart and Costco.

Production Standards:

- Producers and processors must follow production processes consistent with the NOP. Organic farmers must manage soil fertility and crop nutrients through tillage and cultivation practices, crop rotation, and cover crops, supplemented with animal and crop waste materials. This means that synthetic fertilizers, insecticides, herbicides and fungicides are "prohibited substances," but a few synthetic substances are permitted, such as vaccines to prevent livestock diseases, vitamins B_1, C, and E fortification of livestock feed, and the addition of selective soil micronutrients. Farmers can use organic pesticides or naturally occurring minerals but not synthetic pesticides.
- Organic producers must feed their livestock organic feeds, and livestock cannot be given growth hormones or antibiotics. Good management practices, including the use of vaccines, are to be used to reduce animal disease problems. All organically raised animals must be comfortably housed (no total and perpetual indoor confinement) and have access to the outdoors, including access to grazing pasture for ruminants and access to the outdoors for egg laying chickens. However, chickens are not required to be "free-range."

Nutrition, Taste and Appearance:

- A large amount of scientific evidence has concluded that organic foods are no more nutritious than conventional foods, but some antioxidants might be higher. However, pesticide levels on organic fruits and vegetables are lower than on conventional produce, both being below allowable Food and Drug Administration safety limits. In addition, organic chicken, pork and milk have reduced risk of antibiotics-resistant bacteria, but cage-free chickens have greater risk of exposure to bacteria in animal waste that they walk through, raising food safety risks for eggs from these birds.
- Organic foods do not necessarily taste better than conventional foods. Observed taste by a consumer is a result of personal preferences and degree of maturity of a product. Organic produce may have imperfect appearance—odd shapes, varying colors, smaller sizes, and blemishes because of the organic production process. In addition, organic produce may have a shorter shelf life or spoil faster because they are not treated with certain preservatives used on conventional produce.

Fig. A.2. Independent, third-party, verifiable perspective on organic foods.

Please take a few minutes to read the following information.

Natural Food Industry Perspective on "Natural and "All Natural" Foods

The following is a statement on Natural or All Natural foods expressing the views of leaders in the natural foods industry, including Trader Joe's brands, Kashi and Wildtree.

General Statement:

- Natural food websites describe all natural or natural foods as being fresher, better tasting, and containing fewer processed ingredients than conventional foods. In particular, a number of websites tote the health benefits of eating whole grains, fresh fruits and vegetables, grass fed beef and lamb, and free-range pigs and chickens, including the chicken's eggs.

- This emphasis on minimal processing of grains, fruits and vegetables and a natural environment for farm livestock and poultry is represented as a source of especially healthful food and a pleasant environment for farm animals. All natural fruits and vegetables are fresher, tastier and nutritious, and they have fewer food safety concerns than more processed foods. Consequently, many health benefits arise for those consuming natural or all natural food products.

Production Standards:

- To preserve the natural integrity of farm produced food, the only processes that should take place are physical processes that do not fundamentally alter a raw product or that only separate a whole intact food into component parts, i.e., smoking, roasting, freezing, drying, and fermenting.

- All natural free-range eggs are produced by hens that are allowed to roam freely within a farmyard, a shed or chicken coop, which permits large amounts of exercise, rather than being confined to cages. In addition, this free-ranging also allows the hens to consume a more varied diet and to experience natural sunlight, which make eggs taste better.

Nutrition, Taste and Appearance:

- While taste is subjective, many consumers believe that all natural and natural foods taste better than conventional foods. The potential health benefits from consuming an all natural and natural food diet include reduced consumption of potentially harmful substances such as hormones, pesticides, fertilizers and artificial ingredients.

- Consumers may be sensitive to some of these substances such as MSG, a common flavor-enhancer, and dyes used for food coloring. For some consumers, elimination of these sources of sensitivity may improve how they feel.

- In addition, minimally processed all natural or natural foods are more likely to contain whole food sources of beneficial nourishment, including fiber, complex carbohydrates, vitamins, and minerals. To the extent that a consumer's current diet is deficient in fiber, complex carbohydrates or vitamins, incorporating large amounts of all-natural foods can be expected to improve his or her health.

Fig. A.3. Natural food industry perspective on "natural" and "all-natural" foods.

Please take a few minutes to read the following information.

Independent, Third-Party Verifiable Perspective on Natural Foods

The following is a statement on "Natural" or "All Natural" foods by a third-party group consisting of individuals knowledgeable about natural foods, including scientists at several leading land- grant universities.

General Statement:

- In the United States, neither the Food and Drug Administration (FDA) nor the United States Department of Agriculture (USDA) has a formal definition of the term "natural" or "all natural." From a food science perspective, it is difficult to define a food product to be "natural" or "all natural" since food is often processed and no longer a product of the earth.

- Although the Food and Drug Administration (FDA) does not formally define the term "natural" or "all natural" for the food industry, it also does not object to the use of the term for food products that are free of synthetic or artificial ingredients and are minimally processed and without preservatives, artificial additives, growth hormones, antibiotics, hydrogenated oils, stabilizers and emulsifiers. Despite these terms being applied to many foods, the "natural" and "all natural" concept remains ambiguous, and the FDA discourages the food industry from using these terms on food labels in a manner that the agency views as misleading or deceptive.

Production Standards:

- Currently, only very weak production requirements exist for food products labeled as "natural" or "all natural" in the United States. The USDA's Food Safety and Inspection Service (FSIS) suggests the term "natural" or "all natural" for products containing no artificial ingredients or chemical preservatives and with minimal processing, such as traditional processes used to make food edible, safe for human consumption or for preservation, i.e. smoking, roasting, freezing, drying, and fermenting. More severe processing such as acid hydrolysis and chemical bleaching is considered more than minimally processed.

- "Natural" and "all natural" food products do not follow the production protocols of organic food production; rather, they aim to preserve the natural integrity of the food products as they come from the fields and farms. However, several lawsuits have been filed against major national brands that may have used the "natural" or "all natural" label deceptively. These brands include Tostitos, Sun Chips, Kashi, Snapple, Tropicana, Naked Juice, Jamba Juice, Ben and Jerry's, and Campbell Soup. Many of these suits are currently ongoing, although some have been dismissed or partially dismissed by federal judges.

Nutrition, Taste and Appearance:

- While consumers may think the terms "natural" and "all natural" imply "healthy," these food products may be less safe than similar conventional items. In 2002, records from the FDA and Health Canada revealed that some organic and "all natural" products were eight times more likely to be recalled for food safety problems, including bacterial contamination and mislabeling, than conventional foods.

- Furthermore, a large amount of scientific research does not conclude that "natural" or "all natural" foods are more nutritious than conventional foods. However, evidence from the United States FDA found that 6 out of 10 United States shoppers believe that "natural" and "all natural" foods are healthier than conventional foods.

Fig. A.4. Independent, third-party, verifiable perspective on natural food.

Table A.1. Zero bids across commodities and types.

	Conventional	Natural	Organic
Apples	2 (1.9%)	3 (2.9%)	2 (1.9%)
Eggs	3 (2.9%)	4 (3.9%)	1 (1.0%)
Broccoli	1 (1.0%)	2 (1.9%)	0 (0.0%)

Note: The first number in each entry is the count of zero bids. The second number in parentheses expresses each count as a fraction of the 102 bids per commodity and type.

References

Aertsens, J., K. Mondelaers, W. Verbeke, J. Buysse and G. Van Huylenbroeck (2011). The Influence of Subjective and Objective Knowledge on Attitude, Motivations and Consumption of Organic Food. *British Food Journal* 113 (11), 1353–1378.

Anstine, J. (2007). Organic and All Natural: Do Consumers Know the Difference? *Journal of Applied Economics & Policy* 26(1), 15–27.

Ariely, D., G. Loewenstein and D. Prelec (2003). Coherent Arbitrariness: Stable Demand Curves Without Stable Preferences. *Quarterly Journal of Economics* 118(1), 73–106.

Batte, M.T., N.H. Hooker, T.C. Haab and J. Beaverson (2007). Putting their Money Where their Mouths are: Consumer Willingness to Pay for Multi-Ingredient, Processed Organic Food Products. *Food Policy* 32(2), 145–159.

Bazoche, P., P. Combris, E. Giraud-Héraud, A.S. Pinto, F. Bunte and E. Tsakiridou (2014). Willingness to Pay for Pesticide Reduction in the EU: Nothing But Organic? *European Review of Agricultural Economics* 41(1), 87–109.

Becker, G., M. DeGroot and J. Marschak (1964). Measuring Utility by a Single-Response Sequential Method. *Behavioral Science* 9(3), 226–232.

Benny, E. (2012). "Natural" Modifications: the FDA's Need to Promulgate an Official Definition of "Natural" that Includes Genetically Modified Organisms. *The George Washing. Law Review* 80(5), 1504–1526.

Bernard, J.C., C. Zhang and K. Gifford (2006). An Experimental Investigation of Consumer Willingness to Pay for Non-GM Foods When an Organic Option is Present. *Agricultural and Resource Economics Review* 35(2), 374–385.

Bernard, J.C. and D.J. Bernard (2009). What is it About Organic Milk? An Experimental Analysis. *American Journal of Agricultural Economics* 91(3), 826–836.

Bernard, J.C. and D.J. Bernard (2010). Comparing Parts With the Whole: Willingness to Pay for Pesticide-Free, Non-GM, and Organic Potatoes and Sweet Corn. *Journal of Agricultural and Resource Economics* 35(3), 457–475.

Carlson, A. and E. Jaenicke (2016). Changes in retail organic price premiums from 2004 to 2010. Economic Research Report No. 209. United States Department of Agriculture, Economic Research Service.

Chang, J.B. and J.L. Lusk (2009). Fairness and Food Choice. *Food Policy* 34(6), 483–491.

Chang, J.B., J.L. Lusk and F.B. Norwood (2010). The Price of Happy Hens: A Hedonic Analysis of Retail Egg Prices. *Journal of Agricultural and Resource Economics* 35(3), 406–423.

Corrigan, J.R. and M.C. Rousu (2006). Posted Prices and Bid Affiliation: Evidence from Experimental Auctions. *American Journal of Agricultural Economics* 88(4), 1078–1090.

Dettmann, R.L. and C. Dimitri (2009). Who's Buying Organic Vegetables? Demographic Characteristics of U.S. Consumers. *Journal of Food Products Marketing* 16(1), 79–91.

Dimitri, C. and R.L. Dettmann (2012). Organic Food Consumers: What Do We Really Know About Them? *British Food Journal* 114(8), 1157–1183.

FMI (2014). U.S. Grocery Shopper Trends 2014. Report. Food Marketing Institute.

Gifford, K. and J.C. Bernard (2004). The Impact of Message Framing on Organic Food Purchase Likelihood. *Journal of Food Distribution Research* 35(3), 19–28.

Gifford, K. and J.C. Bernard (2006). Influencing Consumer Purchase Likelihood of Organic Food. *International Journal of Consumer Studies* 30(2), 155–163.

Gifford, K. and J.C. Bernard (2011). The Effect of Information on Consumers' Willingness to Pay for Natural and Organic Chicken. *International Journal of Consumer Studies* 35(3), 282–289.

Glennerster, R. and K. Takavarasha (2013). *Running Randomized Evaluations: A Practical Guide.* Princeton, NJ: Princeton University Press.

Govindasamy, R. and J. Italia (1999). Predicting Willingness-To-Pay a Premium for Organically Grown Fresh Produce. *Journal of Food Distribution Research* 30(July), 44–53.

Greene, C., C. Dimitri, B.H. Lin, W. McBride, L. Oberholtzer and T. Smith (2009). Emerging issues in the U.S. organic industry. Economic Information Bulletin No. 55. United States Department of Agriculture, Economic Research Service.

Hamilton, S.F. and D. Zilberman (2006). Green Markets, Eco-Certification, and Equilibrium Fraud. *Journal of Environmental Economics and Management* 52(3), 627–644.

Harris, E.A. and S. Strom (2014). Walmart to Sell Organic Food, Undercutting Big Brands. The New York Times: B9, p. B9.

He, N. and J.C. Bernard (2011). Differences in WTP and Consumer Demand for Organic and Non-GM Fresh and Processed Foods. *Agricultural and Resource Economics Review* 40(2), 218–232.

Howard, P.H. (2009). Organic Industry Structure. *Media-N: J. New Media Caucus* 5(3).

Huang, C.L. (1996). Consumer Preferences and Attitudes Towards Organically Grown Produce. *European Review of Agricultural Economics* 23(3), 331–342.

Huang, C.L. and B.-H. Lin (2007). A Hedonic Analysis of Fresh Tomato Prices Among Regional Markets. *Review of Agricultural Economics* 29(4), 783–800.

Huffman, W.E., J.F. Shogren, M. Rousu and A. Tegene (2003). Consumer Willingness to Pay for Genetically Modified Food Labels in a Market with Diverse Information: Evidence from Experimental Auctions. *Journal of Agricultural and Resource Economics* 28(3), 481–502.

Huffman, W.E., M. Rousu, J.F. Shogren and A. Tegene (2007). The Effects of Prior Beliefs and Learning on Consumers Acceptance of Genetically Modified Foods. *Journal of Economic Behavior and Organization* 63(1), 193–206.

Jaenicke, E., C. Dimitri and L. Oberholtzer (2011). Retailer Decisions About Organic Imports and Organic Private Labels. *American Journal of Agricultural Economics* 93(2), 597–603.

Jolly, D.A., H.G.K. Schultz, V. Diaz-Knauf and J. Johal (1989). Organic Foods: Consumer Attitudes and Use. *Food Technology* 43(11), 60–66.

Klonsky, K. (2012). Comparison of Production Costs and Resource Use for Organic and Conventional Production Systems. *American Journal of Agricultural Economics* 94(2), 314–321.

Kuminoff, N.V. and A. Wossink (2010). Why Isn't More US Farmland Organic? *Journal of Agricultural Economics* 61(2), 240–258.

Lin, B.-H., T.A. Smith and C.L. Huang (2008). Organic Premiums of U.S. Fresh Produce. *Renewable Agriculture and Food Systems* 23(3), 208–216.

Lin, B.-H., S.T. Yen, C.L. Huang and T.A. Smith (2009). U.S. Demand for Organic and Conventional Fresh Fruits: The Roles of Income and Price. *Sustainability* 1(3), 464–478.

Lusk, J.L. and J.F. Shogren (2007). *Experimental Auctions: Methods and Applications in Economic and Marketing Research.* Cambridge, UK: Cambridge University Press.

Maddala, G. (1983). Limited Dependent and Qualitative Variables in Econometrics. *Econometric Society Monographs in Quantitative Economics*, vol. 3. Cambridge, UK: Cambridge University Press.

Maguire, K.B., N. Owens and N.B. Simon (2004). The Price Premium for Organic Babyfood: A Hedonic Analysis. *Journal of Agricultural and Resource Economics* 29(1), 132–149.

Maguire, K.B., N. Owens and N.B. Simon (2006). Focus on Babies: A Note on Parental Attitudes and Preferences for Organic Babyfood. *Journal of Agribusiness* 24(2), 187–195.

McBride, W.D., C. Greene, L. Foreman and M. Ali (2015). The Profit Potential Of Certified Organic Field Crop Production. Economic Research Report No. 188. United States Department of Agriculture, Economic Research Service.

Meas, T., W. Hu, M.T. Batte, T.A. Woods and S. Ernst (2015). Substitutes or Complements? Consumer Preference for Local and Organic Food Attributes. *American Journal of Agricultural Economics* 97(4), 1044–1071.

Menkhaus, D.J., G.W. Borden, G.D. Whipple, E. Hoffman and R.A. Field (1992). An Empirical Application of Experimental Economics in Marketing Research. *Journal of Agricultural and Resource Economics* 17(1), 44–55.

Milgrom, P. and J. Roberts (1986). Relying on the Information of Interested Parties. *The RAND Journal of Economics* 17(1), 18–32.

Misra, S.K., C.L. Huang and S.L. Ott (1991). Consumer Willingness to Pay for Pesticide Free Fresh Produce. West. *Journal of Agricultural Economics* 16(2), 218–227.

Nassauer, S. (2016). Wal-Mart to drop wild oats organic food brand. Wall Street J. (Business Section).

National Organic Program (2002). Organic Production and Handling Requirements. 7 CFR Part 205.

Negowetti, N.E. (2013). A National "Natural" Standard for Food Labeling. *Maine Law Review* 65(2), 581–604.

Nielsen Company (2015). We Are What We Eat: Healthy Eating Trends Around the World, Report. January, 2015.

Non-GMO Project (2016). Non-GMO Project Standard. Version 13 http://www.nongmoproject.org/wp-content/uploads/Non-GMO-Project-Standard.pdf (accessed 5-14-2016).

Onken, K.A., J.C. Bernard and J.D. Pesek Jr. (2011). Comparing Willingness to Pay for Organic, Natural, Locally Grown, and State Marketing Program Promoted Foods in the Mid-Atlantic Region. *Agricultural and Resource Economics Review* 40(1), 33–47.

Organic Trade Association (2016). U.S. Organic Sales Post New Record of $43.3 Billion in 2015. Press Release. May, 2016.

Park, T.A. and L. Lohr (1996). Supply and Demand Factors for Organic Produce. *American Journal of Agricultural Economics* 78(3), 647–655.

Public Law 114-216 (2016). An Act to Reauthorize and Amend the National Sea Grant College Program Act, and For Other Purposes. National Bioengineered Food Disclosure Standard. 130 Stat. 834. July 29, 2016.

Roosen, J., J.A. Fox, D.A. Hennessy and A. Schreiber (1998). Consumers' Valuation of Insecticide Use Restrictions: An Application to Apples. *Journal of Agricultural and Resource Economics* 23(2), 367–384.

Rousseau, S. and L. Vranken (2013). Green Market Expansion by Reducing Information Asymmetries: Evidence for Labeled Organic Food Products. *Food Policy* 40, 31–43.

Rousu, M., W.E. Huffman, J.F. Shogren and A. Tegene (2007). Effects and Value of Verifiable Information in a Controversial Market: Evidence from Lab Auctions of Genetically Modified Food. *Economic Inquiry* 45(3), 409–432.

Shogren, J., M. Margolis, C. Koo and J. List (2001). A Random nth-Price Auction. *Journal of Economic Behavior and Organization* 46(4), 409–421.

Smith, T.A., C.L. Huang and B.-H. Lin (2009). Estimating Organic Premiums in the U.S. Fluid Milk' Market. *Renewable Agriculture and Food Systems* 24(3), 197–204.

Stzrok, J.L. and W.E. Huffman (2015). Willingness to Pay for Organic Food Products and Organic Purity: Experimental Evidence. *AgBioForum* 18(3), 345–353.

Thompson, G.D. (1998). Consumer Demand for Organic Foods: What We Know and What We Need to Know. *American Journal of Agricultural Economics* 80(5), 1113–1118.

Thompson, G.D. and J. Kidwell (1998). Explaining the Choice of Organic Produce: Cosmetic Defects, Prices, and Consumer Preferences. *American Journal of Agricultural Economics* 80(2), 277–287.
Thompson, S. (2014). Artificially “Natural”: Class Action Lawsuits Attack Misleading “Natural” Claims in FDA’s Absence. *Indian Law Review* 47(3), 893–918.
USDA-ERS (2016). Organic Market Overview. http://www.ers.usda.gov/topics/natural-resources-environment/organic-agriculture/organicmarket-overview.aspx (accessed 5-14-2016).
USDA-FSIS (2005). Food Standards and Labeling Policy Book, Report. United States Department of Agriculture, Food Safety and Inspection Service.
US-HHS-FDA (1991). Food Labeling: Nutrient Content Claims, General Principles, Petitions, Definitions of Terms. 56 Fed. Reg. 60,421, 60,466. United States Department of Health and Human Services, Food and Drug Administration.
US-HHS-FDA (2015). Use of the Term “natural” in the Labelling of Human Food Products; Request for Information and Comments.” 80 Fed. Reg. 69,905, 69,909. United States Department of Health and Human Services, Food and Drug Administration.
USDA-NASS (2010). Organic Production Survey (2008). Volume 3, Special Studies, part 2. United States Department of Agriculture, National Agricultural Statistics Service.
USDA-NASS (2012). 2011 Certified Organic Production Survey, Report. United States Department of Agriculture, National Agricultural Statistics Service.
USDA-NASS (2016a). Organic Survey (2014). Volume 3, Special Studies, Part 4. United States Department of Agriculture, National Agricultural Statistics Service.
USDA-NASS (2016b). “Quick Stats 2.0” Website https://quickstats.nass.usda.gov (accessed 5-14-2016).
Vickrey, W.S. (1961). Counterspeculation, Auctions, and Competitive Sealed Tenders. *Journal of Finance* 16(1), 8–37.
Yue, C., F. Alfnes and H.H. Jensen (2009). Discounting Spotted Apples: Investigating Consumers’ Willingness to Accept Cosmetic Damage in an Organic Product. *Journal of Agricultural and Applied Economics* 41(1), 29–46.
Zhang, F., C.L. Huang, B.-H. Lin and J.E. Epperson (2008). Modeling Fresh Organic Produce Consumption with Scanner Data: A Generalized Double Hurdle Model Approach. *Agribusiness* 24(4), 510–522.

Part III

Enhanced Nutrition and Food Safety Achieved Using New Technology

Chapter 11

Improving the Nutrient Content of Food through Genetic Modification: Evidence from Experimental Auctions on Consumer Acceptance*

Gregory J. Colson[†], Wallace E. Huffman[‡], and Matthew C. Rousu[§]

[†] *Department of Agricultural and Applied Economics at the University of Georgia*

[‡] *Agriculture and Life Sciences at Iowa State University*

[§] *Susquehanna University*

This paper assesses consumers' acceptance of nutritionally enhanced vegetables using a series of auction experiments administered to a random sample of adult consumers. Evidence suggests that consumers are willing to pay significantly more for fresh produce with labels signaling enhanced levels of antioxidants and vitamin C achieved by moving genes from within the species, as opposed to across species. However, this premium is significantly affected by diverse information treatments injected into the experiments.

Keywords: Bayesian analysis, experimental auction, food products, genetic modification

1. Introduction

Daily consumption of fruits and vegetables in the United States is significantly lower than national dietary recommendations, despite extensive information campaigns, outreach efforts, and popular-media attention highlighting the relationship between food choices

*This article originally appeared in the **Journal of Agricultural and Resource Economics (2011), 36(2), 343–364**. This is published by permission of **Western Agricultural Economics Association**.

Review coordinated by Gary Brester.

and health outcomes (Guenther *et al.*, 2006; Kimmons *et al.*, 2009). Fruits and vegetables are promoted in the battle against obesity as a superior alternative to high-caloric foods (e.g., sugar based beverages, calorie-dense processed snacks). They are also a natural source of the vitamins, minerals, and antioxidants that play a critical role in determining health outcomes (e.g., Block, Patterson, and Subar, 1992; Joshipura *et al.*, 2001; Epstein *et al.*, 2001), but the proliferation of foods fortified or enriched with vitamins and minerals (e.g., breakfast cereals) and daily multivitamin tablets has diminished this role (Subar *et al.*, 1998; Berner, Clydesdale, and Douglass, 2001).[1] The emergence of non-food sources of nutrients coupled with the burgeoning success of the functional food market has helped motivate the biotechnology industry to develop new crop varieties engineered for superior nutrient levels, such as potatoes with enhanced levels of antioxidants and vitamin C.

While genetic modification has potential for delivering improved product quality by enhancing nutrient content, since their emergence in the market in the mid-1990s genetically modified (GM) foods have remained domestically and internationally a socially and politically sensitive topic due to their perceived unnatural production. Early GM foods were engineered using "transgenic" methods, in which genes from one species are transferred into a different species (e.g., from soil bacteria into corn). Studies have shown that consumers view these transgenic foods as weakly inferior to similar non-GM alternatives (e.g., Huffman *et al.*, 2003; Noussair, Robin, and Ruffieux, 2004; Lusk *et al.*, 2005; Rousu *et al.*, 2007). The transgenic nature of these genetically modified organisms (GMOs) has been one dimension of consumers' and environmental groups' resistance to genetic modification of plants–raising biodiversity, environmental, ethical, and safety concerns–and has been a factor in the larger controversy surrounding GMOs (for reviews of the GM debate see Herdt, 2006; van den Bergh and Holley, 2002).

[1]See Woodside *et al.* (2005) for a review of medical evidence on supplements vs. food sources.

To counter some of these concerns, new "intragenic" bioengineering methods have been developed that only transfer genes from within the species (e.g., from a wild potato variety into a commercial potato variety), exploiting the natural and induced diversity of species that has evolved over time in diverse geo-climates. The traditional method of slicing genes includes the use of antibiotic markers, which mark the location of a transgene on chromosomes, but this practice has raised public concern. A new gene transfer system has been developed that does not infringe on existing gene splicing patents and does not use antibiotic markers (see Rommens *et al.*, 2004). No "foreign" genetic material is used in this new gene transport system, removing one of the perceived negative attributes of earlier transgenic foods. Moreover, the new method has many features in common with traditional plant breeding methods.

This study examines consumers' willingness to pay for foods that have been enhanced with consumer attributes–vitamins and antioxidants–delivered by transgenic and intragenic GM methods controlling for information effects. The study uses an auction market mechanism for revealing willingness to pay and participants randomly chosen from consumers in two urban areas. Building on the methodology of Rousu *et al.* (2007), food labels and information treatments are an important part of the research design. As an additional advance in analyzing experimental auction bid-price data, we construct and estimate a Bayesian econometric model of bid prices that simultaneously controls for bid censoring, commodity fixed effects, bidding-round fixed effects, and correlation of unobserved effects across rounds of bidding.

2. Experimental Design

In spring 2007, an independent survey agency called a random set of telephone numbers in two cities to solicit participants for our experiments. Participants were chosen from two geographically separated cities to obtain some regional balance in our participants, and they were only told that this was a university project on consumers' assessments of household products and foods. Willing participants were then permitted to chose among three or four starting times on the auction days (March 24 in Harrisburg, Pennsylvania, and April 14 in Des Moines, Iowa; both dates were Saturdays), and they were given instructions on how to reach the lab site or classroom. In our experiments, we disseminate several information treatments randomly within the same session to avoid potentially confounding session effects.

Our methodology incorporated and refined established experimental procedures (Hoffman *et al.*, 1993; Shogren *et al.*, 1994; Lusk *et al.*, 2001; Alfnes and Rickertsen, 2003) and advances made by Rousu *et al.* (2007). We used adult consumers from two distinct geographic regions that were drawn from a random phone book sample, ensuring that our results are not artifacts of a single geographic region. We paid participants a fixed amount for participating in the experiments, but do not endow participants with products and have them bid to upgrade to another product, because new evidence by Corrigan and Rousu (2006) and Plott and Zeiler (2007) has shown that session monitors in the past may have induced significant endowment effects in bid prices by emphasizing the personal gift nature of in-kind transfers made to participants as part of the experiment. To assess the impact of interested parties (e.g., biotechnology, environmental, and independent scientific groups) attempting to impact market outcomes by strategically using private information (Milgrom and Roberts, 1986), we injected randomized information treatments into the experiments. We used the nth-price auction mechanism (Shogren *et al.*, 2001), which has been shown to be a demand revealing mechanism that better engages off-margin bidders. To avoid strategic behavior in bidding in sequential price auctions (Bajari and Hortaçsu, 2003; Guerre, Perrigne, and Vuong, 2000; Corrigan and Rousu, 2006), all bids are collected in a session before any information about the bid levels and distributions are revealed. We also randomized all food labels to eliminate sequencing effects. Finally, in many previous experiments where information is disseminated to participants (Lusk *et al.*, 2004; Rousu *et al.*, 2007), each "session" receives the same information treatment. In our experiments, we disseminate several information treatments randomly within the same session, thus avoiding potentially confounding session effects.

Upon arrival at the experiment site, participants were alternately assigned to one of two concurrent sessions (related individuals were assigned to different sessions). Each session consisted of nine to seventeen individuals and lasted approximately ninety minutes. Participants were asked to sign a consent form and were paid $45 dollars for their participation. Next, they were asked to complete a short questionnaire soliciting socioeconomic and demographic information and to answer a few questions about agricultural technologies. A total of fourteen sessions (eight in Des Moines and six in Harrisburg) were conducted. With regards to the location for experiments, it should be noted that many experimental studies are now being conducted in settings that are more familiar to consumers (Lusk, Pruitt, and Norwood, 2006). We also considered the possibility of using an intercept sample in a grocery store in a "framed field experiment" (Harrison and List, 2004), but the length of the experiment prohibited that option.

Participants were informed that they would be engaging in an auction of food products and provided with instructions and examples about the auction method used in the study: the nth-price auction. In this type of auction, all individuals who bid higher than the randomly selected "nth-price" win the auction and pay the nth-price for the commodity. Instead of using the more common Vickrey sealed bid second-price auction mechanism (Vickrey, 1961), the nth-price auction was selected based on evidence that it better engages off-margin bidders while still being a demand-revealing mechanism. The Becker and Marschak (1964) mechanism is also demand-revealing (e.g., Wertenbroch and Skiera, 2002), but the random nth-price auction has been shown to be more accurate at revealing preferences in experiments, potentially due to the endogenous clearing price (Lusk and Rousu, 2007). As part of the instruction process, participants were told that their preferred strategy should be to bid their true preferences.

During a practice phase, participants engaged in a two-round nth-price auction with candy, pens, and pencils to gain experience with the nth-price auction. Participants were then told that the auction would consist of four rounds of bidding, but only one round would be binding and it would be chosen after all bids were submitted.

This format reduces participants' concerns about exceeding their resources (the $45 dollars plus any cash they brought with them to the experiment) and fixes the idea that despite multiple bidding rounds they are bidding on only one unit of each of the auctioned commodities, eliminating potential demand effects associated with multiple purchases.

Three information perspectives on GMOs and a "no information" baseline were used to construct five information treatments for the experiments. The pro-biotech perspective consists of a collection of mainly positive or optimistic statements on GM provided by a group of leading agricultural biotechnology companies. The anti-biotech perspective consists of a collection of mainly negative or pessimistic statements on GM from leading environmental groups. The third-party perspective (or verifiable information) is a collection of statements representing an objective assessment of GM at the time the experiments were conducted. This perspective was linked to scientists, professions, religious leaders and academics, none of whom had a financial stake in GM foods. To ensure that the volume of information contained in the three perspectives on GMOs was not overwhelming to participants, each perspective was limited to one standard sheet of copy paper and organized under five common headings. The three information perspectives are presented in the appendices. The order of pro- and anti-biotech perspectives was randomized in treatments consisting of more than one perspective. The verifiable perspective was always presented last. Throughout this article, the following terms will be used synonymously to refer to types of information: 1) industry, positive, pro-biotech, 2) environmental, negative, and anti-biotech, and 3) verifiable and third party.

In each round of the auction, participants submitted three separate bids: one bid for each of the three auction commodities. The auction commodities were one pound of broccoli, one pound of beefsteak tomatoes, and five pounds of russet potatoes. Products were presented in plain packaging similar to how they are displayed in a grocery store, and a simple label was affixed. In each bidding round, the three commodities with labels were revealed on a table in the front of the lab. In half of the sessions (3 in PA and 4 in IA), the four food labels (one in each round) were: GM Free, Intragenic GM, Transgenic GM, and Plain. Product name (e.g., Russet Potatoes) and weight were listed on each label. The phrase "Plain" describes a label containing only the product name and weight. Figure 1 presents examples of labels. In the other half of the sessions (seven total), the first three rounds of the auction consisted of products with food labels of either GM, Intragenic GM, or Transgenic GM, but also offered additional information: "Enhanced levels of Antioxidants and Vitamin C." Figure 2 presents examples of these labels. A fourth label treatment, not pertinent to this article, always appeared in the final auction round and would not affect bids in the earlier three rounds.

All three products within a round of bidding had the same food label, and the order in which labels were presented was randomized across sessions. After a set of experimental products was revealed, participants were asked to come to the front of the room and view the products before writing their three bids. These bids were then collected by the session monitor before proceeding to the next round of bidding.

Russet Potatoes (5 lb.)	Russet Potatoes (5 lb.) GM Free Product
Russet Potatoes (5 lb.) Intragenic GM Product	Russet Potatoes (5 lb.) Transgenic GM Product

Fig. 1. Examples of auction food labels for products without enhanced nutrients.

Russet Potatoes (5 lb.) Enhanced levels of Antioxidants and Vitamin C GM Product	Russet Potatoes (5 lb.) Enhanced levels of Antioxidants and Vitamin C Intragenic GM Product	Russet Potatoes (5 lb.) Enhanced levels of Antioxidants and Vitamin C Transgenic GM Product

Fig. 2. Examples of auction food labels for products with enhanced nutrients.

After completion of all bidding rounds, the binding bidding round was selected by drawing a number from an envelope. Bids were then posted and ranked on a whiteboard in the front of the lab (no bids were posted prior to this point). Finally, for each of the three commodities a random n was drawn from an envelope to determine the clearing prices. Winners were then identified. All participants were then asked to complete a short exit questionnaire. Upon completing the questionnaire, non-winners were told that they were free to leave, and winners were told to go to an adjacent room to complete their purchases, exchanging money for goods. Given the incomplete regulatory status of the intragenic foods, we were unable to deliver nutrient-enhanced GM fresh vegetables to winners. As an alternative, winners were informed and given plain-labeled food products, which is similar to procedures followed by others in similar circumstances (e.g., Alfnes and Rickertsen, 2003; Corrigan *et al.*, 2009; Tonsor *et al.*, 2005). Participants' reception to these experiments was very good, and we did not receive any complaints.[2]

3. Summary of Data

Table 1 presents basic summary statistics of the auction participants. Participants are 68% female, the mean age is 44 years, mean education is 14.5 years, and mean household income is $51,000. Only 11% of the participants consider themselves well- or extremely well-informed about GM, and 17% report an opinion of GM that

[2]Some of the food labels for fresh vegetables were different from the actual genetic make-up of the product. This discrepancy was not perceptible to participants and no winner was misled about the product that he or she received at the end of the experiment. Some call this practice "deception." There is not much research on the effects of this type of deception, and the few studies that have investigated it find small–if any–impact. Ethically, the benefits from this type of deception need to be weighed against the costs. If university students are the participants and they participate repeatedly within the same lab, then the practice is problematic to good research (e.g., Ortmann and Hertwig, 2002). However, when participants are randomly drawn adults from the general population, the impact on future experimental participants is going to be very small. Some argue for debriefing sessions at the end of experiments (Bonetti, 1998), but our experimental design, which was approved by the ISU Institutional Review Board on Human Subjects, did not lend itself to a complete debriefing of non-winners.

Table 1. Summary statistics for auction participants ($N = 190$).

Variable	Variable Definition	Mean	Stdev
Gender	1 if female	0.68	0.47
Age	Participant's age	44.33	15.8
Income	Household income (in 1000s)	51.09	35.23
Education	Years of schooling	14.47	2.26
Married	1 if married	0.53	0.50
Household	Number of people in household	2.74	1.41
Race	1 if participant is white	0.85	0.36
Informed	1 if well or extremely well informed about GM	0.11	0.31
Opinion	1 if opinion towards GM is supportive	0.17	0.38
Read_Labels	1 if often or always read food labels	0.63	0.48
Envi_Mem	1 if member of environmental group	0.04	0.20
Farm	1 if previously/currently engaged in farming	0.04	0.21
Smoke	1 if smoke	0.23	0.42
Exercise	1 if exercise regularly	0.51	0.51
Health_Diet	Self assessed healthiness of diet (1–10 scale)	6.73	1.61
Health_Phys	Self assessed physical health (1–10 scale)	7.16	1.69

is supportive or strongly supportive. 63% report that they often or always read food labels. Regarding lifestyle indicators, 23% report that they smoke and 51% report exercising regularly. Participants self-assessed the healthiness of diet with a mean score of 6.7 on a scale of 1 to 10 and self-assessed their physical health with a mean of 7.2.

Tables 2 and 3 summarize participants' bid prices for fresh produce in the experimental auctions. Table 2 contains the mean and standard deviation (in parentheses) of bid prices for products without enhanced nutrient labels, conditional on information treatments. Table 3 presents similar summary statistics for products with enhanced nutrient labels. Fewer than 8% of placed bids were zeros. For products without enhanced nutrient labels, the mean bid pooled across all information treatments for products labeled GM Free is greater than for Plain-labeled products; similarly, Plain-labeled products had a higher mean bid than products labeled as Transgenic GM. The ordering of mean bid prices for the Plain label and Intragenic GM label is not the same across products. For products with enhanced nutrient labels, mean bid prices for Intragenic GM

Table 2. Mean bid prices for foods without enhanced nutrient label.

Broccoli (1 lb.)				Tomato (1 lb.)				Potato (1 lb.)			
Plain	GMF	Intra	Trans	Plain	GMF	Intra	Trans	Plain	GMF	Intra	Trans
All Treatments ($N = 92$)											
1.28	1.46	1.42	1.20	1.38	1.52	1.36	1.18	2.16	2.34	2.16	2.00
(0.76)	(0.77)	(0.86)	(0.78)	(0.98)	(0.91)	(0.97)	(0.87)	(1.16)	(1.16)	(1.23)	(1.22)
No Information ($N = 17$)											
1.38	1.69	1.54	1.37	1.39	1.47	1.35	1.25	1.99	2.36	2.03	2.07
(0.61)	(0.82)	(0.70)	(0.66)	(1.08)	(1.04)	(0.88)	(0.83)	(0.84)	(1.05)	(0.79)	(0.88)
Pro-biotech Information only ($N = 20$)											
1.30	1.37	1.74	1.24	1.49	1.46	1.62	1.29	2.33	2.33	2.54	2.21
(0.90)	(0.71)	(0.83)	(0.83)	(1.16)	(0.88)	(1.01)	(1.06)	(1.22)	(1.15)	(0.99)	(1.23)
Anti-biotech Information only ($N = 17$)											
1.21	1.60	1.07	1.14	1.24	1.67	0.95	0.98	2.19	2.71	1.84	1.81
(0.72)	(0.73)	(0.83)	(0.84)	(0.93)	(0.97)	(0.81)	(0.74)	(0.85)	(0.96)	(1.20)	(1.13)
Pro-biotech & Anti-biotech Information ($N = 21$)											
1.45	1.58	1.56	1.22	1.48	1.73	1.60	1.25	2.34	2.43	2.45	2.18
(0.79)	(0.76)	(0.99)	(0.76)	(0.86)	(0.87)	(1.11)	(0.83)	(1.19)	(1.14)	(1.42)	(1.22)
Pro-biotech, Anti-biotech, and Verifiable Information ($N = 17$)											
1.03	1.07	1.08	1.03	1.23	1.22	1.19	1.09	1.86	1.86	1.83	1.66
(0.73)	(0.76)	(0.78)	(0.84)	(0.93)	(0.82)	(0.87)	(0.88)	(1.56)	(1.42)	(1.51)	(1.59)

Notes: Average bid prices are in dollars. Standard deviations are presented in parentheses.

Table 3. Mean bid prices for foods with enhanced nutrient label.

GMF	Intra	Trans	GMF	Intra	Trans	GMF	Intra	Trans
All Treatments ($N = 98$)								
1.51	1.67	1.45	1.42	1.76	1.41	2.45	2.61	2.27
(1.01)	(1.14)	(1.01)	(0.86)	(1.26)	(0.97)	(1.76)	(1.84)	(1.96)
No Information ($N = 20$)								
1.91	1.86	1.83	1.65	1.95	1.73	3.18	3.20	3.23
(1.54)	(1.16)	(1.28)	(1.23)	(1.34)	(1.32)	(3.06)	(2.73)	(3.40)
Pro-biotech Information only ($N = 18$)								
1.63	2.52	1.79	1.81	2.64	1.90	2.73	3.49	2.65
(0.65)	(1.20)	(0.68)	(0.70)	(0.94)	(0.68)	(1.00)	(1.89)	(1.35)
Anti-biotech Information only ($N = 18$)								
1.25	1.07	1.06	1.23	1.10	0.98	2.12	1.92	1.71
(0.82)	(0.89)	(0.89)	(0.69)	(0.66)	(0.63)	(1.46)	(1.34)	(1.37)
Pro-biotech & Anti-biotech Information ($N = 20$)								
1.67	1.84	1.63	1.36	1.74	1.43	2.54	2.64	2.34
(0.87)	(1.15)	(1.04)	(0.67)	(1.32)	(1.04)	(1.14)	(1.45)	(1.24)
Pro-biotech, Anti-biotech, and Verifiable Information ($N = 22$)								
1.10	1.16	0.97	1.11	1.44	1.05	1.74	1.90	1.48
(0.77)	(0.70)	(0.74)	(0.75)	(1.38)	(0.72)	(0.98)	(0.87)	(0.97)

Notes: Average bid prices are in dollars. Standard deviations are presented in parenthesis.

labeled products pooled across all information treatments are greater than for generic GM, which in turn are higher than for Transgenic GM labeled products.

Tables 4, 5, and 6 present differences in mean bid prices across a selection of different labels. Table 4 presents a comparison between bid prices for GM Free vs. Intragenic GM and GM Free vs. Transgenic GM labeled products, without enhanced nutrient labels. Consumers are willing to pay a premium for GM Free over Transgenic GM labeled products under all information treatments. Average premiums under different information treatments range from \$0.03 to \$0.46 per pound of broccoli, \$0.13 to \$0.69 per pound of tomatoes, and \$0.13 to \$0.90 per five pounds of potatoes. In general, participants are willing to pay premiums for GM Free over Intragenic GM

Table 4. Difference in mean bid prices for foods without enhanced nutrient label.

	GM Free w/o EN vs. Intragenic w/o EN			GM Free w/o EN vs. Transgenic w/o EN		
Info Treatment	Broccoli	Tomato	Potato	Broccoli	Tomato	Potato
No info	\$0.15	\$0.12	\$0.33	\$0.32	\$0.22	\$0.29
Pro	\$−0.36	\$−0.16	\$−0.21	\$0.13	\$0.17	\$0.13
Anti	\$0.53**	\$0.72**	\$0.87**	\$0.46**	\$0.69**	\$0.90**
Pro & Anti	\$0.01	\$0.13	\$−0.01	\$0.36	\$0.49*	\$0.25
Pro, Anti, & Ver	\$−0.01	\$0.02	\$0.03	\$0.03	\$0.13	0.20

Notes: Asterisk (*) and double asterisk (**) denote variable significant at 10% and 5% respectively.

Table 5. Difference in mean bid prices for foods with and without enhanced nutrient label.

	Intragenic w/EN vs. Intragenic w/o EN			Transgenic w/EN Transgenic w/o EN		
Info Treatment	Broccoli	Tomato	Potato	Broccoli	Tomato	Potato
No info	\$0.32	\$0.60	\$1.17*	\$0.46	\$0.48	\$1.16
Pro	\$0.78**	\$1.02**	\$0.95*	\$0.55**	\$0.61**	\$0.44
Anti	\$0.00	\$0.14	\$0.08	\$−0.08	\$−0.01	\$−0.10
Pro & Anti	\$0.27	\$0.14	\$0.19	\$0.41	\$0.18	\$0.16
Pro, Anti, & Ver	\$0.09	\$0.25	\$0.06	\$−0.06	\$−0.04	\$−0.18

Notes: Asterisk (*) and double asterisk (**) denote variable significant at 10% and 5% respectively.

labeled products that are smaller, or even negative (e.g., under the pro-biotechnology treatment). Participants receiving anti-biotech information are willing to pay the greatest premium for GM Free labeled products.

A comparison of average bid prices for GM labeled products with and without enhanced nutrient labels are presented in Table 5. The first part of Table 5 considers products with intragenic labels; differences are positive, implying that consumers value enhanced

Table 6. Difference in mean bid prices for foods with enhanced nutrient label and plain label foods.

	Intragenic w/EN vs. Plain Label w/o EN			Transgenic w/EN vs. Plain Label w/o EN		
Info Treatment	Broccoli	Tomato	Potato	Broccoli	Tomato	Potato
No info	\$0.48	\$0.56	\$1.21*	\$0.45	\$0.34	\$1.24
Pro	\$1.22**	\$1.15**	\$1.16**	\$0.49*	\$0.41	\$0.32
Anti	\$−0.14	\$−0.14	\$−0.27	\$−0.15	\$−0.26	\$−0.48
Pro & Anti	\$0.39	\$0.26	\$0.30	\$0.18	\$−0.05	\$0.00
Pro, Anti, & Ver	\$0.13	\$0.21	\$0.04	\$−0.06	\$−0.18	\$−0.38

Notes: Asterisk (*) and double asterisk (**) denote variable significant at 10% and 5% respectively.

antioxidants and vitamin C. Premiums are the greatest under the pro-biotech information treatment and least under the antibiotech treatment. The second part of the table presents mean differences in bid prices for Transgenic GM labeled products with and without enhanced nutrient labels. Here the differences are smaller and– in the case of anti-biotech information and the pro, anti and third-party information treatments–differences are slightly negative but not significantly different from zero. Combined, these results show that consumers value enhanced nutrition in fresh vegetables but are willing to pay a greater premium when these traits are obtained through intragenics instead of transgenics. The first part of Table 6 compares bid prices for Intragenic GM labeled products with enhanced nutrient labels versus Plain Label products without enhanced nutrient labels. Across all treatments, except for anti-biotech, consumers are willing to pay a premium for the Intragenic GM labeled products with enhanced nutrient labels. The second part of Table 6 is a comparison between Transgenic GM labeled products with enhanced nutrient labels versus Plain Label products. Here the differences are smaller and negative in the conflicted information setting with anti-, pro-, and verifiable-information. Table 6 provides additional evidence on the positive value placed by consumers on nutrition derived through intragenics.

4. Impacts of Controversial and Verifiable Information on WTP

Although the data and unconditional analysis are suggestive of the impact of information on the valuation of various types of GM products, a more rigorous analysis of bid prices is necessary to identify label and information effects. In this section, a multivariate regression model is constructed for this task.

Before deriving the econometric model, it is useful to summarize the issues that need to be incorporated in a model of bid prices. Bids for a given type of produce may be correlated in successive rounds of bidding. While it is common in the experimental economics literature to ignore correlation across rounds, efficiency can be gained by incorporating this information into the estimation procedure. Given the potential diversity of relative preferences for the experimental products, a general error specification is a natural starting place for estimation. Hence, a seemingly unrelated regression (SUR) model (Zellner, 1962) is selected to account for correlation across rounds of bidding.

Zero bids raise special problems in bid price models. A zero bid could represent a participant's protest of GMOs; but more generally, a zero bid for a product presents a censoring problem (i.e., bid prices are restricted to the non-negative interval). In the case of single equation models, censoring can be easily managed (e.g., a Tobit model). In the case of a system of equations with censoring, there are a number of classical estimation techniques that have been proposed, but they suffer from a variety of econometric issues and general intractability, particularly for models with larger numbers of equations and cross-equation correlation of disturbances (e.g., a SUR-Tobit model). A SUR-Tobit model is difficult to estimate because multiple integrals are required for maximum likelihood estimation, requiring simulation algorithms (Huang, Sloan, and Adamache, 1987; Meng and Rubin, 1996; Huang, 1999).

An alternative, selected for this study, is to estimate the model via Bayesian techniques with data augmentation (Albert and Chib, 1993). In this approach the complication of multiple integrals in the SUR-Tobit model is eliminated and replaced with a straightforward

Gibbs sampler, which is fast to estimate and easily scaled to larger order systems. To simultaneously address these econometric issues, Huang's (2001) model is adapted and extended to create a Bayesian-SUR-Tobit model of individual bid prices with commodity-specific-fixed effects.

Let y_{ij} denote the bid price by an individual, $i = 1, 2, \ldots, N$, for a food product with label $j = 1, 2, \ldots, J (J = 4$ for the auctions without enhanced nutrient labels and $J = 3$ for the auctions with enhanced nutrient labels). The latent WTP of the ith individual for the food product under label j can be expressed as:

$$y^*_{ij} = x'_{ik}\beta_j + \varepsilon_{ij}, \quad i = 1, 2, \ldots, N, \quad j = 1, 2, \ldots, J, \tag{1}$$

where:

$$y_{ij} = \begin{cases} y^*_{ij} & \text{if } y^*_{ij} > 0 \\ 0 & \text{if } y^*_{ij} \leq 0, \end{cases} \tag{2}$$

where y_{ij} is the observed bid price, y^*_{ij} is the latent bid price, and $\varepsilon = (\varepsilon_{i1}, \ldots, \varepsilon_{iJ})'^{iid} \sim N(0, \Omega)$. For individual i, we can express a system of equations, one equation for each label $j = 1, 2, \ldots, J$, as:

$$\begin{bmatrix} y^*_{i1} \\ y^*_{i2} \\ \vdots \\ y^*_{iJ} \end{bmatrix} = \begin{bmatrix} x'_{i1} & 0 & \cdots & 0 \\ 0 & x'_{i2} & \cdots & 0 \\ \vdots & \vdots & \ddots & \vdots \\ 0 & 0 & \cdots & x'_{iJ} \end{bmatrix} \begin{bmatrix} \beta_1 \\ \beta_2 \\ \vdots \\ \beta_J \end{bmatrix} + \begin{bmatrix} \varepsilon_{i1} \\ \varepsilon_{i2} \\ \vdots \\ \varepsilon_{iJ} \end{bmatrix}. \tag{3}$$

In stacked notation, for each individual, i, we can express the system of WTP equations as $y^*_i = x_i\beta + \varepsilon_i, i = 1, 2, \ldots, N$, where $y^*_i = (y^*_{i1}, \ldots, y^*_{iJ})$ is a $J \times 1$ vector, $x_i = diag(x_{i1}, \ldots, x_{iJ})$ is a $J \times k$ matrix, and $\beta = (\beta'_1, \ldots, \beta'_J)$ is a $Jk \times 1$ vector. Finally, stacking over all N individuals we have a complete system of equations $y^* = X\beta + \varepsilon$.

Following Albert and Chib (1993), latent bid prices are model parameters, and the augmented posterior density function for unknown model parameters is proportional to the product of two conditional distributions and the prior:

$$p(\beta, \Omega, y^*|y) \propto p(y|y^*, \beta, \Omega)p(y^*|\beta, \Omega)p(\beta, \Omega). \tag{4}$$

The conditional distribution function for y is directly predicted by the latent bid price outcomes:

$$p(y|y^*, \beta, \Omega) = \prod_{i=1}^{N}\prod_{j=1}^{J}\{I(y_{ij}^* > 0)I(y_{ij} = y_{ij}^*) + I(y_{ij}^* \leq 0)I(y_{ij} = 0)\}, \tag{5}$$

and the conditional distribution function for y^* is proportional to:

$$p(y^*|\beta, \Omega) \propto \prod_{i=1}^{N}\left\{|\Omega|^{-N/2}\exp\left[-\frac{1}{2}\sum_{i=1}^{N}(y_i^* - x_i\beta)\Omega^{-1}(y_i^* - x_i\beta)\right]\right\}. \tag{6}$$

Priors for the unknown model parameters, β and Ω, are assumed to be independent:

$$\begin{aligned} \beta &\sim N(\beta_0, V_\beta) \\ \Omega^{-1} &\sim W(a_\varepsilon, V_\varepsilon), \end{aligned} \tag{7}$$

where N and W denote the multivariate normal and Wishart distributions. The conditional posterior distributions for β and Ω^{-1} are given by standard results:

$$\beta|\Omega^{-1}, y \sim N(D_\beta d_\beta, D_\beta), \tag{8}$$

where

$$\begin{aligned} D_\beta &= (X'(\Omega^{-1} \otimes I_N)X + V_\beta^{-1})^{-1} \\ d_\beta &= X'(\Omega^{-1} \otimes I_N)y + V_\beta^{-1}\beta_0, \end{aligned} \tag{9}$$

and

$$\Omega^{-1}|\beta, y \sim W\left(N + a_\varepsilon, \left[V_\varepsilon^{-1} + \sum_{i=1}^{N}(y_i - X_i\beta)(y_i - X_i\beta)\right]^{-1}\right). \tag{10}$$

Finally, the conditional posterior for latent bids is a multivariate truncated normal given by:

$$y_{ij}^*|\beta, \Omega^{-1} \sim TN_{[-\infty,0]}(\mu_{j|-j}, \omega_{j|-j}^2)\forall ijs.t.y_{ij} = 0, \tag{11}$$

where:

$$\begin{aligned} \mu_{j|-j} &= \mu_j + \Omega'_{j-j}\Omega_{-j-j}^{-1}(y_{-j}^{*'} - \mu_{-j}) \\ \omega_{j|-j}^2 &= \omega_{jj}^2 - \Omega'_{j-j}\Omega_{-j-j}^{-1}\Omega'_{j-j}, \end{aligned} \tag{12}$$

where $\mu = X_i\beta$, μ_j is the jth row element of μ, and μ_{-j} is obtained by deleting the jth row element of μ. The matrix Ω_{-j-j} is derived from Ω by eliminating the jth column and row and Ω_{j-j} is the vector derived from the jth column of Ω by removing the jth row term. Iteratively sampling from the conditional posterior distributions for β, Ω^{-1}, and y^* yields a set of draws from the joint pdf.

5. Econometric Results

We separate observations of bid prices into one set without enhanced nutrient labels and one with enhanced nutrient labels and fit the Bayesian-SUR-Tobit model developed in the previous section to these two groups separately. Instead of estimating a separate model for each food product (fresh broccoli, tomatoes, and potatoes), the model is further stacked over the three commodities, and a commodity-specific fixed effects is added for each commodity. These dummy variables are accompanied in each of the models by the socio-demographic variables presented in Table 1, including individual characteristics (e.g., income, education), personal opinions and knowledge regarding genetic modification, and several health status indicators. Key values of parameters in the priors $\beta_0 = 0$, $V_\beta = (10\text{e}4)I_j$, $a_\varepsilon = J^*k$, and $V_\varepsilon = (10\text{e}4)I_j$, allowing the econometric results to be dominated by the sample information and not the priors. Following a 1,000 iteration burn-in, a total of 10,000 draws from the Gibbs sampler were used in estimation.

Following the tradition of Bayesian econometrics, we present empirical results for the posterior mean of key parameters and their standard deviations and posterior probability of the estimated parameter being greater than zero (Koop, Poirier, and Tobias, 2007). Table 7 presents estimates of the bid price model for commodities without enhanced nutrients. A table of corresponding marginal effects (abbreviated M.E.) is presented in the appendices as Table A.1.

The signs of the estimated posterior means of the information treatment dummy variables are consistent with expectations. Individuals who receive anti-biotech information are willing to pay a premium for the GM Free label (M.E. \$0.054) and discount both the Intragenic GM and Transgenic GM labels (M.E. \$−0.120 and \$−0.228). For individuals who receive only pro-biotech information, the situation is reversed, with higher WTP for Intragenic and Transgenic GM labels (M.E. \$0.120 and \$0.082) and lower WTP for the GM-Free label (M.E. \$−0.066).

Individuals who receive pro- and anti-biotech information treatments have greater WTP for all four types of labels. However, the impact on relative WTP for Intragenic GM vs. GM Free and Transgenic GM vs. GM Free labels is less than when participants receive pro-biotech information in isolation. This indicates that, in combination, the anti-biotech information dampens the augmenting impact that pro-biotech information has on WTP for GM labels relative to GM Free. Individuals who received the combined pro, anti, and third party perspectives show reduced WTP for all four labels. While the marginal impact is similar for each of the four labels, the largest reduction occurs for the GM Free label.

We can also see that individuals who are older, white, have larger households, and have higher household incomes are willing to pay more for foods with each of the four labels. Individuals with affiliations to environmental groups have a negative posterior mean for each label; as expected, and the decrease in WTP among this group is most pronounced for the Transgenic GM label and least for the GM Free label (M.E. \$−0.592 and \$−0.174).

Results for the opinion variables present an interesting picture. Individuals who typically read food labels–signaling an interest in the nutrient content of foods–have lower WTP for all four labels,

Table 7. Bayesian estimates of bid price equations for products without enhanced nutrient label ($N = 92$, Obs $= 1,104$).

	$Y^{PlainLabel}$			Y^{GMFree}			$Y^{Intragenic}$			$Y^{Transgenic}$		
Variables	Mean	Stdev	Pr	Mean	Stdev	Pr	Mean	Stdev	Pr	Mean	Stdev	Pr
	Information Treatment Dummy Variables (No Information Dummy Omitted)											
Pro	0.116	0.185	0.74	−0.140	0.187	0.23	0.284	0.195	0.93	0.105	0.183	0.79
Anti	−0.119	0.209	0.28	0.191	0.212	0.82	−0.171	0.219	0.21	−0.257	0.207	0.11
Pro & Anti	0.252	0.169	0.93	0.212	0.170	0.89	0.369	0.180	0.98	0.112	0.169	0.75
Pro, Anti, & Ver	−0.108	0.195	0.29	−0.243	0.197	0.11	−0.013	0.205	0.53	−0.112	0.194	0.28
	Demographic Variables											
Gender	−0.130	0.146	0.19	−0.067	0.146	0.33	0.058	0.153	0.65	−0.228	0.145	0.06
Race	0.432	0.176	0.99	0.240	0.179	0.91	0.292	0.183	0.95	0.239	0.173	0.92
Age	0.021	0.004	1.00	0.019	0.004	1.00	0.017	0.004	1.00	0.016	0.004	1.00
Income	0.004	0.002	0.98	0.005	0.002	1.00	0.005	0.002	0.99	0.005	0.002	1.00
Educ	−0.004	0.020	0.42	−0.010	0.021	0.32	0.028	0.022	0.90	0.028	0.020	0.92
Married	−0.137	0.131	0.14	−0.071	0.131	0.30	−0.180	0.136	0.09	−0.159	0.130	0.11
Household	0.139	0.045	1.00	0.178	0.045	1.00	0.095	0.047	0.98	0.086	0.044	0.98
lowa	−0.182	0.118	0.06	−0.169	0.121	0.08	−0.115	0.125	0.17	0.127	0.118	0.86
Farm	0.014	0.138	0.54	0.105	0.139	0.78	0.095	0.144	0.75	0.131	0.137	0.83
Envi_Mem	−0.495	0.313	0.06	−0.250	0.318	0.22	−0.315	0.328	0.17	−0.628	0.310	0.02

(Continued)

Table 7. (*Continued*)

Variables	$Y^{PlainLabel}$			Y^{GMFree}			$Y^{Intragenic}$			$Y^{Transgenic}$		
	Mean	Stdev	Pr	Mean	Stdev	Pr	Mean	Stdev	Pr	Mean	Stdev	Pr
	Opinion Variables											
Informed	0.392	0.213	0.97	0.236	0.214	0.86	0.025	0.220	0.54	0.263	0.212	0.89
Opinion	−0.046	0.160	0.39	0.055	0.162	0.63	0.064	0.166	0.65	−0.114	0.159	0.24
Read_Labels	−0.289	0.154	0.03	−0.091	0.153	0.28	−0.183	0.155	0.12	−0.436	0.148	0.00
	Health Variables											
Smoke	0.381	0.147	1.00	0.437	0.149	1.00	0.274	0.152	0.96	0.262	0.145	0.97
Exercise	0.043	0.140	0.63	0.105	0.141	0.78	0.386	0.146	1.00	0.162	0.138	0.88
Health_Diet	0.055	0.047	0.88	0.075	0.047	0.95	−0.016	0.048	0.37	0.071	0.046	0.94
Health_Phys	−0.047	0.041	0.13	−0.053	0.041	0.10	−0.052	0.043	0.11	−0.080	0.041	0.03
Round 1 Label	0.115	0.066	0.96	0.148	0.082	0.97	0.282	0.095	1.00	0.218	0.061	1.00

Inter-Round Correlation Coefficients

$\rho_{Plain,GMF}$	$\rho_{Plain,Intra}$	$\rho_{Plain,Trans}$	$\rho_{GMF,Intra}$	$\rho_{GMF,Trans}$	$\rho_{Intra,Trans}$
0.46	0.51	0.31	0.51	0.13	0.24

Notes: Mean, Stdev, and Pr respectively denote the posterior mean, $E(\cdot|y)$, posterior standard deviation, $Std(\cdot|y)$, and posterior probability of being greater than zero, $Pr(\cdot > 0|y)$

but the marginal effect is most pronounced for Intragenic and Transgenic GM labels. Individuals who are more informed about GM coming into the experiments are willing to pay more than their counterparts for Transgenic GM label and Plain label foods in particular. However, individuals with a positive opinion of GM coming into the experiments have a posterior mean close to zero across the four labeling treatments, indicating that the information treatments, in part, confounded prior opinions of GM.

Results for the healthy attitude proxies are mixed. While the signs of the posterior means for smoking (a negative attitude), regular exercise (a positive attitude), and highly rated healthiness of diet (a positive attitude) are consistent across the four food labels, there is little variation in the magnitude of the marginal effect across the different products.

Finally, results show a significant first-round bidding or framing effect. Specifically, participants bid relatively more in the first round for some labels. Our results also show that error terms in the bid price equations are positively correlated. For example, the estimated correlation between error terms for rounds with the Plain and Intragenic GM labels is 0.51 and between rounds with Plain and Transgenic GM labels is 0.31. Despite the greater modeling burden, taking account of the cross-label/round error correlations raises the efficiency of estimation.

Table 8 summarizes the econometric model of bid prices for food products with enhanced nutrient labels (corresponding marginal effects are presented in Table A.2 in the appendices).

The posterior mean is positive across all three labels for individuals receiving the pro-biotech information treatment, with the greatest relative increase for the Intragenic GM label. Individuals who receive anti-biotech information have reduced WTP for all three labels. In combination, individuals receiving both the pro- and anti-biotech information have a lower WTP for the GM and Transgenic GM labels, but higher for the Intragenic GM label. This indicates that these two perspectives in combination largely counterbalance each other in terms of their impact on valuations, but the positive impact on WTP for the Intragenic GM label still holds marginally.

Table 8. Bayesian estimates of bid price equations for products with enhanced nutrient label ($N = 98$, Obs = 882).

	Y^{GM}			$Y^{Intragenic}$			$Y^{Transgenic}$		
Variables	Mean	Stdev	Pr	Mean	Stdev	Pr	Mean	Stdev	Pr
	Information Treatment Dummy Variables (No Information Dummy Omitted)								
Pro	0.198	0.256	0.78	0.982	0.267	1.00	0.479	0.341	0.93
Anti	−0.109	0.266	0.34	−0.478	0.279	0.05	−0.387	0.355	0.13
Pro & Anti	−0.085	0.258	0.36	0.152	0.270	0.72	−0.250	0.349	0.24
Pro, Anti, & Ver	−0.718	0.243	0.00	−0.566	0.256	0.01	−0.506	0.325	0.06
	Demographic Variables								
Gender	−0.045	0.164	0.39	0.224	0.173	0.91	0.252	0.220	0.87
Race	−0.419	0.232	0.03	−0.488	0.243	0.02	−0.476	0.305	0.06
Age	0.023	0.006	1.00	0.016	0.006	0.99	0.017	0.008	0.98
Income	−0.008	0.003	0.01	−0.008	0.003	0.01	−0.012	0.004	0.00
Educ	−0.022	0.035	0.27	0.073	0.036	0.98	−0.001	0.048	0.50
Married	−0.072	0.199	0.35	−0.413	0.207	0.03	0.021	0.264	0.53
Household	0.265	0.062	1.00	0.215	0.065	1.00	0.156	0.083	0.97
Iowa	−0.170	0.161	0.14	−0.265	0.170	0.06	0.352	0.224	0.95
Farm	−0.473	0.206	0.01	−0.518	0.221	0.01	−0.298	0.282	0.14
Envi_Mem	−0.323	0.419	0.22	−0.458	0.445	0.15	−0.213	0.562	0.35
	Opinion Variables								
Informed	0.169	0.291	0.72	0.178	0.304	0.72	0.284	0.380	0.77
Opinion	0.448	0.262	0.95	−0.025	0.282	0.46	0.804	0.348	0.99
Read_Labels	0.483	0.184	1.00	0.348	0.194	0.97	0.068	0.249	0.60
	Health Variables								
Smoke	0.308	0.189	0.94	0.201	0.200	0.84	0.154	0.244	0.73
Exercise	0.146	0.164	0.81	0.551	0.176	1.00	0.408	0.224	0.97
Health_Diet	−0.046	0.074	0.27	−0.134	0.078	0.05	−0.112	0.098	0.13
Health_Phys	0.179	0.074	0.99	0.132	0.079	0.95	0.171	0.100	0.96
Round 1 Label	−0.210	0.110	0.02	0.342	0.138	1.00	0.626	0.159	1.00
	Inter-Round Correlation Coefficients								
	$\rho_{GM,Intra}$			$\rho_{GM,Trans}$			$\rho_{Intra,Trans}$		
	0.49			0.46			0.54		

Notes: Mean, Stdev, and Pr respectively denote the posterior mean, $E(\cdot|y)$, posterior standard deviation, $Std(\cdot|y)$, and posterior probability of being greater than zero, $Pr(\cdot > 0|y)$.

Finally, when verifiable information is introduced, valuations for all three labels are lower, indicating that verifiable information bolsters the negative impact of anti-biotech information on WTP for GM food products with enhanced consumer attribute labels.

Individuals who are members of environmental groups, have experience in farming, are white, or have higher household incomes have lower WTP for all three GM labels. Consumers who are older or have larger households have a higher WTP. Consistent with the results for products without enhanced nutrient labels, individuals who were informed about GM before the experiments or typically read food labels are willing to pay more for each of the three labels. Interestingly, individuals with a favorable prior opinion towards GM are willing to pay more for both the GM and Transgenic GM labels, but the posterior is flat and centered at zero for the Intragenic GM label. This indicates that prior perceptions toward GM did not carry over into valuations of the Intragenic GM with enhanced nutrition label.

As in the estimates for products without enhanced nutrient labels, the signs of the posterior estimates for the healthy attitude variables do not present a clear relation with WTP. Surprisingly, individuals who regularly exercise or smoke are willing to pay more for each of the three labels, but those with self-assessed healthier diets are willing to pay less. However, individuals who have higher self-assessed physical healthiness are willing to pay more under each of the three labels.

Finally, as in the case of products without enhanced nutrient labels, a significant first-round label effect and correlation of error terms across rounds of bidding/labels occurs. The correlation coefficients across the different labels is approximately 0.5, which is large and consistent with the cross-label correlations in bid prices for products without enhanced nutrient labels.

6. Discussion

In the effort to improve health outcomes in the United States, current policy objectives focus on encouraging consumption of healthy alternatives, including fruits, vegetables, and whole grains. In addition to typically being lower-calorie options compared to alternatives, fruits and vegetables in particular make important dietary contributions to overall health quality and may reduce potential health risks. One potential strategy for increasing both potency and consumption of these foods is improving the nutritional value and appeal of fruits and vegetables. By departing from fortification approaches, biotechnology companies have developed new "healthier" varieties via controversial genetic methods that would not be feasible using standard plant breeding techniques. Our experimental evidence indicates that potential exists for genetically modified foods containing enhanced vitamin and antioxidant content to find acceptance among consumers, with several important caveats.

Our experiments indicate that information brought to the public's attention by interested parties attempting to influence market outcomes plays an important role in determining acceptance of GM foods with enhanced nutrients. While pro-biotechnology information in isolation has a strong augmenting impact on WTP, the premium consumers are willing to pay declines significantly in a crowded information environment with positive, negative, and verifiable information. Additionally, experiments indicate that the very nature of the engineering process has an effect on consumer WTP for nutritionally enhanced foods. Consumers appear to be more accepting of nutrition enhancements achieved using intragenics, which is arguably closer to conventional plant breeding methods, compared to transgenics, where "foreign" genetic material outside of the species is used, indicating that intragenics may be a more appealing product development path for biotechnology companies attempting to bring enhanced food products to market.

Acknowledgments

We acknowledge helpful advice from Justin Tobias, Phil Dixon, and Alicia Carriquiry. We also acknowledge financial assistance from J.R. Simplot Corporation, the BIGMAP program, and the Iowa Agriculture Experiment Station.

References

Albert, J.H. and S. Chib (1993). Bayesian Analysis of Binary and Polychotomous Response Data. *Journal of the American Statistical Association* 88, 669–679.

Alfnes, F. and K. Rickertsen (2003). European Consumers' Willingness to Pay for U.S. Beef in Experimental Auction Markets. *American Journal of Agricultural Economics* 85, 396–405.

Bajari, P. and A. Hortaçsu (2003). TheWinner's Curse, Reserve Prices, and Endogenous Entry: Empirical Insights from eBay Auctions. *RAND Journal of Economics* 34, 329–355.

Becker, M.H.D. and J. Marschak (1964). Measuring Utility by a Single-Response Sequential Method. *Behavioral Science* 9, 226–232.

Berner, L.A., F.M. Clydesdale and J.S. Douglass (2001). Fortification Contributed Greatly to Vitamin and Mineral Intakes in the United States, 1989–1991. *Journal of Nutrition* 131, 2177–2183.

Block, G., B. Patterson and A. Subar (1992). Fruit, Vegetables, and Cancer Prevention: A Review of the Epidemiological Evidence. *Nutrition and Cancer* 18, 1–29.

Bonetti, S. (1998). Experimental Economics and Deception. *Journal of Economic Psychology* 19, 377–395.

Corrigan, J.R., D.P.T. Depositario, R.M. Nayga, X. Wu and T. P. Laude (2009). Comparing Open-Ended Choice Experiments and Experimental Auctions: An Application to Golden Rice. *American Journal of Agricultural Economics* 91, 837–853.

Corrigan, J.R. and M.C. Rousu (2006). The Effect of Initial Endowments in Experimental Auctions. *American Journal of Agricultural Economics* 88, 448–457.

Epstein, L.H., C.C. Gordy, H.A. Raynor, M. Beddome, C.K. Kilanowski and R. Paluch (2001). Increasing Fruit and Vegetable Intake and Decreasing Fat and Sugar Intake in Families at Risk for Childhood Obesity. *Obesity Research* 9, 171–178.

Guenther, P.M., K.W. Dodd, J. Reedy and S.M. Krebs-Smith (2006). Most Americans Eat Much Less than Recommended Amounts of Fruits and Vegetables. *Journal of the American Dietetic Association* 106, 1371–1379.

Guerre, E., I. Perrigne and Q. Vuong (2000). Optimal Nonparametric Estimation of First-Price Auctions. *Econometrica* 68, 525–574.

Harrison, G.W. and J.A. List (2004). Field Experiments. *Journal of Economic Literature* 42, 1009–1055.

Herdt, R.W. (2006). Biotechnology in Agriculture. *Annual Review of Environment and Resources* 31, 265–295.

Hoffman, E., D.J. Menkhaus, D. Chakravarti, R.A. Field and G.D. Whipple (1993). Using Laboratory Experimental Auctions in Marketing Research: A Case Study of New Packaging for Fresh Beef. *Marketing Science* 12, 318–338.

Huang, C.J., F.A. Sloan and K.W. Adamache (1987). Estimation of Seemingly Unrelated Tobit Regressions via the EM Algorithm. *Journal of Business and Economic Statistics* 5, 425–430.

Huang, H. (1999). Estimation of the SUR Tobit Model via the MCECM Algorithm. *Economics Letters* 64, 25–30.

Huang, H. (2001). Bayesian Analysis of the SUR — Tobit Model. *Applied Economics Letters* 8, 617– 622.

Huffman, W.E., J.F. Shogren, M.C. Rousu and A. Tegene (2003). Consumer Willingness to Pay for Genetically Modified Food Labels in a Market with Diverse Information: Evidence from Experimental Auctions. *Journal of Agricultural and Resource Economics* 28, 481–502.

Joshipura, K.J., F.B. Hu, J.E. Manson, M.J. Stampfer, E.B. Rimm, F.E. Speizer, G. Colditz, A. Ascherio, B. Rosner, D. Spiegelman and W.C. Willett (2001). The Effect of Fruit and Vegetable Intake on Risk for Coronary Heart Disease. *Annals of Internal Medicine* 134, 1106–1114.

Kimmons, J., C. Gillespie, J. Seymour, M. Serdula and H. M. Blanck (2009). Fruit and Vegetable Intake Among Adolescents and Adults in the United States: Percentage Meeting Individualized Recommendations. *Medscape Journal of Medicine* 11, 26.

Koop, G., D.J. Poirier and J.L. Tobias (2007). *Bayesian Econometric Methods.* New York, NY: Cambridge University Press.

Lusk, J.L., M.S. Daniel, D.R. Mark and C.L. Lusk (2001). Alternative Calibration and Auction Institutions for Predicting Consumer Willingness to Pay for Nongenetically Modified Corn Chips. *Journal of Agricultural and Resource Economics* 26, 40–57.

Lusk, J.L., L.O. House, C. Valli, S.R. Jaeger, M. Moore, J. Morrow and W.B. Traill (2004). Effect of Information about Benefits of Biotechnology on Consumer Acceptance of Genetically Modified Food: Evidence from Experimental Auctions in the United States, England, and France. *European Review of Agricultural Economics* 31, 179–204.

Lusk, J.L., M. Jamal, L. Kurlander, M. Roucan and L. Taulman (2005). A Meta-Analysis of Genetically Modified Food Valuation Studies. *Journal of Agricultural and Resource Economics* 30, 28–44.

Lusk, J.L., J. Pruitt and B. Norwood (2006). External Validity of a Framed Field Experiment. *Economics Letters* 93, 285–290.

Lusk, J. L. and M.C. Rousu (2007). "Market Price Endogeneity and Accuracy of Value Elicitation Mechanisms." In J.A. List (Ed.), *Using Experimental Methods in Environmental and Resource Economics.* Cheltenham, UK: Edward Elger Publishing.

Meng, X.-L. and D.B. Rubin (1996). "Efficient Methods for Estimation and Testing with Seemingly Unrelated Regressions in the Presence of Latent Variables and Missing Observations." In D.A. Berry, K.M. Chaloner and J.K. Geweke (Eds.), *Bayesian Analysis in Statistics and Econometrics.* New York: John Wiley & Sons.

Milgrom, P. and J. Roberts (1986). Relying on the Information of Interested Parties. *RAND Journal of Economics* 17, 18–32.

Noussair, C., S. Robin and B. Ruffieux (2004). Do Consumers Really Refuse to Buy Genetically Modified Food? *Economic Journal* 114, 102–120.

Ortmann, A. and R. Hertwig (2002). The Costs of Deception: Evidence from Psychology. *Experimental Economics* 5, 111–131.

Plott, C.R. and K. Zeiler (2007). Exchange Asymmetries Incorrectly Interpreted as Evidence of Endowment Effect Theory and Prospect Theory? *American Economic Review* 97, 1449–1466.

Rommens, C.M., J.M. Humara, J. Ye, H. Yan, C. Richael, L. Zhang, R. Perry and K. Swords (2004). Crop Improvement through Modification of the Plant's Own Genome. *Plant Physiology* 135, 421–431.

Rousu, M.C., W.E. Huffman, J.F. Shogren and A. Tegene (2007). Effects and Value of Verifiable Information in a Controversial Market: Evidence from Lab Auctions of Genetically Modified Food. *Economic Inquiry* 45, 409–432.

Shogren, J.F., M. Margolis, C. Koo and J.A. List (2001). A Random nth-Price Auction. *Journal of Economic Behavior & Organization* 46, 409–421.

Shogren, J.F., S.Y. Shin, D.J. Hayes and J.B. Kliebenstein (1994). Resolving Differences in Willingness to Pay and Willingness to Accept. *American Economic Review* 84, 255–270.

Subar, A.F., S.M. Krebs-Smith, A. Cook and L.L. Kahle (1998). Dietary Sources of Nutrients among US Adults, 1989 to 1991. *Journal of the American Dietetic Association* 98, 537–547.

Tonsor, G.T., T.C. Schroeder, J.A. Fox and A. Biere (2005). European Preferences for Beef Steak Attributes. *Journal of Agricultural and Resource Economics* 30, 367–380.

van den Bergh, J.C.J.M. and J.M. Holley (2002). An Environmental-Economic Assessment of Genetic Modification of Agricultural Crops. *Futures* 34, 807–822.

Vickrey, W. (1961). Counterspeculation, Auctions, and Competitive Sealed Tenders. *Journal of Finance* 16, 8–37.

Wertenbroch, K. and B. Skiera (2002). Measuring Consumers? Willingness to Pay at the Point of Purchase. *Journal of Marketing Research* 39, 228–241.

Woodside, J.V., D. McCall, C. McGartland and I.S. Young (2005). Micronutrients: Dietary Intake v. Supplement Use. *Proceedings of the Nutrition Society* 64, 543–553.

Zellner, A. (1962). An Efficient Method of Estimating Seemingly Unrelated Regressions and Tests for Aggregation Bias. *Journal of the American Statistical Association* 57, 348–368.

Appendix A: Environmental Group Perspective on GM

General information

Genetic modification (GM) takes genes from one organism and places then into another. The process lets scientists manipulate genes in an unnatural way. Inadequate safety testing of GM plants and food products has occurred. Humans and the Earth are being used as guinea pigs for testing whether "Frankenfoods" are safe. GM foods should be banned because their effect on consumers and the environment is unknown and potentially catastrophic! Genetic modification is one of the most risky things being done to your food sources today and should be stopped before more damage is done.

Scientific impact

All genetic modifications of plants are risky. All GM techniques are relatively new and no one can guarantee that consumers or the environment will not be harmed. The biggest potential hazard of GM foods is the unknown.

Human impact

Genetically modified foods could pose serious risks to human health. Some foods contain allergens, and the potential exists for allergens to be transferred into a GM food product that no one would suspect. For example, if the genes from a peanut were transferred into a tomato, and someone who is allergic to peanuts eats this GM tomato, he could display a peanut allergy.

Another problem with transgenic foods is a moral issue. Many GM techniques transfer genes across species. We believe it is morally wrong to alter life forms on such a fundamental level.

Financial impact

GM foods are being pushed onto consumer by big businesses which only care about their own profits and ignore possible negative side effects. These groups are actually patenting new life forms they create

Table A.1. Marginal effects of explanatory variables on bid prices for products without enhanced nutrient label ($N = 92$, Obs $= 1,104$).

	$Y^{PlainLabel}$		Y^{GMFree}		$Y^{Intragenic}$		$Y^{Transgenic}$	
Dep Var	Mean	Stdev	Mean	Stdev	Mean	Stdev	Mean	Stdev
	Information Treatment Dummy Variables (No Information Dummy Omitted)							
Pro	0.048	0.090	−0.066	0.090	0.120	0.084	0.082	0.132
Anti	−0.084	0.132	0.054	0.066	−0.120	0.150	−0.228	0.198
Pro & Anti	0.108	0.072	0.066	0.054	0.156	0.072	0.060	0.108
Pro, Anti, & Ver	−0.072	0.120	−0.120	0.114	−0.006	0.114	−0.102	0.156
	Demographic Variables							
Gender	−0.060	0.066	−0.018	0.054	0.036	0.090	−0.132	0.084
Race	0.330	0.174	0.120	0.102	0.204	0.150	0.210	0.168
Age	0.012	0.000	0.006	0.000	0.006	0.000	0.012	0.000
Income	0.000	0.000	0.000	0.000	0.000	0.000	0.006	0.000
Educ	0.000	0.012	−0.006	0.006	0.012	0.012	0.018	0.012
Married	−0.072	0.072	−0.024	0.048	−0.096	0.078	−0.108	0.090
Household	0.060	0.024	0.060	0.018	0.048	0.024	0.054	0.030
Iowa	−0.090	0.060	−0.060	0.042	−0.060	0.066	0.090	0.090
Farm	0.000	0.072	0.030	0.048	0.042	0.072	0.078	0.084
Envi_Mem	−0.474	0.390	−0.174	0.222	−0.288	0.324	−0.592	0.522
	Opinion Variables							
Informed	0.138	0.066	0.060	0.060	−0.006	0.120	0.132	0.108
Opinion	−0.036	0.090	0.012	0.060	0.024	0.084	−0.096	0.132
Read_Labels	−0.138	0.078	−0.030	0.054	−0.090	0.078	−0.270	0.096
	Health Variables							
Smoke	0.150	0.054	0.120	0.042	0.120	0.060	0.150	0.078
Exercise	0.024	0.072	0.036	0.054	0.198	0.084	0.108	0.096
Health_Diet	0.024	0.024	0.024	0.018	−0.006	0.024	0.042	0.030
Health_Phys	−0.024	0.018	−0.018	0.012	−0.024	0.024	−0.048	0.024
Round 1 Label	0.054	0.030	0.048	0.030	0.114	0.036	0.138	0.036

Notes: Mean and Stdev denote the posterior mean, $E(\cdot|y)$, and posterior standard deviation, $Std(\cdot|y)$.

with plans to sell for profits. Studies have shown that GM crops may even get lower yields than conventional crops.

Environmental impact

GM foods could pose major environmental hazards. Little testing of GM plants for environmental impacts has occurred. One potential

Table A.2. Marginal effects of explanatory variables on bid prices for products with enhanced nutrient label ($N = 98$, Obs = 882).

	Y^{GM}		$Y^{Intragenic}$		$Y^{Transgenic}$	
Dep Var	Mean	Stdev	Mean	Stdev	Mean	Stdev
	Information Treatment Dummy Variable (No Information Dummy Omitted)					
Pro	0.040	0.056	0.140	0.034	0.136	0.094
Anti	−0.034	0.072	−0.122	0.080	−0.142	0.132
Pro & Anti	−0.026	0.068	0.026	0.050	−0.092	0.122
Pro, Anti & Ver	−0.226	0.092	−0.142	0.074	−0.182	0.122
	Demographic Variables					
Gender	−0.010	0.040	0.048	0.038	0.084	0.074
Race	−0.086	0.046	−0.080	0.038	−0.138	0.084
Age	0.008	0.002	0.004	0.002	0.010	0.004
Income	−0.002	0.000	−0.002	0.000	−0.006	0.002
Educ	−0.006	0.012	0.020	0.010	−0.002	0.026
Married	−0.018	0.050	−0.082	0.042	0.008	0.086
Household	0.082	0.022	0.058	0.020	0.082	0.044
Iowa	−0.042	0.040	−0.052	0.034	0.118	0.076
Farm	−0.140	0.070	−0.128	0.064	−0.106	0.100
Envi_Mem	−0.118	0.146	−0.140	0.144	−0.098	0.202
	Opinion Variables					
Informed	0.030	0.066	0.026	0.056	0.078	0.112
Opinion	0.088	0.048	−0.012	0.060	0.216	0.084
Read_Labels	0.132	0.056	0.076	0.046	0.024	0.082
	Health Variables					
Smoke	0.068	0.040	0.036	0.038	0.046	0.076
Exercise	0.038	0.042	0.116	0.042	0.136	0.076
Health_Diet	−0.014	0.024	−0.036	0.022	−0.060	0.054
Health_Phys	0.056	0.024	0.036	0.022	0.092	0.056
Round 1 Label	−0.054	0.030	0.064	0.026	0.188	0.048

Notes: Mean and Stdev denote the posterior mean, $E(\cdot|y)$, and posterior standard deviation, $Std(\cdot|y)$.

risk of GM crops is their impact on wildlife, including wild species of plants and insects. A study showed that one type of GM plant killed Monarch butterflies. Another potential environmental hazard could come from pests that become resistant to new naturally occurring toxic substances engineered into plants to kill pests–insects and worms–or to make a plant resistant to a particular herbicide application. The target pests that get exposed to these new GM crops could quickly develop tolerances and wipe out many of the potential advantages of GM pest resistance.

Appendix B: Biotechnology Industry Perspective on GM

General information

Genetically modified (GM) plants have the potential to be one of the greatest discoveries in the history of farming. GM crops have lowered food production costs by improving insect and disease resistance and weed control in plants. New genetic engineering techniques could dramatically enhance consumer benefiting attributes of food such as vitamins, antioxidants, flavor, and shelf life. These improvements to plant quality can only be attained through GM, not conventional breeding.

The process of genetic modification takes genes from one organism and places them into another. There are two distinct types of GM used by biotechnology companies. Transgenic GM transfers genes between two unrelated organisms, for example from soil bacteria to corn. Intragenic GM involves transferring genes between two breeds of the same organism, for example, from wild species of corn to a commercial variety of corn.

Scientific impact

Both transgenic and intragenic techniques are used to produce food products that are approved by the Food and Drug Administration (FDA). Intragenic modification is a genetic technique for significantly speeding up the conventional process of plant cross-breeding, which has been undertaken by farmers and plant breeders for thousands of years. Many industry groups believe intragenics should require minimal FDA testing because no foreign genes or proteins are added to the GM plant. We have only seen the tip of the iceberg of the future potential of GM for improving worldwide health and nutrition through enhanced plants.

Human impact

The potential exists for GM to dramatically enhance traits that have direct value to consumers, such as increased vitamins and antioxidants, more flavor, longer shelf life, lower pesticide use, and reduced cost of production. Superior GM plants will help reduce worldwide malnutrition and improve the healthiness of foods. The FDA has approved GM food for human consumption, and Americans have been consuming GM foods for a decade. While every food (modified or not) poses some risks, there has never been a documented case of a person getting sick from GM food.

Financial impact

With the introduction of enhanced nutrition, antioxidants, shelf life, flavors, and other consumer desired attributes using GM technology, consumers will for the first time enjoy the direct benefits of genetic engineering. GM plants have reduced farmers' costs, which mean lower food prices. Worldwide the number of hungry people is declining. GM technology is helping to feed the world and improve worldwide nutrition.

Environmental impact

Genetic modification of plants has the potential to be one of the most environmentally helpful discoveries ever. GM technology has produced new methods of insect control that reduce chemical insecticide application by 50% or more. GM weed control is providing new methods to control weeds, which are a problem in no-till farming. This means greater crop yields and less environmental damage.

Appendix C: Independent, 3rd Party, Verifiable Perspective on GM

General information

The process of genetic modification (GM) takes genes from one organism and places them into another. There are two distinct types of GM used by biotechnology companies. Transgenic GM transfers genes between two unrelated organisms, for example, from soil bacteria to corn. Intragenic GM involves transferring genes between two varieties of the same organism, for example, from wild species of corn to a commercial variety. Hence, intragenic modification has much in common with conventional plant breeding.

Scientific impact

The Food and Drug Administration (FDA) standard for GM food products is based on the principle that they have essentially the same ingredients, although modified from the original plant. Almost all GM crops meet the FDA's substantive equivalent requirement. Hence, they do not require special testing before commercial marketing can occur.

Human impact

Many scientists see intragenics as having real potential for enhancing consumer attributes of plants such as dramatically increasing vitamin and antioxidant levels, extending shelf life, and reduced chemical pesticide application without concerns about gene transfer across species. These improvements to plants are only possible using genetic modification and not conventional breeding.

All foods present a risk of an allergic reaction to a small fraction of the population. No FDA approved GM food poses any known unique human health risks, but when genes are transferred across species, a new allergen is possible. This is more likely with transgenics than intragenics. While GM crops can result in higher yields and enhanced nutrition, there is no consensus whether GM foods have or will reduce worldwide hunger.

Many people have moral or religious objections to GM. Some groups see intragenics as being more acceptable because genes are transferred between two breeds of the same species.

Financial impact

GM seeds and other organisms are produced by businesses that seek profits. For farmers to switch to GM crops, they must see benefits from making a change. Consumers must also see benefits from consuming GM foods–lower price or enhanced consumer attributes. However GM technology may lead to changes in the organization of the agri-business industry and farming.

Environmental impact

The long-term effects of GM on the environment are largely unknown. Bioengineered insect resistance has reduced farmers' applications of environmentally hazardous insecticides, but resistance to this bio-control system will increase over time. More studies are occurring to help assess the impact of bioengineered plants on the environment. Some studies reported harm to Monarch butterflies from GM crops, but other scientists were not able to recreate the results.

Enhanced consumer attributes, such as vitamins, antioxidants, and longer shelf life due to intragenics pose no known environmental hazards.

Chapter 12

Consumer Valuation of Information about Food Safety Achieved Using Biotechnology: Evidence from New Potato Products*,†

Jonathan R. McFadden‡,¶,** and Wallace E. Huffman§,‖,††

‡*Economic Research Service, U.S. Department of Agriculture,*
1400 Independence Ave., SW Mail Stop 1800,
Washington, DC 20250, United States
§*Department of Economics, Iowa State University,*
460E Heady Hall, Ames,
IA 50011, United States
¶*jonathan.mcfadden@ers.usda.gov*
‖*whuffman@iastate.edu*

This study focuses on the role of food labels and information for affecting consumers' valuation of food safety achieved through application of biotechnological (biotech) methods. In 2002, potato products cooked to a high temperature were first reported to contain the human carcinogen acrylamide. Research discoveries using genetic engineering can substantially reduce carcinogenic-forming potential, and thereby increase food safety of potato products. Adult consumers from three distant regions of the U. S. were the subjects in lab auctions of potato products. They engaged in distinct rounds of bidding, first without packaged information and again after receiving information about the food safety risks and benefits of new biotech potato products. The study finds that willingness-to-pay (WTP) for these new potato products are not significantly different from conventional potato products under no

*This article originally appeared in the **Food Policy (2017), 69, 82–96**. This is published by permission of **Elsevier**.
†The views expressed are those of the authors and should not be attributed to the Economic Research Service or the U.S. Department of Agriculture (USDA).
**The experiment was completed while at Iowa State University.
††Corresponding author.

information. However, exposure to a scientific perspective and scientific plus industry perspectives increases participants' willingness-to-pay for the new potato products and reduces willingness-to-pay for conventional products. Exposure to the negative perspective on the new technology significantly reduces willingness-to-pay. Consumer valuation of the new potato products is affected by food labels, information, and consumer attributes. A consumer information program could be needed to gain consumer acceptance of these potato products or other foods that have been genetically modified to increase certain food safety dimensions.

Keywords: Experimental auctions, Potato products, Information and labeling effects, Biotechnology, Food safety

1. Introduction

Potatoes are native to South America but spread to North America, Northern and Eastern Europe, and beyond in the seventeenth- and eighteenth centuries as explorers and travelers returned from Peru. The potato had several advantages over other food staples: convenience in storability, ample sources of energy and fiber, and ease of growing on small plots. Initial introductions into cropping and food systems were not without problems arising from low genetic diversity under unsatisfactory growing conditions, such as potato blight (Hirsch *et al.*, 2013). Today, the potato is the most important vegetable in the United States (Jansky *et al.*, 2014) and is widely consumed in the American and European diets.[1,2]

The popularity of the potato for human consumption is being challenged by the recent discovery that potato products cooked at high temperatures (>250°F) contain acrylamide (Tareke *et al.*, 2002), which the International Agency for Research on Cancer (IARC) classifies as "probably carcinogenic to humans (Group 2A)" (Lineback *et al.*, 2012). Acrylamide is also a neuro-toxin (OEHHA, 2014). Affected potato products include French fries, hash browns, and potato chips. Roughly, the darker is the browning, the higher is the acrylamide concentration.[3]

[1]The potato is a good source of potassium (Institute of Medicine, 2015).
[2]In 2013, FAO estimates world production of potatoes at 368 million tons. Roughly two thirds are eaten directly by humans (FAO, 2015).
[3]Acrylamide is a by-product of the Maillard chemical reaction and is a stable compound after formation. After controlling for water content, a positive relationship exists between

In 2005, the California Office of the Attorney General sued the U. S. potato industry under Proposition 65 (California Safe Drinking Water and Toxic Enforcement Act of 1986), necessitating substantial reductions in acrylamide levels in potato products. The settlement of the suit required many California restaurant chains to post warnings about acrylamide exposure from eating cooked potato products (OAG, 2008). The goal of these warnings is to promote public health. They also signal to the U.S. potato industry a need to reduce the acrylamide content of potato products cooked to high temperatures.

Biological researchers have been searching for a method to substantially reduce acrylamide levels in cooked potato products. However, conventional plant breeding and new field- and potato-storage management strategies have been unsuccessful. In contrast, new genetic engineering methods provide viable opportunities. Bethke and Bussan (2013) report that biotech gene-silencing methods significantly reduce asparagine and sugar content of cooked potato products. These are the two naturally-occurring compounds in potatoes that react to form acrylamide in potato products cooked at high temperatures. As such, in 2016, the FDA indicated that biotech varieties may ultimately be the most effective solution for acrylamide reduction (US-FDA, 2016). Hence, genetic modification (GM) can be employed to significantly improve the food safety of an important food crop — the white potato.

This article reports the results of applying an innovative approach to valuing and comparing various information treatments on risk reduction. The risk, food-borne acrylamide exposure, is an emerging food safety issue that has attracted worldwide attention. This study focuses on the effect of food labels and packaged information on consumers' valuation of low-acrylamide biotech potato products. Adult consumers from three distant regions of the U.S. were chosen as subjects. They were from California, Iowa and Massachusetts. Iowa and California are large producers of food and agricultural products,

---

chip and fry color and acrylamide content. Processing conditions, environmental factors during plant growth, storage, and field management practices influence acrylamide-forming potential in potatoes. Acrylamide concentrations in potatoes vary widely as a function of reducing sugar content and cooking temperatures (Bethke and Bussan, 2013).

while farmers in Massachusetts tend to be engaged in small-scale farming and agro-tourism. In addition, three New England states voted over 2013–14 to require some type of mandatory labeling of GM foods, but state-level votes in other U.S. regions have failed. Hence, our sample contains significant regional diversity. The subjects participated in laboratory auctions, which created a unique data set.

The experimental design included three white potato products (fresh whole potatoes, frozen French fries, and potato chips), two types of products (low-acrylamide biotech and conventional), and six information treatments. The six information treatments were constructed from three different perspectives — a scientific perspective on health risks from acrylamide exposure, an industry perspective on low-acrylamide potatoes obtained by using biotechnology, and environmental groups' perspective on biotechnology. Subjects participate in private-information-revealing experimental auction markets, where they place bids under both the "treatment" and "control," i.e., a with-in person statistical experimental design (List, 2009). Each participant bids on both types of the three products, both before receiving one of the randomly-assigned information treatments and after receiving information.

Prior to receiving packaged information, subjects' average willingness-to-pay for new biotech potato products is not significantly different from that of conventional potato products. However, when subjects receive packaged information, willingness-to-pay differences emerge. Under the scientific information treatment, all average treatment effects (ATEs) are statistically significant at the 10 percent level, being positive for biotech commodities but negative for conventional commodities. When subjects receive environmental groups' information, average WTP is reduced for biotech and conventional products, but the reduction is larger and statistically stronger for the biotech products. When subjects receive industry and scientific perspectives, the mean bids for the biotech products increase significantly, and bids for conventional products decline significantly.

2. Related Research

The current study stems from past applied work on incentive-compatible elicitation mechanisms, food experiments, and experiments involving information and labeling treatments. For the past half-century, economists have used the Vickrey (1961) second-price auction to learn about consumer WTP. Participants submit private bids, and the highest bidder wins but pays the second-highest price. Since the weakly dominant strategy is sincere bidding, the auction is said to be incentive compatible (Hurwicz, 1972). Second-price auctions remain widely used because of their incentive-compatible properties and ease of implementation, including identification of a single winner.

Another common method of eliciting consumer valuations is the Becker *et al.* (1964) mechanism (hereafter denoted as BDM). In this value elicitation procedure, participants are asked to report their WTP for a good (or group of goods) and then a random price is drawn, typically from a uniform distribution. Those whose valuations are higher than the random price receive the good and pay the random price, whereas participants with a WTP less than the random price do not receive the good and pay nothing. The BDM device is therefore incentive compatible.

Shogren *et al.* (2001) build on the BDM mechanism and Vickrey auction to create a mechanism known as a random nth-price auction. In this method, participants bid on the experimental product, bids are ranked from highest to lowest, and a random number, n, is drawn uniformly on $[2, k]$, where k is the number of session participants. Those who bid higher than the nth bid are winners but pay the nth bid, not their submitted bid. As in the BDM, the bidders' dominant strategy is to truthfully reveal their valuations. Shogren *et al.* (2001) argue that random nth-price auctions are desirable because of engaging "low-margin" bidders, i.e., low-value participants still have a positive probability of winning. Using an experiment with induced values, the authors find more sincere bidding from low-value individuals in the random nth-price auction than the Vickrey auction.[4] This auction is especially useful for assessing valuations for all subjects in an auction.

The literature on the use of experimental auctions to examine the effects of information treatments on WTP for food products has evolved since the 1990s (Menkhaus *et al.*, 1992; Hoffman *et al.*, 1992; Schmitz *et al.*, 1993; Hayes *et al.*, 1995; Fox *et al.*, 2002). In the first three studies, fifth-price auctions with repeated trials were conducted in the Denver and Los Angeles areas to assess WTP for beef steaks with two types of packaging. They found that information treatments, income, household size, employment, beef shape, cholesterol concerns, and "all natural" labels significantly influence the WTP difference between vacuum-sealed and plastic overwrapped steaks. The latter two studies employ Vickrey auctions to assess valuations for meat sandwiches with varying levels of food safety risks.

Since the wider adoption of laboratory experimental methods in agricultural economics, there has been an increase in studies examining information and labeling effects (Rousu *et al.*, 2007; McFadden and Lusk, 2015). Although these studies use different experimental methods, one commonality is that prior beliefs, information treatments, labels and seals, and framing have substantial effects on consumer values. For example, Hayes *et al.* (2002) find that information treatments have asymmetric effects on WTP for irradiated pork sandwiches, with negative information having more impact than positive information. Similarly, Marette *et al.* (2010) show that information about the benefits of consuming yogurt with plant sterols to reduce cholesterol increases consumers' WTP. In certain studies, information and labeling effects can be isolated to look at very specific influences on WTP. Randomization of information treatments is central to identification of information treatment effects.

[4]There is a growing literature examining potential divergence between incentive compatibility in theory and in practice (Lusk and Shogren, 2007). In recent work, Malmendier and Lee (2011) document irrational bidding behavior in eBay auctions. We use random *n*th-price auctions because of their frequent use in food auction research and their desirable properties in practice.

3. Experimental Design

3.1. *Overview*

Due to cost considerations and accessibility, three different agencies were used to obtain participants for our experiments. Survey and Behavioral Research Services (SBRS) at Iowa State University developed the protocol for recruiting participants. A random sample of households was contacted to determine if there was an eligible individual to participate in group sessions in a project by Iowa State University on how people select food and household products. Recruiters were seeking adults who were 18–65 years of age who could follow directions and write in English, and could come on a target date to a laboratory site. In particular, they were told that the sessions involved a food preference experiment set up, like an auction or bidding activity, and they would be paid $65 for about an hour of their time. Willing adults were asked to pick among the three available starting times. They were told that there would be follow-up communication confirming time and location, with directions to the lab site.

SBRS undertook recruitment for the Des Moines, Iowa, area. They started with a set of random telephone numbers. When a willing participant was obtained, the individual was also permitted to suggest the name of another family member or friend who might be willing to participate in the experiments. Family and friends were separated into different concurrent sessions, except in two cases. This variation of random sampling was only undertaken in the Des Moines area and contributed to only 25% of that area's sub-sample. In the Los Angeles area, we worked with a marketing research company — Focus and Testing. In the Boston area, we worked with Answer Quest. Focus and Testing drew participants from a subject pool of 100,000 adults, and Answer Quest drew participants from a subject pool of 60,000 adults. These recruits had participated in one or more prior marketing research projects undertaken by these companies but not experimental auctions. Recruitment at the latter two sites was similar to the Des Moines area, except that willing participants were not permitted to suggest names of other possible recruits, and there was a 60–40 female-male division of recruits. This was to insure that subjects matched the gender ratio of the population of U.S. grocery store shoppers.[5] A total of 304 adults, split roughly equally across the three areas, were chosen and were paid $65 for participating. For reasons summarized above, we chose the random nth-price auction over other auction mechanisms.

Our experimental lab products are: a 5 lb. bag of Russet potatoes, 12 oz. bag of classic-cut potato chips, and 2 lb. bag of frozen French fries. We offer three potato products because of the experiment's high fixed costs and participants' heterogeneous preferences.[6] Each product has two types, conventional (available in grocery store) and low-acrylamide using biotechnology. Low-acrylamide products using biotechnology had not been deregulated by the USDA for commercial sale at the time of the experiment.[7] However, one company had low-acrylamide, biotech potato products under development that met our display needs, which they provided as display products for the bidding rounds. At the end of each session, all participants were told that only conventional potato products were available to take home on the day of the experiment.

Our experiments provide two types of information to participants: food labels (two types) and information treatments (six). All commodities are packaged in transparent plastic, rather than commercial packaging, to focus attention on the food labels. Conventional products carry a plain label briefly describing the product and net weight. The low-acrylamide biotech products have labels that briefly describe the product, including "low acrylamide," and net weight but also include the added information: "Product Made Using

[5]Approximately 71% of Des Moines subjects are female, which is somewhat larger than the proportion of women in the U.S. grocery shopper population. However, 62% of our total sample are women.

[6]These diverse attributes suggest that fresh potatoes, potato chips, and frozen French fries are not strong substitutes or complements. Given that the commodities serve different needs, understatements or overstatements of WTP arising from complementary or substitutable commodities are very unlikely (Rousu *et al.*, 2008).

[7]The USDA deregulated biotech potatoes used in these experiments on November 7, 2014. The Innate™ 1.0 biotech potato is resistant to black-spot bruising, will not brown when cut and exposed to air, and has low sugar content and acrylamide forming potential (O'Connell, 2014).

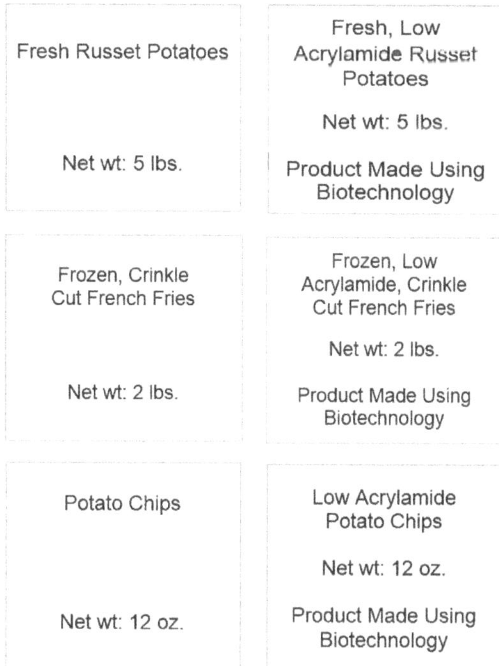

Fig. 1. Commodity labels: Conventional (left) and low-acrylamide (right).

Biotechnology." These labels are affixed to the front of each product's packaging. Fig. 1 depicts the labels for all products.

Information treatments are constructed from three perspectives, each limited to one 8.5″ × 11″ sheet of paper. These perspectives are (i) an industry perspective on using biotechnology to reduce acrylamide levels in food products, (ii) a scientific perspective on health risks of acrylamide exposure, and (iii) environmental groups' (negative) perspective on biotechnology.[8] Each of these perspectives is organized under the same three headings (General Statement,

[8]We do not suggest that all environmental groups view GM products negatively. Environmental non-profit organizations have very diverse objectives, such as conservation of wetlands, freshwater rivers, and bird habitats. This perspective only refers to groups that are opposed to biotechnology. Our choice of the term "environmental groups" was guided by similar studies in the food auctions literature.

Nutrition and Health, and Environmental Impacts and Food Security; see Appendix) to aid participants in reading and processing information contained in them.

The information in these perspectives was adapted from websites, medical research, government reports, non-governmental organization (NGO) reports, and statements from anti-biotechnology advocacy groups. Some statements in the industry perspective are adapted from Rasmussen *et al.* (2013) and the J.R. Simplot Company's website, www.simplot.com. Information about the harmful impacts of acrylamides exposure used in the scientific perspective are from various medical journal articles and government reports (Garland and Patterson, 1967; Shiraishi, 1978; Naruszewicz *et al.*, 2009; JECFA, 2011; Nixon *et al.*, 2012). For the environmental groups' perspective, many statements are based on claims from Antoniou *et al.* (2012) and the Non-GMO Project's website, www.nongmoproject.org. As with other design features, the information perspectives were pretested and subsequently revised on several occasions prior to implementation in the experiment.

The three different information perspectives are used to create six information treatments: (1) industry perspective only, (2) scientific perspective only, (3) environmental groups' perspective only, (4) industry and scientific perspective, (5) industry and environmental perspective, and (6) scientific and environmental perspective. As in Rousu *et al.* (2007) and Colson *et al.* (2011), when participants receive two perspectives, the order of the perspectives is randomized.[9] A randomized design permits simpler identification of treatment effects and obviates controlling for sample selection (List, 2009; Lusk and Shogren, 2007).

3.2. *Flow of subject activities in the experiments*

The steps of a typical auction session are provided in the flowchart of Fig. 2. Upon arriving at the experimental site, participants read

[9]No participant received a treatment consisting of all three information perspectives. This reduces subject burden and session lengths.

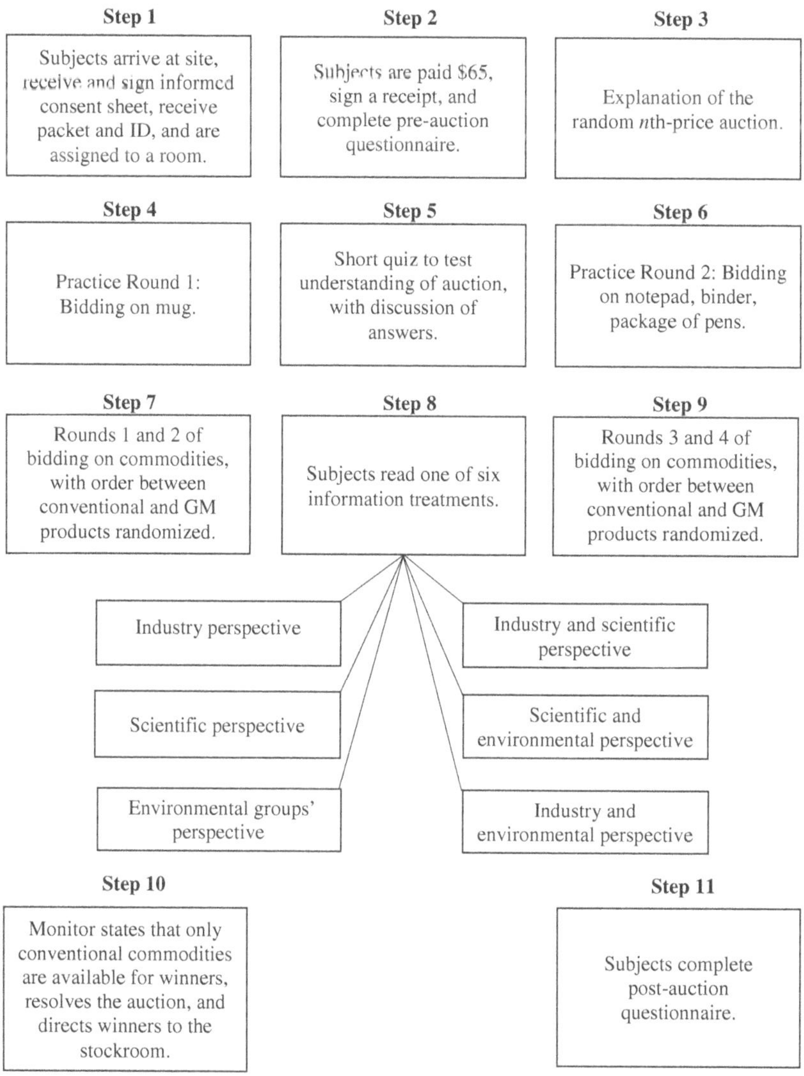

Fig. 2. The flow of subject activities in the experiments.

and sign an informed consent sheet, required for human subjects' research. They are then given an ID number and randomly assigned to one of two concurrent sessions (sessions that occur at the same time but were administered by different monitors). They are also given a packet of material used in the experiment, which contains a randomly-assigned information treatment.[10] They are then instructed to begin completing the pre-auction questionnaire. The 19 questions in the pre-auction questionnaire mainly relate to demographics, prior knowledge about acrylamide, biotech foods, and non-biotech foods, and health. The breadth of the food-related questions are not likely to cause problems in the following auctions. We distribute the $65 payment for participating in our experiments before starting the practice rounds.

Participants receive instructions on the auction mechanism. After an initial introduction, they are told that they will be participating in an experimental auction and will gain experience with the auction mechanism. They are also notified that they can win at most one unit of each commodity. At this stage, they are encouraged to ask clarifying questions. After questions have been addressed, we proceed to one of two practice rounds. The first practice round trains participants on the experimental auction methods, while the second practice round is designed to familiarize participants with bidding simultaneously on more than one product. The product in the first practice round is a ceramic mug. Participants are asked to come in a single file to the front of the room, view the mug, return to their seats, and write down bids. No talking is permitted. The session monitor and assistant then collect and rank the bids, draw the random n, and identify winners. A short quiz is then administered to check participants' understanding of the auction rules, and then we review the answers and encourage questions. In the second practice round, participants gain experience placing three bids at one time on a notepad, binder, and a package of pens. Bids for the three products are sorted separately and displayed, and "winner" IDs are revealed. Any remaining questions from subjects are answered.

Bidding on experimental products starts next. There are four distinct rounds of bidding, with one round on each set of three

[10]Individuals who arrive together (e.g., friends and family members) are placed in separate sessions, unless special circumstances require assignment to the same session. In this case, we ask that they do not sit next to each other, unless this is also necessary for extenuating reasons. There were two instances of this in our sample.

potato products: fresh potatoes, fries, and chips. Each set of products is covered by a blue bin before a session begins. Both sets of conventional products have identical packaging and labeling, and the products look nearly identical. Likewise, both sets of biotech products also have identical packaging and labeling and look very similar. In rounds one and two, subjects have not yet read their information treatment. They bid on a set of conventional and biotech products, with the type (conventional or biotech) in the first round determined randomly. As in the practice rounds, participants are asked to come in a single file without talking to the front of the room, view the products, return to their seats, and write their bids on a bid sheet. Participants are not allowed to pick up the products.[11] Bid sheets are collected from the participants after each round of bidding. After bid sheets have been collected from the second round of bidding, participants are asked to turn to the information treatment included in their packet and are given approximately ten minutes to read it.[12] Participants then engage in rounds three and four of bidding. The order is randomized as to whether they see the conventional or biotech products first.

After the bid sheets have been collected from the fourth and final round of bidding in a session, participants are told that only the conventional products are available to the winners. Next, the binding round, random n, winning prices, and winning IDs are revealed for a session.[13] Session monitors input bids for the binding round into a spreadsheet, the bids for this round are projected at the front of

[11]Subjects are permitted to touch the commodities' packaging. Peck and Shu (2009) find that, across a wide array of household objects, simply touching an item can increase perceived ownership, which raises valuations if the object does not provide negative sensory feedback. We expect that prohibiting subjects from picking up the products will not influence our results, given consumers' widespread experience with conventional potato products.

[12]To ensure that outside information does not contaminate any specific treatment, session monitors and assistants do not answer questions related to informational content. To avoid confusion, we state that subjects will have either one or two pages of information. This added piece of information about the design is not expected to substantially influence the treatments' efficacy or bidding behavior.

[13]For purposes of product control, we randomly choose the binding conventional round and random n prior to all sessions.

each session's room, and participants watch as the session monitor ranks the bids and reveals the winning IDs.[14] A short post-auction questionnaire is administered; winners of products are then directed to a "stockroom" to purchase the won items(s) and non-winners are excused.

3.3. *Deception in economics experiments*

Interesting research questions may arise before new biological products have been completely de-regulated. Low-acrylamide biotech potato products were not deregulated for commercial sale at the time of the experiment. Hence, it was infeasible for subjects to take home these products. A challenge then arises of how to minimize deception. Rousu *et al.* (2015) examine the cases for and against deception in applied economics research. The second table of their paper lists examples of published papers by agricultural economists that have used some form of deception, and furthermore, they describe some of the ways in which deception might be used successfully in research.

Colson *et al.* (2016) report rankings of severity of deception found in a survey of scholars engaged in applied economics research. They identify the most severe forms of deception as cases where: (1) subjects are deceived in a way that causes physical or physiological trauma, (2) subjects are promised payments for participating in an experiment but receive nothing, or (3) subjects leave an experiment with a different product than what they thought that they had won. Deception of lesser consequences are identified as: (1) providing subjects with no or misleading information about the main purpose of a study, (2) failing to inform subjects that they are part of a study, (3) using confederates or bogus devices,[15] and

[14]Demand reduction is an unlikely problem in our setting. This is because subjects are not bidding on multiple units of the same product. Rather, they are bidding on products that are not strong complements or substitutes. Although demand reduction can occur when auctioning many units of the same good (Corrigan and Rousu, 2006), this is not problematic when the goods are unrelated in consumption (Lusk and Shogren, 2007).
[15]Confederates are subjects who appear to be participating in a study but are actually following the experimenter's instructions. Bogus devices are items in an experiment that do not function as one would expect.

(4) providing subjects with incomplete information about a product for sale in an experiment.

List (2008) emphasized that a strict adherence to a no-deception policy is contrary to the goal of conducting and publishing research on important topics. For example, in some cases, research can only be undertaken if deception is used (e.g., testing for hiring discrimination by using resumes of individuals differing by gender- or ethnic-associated names). He concluded that the benefits and costs of deception (and informed consent) should be carefully considered in each situation, and blanket rules against deception in economic experiments should be avoided.[16]

Moreover, the U.S. Office of Human Research Protection does not forbid deception in research (US-HHS-OHRP, 2009), and in some other fields, deception in experiments is common and widely accepted. Psychologists have regularly used deception in laboratory experiments for more than 100 years (e.g., Solomons, 1897). Hertwig and Ortmann (2008) report that more than half of the papers published in top social psychology journals in 2002 were based on research that involves some form of deception. Deception is widely used and accepted in some other fields, including marketing, nutrition, and medicine (e.g., Nunes and Boatwright, 2004; Wansink *et al.*, 2005; Wansink and Cheney, 2005).

Regarding our research, we designed the experiments to minimize deception at each step, given the circumstances. At the initiation of each session, we told subjects they would be participating in an auction of food products. We told them that winners would execute winning bids, but that they would at most be a winner of only one unit of each type of product. In rounds of bidding with GM potato products, all labels were accurate, i.e., the label on the

[16]His conclusion notwithstanding, a leading agricultural economics journal, the *American Journal of Agricultural Economics*, imposed a policy in 2011 banning the publication of articles using deceptive practices. However, in 2015, the journal changed the policy. Authors are now asked to describe the extent and nature of deception involved in submitted papers and to provide justification for deception. The editor then decides whether deception is warranted or excessive given the research justification.

product correctly identified the product. Each subject participated in four rounds of bidding, and no information on bids in a session was revealed until all bids were placed and collected. The monitor then randomly selected one of the two rounds of bidding on the conventional products to be binding.

Hence, our procedures insured that products on display in the auctions were correctly labeled (authentic), no winner paid a price different from the rules in the auction, and no subject left the auction with a product that was different from what he or she thought had been won. We believe that our experimental procedures resulted in a low level of risk to human subjects, given the objective of the research project and state of biotech product deregulation at the time of the experiments.

4. Data Analysis Methods, Econometric Model, and Data

We provide two methods of data analysis. The first analyzes average treatment effects in subjects' bids before and after they received a packaged information treatment. The second is a regression analysis of subjects' bid differences within and across technology types and information treatments.

4.1. *Estimating average treatment effects (ATEs)*

For a given commodity, the (information) treatment effect for the ith participant is $E(WTP_{iT} - WTP_{iC})$, where "subscript" T refers to treatment and C to control and $E(\cdot)$ is the expectation operator. The ATE over all participants is then $\overline{WTP_T} - \overline{WTP_C}$, where the "overbar" notation represents the sample mean over all participants. With random assignment of treatment (control), unobservable of participants are balanced between participants assigned to the treatment and control. It is the random assignment of subjects to treatment that becomes the instrumental variable permitting identification of treatment effects (List, 2009).

4.2. Econometric model

We conduct further empirical investigation by using regression analysis of WTP differences. Our econometric model primarily uses within-individual WTP differences: WTP post-information treatment less WTP pre-information treatment, which differences out unchanged unobservables. We also examine WTP differences post-information treatment.

For each of the three products, we estimate the following econometric models of WTP differences:

$$Bid_{i,\,post}^{GM} - Bid_{i,\,pre}^{GM} = \boldsymbol{x'_i\beta_1} + \boldsymbol{T'_i}\gamma_1 + \varepsilon_{1,i} \quad (1)$$

$$Bid_{i,\,post}^{GM} - Bid_{i,\,pre}^{Conv} = \boldsymbol{x'_i\beta_2} + \boldsymbol{T'_i}\gamma_2 + \varepsilon_{2,i} \quad (2)$$

$$Bid_{i,\,post}^{GM} - Bid_{i,\,post}^{Conv} = \boldsymbol{x'_i\beta_3} + \boldsymbol{T'_i}\gamma_3 + \varepsilon_{3,i}. \quad (3)$$

In the above equations, $\boldsymbol{x'_i}$ is a vector of demographic and other controls for person i, $\boldsymbol{T'_i}$ are indicator variables for information treatments, and $\varepsilon_{.,i}$ is a random error. Further, the term $Bid_{i,\,post}^{GM} - Bid_{i,\,pre}^{GM}$ is the difference in WTP for biotech foods before and after the information treatment, while $Bid_{i,\,post}^{GM} - Bid_{i,\,pre}^{Conv}$ is WTP for biotech foods post-information treatment less WTP for conventional foods without information. The latter is designed to examine information impacts in transitioning from conventional to low-acrylamide biotech potato products.[17] This is analogous to examining expected price differences before and after introduction, which is of practical importance given recent USDA deregulation. Last, we analyze $Bid_{i,\,post}^{GM} - Bid_{i,\,post}^{Conv}$, the difference in WTP for biotech and conventional foods post-information. This is designed to capture the premium a shopper would be willing to pay for biotech potatoes after learning about using GM to reduce acrylamide-forming potential. The estimates of the γ parameters indicate treatment effects.

[17]Interpreting this dependent variable is more challenging. It may not strictly represent a premium for food safety because it confounds sources of perceived risk: the biotech products reduce acrylamide potential but use biotechnologies that are not universally accepted by consumers.

The WTP data have two potential problems. First, some zero bids are truncated at zero, i.e., subjects would have preferred to bid a negative value but felt constrained to bid zero instead. Other zero bids are legitimate bids with no truncation, i.e., zero was the preferred bid (Maddala, 1983). Second, valuations, although sincere, might be arbitrary (Ariely *et al.*, 2003). We ignore the second and focus on the first. Hence, the WTP differences in Eqs. (1)–(3) are sometimes censored. To accommodate censoring, we apply a general version of a differenced Tobit model (Rousu *et al.*, 2007). The general nature of this model arises from multiple sources of truncation of WTP at zero. Coefficients are obtained by maximizing each equation's log-likelihood, which is the sum of continuous and discrete distributions.

There are four censoring cases to consider, depending on the location of zeros in the dependent variables of Eqs. (1)–(3). *Case 1*: both bids are positive. The dependent variable is uncensored and the likelihood is continuous. *Case 2*: the post-information bid is zero because the subject preferred to bid a negative value but felt constrained to bid zero, and the pre-information bid is positive. In this case, the true WTP difference is left-censored at the value of subject i's pre-information bid. *Case 3*: the post-information bid is positive and the pre-information bid is zero because the subject preferred to bid a negative value but bid zero instead. In this case, the true WTP difference is right-censored at the value of subject i's post-information bid. *Case 4*: both bids are truncated at zero because a subject preferred a negative value but bid zero instead. Unfortunately, when we observe a zero value for WTP, we do not know whether it is a legitimate zero or a "constrained" zero. As such, these double-zero observations are not included in the regression analysis.

4.3. *Data*

We draw on past food experiment research, economic theory, and practical considerations to guide question choice in the questionnaires. Data from the pre-auction questionnaires concern demographics, including gender, age, years of education, race, marital status, household size and income, religion, residential location, and occupation. We also ask participants about their degree of knowledge about acrylamide, biotechnology, and food safety. Post-auction questions include previous participation in an economics experiment, distance traveled to the site, and accompaniment by another participant to the site.

Of the 304 participants, 101 are from Des Moines, 106 are from Los Angeles, and 97 are from Boston. The average number of participants per session is 16.9. Four participants had missing data and were excluded from the analysis, leaving a sample of 300. The average number of participants per treatment is 50 (300/6). Of these 300, there are 3–5 double-zero bids for potatoes, 9–13 double-zero bids for French fries, and 6–10 double-zero bids for potato chips.

The average age of the participants is 43 years, with 14.4 years of schooling, living with one other adult in the household, and having a per-person annual pre-tax income of $28,550 (see Table 1).[18] Roughly 42% and 49% of the participants were at least somewhat informed about biotech and non-biotech food products prior to the experiment ("Previously informed about biotech" and "Previously informed about non-biotech"), and 88% reported reading labels at least some of the time when buying new food products ("Looks at labels").[19] In general, there is considerable variation in participants' attributes.[20]

[18]Annual incomes are coded as the midpoint of the income bracket chosen by the respondent in the pre-auction questionnaire.

[19]Subjects in the Des Moines area consisted of randomly-dialed participants, returning participants from a previous food experiment, and contacts of randomly-dialed participants. This substantially reduced recruitment expense. However, we are not concerned about correlated bids, dependent data, or an unrepresentative sample for the following reasons: (i) except in two cases, subjects who arrive with other subjects are assigned to different concurrent sessions, (ii) the auctions are incentive compatible, regardless of subjects' familiarity with other subjects, (iii) the sample matches several key dimensions of the population of US grocery shoppers, and (iv) these types of recruits comprise less than 25% of our full sample.

[20]Roughly 76% of the sample is white, a consequence of the eligibility requirement that subjects can read and write English. This has some bearing on average demographics, but Table 1 confirms there is ample variation in our data.

Table 1. Summary statistics for bid differences and regressors ($N = 300$).

Variable	Mean	Std. Dev.	Min	Max
Dependent variables				
GM Pot Bid Post − GM Pot Bid Pre ($)	−0.11	1.56	−8.95	7.00
GM Pot Bid Post − Conv Pot Bid Pre ($)	−0.13	1.67	−13.95	6.00
GM Pot Bid Post − Conv Pot Bid Post ($)	−0.05	1.54	−8.00	4.00
GM Fry Bid Post − GM Fry Bid Pre ($)	−0.20	1.37	−9.55	4.65
GM Fry Bid Post − Conv Fry Bid Pre ($)	−0.13	1.39	−11.55	4.50
GM Fry Bid Post − Conv Fry Bid Post ($)	0.02	1.34	−8.00	3.50
GM Chip Bid Post − GM Chip Bid Pre ($)	−0.07	1.19	−7.32	4.90
GM Chip Bid Post − Conv Chip Bid Pre ($)	−0.08	1.17	−8.32	4.70
GM Chip Bid Post − Conv Chip Bid Post ($)	0.01	1.24	−8.00	4.00
Demographics and other regressors				
Age (years)	42.82	14.08	18	65
Age^2 (years)	2,031.04	1,191.22	324	4225
Adults in household (integer)	2.08	0.85	1	6
Per capita income ($ thousand)	28.55	19.53	1.50	112.50
Per capita $income^2$ ($ thousand)	2,792.60	3,322.49	9.00	24,200.00
Looks at labels (0–1)	0.88	0.33	0	1
Previously informed about biotech foods (0–1)	0.42	0.49	0	1
Previously informed about non-biotech foods (0–1)	0.49	0.50	0	1
Arrived with someone else (0–1)	0.22	0.42	0	1
Distance traveled (miles)	9.32	7.99	1.00	100.00
Exercises regularly (0–1)	0.75	0.43	0	1
Information treatments				
Industry information	0.19	0.39	0	1
Scientific information	0.18	0.38	0	1
Industry and scientific information	0.17	0.38	0	1
Industry and environmental information	0.15	0.36	0	1
Scientific and environmental information	0.15	0.34	0	1

5. Empirical Results

5.1. *Mean bids by commodity and information treatment*

Table 2 provides summary statistics on bids for the three potato products by type, before and after subjects received an information treatment. Mean bids across all participants pre-information treatment are \$2.60 per bag of conventional and GM potatoes, \$2.50 and \$2.41 per bag of biotech and conventional fries, and \$2.10 and \$2.07 per bag of biotech and conventional chips. The differences in mean WTP are small and not significantly different from zero prior to information.[21] Hence, we find no labeling effect on WTP for potato products pre-information treatment.

The ATE calculations and tests of zero ATEs, by information treatment, are presented in Table 3. For the roughly 54 participants who received the scientific perspective only treatment, the mean bid of participants for biotech potatoes is \$2.34 pre-information and \$2.88 post-information, an ATE of \$0.55. In contrast, the mean bid for conventional potatoes declined from \$2.68 to \$2.38 after information, an ATE of –\$0.30. The mean bid for biotech fries is \$2.41 pre-information and \$2.73 post-information (ATE of \$0.32), whereas for conventional fries, the mean bid is \$2.40 pre-information and \$2.15 post-information (ATE of –\$0.25). Last, the ATE for biotech chips is \$0.38, but for conventional chips, the ATE is –\$0.29. Hence, under the scientific information treatment, all ATEs are significantly different from zero at least at the 10% level, being positive for biotech commodities but negative for conventional commodities.[22]

When participants receive the information treatment containing only the environmental groups' perspective, mean bids for both biotech and conventional products are reduced, but the reduction is

[21] In separate t-tests, we cannot reject at the 5% level the null hypotheses that the mean of the bid differences are zero between GM and conventional potatoes (t-statistic $= -0.27$), GM and conventional fries (t-statistic $= 1.15$), and between GM and conventional chips (t-statistic $= -0.12$), all prior to subjects receiving an information treatment.

[22] When participants receive the industry perspective only, the pre- vs. post-information change in bids is smaller in magnitude and tends to decrease after exposure to information.

Table 2. Mean bids for subjects, excluding double-zero bids.

Commodity type	N	Mean	Std Dev	Min	Max	Mean	Std Dev	Min	Max
Before information						**After information**			
Mean bids – all bidders and treatments									
GM potatoes	295	2.60	1.77	0	11.00	2.48	1.91	0	13.99
Conv. potatoes	297	2.60	1.65	0	15.00	2.52	1.70	0	10.99
GM French fries	287	2.50	1.68	0	12.00	2.29	1.71	0	11.99
Conv. fries	290	2.41	1.46	0.01	12.00	2.25	1.52	0	9.00
GM chips	290	2.10	1.38	0	8.00	2.02	1.50	0	8.99
Conv. chips	294	2.07	1.23	0	9.00	1.98	1.39	0	9.00
A. Bidders receiving industry-only perspective									
GM potatoes	56	2.51	1.49	0.20	6.00	2.51	1.86	0	7.00
Conv. potatoes	56	2.49	1.27	0.10	7.00	2.24	1.32	0	5.50
GM French fries	53	2.55	1.59	0.01	6.99	2.35	1.66	0	7.00
Conv. fries	54	2.19	1.13	0.01	5.00	1.98	1.18	0	4.00
GM chips	54	2.15	1.40	0.01	5.99	2.09	1.61	0	6.50
Conv. chips	55	1.98	1.08	0.01	4.00	1.92	1.34	0	6.00
B. Bidders receiving scientific-only perspective									
GM potatoes	54	2.34	1.86	0	10.99	2.88	2.39	0	13.99
Conv. potatoes	54	2.68	1.91	0.50	11.99	2.38	2.02	0	10.99
GM French fries	52	2.41	1.70	0	7.99	2.73	2.21	0	11.99
Conv. fries	53	2.40	1.64	0.01	9.99	2.15	1.75	0	8.99
GM chips	52	2.09	1.33	0	5.00	2.47	1.77	0	8.99
Conv. chips	52	2.16	1.34	0.45	7.99	1.87	1.34	0	6.00
C. Bidders receiving environmental-only perspective									
GM potatoes	52	3.48	2.28	0.25	11.00	2.22	1.68	0	5.00
Conv. potatoes	53	3.19	2.33	0.10	15.00	3.03	1.72	0.10	6.00
GM French fries	52	3.10	2.17	0.15	12.00	2.02	1.63	0	8.00
Conv. fries	52	3.02	1.95	0.02	12.00	2.75	1.48	0.10	9.00
GM chips	53	2.54	1.64	0.10	8.00	1.75	1.32	0	5.00
Conv. chips	54	2.46	1.52	0.20	9.00	2.29	1.29	0	5.00
D. Bidders receiving industry and scientific information									
GM potatoes	51	2.54	1.68	0.25	9.00	2.98	1.71	0.05	9.00
Conv. potatoes	51	2.57	1.36	0	6.00	2.33	1.37	0.01	6.00
GM French fries	50	2.30	1.41	0.10	8.00	2.68	1.33	0.05	7.00
Conv. fries	49	2.36	1.07	0.05	5.00	2.13	1.19	0.05	5.00
GM chips	51	1.93	1.32	0.15	6.00	2.40	1.31	0.05	6.00
Conv. chips	50	2.06	1.09	0.15	4.00	1.85	1.16	0.10	5.00

Table 2. (*Continued*)

Commodity type	N	Mean	Std Dev	Min	Max	Mean	Std Dev	Min	Max
Before information						**After information**			
E. Bidders receiving industry and environmental information									
GM potatoes	44	2.35	1.39	0.50	5.50	2.07	1.51	0	5.99
Conv. potatoes	44	2.28	1.11	0.25	5.00	2.68	1.86	0	9.00
GM French fries	43	2.40	1.48	0	6.99	1.95	1.48	0	5.00
Conv. fries	44	2.36	1.18	0.50	5.25	2.41	1.92	0	9.00
GM chips	43	1.87	1.13	0.25	4.00	1.65	1.22	0	5.00
Conv. chips	45	1.89	1.04	0	4.25	2.09	1.82	0	9.00
F. Bidders receiving scientific and environmental information									
GM potatoes	38	2.25	1.40	0.01	5.00	2.05	2.05	0	10.00
Conv. potatoes	39	2.21	1.30	0.01	5.00	2.47	1.82	0	8.10
GM French fries	37	2.12	1.43	0.01	6.00	1.89	1.65	0	6.00
Conv. fries	38	2.02	1.43	0.01	7.00	2.03	1.46	0	5.00
GM chips	37	1.89	1.32	0.01	4.69	1.61	1.49	0	5.19
Conv. chips	38	1.78	1.09	0.01	4.29	1.83	1.36	0	4.79

Table 3. Average treatment effects by commodity and information treatment.

Information	Bid^{GM}_{Pot}	Bid^{GM}_{Fry}	Bid^{GM}_{Chip}	Bid^{Conv}_{Pot}	Bid^{Conv}_{Fry}	Bid^{Conv}_{Chhip}
Industry (I)	\$0.01	−\$0.20	−\$0.06	−\$0.25***	−\$0.21**	−\$0.07
Scientific (S)	\$0.55***	\$0.32*	\$0.38**	−\$0.30*	−\$0.25*	−\$0.30**
Environmental (E)	−\$1.26***	−\$1.08***	−\$0.79***	−\$0.17	−\$0.27	−\$0.17
I and S Info	\$0.43***	\$0.38***	\$0.47***	−\$0.24**	−\$0.23*	−\$0.21**
I and E Info	\$0.28*	−\$0.45**	−\$0.22*	\$0.40	\$0.04	\$0.21
S and E Info	−\$0.20	−\$0.24	−\$0.28	\$0.26	\$0.02	\$0.06

Note: Table entries are average treatment effects (ATEs) for the three GM products and three conventional products. For each entry, we perform a one-sample, two-sided t-test of the null hypothesis of a zero ATE. Double zero bids are excluded. Sample sizes by commodity and information treatment are the same as those reported in panels A-F of Table 1. $^{*}p < 0.1$, $^{**}p < 0.05$, $^{***}p < 0.01$.

larger and only statistically significant for the biotech products. The mean bid for biotech potatoes is \$3.48 pre-information and \$2.22 post-information (ATE of −\$1.26). The pre-information mean bid for biotech French fries is \$3.10 and \$2.02 post-information (ATE of −\$1.08), with a similar ATE of −\$0.79 for biotech chips.

For participants who receive the industry and scientific perspectives, the mean bids for the biotech products post-information increases significantly, but bids for conventional products decline significantly. For the biotech products, ATEs range in \$0.38–\$0.47, while ATEs range between −\$0.21 and −\$0.24 for conventional products. These are similar to those for participants receiving only the scientific perspective.

For those who receive the industry and environmental groups' perspectives, mean bids for the biotech products decrease post-information, while mean bids for the conventional products increase. For the former group of products, mean bids decrease by \$0.22–\$0.45 after information. However, mean bids for the conventional products increase between \$0.04 and \$0.40 after information, though these ATEs are not statistically different from zero. As such, the negative perspective toward biotech potato products dominates the positive perspective in affecting mean bids under this information treatment.

For participants who receive the information treatment containing the scientific and environmental perspectives, none of the ATEs are significantly different from zero. Hence, the positive scientific perspective for low-acrylamide potato products effectively neutralizes the negative perspective of environmental groups.

5.2. *Regression analysis of bid differences and effects of information*

Summary statistics for variables included in the regression analysis are presented in Table 1. Mean WTP across all information treatments is lower than mean WTP pre-information. For example, average WTP for the biotech potatoes is \$0.11 lower after subjects received packaged information. This trend is similar for French fries and potato chips. Post-information GM premiums range between −\$0.05 and \$0.01, depending on product.

Table 4 presents results of the three econometric models of WTP differences for the fresh potatoes.[23] In column 1, relative to

[23]In addition to the regressors listed in Table 1, we included participant's education, health variables, number of pre-teen children, number of teenage children, and fixed effects for experiment location, session times, and monitor. These latter variables were not individually or jointly significant, so they have been deleted from the models presented in Tables 4–6.

Table 4. Estimates of model of WTP differences with information effects and possible censoring: Biotech and conventional potatoes.

Variable	(1) $Bid_{i,post}^{GMPot} - Bid_{i,pre}^{GMPot}$	(2) $Bid_{i,post}^{GMPot} - Bid_{i,pre}^{ConvPot}$	(3) $Bid_{i,post}^{GMPot} - Bid_{i,post}^{ConvPot}$
Age	0.037	0.025	−0.034
Age^2	−0.0004	−0.0001	0.0004
Adults in household	0.213*	0.263*	0.057
Per capita income	0.009	0.010	−0.007
Per capita income^2	−0.000	−0.000	0.000
Looks at labels	−0.201	−0.256	−0.034
Previously informed about biotech foods	0.500**	0.368	0.299
Previously informed about non-biotech foods	−0.347	−0.491**	−0.448**
Arrived with someone else	−0.332	−0.233	−0.192
Distance traveled	0.022*	0.022*	0.033***
Exercises regularly	−0.536***	−0.459**	−0.464**
Industry information	1.180***	0.974***	1.113***
Scientific information	1.972***	1.428***	1.477***
Industry and scientific information	1.759***	1.589***	1.599***
Industry and environmental information	1.063***	0.972***	0.411
Scientific and environmental information	1.050***	0.747**	0.419
Constant	−2.087*	−2.035*	−0.090
Observations	295	297	292
LR χ^2 statistic	68.9***	43.1***	64.3***

Note: The 5% critical value of $\chi^2(16)$ is 26.3. $^*p < 0.1$, $^{**}p < 0.05$, $^{***}p < 0.01$.

Table 5. Estimates of model of WTP differences with information effects and possible censoring: Biotech and conventional French fries.

Variable	(4) $Bid_{i,post}^{GMFry} - Bid_{i,pre}^{GMFry}$	(5) $Bid_{i,post}^{GMFry} - Bid_{i,pre}^{ConvFry}$	(6) $Bid_{i,post}^{GMFry} - Bid_{i,post}^{ConvFry}$
Age	0.052	0.060	−0.008
Age^2	−0.001	−0.001	0.000
Adults in household	0.215*	0.164	−0.013
Per capita income	0.018*	0.012	−0.008
Per capita income^2	−0.0001*	−0.0001	0.0001
Looks at labels	−0.190	−0.253	−0.056
Previously informed about biotech foods	0.168	0.103	0.135
Previously informed about non-biotech foods	−0.180	−0.440**	−0.278
Arrived with someone else	−0.151	−0.164	−0.072
Distance traveled	0.012	0.012	0.026***
Exercises regularly	−0.459**	−0.289	−0.324*
Industry information	0.824***	1.000***	1.049***
Scientific information	1.550***	1.391***	1.350***
Industry and scientific information	1.543***	1.468***	1.416***
Industry and environmental information	0.692**	0.601**	0.365
Scientific and environmental information	0.793***	0.678**	0.557*
Constant	−2.368**	−2.463**	−0.360
Observations	287	291	283
LR χ^2 statistic	55.3***	51.8***	56.4***

Note: The 5% critical value of $\chi^2(16)$ is 26.3. $^*p < 0.1$, $^{**}p < 0.05$, $^{***}p < 0.01$.

Table 6. Estimates of model of WTP differences with information effects and possible censoring: Biotech and conventional potato chips.

Variable	(7) $Bid_{i,post}^{GMChip} - Bid_{i,pre}^{GMChip}$	(8) $Bid_{i,post}^{GMChip} - Bid_{i,pre}^{ConvChip}$	(9) $Bid_{i,post}^{GMChip} - Bid_{i,post}^{ConvChip}$
Age	0.084**	0.071**	0.013
Age^2	−0.001**	−0.001*	−0.000
Adults in household	0.343***	0.241**	0.130
Per capita income	0.014*	0.009	0.005
Per capita income^2	−0.0001*	−0.0001	−0.0001
Looks at labels	−0.005	−0.063	0.135
Previously informed about biotech foods	0.200	0.010	0.121
Previously informed about non-biotech foods	−0.197	−0.289*	−0.246
Arrived with someone else	−0.188	−0.211	0.051
Distance traveled	0.017*	0.013	0.025***
Exercises regularly	−0.327**	−0.327**	−0.294*
Industry information	0.602***	0.696***	0.687***
Scientific information	1.301***	1.161***	1.290***
Industry and scientific information	1.274***	1.199***	1.229***
Industry and environmental information	0.561**	0.447*	0.109
Scientific and environmental information	0.406	0.436*	0.279
Constant	−3.198***	−2.593***	−1.204
Observations	290	294	287
LR χ^2 statistic	62.9***	53.1***	63.8***

Note: The 5% critical value of $\chi^2(16)$ is 26.3. $^{*}p < 0.1$, $^{**}p < 0.05$, $^{***}p < 0.01$.

participants who viewed the environmental groups'-only treatment (the excluded treatment), the industry-only treatment increases WTP for biotech potatoes by $1.18, and the scientific-only treatment increases WTP by $1.97. The relative impact of a subject receiving the industry and scientific perspectives is $1.76, a weighted average of the two separate effects. The impacts of the industry and environmental groups' treatment and the scientific and environmental groups' treatment are significantly lower but similar in size at $1.06 and $1.05. This suggests that the environmental groups' perspective decreases subjects' positive WTP relative to the positive effects of industry or scientific information.

Participants who indicated that they were well-informed or somewhat well informed about biotech food products coming into the experiments were willing to pay $0.50 more for biotech potatoes after information than the uninformed. Those who exercise regularly are willing to pay $0.54 less for biotech potatoes after information. However, an additional adult in the household increases post-information WTP for potatoes by $0.21.

Although the dependent variables in models (2) and (3) differ in terms of the reference commodity (conventional potatoes pre- vs. post-information), the estimates of the information effects are similar to those in (1). The main difference is that the information treatments with the environmental groups' perspectives do not influence post-information GM premiums (model 3).

Regressions explaining WTP differences for biotech French fries are reported in Table 5. The industry-only treatment increases subjects' relative WTP for biotech fries by $0.82, and the scientific-only treatment increases WTP by $1.55 (column 4). The impact of a subject viewing the industry and scientific perspectives is $1.54, approximately the same as the estimate of the scientific-only treatment effect. The relative effects on a subject's WTP after viewing the industry and environmental groups' treatment and the scientific and environmental groups' treatment are significantly smaller and similar at $0.69 and $0.79. This again suggests that the environmental groups' perspective decreases WTP relative to the positive effects of viewing scientific or industry information only.

There are a number of other interesting results for fries apart from the information treatment effects. A subject's information premium for biotech fries is increasing in the number of adults in the household and annual per-capita household income up to $90,000 (column 4). Participants who exercise regularly have lower WTP differences. Information treatment effects in columns (5) are similar to those in column (4). However, the industry and environmental groups' information treatment has no explanatory power for the post-information GM fry premium.

Regressions of WTP differences for biotech potato chips are reported in Table 6. The signs and magnitudes of the single-perspective information treatment effects are similar to those for potatoes and fries. The information effects are somewhat smaller and are insignificant in three instances across two of the models (column 7 and column 9).

Participants' characteristics contribute more to explaining the premiums for biotech chips than the other two products. The WTP differences increase with subjects' age up to 36–42 years. Up to $70,000, the marginal effect of additional household income is positive (column 7), and an additional adult in a subject's household increases the premium by $0.24–$0.34 (columns 7–8). Similar to the other two products, those who exercise regularly have decreased premiums by $0.29–$0.33.

Last, we provide evidence of additive information effects on WTP differences for the biotech products. We test the null hypothesis that $\beta_1 + \beta_S = \beta_{SI}$, where β_1 is the coefficient of the industry-only treatment, β_S is the coefficient of the scientific only treatment, and β_{SI} is the coefficient of the scientific and industry treatment. Except for biotech potatoes, we cannot reject the null hypothesis of simple additivity of separate industry only and scientific only effects equaling the estimate of the combined effect.[24] For biotech potatoes, the sum of the two separate information effects ($3.15)

[24]The likelihood ratio test statistic for the results in columns (1), (4), and (7) are 11.77, 4.89, and 3.90. The 1% critical value of the $\chi^2(1)$ distribution is 6.63.

is nearly double the estimate for the combined effect ($1.76). This suggests there is some type of complementary interaction between the two perspectives in the treatment. One reason for this might be that single-perspective information obtained in the experiment could help subjects, in the future, prepare fresh potatoes so as to eliminate acrylamide formation (e.g., boiling or steaming). This is not possible for processed fries and chips.

6. Conclusions

Policy implications of our results arise from finding that consumers respond to information in food labels and perspectives of the potato industry, scientific community, and environmental groups on food-borne acrylamide and new low-acrylamide biotech potato products, and that this response is also conditioned somewhat by heterogeneity of consumers' attributes. Households with a larger number of adults are willing to pay more for biotech than conventional potato products. Larger household per capita income increases WTP for biotech fries and chips post-information at low income levels, but the effect diminishes as income increases. Those consumers who indicated that they were informed about biotech foods coming into the experiment were willing to pay more for biotech fresh potatoes, but not for the processed fries and chips. Consumers who exercise regularly have reduced WTP for biotech products after information, reduced GM premiums post-information, and reduced WTP for biotech products post-information relative to conventional products pre-information.

Perhaps surprising, we find no evidence of a labeling effect on subjects' valuations. That is, prior to subjects viewing a randomly-assigned information treatment, there is no statistically significant difference in mean WTP for conventional potato products with plain labels and biotech potato products with biotech-indicating labels. Further, there is no significant effect on WTP due to a consumer indicating that he or she reads food labels when buying a food product for the first time. This is suggestive evidence that, in isolation, labels for biotech potato products do not influence consumer valuations any more or less than labels for conventional (non-GM) potato products.

The information effects suggest that consumers' WTP decisions may be malleable. Those receiving the industry information (only), industry and scientific information, scientific information (only), industry and environmental groups' information, and scientific and environmental groups' information show sizeable and statistically significant increased WTP relative to those who received environmental groups' information (only). However for fresh potatoes, the largest increases occur when consumers receive scientific (only) information. For fries, the largest increases occur for industry and scientific information. For chips, the increase in WTP for scientific (only) information and for industry and scientific information are roughly equal. To provide context, the size of these large information treatment effects are roughly equal to 50% of the mean price of the corresponding conventional potato products pre-information.

Low-acrylamide biotech potatoes have been recently approved in the U.S. for commercial sale. Consumers' adoption of new low-acrylamide potato products would reduce a principal source of carcinogens in diets. Consumer and producer acceptance of these biotech foods will, in large part, be determined by the relative success of the pro-GM and anti-GM groups in influencing public opinion. Given that we do not find a labeling effect, the results demonstrate these new biotech potato products will be most successful if there is an effective information campaign describing the objective benefits of these products.

Since the implementation of our experiment, there have been more than two years of White Russet™ potato (biotech potatoes with low-acrylamide-forming potential) sales in the U.S. The seed potatoes are sold directly to farmers, who then contract with one of two licensed intermediaries. These intermediaries market to grocery stores, restaurants, and food service providers. To date, roughly 40 million pounds of these biotech potatoes have been sold. During this early stage of introduction, independent grocery stores have been selling them at the same price as conventional potatoes. To the extent that early introduction represents a low-information setting, this externally confirms our finding of no significant difference in mean WTP for conventional and biotech fresh potatoes prior to information.

Our experimental design and results offer opportunities for future research. Labeling of GM foods (mandatory or voluntary) remains an important economic issue in the U.S. and European Union. To the extent that subjects have access to somewhat similar information, an experimental design that elicits valuations for other potential labels could be interesting. Alternatively, one could analyze "authority" effects on valuation, e.g., given identical information treatments, subjects may respond differently based on their preferred information source. Our results suggest that some demographic variables mediate information effects, which could inform future experimental designs.

Acknowledgments

We thank Gary Laney, Katie Lacy, Sonya Huffman, Nadia Huffman, Nicole Caviris, Zach Martin, Jesse Strzok, and Jared Martin for valuable assistance. We thank John List, Elizabeth Hoffman, Kyrre Rickertsen, Matthew Rousu, Paul Bethke, Paul Mitchell, Justus Wesseler, and participants at the ISU Human Resources Workshop for helpful comments. We acknowledge funding and assistance from the USDA through the University of Wisconsin, the J.R. Simplot Company, and the Iowa Agriculture and Home Economics Experiment Station. This research was approved by the ISU Institutional Review Board (IRB) in ISU IRB #14-115.

Appendix A

See Figs. A.1–A.3.

Please take a few minutes to read the following.

An industry perspective on low acrylamide potatoes using biotechnology

General Statement:

- In 2002, acrylamide was discovered in foods containing certain natural sugars when cooked at high temperatures (above roughly 250° F), such as French fries, hash browns and potato chips. They are also formed in the baking of bread (crust) and cookies and roasting of coffee beans.
- Acrylamide, a chemical compound, is formed in foods from naturally occurring sugars and asparagine (a building block of a protein). It is a toxin and possible carcinogen in humans.
- Conventional plant breeding efforts have resulted in only small reductions of acrylamide in processed potato products.
- In contrast, biotech methods have significantly reduced the acrylamide level in processed potato products, thereby improving food safety.
- Biotech plant breeding methods take genes from one organism and transfer them to another. In new biotech potatoes, all genes come from wild and domesticated potato varieties. Hence, this biotech method has much in common with conventional plant breeding.

Nutrition and Health:

- Long-term, low-level intake of acrylamide by lab animals has been shown to create serious health problems. One study found that consuming conventional potato chips regularly for a month caused some health changes in humans.
- Under Proposition 65, California requires certain restaurants, including Applebee's, Chili's, McDonald's, Burger King, Wendy's, and KFC, to post a warning stating that "cooked potatoes that have been browned, such as French fries, hash browns and baked potatoes, contain acrylamide, a chemical known to the State of California to cause cancer."
- A major biotech accomplishment has been the development of new potato varieties that have approximately a 70% reduction in acrylamide levels in processed potato products relative to conventional.
- Biotech methods were also used to develop golden rice, which enhanced vitamin A content. Other opportunities exist for enhancing consumer attributes, such as, antioxidants and vitamins in food.
- All Food and Drug Administration (FDA)-approved biotech-foods have been assessed to be substantially equivalent in nutrient content to conventional foods. As such, labeling in the U.S. is voluntary.
- Biotech foods, relative to conventional foods, have similar low allergy potential.

Environmental Impacts and Food Security:

- Acrylamide is used in industrial processes to make paper, dyes, and plastics, but food and cigarette smoke are the two main sources of human exposure to acrylamide.
- Commercial potatoes are grown from pieces of whole potato and not seed. In addition, many commercial potatoes are either sterile or not sexually compatible with wild potatoes. Hence, there is very low risk of biotech potatoes crossing with other potato varieties or other plants.
- There are no adverse impacts on the environment of the new biotech potatoes.
- In general, biotech crops have reduced the use of toxic insecticides and increased the use of other environmentally friendly farming practices.
- Food security can be greatly improved with continued advancement and adoption of genetically modified crops.
- Approval is being sought for biotech potato exports to Canada, Mexico, South Korea, and Japan.

Fig. A.1. Industry perspective on low acrylamide potatoes using biotechnology.

Please take a few minutes to read the following.

A scientific perspective on acrylamide exposure

General Statement:

- In 2002, acrylamide was discovered in foods containing certain natural sugars when cooked at high temperatures (above roughly 250° F), such as French fries, hash browns and potato chips. Acrylamide is also formed in the baking of bread (crust) and cookies and roasting of coffee beans.
- Acrylamide, a chemical compound, is formed in these foods from naturally occurring sugars and asparagine (a building block of a protein).
- Compared to other foods in the American diet, French fries and potato chips are a major source of acrylamide.
- The amount of acrylamide increases with longer frying or baking times and higher cooking temperatures. However, boiling, steaming, and microwaving potatoes produces negligible acrylamide.
- Under Proposition 65, California requires certain restaurants, including Applebee's, Chili's, McDonald's, Burger King, Wendy's, and KFC, to post a warning stating that "cooked potatoes that have been browned, such as French fries, hash browns and baked potatoes, contain acrylamide, a chemical known to the State of California to cause cancer."
- However, the FDA has not advised people to stop eating these potato products.

Nutrition and Health:

- The International Agency for Research on Cancer (IARC) designates acrylamide as a "probable human carcinogen," and the US National Toxicology Program classifies acrylamide as "reasonably anticipated" to be capable of causing cancer in humans.
- The first health studies of acrylamide occurred in the 1960s and explored health hazards with on-the-job contact. Chronic exposure produces toxicity that weakens muscles and reflexes and leads to sensory loss.
- Evidence from studies in the 1970s and 1980s advanced the linkages between cancer and acrylamide.
- In a recent study, lab animals that consumed low doses of acrylamide placed in drinking water over a long period of time (roughly equal to annual acrylamide consumption of an adult) experienced gene damage.
- Another study found that humans who consumed conventional potato chips daily for four weeks had higher incidence of early development of cardio-vascular diseases.
- Additional studies suggest linkages between high acrylamide diets and development of breast cancer. High levels of acrylamide consumption by pregnant women have been associated with increased frequency of low birth weight babies.

Environmental Impacts and Food Security:

- Acrylamide is used in industrial processes to make paper, dyes, and plastics, but food and cigarette smoke are the two main sources of human exposure to acrylamide.
- If the occupational and health risks associated with acrylamide were to be substantially reduced or eliminated, food security would improve, especially in the United States and some other developed countries where a majority of the population regularly consumes foods that contain acrylamide.

Fig. A.2. Scientific perspective on acrylamide exposure.

Please take a few minutes to read the following.

An environmental group perspective on biotechnology

General Statement:

- Biotech plant breeding takes genes from one organism and places them into another. This process manipulates genes and alters genetic makeup and properties. The cutting of genetic material from one organism and inserting it into another is quite imprecise and can cause mutations. There has been inadequate testing of these products.
- Biotech methods frequently use antibiotic-resistant gene segments in soil bacteria or viruses and transfer them into plants. This process is risky, leading to unanticipated outcomes.
- Biotech plants are regulated by the federal government, but federal regulation relies heavily on data collected from field trials and other testing by the biotech industry.
- Biotech seeds were first marketed to U.S. farmers in the mid-1990s, and rapid farmer adoption occurred in field crops (corn, soybean, cotton and canola). Later developments have been in papaya and vegetables (sugar beet, squash and potato).
- In the United States, the sales of biotech foods have grown very rapidly. This growth is driven by self-interested producers and marketers seeking to maximize crop yields and minimize production costs.

Nutrition and Health:

- Of the laboratory plants that are successfully modified to express the "right" traits, genetic engineers select among those that look strong, healthy, and capable of further breeding. There is poor screening to eliminate varieties that produce harmful substances or low nutrient quality.
- New allergens are likely to be introduced into the food supply.
- Early research reported some health problems in laboratory animals consuming first-generation insect-resistant potatoes.
- Several scientific studies show that laboratory animals that have been fed biotech food developed one or more toxic effects on vital and/or reproductive organ functioning, relative to a control group.
- A recent Canadian study found a common protein from insect-resistant corn in the bloodstream of pregnant women and their fetuses. Another recent study found that high concentrations of this protein resulted in severe damage to human embryonic kidney cells.
- The nutritional content of biotech foods, relative to conventional foods, is variable.

Environmental Impacts and Food Security:

- Private companies are not capable of screening new biotech materials for every possible pathogen or environmental stress. Unnoticed and unsafe mutations could strike after the occurrence of extreme stress, such as plant disease outbreaks, droughts, floods, and heat waves.
- New biotech crops may cross-pollinate with other plants and are likely to cause super weeds.
- Some herbicides used on biotech crops diffuse into the air and leach into streams and waterways in some areas. These herbicides are toxic to amphibians (e.g., frogs, salamanders) and earthworms, which impacts bird populations.
- Biotech crops are doing little to help international food security or relieve hunger in poor countries. The major biotech crops, corn and soybeans, are mainly used for animal feed, biofuels, and processed human food in developed countries.

Fig. A.3. Environmental group perspective on biotechnology.

References

Antoniou, M., C. Robinson and J. Fagan (2012). GMO Myths and Truths. Earth Open Source, London.

Ariely, D., G. Loewenstein and D. Prelec (2003). Coherent Arbitrariness: Stable Demand Curves Without Stable Preferences. *Quarterly Journal of Economics* 118(1), 73–106.

Becker, G., M. DeGroot and J. Marschak (1964). Measuring Utility by a Single-Response Sequential Method. *Behavioral Science* 9(3), 226–232.

Bethke, P.C. and A.J. Bussan (2013). Acrylamide in Processed Potato Products. *American Journal of Potato Research* 90(5), 403–424.

Colson, G.J., W.E. Huffman and M.C. Rousu (2011). Improving the Nutrient Content of Food Through Genetic Modification: Evidence from Experimental Auctions on Consumer Acceptance. *Journal of Agricultural and Resource Economics* 36(2), 343–364.

Colson, G.J., J.R. Corrigan, C. Grebitus, M.L. Loureiro and M.C. Rousu (2016). Which Deceptive Practices, if Any, Should be Allowed in Experimental Economics Research? Results from Surveys of Applied Experimental Economists and Students. *American Journal of Agricultural Economics* 98(2), 610–621.

Corrigan, J.R. and M.C. Rousu (2006). The Effect of Initial Endowments in Experimental Auctions. *American Journal of Agricultural Economics* 88(2), 448–457.

FAO (Food and Agriculture Organization) (2015). Data download. URL: faostat.fao.org.

Fox, J.A., D.J. Hayes and J.F. Shogren (2002). Consumer Preferences for Food Irradiation: How Favorable and Unfavorable Descriptions Affect Preferences for Irradiated Pork in Experimental Auctions. *Journal of Risk Uncertainty* 24(1), 75–95.

Garland, T. and M. Patterson (1967). Six Cases of Acrylamide Poisoning. *British Medical Journal* 4, 134–138.

Hayes, D.J., J.F. Shogren, S. Youll Shin and J.B. Kliebenstein (1995). Valuing Food Safety in Experimental Auction Markets. *American Journal of Agricultural Economics* 77(1), 40–53.

Hayes, D.J., J.A. Fox and J.F. Shogren (2002). Experts and Activists: How Information Affects the Demand for Food Irradiation. *Food Policy* 27(2), 185–193.

Hertwig, R. and A. Ortmann (2008). Deception in Experiments: Revisiting the Arguments in Its Defense. *Ethics Behavior* 18(1), 59–92.

Hirsch, C.N., C.D. Hirsch, K. Felcher, J. Coombs, D. Zarka, A. Van Deynze, W. De Jong, R.E. Veilleux, S. Jansky, P. Bethke, D.S. Douches and C.R. Buell (2013). Retrospective View of North American Potato (*Solanum tuberosum* L.) Breeding in the 20th and 21st Centuries. *Genes, Genomes, Genetics* 3(1), 1003–1013.

Hoffman, E., D.J. Menkhaus, D. Chakravarti, R.A. Field and G.D. Whipple (1992). Using Laboratory Experimental Auctions in Marketing Research: A Case Study of New Packaging for Fresh Beef. *Marketing Science* 12(3), 318–338.

Hurwicz, L. (1972). "On informationally decentralized systems." In C.B. McGuire and R. Radner (Eds.), *Decision and Organization: A Volume in Honor of Jacob Marschak*, vol. 12. Amsterdam: North-Holland.

Institute of Medicine (2015). Review of WIC food packages: An evaluation of white potatoes in the cash value voucher. Letter Report. Institute of Medicine of the National Academies.

Jansky, S.H., D.M. Spooner and P.C. Bethke (2014). "Potato." In S. Smith, B. Diers, J. Specht and B. Carver (Eds.), *Yield Gains in Major U.S. Field Crops, ASA, CSSA*, SSSA: Madison, WI, pp. 195–217.

JECFA (Joint FAO/WHO Expert Committee on Food Additives) (2011). Safety evaluation of certain contaminants in food. World Health Organization: Geneva.

Lineback, D.R., J.R. Coughlin and R.H. Stadler (2012). Acrylamide in Food: A Review of the Science and Future Considerations. *Annual Review of Food Science and Technology* 3, 15–35.

List, J.A. (2008). Informed Consent in Social Science: Response. *Science* 322(5902), 672.

List, J.A. (2009). An Introduction to Field Experiments in Economics. *Journal of Economic Behavior and Organization* 70(3), 439–442.

Lusk, J.L. and J.F. Shogren (2007). *Experimental Auctions: Methods and Applications in Economic and Marketing Research.* Cambridge, UK: Cambridge University Press.

Maddala, G. (1983). Limited dependent and qualitative variables in econometrics. *Econometric Society Monographs in Quantitative Economics*, vol. 3. Cambridge, UK: Cambridge University Press.

Malmendier, U. and Y.H. Lee (2011). The Bidder's Curse. *American Economic Review* 101(2), 749–787.

Marette, S., J. Roosen, S. Blanchemanche and E. Feinblatt-Mélèze (2010). Functional Food, Uncertainty and Consumers' Choices: A Lab Experiment with Enriched Yoghurts for Lowering Cholesterol. *Food Policy* 35(5), 419–428.

McFadden, B.R. and J.L. Lusk (2015). Cognitive Biases in the Assimilation of Scientific Information on Global Warming and Genetically Modified Food. *Food Policy* 54, 35–43.

Menkhaus, D.J., G.W. Borden, G.D. Whipple, E. Hoffman and R.A. Field (1992). An Empirical Application of Experimental Economics in Marketing Research. *Journal of Agricultural and Resource Economics* 17(1), 44–55.

Naruszewicz, M., D. Zapolska-Downar, A. Kosmider, G. Nowicka, M. Kozlowska-Wojciechowska, A. Vikstrom and M. Tornqvist (2009). Chronic Intake of Potato Chips in Humans Increases the Production of Reactive Oxygen Radicals by Leukocytes and Increases Plasma C-Reactive Protein: A Pilot Study. *American Journal of Clinical Nutrition* 89(3), 773–777.

Nixon, B.J., S.J. Stanger, B. Nixon and S.D. Roman (2012). Chronic Exposure to Acrylamide Induces DNA Damage in Male Germ Cells of Mice. *Toxicological Sciences* 129(1), 135–145.

Nunes, J.C. and P. Boatwright (2004). Incidental Prices and Their Effect on Willingness to Pay. *Journal of Marketing Research* 41(4), 457–466.

OAG (California Department of Justice, Office of the Attorney General) (2008). *Atty. Gen. Brown Settles Potato Chip Lawsuit With Heinz, Frito-Lay & Kettle Foods.* Website. URL: http://oag.ca.gov/news/press-releases/atty- gen-brown-settles-potato-chip-lawsuit-heinz-frito-lay-kettle-foods (accessed on 05/25/2012).

O'Connell, J. (2014). *USDA Deregulates Biotech Potato.* Capital Press. URL: http://www.capitalpress.com/Nation_World/Nation/20141107/usda-deregulates-biotech-potato.

OEHHA (California Office of Environmental Health Hazard Assessment) (2014). *Proposition* 65. Website. URL: http://www.oehha.ca.gov/prop65.html (accessed on 02/08/2014).

Peck, J. and S.B. Shu (2009). The Effect of Mere Touch on Perceived Ownership. *Journal of Consumer Research* 36(3), 434–447.

Rasmussen, J., K. Bradley and H. Baker (2013). Innate^{TM} Varieties: 1.0 and Beyond. Working Paper. J.R. Simplot Company.

Rousu, M., W.E. Huffman, J.F. Shogren and A. Tegene (2007). Effects and Value of Verifiable Information in a Controversial Market: Evidence from Lab Auctions of Genetically Modified Food. *Economic Inquiry* 45(3), 409–432.

Rousu, M.C., R.H. Beach and J.R. Corrigan (2008). The Effects of Selling Complements and Substitutes on Consumer Willingness to Pay: Evidence from a Laboratory Experiment. *Canadian Journal of Agricultural Economics* 56(2), 179–194.

Rousu, M.C., G. Colson, J.R. Corrigan, C. Grebitus and M.L. Loureiro (2015). Deception in Experiments: Towards Guidelines on Use in Applied Economics Research. *Applied Economic Perspectives and Policy* 37(3), 524–536.

Schmitz, J.D., D.J. Menkhaus, G.D. Whipple, E. Hoffman and R.A. Field (1993). Impact of Changing Consumer Preferences on Willingness-To-Pay for Beef Steaks in Alternative Retail Packaging. *Journal of Food Distribution Research* 24(2), 23–35.

Shiraishi, Y. (1978). Chromosome Aberrations Induced by Monomeric Acrylamide in Bone Marrow and Germ Cells of Mice. *Mutation Research* 57(3), 313–324.

Shogren, J., M. Margolis, C. Koo and J. List (2001). A Random *n*th-Price Auction. *Journal of Economic Behavior and Organization* 46(4), 409–421.

Solomons, L. (1897). Discrimination in Cutaneous Sensations. *Psychological Review* 4, 246–250.

Tareke, E., P. Rydberg, P. Karlsson, S. Eriksson and M. Tornqvist (2002). Analysis of Acrylamide, a Carcinogen Formed in Heated Foodstuffs. *Journal of Agricultural and Food Chemistry* 50(17), 4998–5006.

US-FDA (U.S. Food and Drug Administration) (2016). Guidance for Industry: Acrylamide in Food. Final Guidance. U.S. Department of Health and Human Services, Center for Food Safety and Applied Nutrition.

US-HHS-OHRP (U.S. Department of Health and Human Services, Office for Human Research Protections) (2009). 45 C.F.R. Part 46. URL: http://www.hhs.gov/ohrp/regulations-and-policy/regulations/45-cfr-46/index.html.

Vickrey, W.S. (1961). Counterspeculation, Auctions, and Competitive Sealed Tenders. *Journal of Finance* 16 (1), 8–37.

Wansink, B. and M.M. Cheney (2005). Super Bowls: Serving Bowl Size and Food Consumption. *Journal of the American Medical Association* 293(14), 1727–1728.

Wansink, B., J.E. Painter and J. North (2005). Bottomless Bowls: Why Visual Cues of Portion Size May Influence Intake. *Obesity Research* 13(1), 93–100.

Chapter 13

Consumer Demand for Low-Acrylamide-Forming Potato Products: Evidence from Lab Auctions*

Jonathan McFadden† and Wallace Huffman‡,§

†*Economic Research Service, U.S., Department of Agriculture, Washington, DC 20024, USA*

‡ *Department of Economics, Iowa State University, Ames, IA 50011, USA*

§*whuffman@iastate.edu*

A probable human carcinogen — acrylamide — was discovered about a decade ago to form naturally in potato products cooked at high temperatures. Using incentive compatible lab auctions and adult consumers from three distant U.S. locations, we test food labeling and information effects on willingness-to-pay (WTP) for conventional and genetically engineered potatoes, fries, and chips to reduce acrylamide levels. These new potato products are the first biotech products to have enhanced food safety for a popular food in the American and European diets. Relative to an environmental group's perspective, scientific and industry information significantly increase WTP for biotech relative to conventional potato products. In addition, we find significant location-specific fixed effects of information treatments. Scientific and industry information can be used to nudge consumers towards improved decision making and healthy diets.

Resumen Un probable carcinógeno humano — acrilamida — se descubrió hace como una década; se forma naturalmente en productos de papa cocinados a alta temperatura. Mediante el uso de subastas de laboratorio compatibles con incentivos y consumidores adultos de tres localidades distantes en los EUA, probamos etiquetado de alimentos y efectos de información sobre la disponibilidad de pagar (WTP) papas

---

*This article originally appeared in **American Journal of Potato Research (2017), 94, 465–480**. This is published by permission of **Springer Nature**.
The views expressed are those of the authors and should not be attributed to the Economic Research Service or the USDA.

convencionales y las obtenidas por ingeniería genética, a la francesa y en hojuelas, para reducir los niveles de acrilamida. Estos nuevos productos de papa son los primeros biotecnológicos que han aumentado la seguridad alimentaria para un alimento popular en las dietas americana y europea. En relación a la perspectiva de grupos ambientalistas, la información científica y de la industria aumenta significativamente la WTP para los productos de la papa biotecnológica en relación a los de la convencional. Además, encontramos efectos fijos por localidad significativos de tratamientos a la información. La información científica y de la industria puede usarse para empujar a los consumidores hacia un mejoramiento en la toma de decisiones y dietas sanas.

Keywords: Potato products, Information and labeling effects, Acrylamide, Biotechnology, Food safety

1. Introduction

A relatively new foodborne risk has been discovered in starchy foods cooked at high temperatures. Varying amounts of acrylamide have been found in many carbohydrate-rich foods cooked at high temperatures (Tareke *et al.*, 2002; Mottram *et al.*, 2002). This includes potato products cooked at high temperatures. Baked potatoes, potato chips, French fries, and hash browns frequently contain relatively high concentrations of acrylamide.[1] Roasted coffee beans, baked breads and cookies, and other foods with high levels of asparagine (an amino acid) and certain sugars also contain acrylamide (Bethke and Bussan 2013).

Acrylamide has been classified as a toxin and potential carcinogen based on the results of several medical studies (Garland and Patterson 1967; JECFA 2011; Naruszewicz *et al.*, 2009; Nixon *et al.*, 2012; Shiraishi 1978). This research used clinical, epidemiological, and laboratory data, but most were laboratory studies of rats and mice that have been exposed to acrylamide. These studies have consistently found links between acrylamide exposure and the formation of cancers, benign tumors, and developmental defects in lab animals. Other studies analyze the short-term effects of dietary acrylamide on human cancer risk. The International Agency for

[1] Raw, boiled, and steamed potatoes do not contain acrylamide.

Research on Cancer designates acrylamide as "probably carcinogenic to humans (Group 2A)" (Lineback *et al.*, 2012).

Acrylamide is a stable compound formed from the Maillard chemical reaction (Mottram *et al.*, 2002). There is a general positive relationship between fry color, chip color, and acrylamide content, holding water volume fixed. Growing factors, storage conditions, other field management practices, and processing attributes affect the potential for acrylamide to form (Bethke and Bussan 2013).[2] The potato industry is targeting many of these aspects to mitigate health concerns. Recent biotechnology (biotech) solutions include gene silencing to reduce asparagine and sugar content (Bethke and Bussan 2013). One leading food company has currently achieved lower levels of acrylamide-forming potential using genetic engineering (GE) than with changes in field and storage management and processing methods. Moreover, the U.S. Food and Drug Administration (USFDA) has expressed that GE could be, currently, the most effective solution for acrylamide reduction (USFDA 2016). Tran *et al.* (2016) estimate an average 65% decrease in dietary acrylamide exposure from potato chips and French fries made from low-acrylamide biotech potatoes relative to conventional potatoes.

As much as one-third of daily food calories consumed in the U.S. and Europe contains detectable levels of acrylamide (Mucci and Adami 2009), which has direct implications for dietary food safety and the economic viability of the potato industry. Under the Safe Drinking Water and Toxic Enforcement Act of 1986 (Proposition 65), the California Office of Environmental Health Hazard Assessment added acrylamide as a carcinogen in 1990 and again as a male developmental toxin in 2011 (OEHHA 2014). As part of the settlement of a broad 2005 lawsuit under Proposition 65 initiated by California's Office of the Attorney General, many restaurant chains are now posting acrylamide warnings on their

[2]Acrylamide concentrations in potatoes vary widely as a function of reducing sugar content and cooking temperatures, though concentrations in chips and fries are 100 to 1000 μg kg^{-1} and 100 to 700 μg kg^{-1}, respectively. U.S. per capita use of white potatoes in 2013 is approximately the same as in 1980, about 52.2 kg, but the share going to processed foods increased from 53% to 69% (Institute of Medicine 2015).

premises (OAG 2008). These warnings were designed to promote public health and may reduce consumption of conventional potato products. Thus, it is economically interesting to estimate consumers' valuations for conventional potatoes (perceived by some to have risk from potential carcinogens but no risk from GE) and biotech potatoes (perceived by some to have risk from GE but no risk from potential carcinogens), and the impacts of information and labeling on these valuations.

Early studies that use economics experiments to examine the effects of information and labeling on WTP for food products used more general versions of second-price auctions with repeated rounds (Hoffman *et al.*, 1992; Menkhaus *et al.*, 1992; Schmitz *et al.*, 1993). In these studies, authors use fifth-price auctions to investigate WTP for beef steaks in samples from Los Angeles and Denver. The research concerned the economic viability of new beef steak packaging, in addition to the impacts of different types of information. Apart from significant information effects, this set of studies finds significant effects of income, employment, household size, food product attributes, and "all natural" labels on WTP between two foods with distinct packaging.

In a related study, Hayes *et al.* (1995) use second-price auctions to study WTP for meat sandwiches with varying levels of risk. After 10 rounds of bidding, subjects were given information on one of five foodborne pathogens, and then bid in an additional 10 rounds. They find that subjects place a $0.70 per-meal premium on food safety and weigh their own prior beliefs about the odds of illness more heavily than new information. Similarly, Hayes *et al.* (2002) find asymmetric effects of information on WTP for irradiated pork sandwiches, with positive information about irradiation having much less impact than negative information.

Most closely related to our study objectives and research methodology is a set of studies using random *n*th-price auctions to analyze information effects on conventional and GE food products (Huffman *et al.*, 2003; Rousu *et al.*, 2007; Huffman *et al.*, 2007; Colson *et al.*, 2011; Lacy and Huffman 2016). In the first three studies, the authors estimate the impacts of pro-biotech, anti-biotech, and third-party verifiable information on conventional and biotech vegetable oil, fresh potatoes, and tortilla chips. They found that participants discount biotech-labeled foods by 14% relative to conventional-labeled foods. In regression analysis, they found large and significant effects of anti-biotech information, with generally insignificant effects of subject demographics. Colson *et al.* (2011) have similar findings in experimental auctions research of using GE methods to enhance the nutritional quality of broccoli, tomatoes, and potatoes. Lacy and Huffman (2016) use data from a different set of experiments performed on the same day as our experiments. Their study examines consumer demand for sulfite-free fresh potatoes and fresh potato dices.

This article applies an innovative approach to valuing and comparing information treatments on a food safety risk, acrylamide. The objective is to estimate the effects of food labels and information treatments on consumers' valuations of both conventional and low-acrylamide biotech potato products (fresh potatoes, frozen fries, and chips), including testing for city-specific information effects. In April and May, 2014, adult consumers from three distinct and distant U.S. cities participated in an experiment using an auction method. These cities were in Iowa, California, and Massachusetts. Large-scale field crop and specialty crop production is a significant component of the Iowa and California economies. The agricultural economy of Massachusetts is relatively smaller and tends to involve agro-tourism and small-scale farming. Thus, our sample contains significant regional variation and potentially important diversity in viewpoints about food safety and consumption of foods produced with GE ingredients.

The balance of the article is organized as follows. The next section provides a general overview of the experimental design, including participant recruitment and choice of auction method. In the following subsections, we detail the steps undertaken by subjects in a typical auction session, present the econometric model, and describe demographics and other characteristics of our sample. The results section presents and discusses average treatment effects (ATEs) and regression results. The final section concludes. All tables and figures of results, including information perspectives used in the experiment, are contained at the end.

2. Materials and Methods

2.1. *Experimental design: Synopsis*

Three organizations were employed to sample participants for our experiments that occurred during April and May, 2014. Iowa State University's Survey and Behavioral Research Services (SBRS) developed the initial procedures for participant recruitment. Households were randomly contacted to assess interest from eligible individuals to participate in a food-related project by Iowa State University researchers. Recruiters sought adults between 18 and 65 years of age who could follow directions, write in English, and arrive on a specific date at the experiment site. Potential subjects were told that the research sessions were about food preferences, similar to an auction. They were also informed that they would receive $65 for about an hour of their time. Willing adults were asked to choose from among three starting times. They were told that there would be a follow-up letter with directions to the lab site and with a reminder of the time and location.

SBRS recruited the sample for the Des Moines area, beginning with a set of random telephone numbers. In Des Moines only, willing participants were also allowed to suggest the name of another friend or family member who might be willing to participate. For recruitment of participants in Woburn, which is near Boston, and Los Angeles, we employed two marketing research companies — Answer Quest and Focus & Testing. Answer Quest randomly sampled from its subject pool of 60,000 adults in the broader Boston area, while Focus & Testing randomly sampled from its subject pool of 100,000 adults in the Los Angeles area. These individuals had been part of one or more prior private research projects (e.g., focus groups) but had not previously participated in university research involving experimental auctions. Recruitment in these cities was similar to recruitment in the Des Moines area, except that participants were not allowed to offer contact information of other possible recruits.

We also targeted 60% female samples in Los Angeles and Woburn in order to match the gender ratio of the U.S. grocery store shopper population.[3]

Laboratory auctions have become a preferred research method for assessing consumers' willingness-to-pay (WTP) for new food products (Lusk and Shogren 2007). A Vickrey second-price auction is an auction in which the highest bidder pays the second-highest price. It eliminates the winner's curse of over-bidding and performs well when there is a single winner (Shogren *et al.*, 2001). The random nth-price auction has been shown to be superior in many food auctions. In this auction, the winners are those that bid strictly higher than a random n lying between 2 and the number of subjects in the auction, say k (Shogren *et al.*, 2001; Lusk and Shogren 2007). The random nth-price auction is superior to the second-price auction for revealing private valuations when bidders have diverse food preferences for new goods. In the nth-price auction, every participant has a positive probability of being a winner for each product (potato, fries, chips), and it is common to have multiple winners.[4]

The potato products included in our experiments are a 2.27 kg bag of Russet potatoes, 0.34 kg bag of classic-cut potato chips, and 0.91 kg bag of frozen French fries. We offer a diverse set because the experiment had high fixed costs and subjects can have diverse preferences.[5] Each product has two technology types: conventional and low-acrylamide using biotechnology. At the time of the experiment, these low-acrylamide biotech products had not been deregulated for commercial sale.[6] One company had low-acrylamide biotech potato

[3]Female subjects comprise 71% of the Des Moines sample, slightly larger than the percent of women in the U.S. grocery shopper population. In our full sample, 62% are women.

[4]The second-price auction somewhat outperforms the random nth-price auction when a single winner is needed but is less suitable for our analysis of the underlying distribution of WTP for potato products.

[5]It is unlikely that the potato products used in the experiments are strong substitutes or complements. Given that the commodities serve different purposes and differ according to preparation and convenience, understatements or overstatements of WTP arising from relatedness in consumption is implausible (Rousu *et al.*, 2008).

[6]These biotech potatoes were deregulated on November 7, 2014. The Innate™ 1.0 biotech potato is resistant to black-spot bruising, will not brown when cut and exposed to air, and has low sugar content and acrylamide-forming potential (O'Connell 2014).

products under development and in late-stage regulatory approval, and they agreed to provide us with correctly labeled display products. All subjects were told at the end of each auction session that only one type of potato product was available to take home — the conventional type.

There are two kinds of information presented to subjects in the experiment: food labels (two) and information treatments (six). The commodities are packaged in transparent plastic rather than commercial packaging, which tends to be nontransparent. The conventional products have a plain label that describes the product and net weight. The low-acrylamide biotech products have labels describing the product, including "Low Acrylamide" wording, net weight, and the phrase "Product Made Using Biotechnology." These labels are on the front of each product's packaging (Fig. 1).

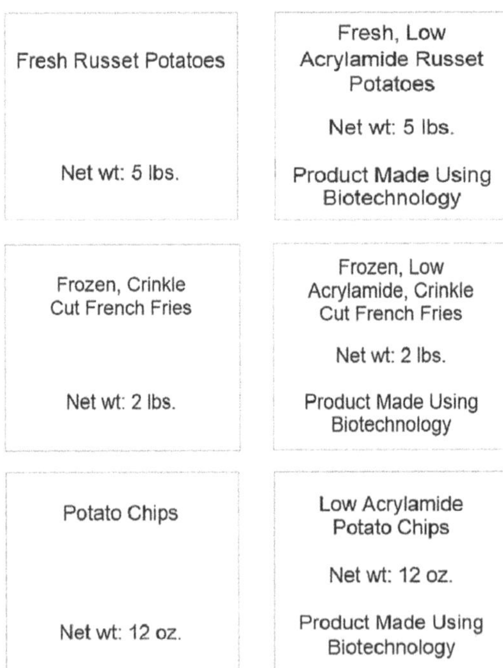

Fig. 1. Commodity labels: Conventional (*left*) and low-acrylamide (*right*).

Following Rousu *et al.* (2007) and Colson *et al.* (2011), we develop three diverse perspectives related to potato products. These perspectives are (1). an industry perspective on using biotechnology to reduce acrylamide levels in food products, (2). a scientific perspective on health risks of acrylamide exposure, and (3). an environmental group's (negative) perspective on biotechnology. Each perspective has the same three general headings (General Statement, Nutrition and Health, and Environmental Impacts and Food Security). The total length of each perspective is limited to one 8.5×11 sheet of paper to reduce the probability that subjects feel information overload.

These three information perspectives were used to construct six information treatments: (1) industry perspective only, (2) scientific perspective only, (3) environmental group's perspective only, (4) industry and scientific perspective, (5) industry and environmental group's perspective, and (6) scientific and environmental group's perspective. The order of perspectives is randomized within experimental packets containing two perspectives. In order to reduce burden and session length, no subject received a treatment comprised of all information perspectives.

Statements in the perspectives were adapted from medical research, government reports, websites, information from anti-biotechnology advocacy groups, and non-governmental organization (NGO) reports. Information about harmful acrylamide exposure used in the scientific perspective are from medical journal articles and government reports (Garland and Patterson 1967; JECFA 2011; Naruszewicz *et al.*, 2009; Nixon *et al.*, 2012; Shiraishi 1978). Certain statements in the industry perspective are adapted from the J.R. Simplot Company website and Rasmussen *et al.* (2013). Sentences in the environmental group's perspective are based on claims from the Non-GMO Project website and Antoniou *et al.* (2012). All information perspectives were pretested and revised before being used in the experiment.

The experimental design introduces randomness at several levels. First, participants are recruited generally randomly (without extensive selection criteria). Second, they are randomly assigned to a packet containing the instructions, questionnaires, and information treatment. Third, they are alternately assigned to one of two concurrent sessions at each starting time. Fourth, the order of bidding on products across four rounds of bidding is randomized (e.g., we randomly select the order of conventional versus biotech products). Fifth, we randomly select one of the two rounds of bidding on conventional products as the binding round. This randomized design permits simpler identification of treatment effects and eliminates the need to correct for sample selection (Lusk and Shogren 2007; List 2009).

3. Steps in each Experiment Session

Figure 2 depicts the steps of the auction sessions. After arriving at the experimental site, participants sign an informed consent sheet. Once agreeing to participate, adults receive an ID number and are randomly assigned to one of two concurrent sessions and a packet (random assignment to treatment).[7] They then complete the preauction questionnaire. Questions in the pre-auction questionnaire relate to demographics, previous knowledge about biotech foods, non-biotech foods, and acrylamide, and health (subjective risk perception of developing cancer and current level of hunger). While they are completing the pre-auction questionnaire, we hand out a $65 participation fee and have participants sign a receipt.

After the pre-auction questionnaires are completed, session monitors announce that subjects will be participating in an experimental auction and explain how the auction works. Subjects are told that they can win at most one unit of each auctioned product in the experiment. They are allowed and encouraged to ask questions for clarification at this point in the session. We next have one of two practice rounds after answering questions. The first practice round teaches participants about the auction methods. The second practice round introduces participants to bidding jointly on three items.

[7]Unless there are special circumstances, subjects arriving together are put in different sessions. There were only two instances of this. These subjects were asked to not sit next to each other.

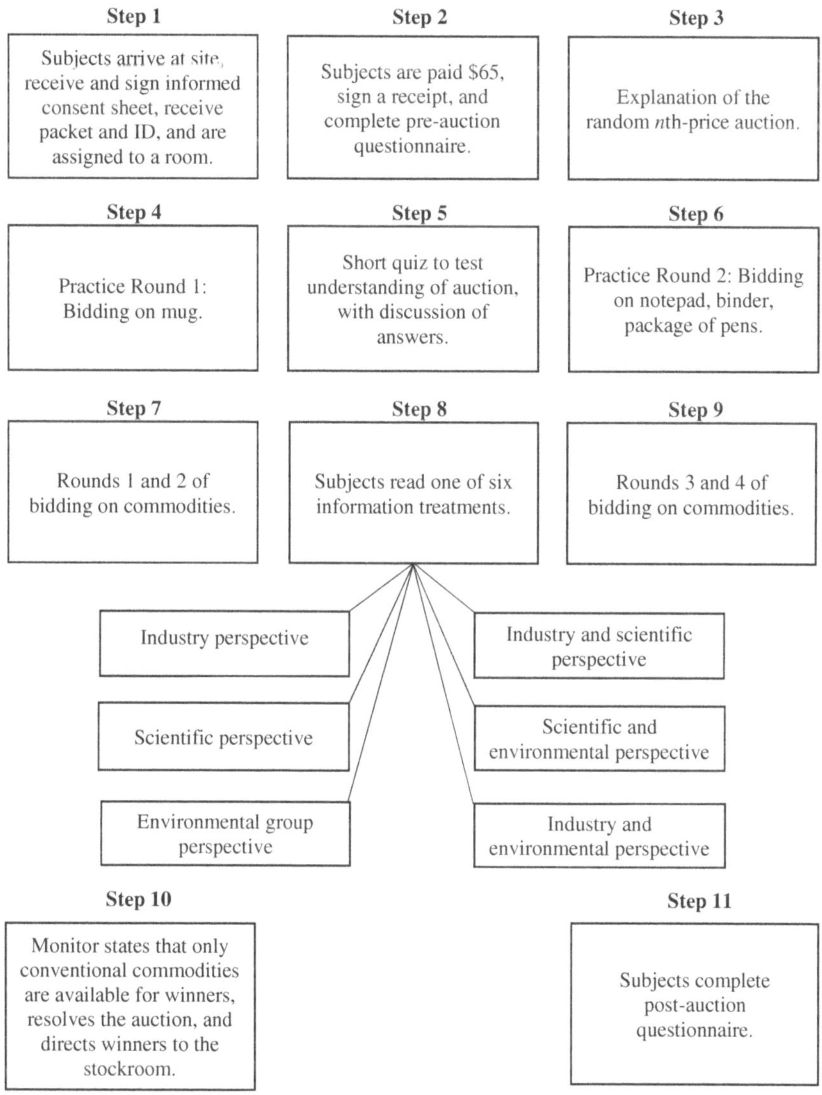

Fig. 2. The flow of subject activities in the experiments.

The item in the first practice round is a ceramic mug. Subjects are asked to come to the front of the room in single file without talking, view the mug, return to their seats, and write down their bids on a bid sheet. The session monitor and assistant then collect and rank the bids, draw the random n, and identify winners. We then give a short quiz to check for understanding of the auction rules. Participants are given four to five minutes to finish the quiz, followed by a review and explanation of the answers. In the second practice round, subjects place three bids at one time on a notepad, binder, and a package of pens. Bids for the three items are sorted and displayed, and "winner" IDs are presented.

Next, there are four separate rounds of bidding on sets of three experimental products, and participants are told that only one round will be binding. Similar to the practice rounds, participants are asked to come to the front of the room without talking, view the products, return to their seats, and write their bids on a bid sheet. Participants are not allowed to pick up the products. After collection of bid sheets, participants are given ten minutes to read their randomly-assigned information treatment.[8] Participants then bid on two separate sets of products (conventional products and biotech products), and bid sheets are collected. The only posted bids are those for the binding type (conventional) in the binding round at the end of the auction. Note that participants bid on different sets of products across the four rounds. However, both sets of conventional products have identical packaging, labeling, and appearance, as do both sets of biotech products.

Participants are then asked to fill out the post-auction questionnaire. We ask about monthly purchases, weekly consumption, and household inventories of potato and potato-related products. Additional questions concern health-related indicators and participants' level of understanding and agreement with the information perspectives.

After all bids have been placed and collected, participants are told that only conventional products are available for the winners.[9] Then, the binding round, random n, winning prices, and winning

[8]So that outside information does not confound our experiment's information treatments, we do not answer questions related to information content. We also state that subjects will have either one or two pages of information to read.

[9]Given this sequence of steps, bids will not exhibit hypothetical bias. To our knowledge, there were no direct consequences of not selling the GE potato products.

IDs are revealed. For budgeting purposes, we randomly chose the binding conventional round and random n before the experiments. After entering bids for the binding round into a spreadsheet, the bids for all products are projected at the front of each session's room, and participants watch the session monitor rank the bids and reveal the winning IDs. Winners are then directed to a "stockroom" to purchase the won item(s), and questionnaires are quickly checked to ensure completion prior to each participant's departure.

4. Data Analysis Methods, Econometric Model, and Data

We conduct the main empirical investigation by using regression analysis of WTP differences. Prior to regression analysis, we estimate average treatment effects (ATEs) resulting from subjects receiving randomly-assigned information. For a given product, the ATE is estimated as the mean bid post-information less mean bid pre-information for subjects in a given information treatment. Since we have random assignment to treatment, subjects' unobservables are balanced across treatments. This random assignment is the instrumental variable allowing identification of treatment effects (List 2009).

The econometric model exploits within-individual WTP differences (which can be positive, zero, or negative): WTP post-information treatment less WTP pre-information treatment. It also differences out subject-specific unobservables. The econometric model for WTP differences is:

$$Bid_{i,past}^{GM\,Pot} - Bid_{i,pre}^{Conv\,Pot} = \boldsymbol{z}'_{\boldsymbol{i}}\boldsymbol{\theta}_1 + \boldsymbol{IT}'_{\boldsymbol{i}}\boldsymbol{\delta}_1 + \boldsymbol{LT}'_i\boldsymbol{\beta}_1 + \mu_{1,i} \quad (1)$$

$$Bid_{i,past}^{GM\,Fry} - Bid_{i,pre}^{Conv\,Fry} = \boldsymbol{z}'_{\boldsymbol{i}}\boldsymbol{\theta}_2 + \boldsymbol{IT}'_{\boldsymbol{i}}\boldsymbol{\delta}_2 + \boldsymbol{LT}'_i\boldsymbol{\beta}_2 + \mu_{2,i} \quad (2)$$

$$Bid_{i,past}^{GM\,Chip} - Bid_{i,pre}^{Conv\,Chip} = \boldsymbol{z}'_{\boldsymbol{i}}\boldsymbol{\theta}_3 + \boldsymbol{IT}'_{\boldsymbol{i}}\boldsymbol{\delta}_3 + \boldsymbol{LT}'_{\boldsymbol{i}}\boldsymbol{\beta}_3 + \mu_{3,i} \quad (3)$$

In specification (1)–(3) above, $\boldsymbol{z}'_{\boldsymbol{i}}$ is a vector of demographic, location, and other subject-specific controls for person i, $\boldsymbol{IT}'_{\boldsymbol{i}}$ are indicator variables for information treatments, $\boldsymbol{LT}'_{\boldsymbol{i}}$ are indicators of the Woburn and Los Angeles sites interacted with information treatments, and $\mu_{j,i}$ are normally-distributed random disturbance terms. The term $Bid_{i,post}^{GM\,Pot} - Bid_{i,pre}^{Conv\,Pot}$ is the WTP for biotech potatoes post-information treatment less WTP for conventional potatoes pre-information. This is analogous to analyzing information impacts in markets shifting from conventional potatoes to low-acrylamide biotech potatoes, which is likely to be important in light of recent deregulation. Econometric estimates of the δ parameters measure the information treatment effects, though they must be considered in conjunction with the β estimates of interaction terms. Given the distance across markets, and that viewpoints about biotechnology and food safety can vary geographically, we test the hypothesis that the impact of information treatment effects are different across the three experimental sites — Woburn and Los Angeles relative to Des Moines. This motivates our inclusion of the interaction terms in our empirical model.

One complication with the WTP data is that some zero bids are truncated at zero. That is, subjects can prefer to bid a negative amount but feel constrained to bid zero. In principle, consumers with negative WTP would require a subsidy to take home the product. Other zero bids are legitimate bids without truncation, i.e., zero was the preferred bid (Maddala 1983). Thus, the WTP differences in eqs. (1)–(3) are censored for some observations. To incorporate censoring, a general version of a differenced Tobit model is developed to account for multiple sources of truncation of WTP at zero.

Based on location of zeros in the bids, there are four censoring cases. *Case 1*: both bids are positive. The dependent variable is uncensored. *Case 2*: the post-information bid is zero (i.e., the subject preferred to bid a negative value but felt constrained to bid zero), and the pre-information bid is positive. The true WTP difference is left-censored at the value of the pre-information bid. *Case 3*: the post-information bid is positive and the pre-information bid is zero. The true WTP difference is right-censored at the value of the post-information bid. *Case 4*: both bids are truncated at zero because a subject preferred a negative value but bid zero instead. When WTP is zero, we do not know whether the observation is a sincere zero or a "constrained" zero.

This estimation procedure is inefficient if WTP among potato products is jointly determined. Since participants are informed at the beginning of the experiments that they will be bidding on fresh potatoes, French fries, and potato chips, this could condition them to think about their WTP simultaneously. We could impose cross-equation correlation of disturbances (Zellner 1962), but the auction commodities differ greatly with respect to preparation times, convenience, and storage. Thus, we assume that valuations are formed independently.

5. Data

Our pre- and post-auction questionnaires provide information that results in a more detailed dataset than those underlying most experimental auction studies. This arises, in part, due to the literature's limited knowledge of factors affecting consumers' valuations of biotech foods containing consumer traits. We built on past food economics research, economic theory, and other considerations in choosing the questions to include in the questionnaires.[10]

The pre-auction questionnaire solicits subjects' demographic information, including age, gender, race, years of education, residential location, household size and income, and marital status.[11] Also, participants are asked about their degree of knowledge about biotechnology, acrylamide, and food safety.[12] The post-auction questionnaire consists of several related questions, including frequency of exercise.

[10]A consultation with the ISU Department of Statistics also provided assistance on sampling, question arrangement, and question wording in the questionnaires.

[11]We compared the demographics of subjects in each city to city demographics reported in 2014 by the Census Bureau's American Community Survey. Our city-specific subsamples are generally representative of each city's population regarding marital status, age, household size, schooling, and income. Further details are available upon request.

[12]We follow the food experiments literature in asking general questions in the pre- and post-auction questionnaires (Huffman *et al.*, 2003; Colson *et al.*, 2011). These studies ask how well-informed participants are about genetic modification, both before and after the experiment, which directly relates to the study objectives. Given the general nature of these questions, as well as the fact that we use an incentive-compatible method, we do not expect this to bias subjects' bids.

There are 304 participants in the total sample: 101 from Des Moines, 106 from Los Angeles, and 97 from Woburn. Mean number of participants per session is 16.9. Four participants had missing data, so the sample size for our analysis is 300. This implies an average number of participants per treatment of 50 (300/6). Of these observations, there are 3 double-zero bids for potatoes, 9 double-zero bids for French fries, and 6 double-zero bids for potato chips.[13]

The average age of the participants is 43 years, living with one other adult in the household, and having a per-person annual pretax income of $28,550.[14] Approximately 42% and 49% of the participants were at least somewhat informed about biotech and non-biotech food products prior to the experiment. Note that 75% of the sample exercises at least twice per week. Indicator variables for the city in which the sessions occurred control for location-specific effects, e.g., cost of living and local food cultures. In general, there is ample variability in subject demographics, particularly in per-person incomes and previous knowledge of biotech and non-biotech food products (see Table 1).[15]

6. Results and Discussion

6.1. *Mean bids by commodity and information treatment*

Table 2 presents average bids across the three products and two technology types before and after subjects view their randomly-assigned information treatment. Since each subject bids on all products before information and again after receiving information, we can compute average treatment effects for each information

[13]Nearly 4.7% of the potato bid differences have one zero bid (either before or after information), and 5.7% of the French fry and chip bid differences have one zero bid. The number of single-zero bids represents the maximum number of possible censored bid differences.

[14]Annual incomes are entered as the midpoint of the income bracket chosen by the respondent.

[15]The sample is 76% white. This likely stems from the eligibility requirement that subjects be able to read and write English. Although our sample is quite diverse, our results generalize mainly to the population of English-speaking U.S. grocery shoppers.

Table 1. Summary statistics for bid differences and regressors ($N = 300$).

Variable	Mean	St.Dev	Min	Max
Dependent variable				
GM Pot Bid Post – Conv Pot Bid Pre ($)	−0.13	1.67	−13.95	6.00
GM Fry Bid Post – Conv Fry Bid Pre ($)	−0.13	1.39	−11.55	4.50
GM Chip Bid Post – Conv Chip Bid Pre ($)	−0.08	1.17	−8.32	4.70
Information treatments				
Industry perspective	0.19	0.39	0	1
Scientific perspective	0.18	0.38	0	1
Environmental group's perspective	0.16	0.38	0	1
Industry and scientific perspectives	0.17	0.38	0	1
Industry and environmental group's perspectives	0.15	0.36	0	1
Scientific and environmental group's perspectives	0.15	0.34	0	1
Subject demographics				
Age(years)	42.82	14.08	18	65
Age^2(years)	2,031.04	1,191.22	324	4,225
Per capita income ($ thousand)	28.55	19.53	1.50	112.50
Adults in household (integer)	2.08	0.85	1	6
Previously informed about biotech food products (0–1)	0.42	0.49	0	1
Previously informed about non-biotech food products(0–1)	0.49	0.50	0	1
Exercises regularly (0–1)	0.75	0.43	0	1
Experimental design				
Los Angeles location (0–1)	0.35	0.48	0	1
Woburn, MA location(0–1)	0.32	0.47	0	1
Des Moines location (0–1)	0.33	0.47	0	1

treatment. We first perform t-tests of the null hypothesis that the mean of bid differences are zero between the biotech and conventional products prior to subjects reading an information treatment. We cannot reject the null hypothesis at the 5% significance level for potatoes (t-statistic $= -0.27$), French fries (t-statistic $= 1.15$), and chips (t-statistic $= -0.12$). Therefore, we find no evidence of

Table 2. Mean bids for subjects and average treatment effects, excluding double-zero bids.[a]

Commodity Type	N	Mean	Min	Max	Mean	Min	Max	Average Treatment Effect (ATE)
		Before Information			After Information			
A. Industry-only perspective								
GM potatoes	56	2.51	0.20	6.00	2.51	0	17.00	\$0.01
Conv. potatoes	56	2.49	0.10	7.00	2.24	0	5.50	−\$0.25***
GM French fries	53	2.55	0.01	6.99	2.35	0	7.00	−\$0.20
Conv. fries	54	2.19	0.01	5.00	1.98	0	4.00	−\$0.21**
GM chips	54	2.15	0.01	5.99	2.09	0	6.50	−\$0.06
Conv. chips	55	1.98	0.01	4.00	1.92	0	6.00	−\$0.07
B. Scientific-only perspective								
GM potatoes	54	2.34	0	10.99	2.88	0	13.99	\$0.55***
Conv. potatoes	54	2.68	0.50	11.99	2.38	0	10.99	−\$0.30*
GM French fries	52	2.41	0	7.99	2.73	0	11.99	\$0.32*
Conv. fries	53	2.40	0.01	9.99	2.15	0	8.99	−\$0.25*
GM chips	52	2.09	0	5.00	2.47	0	6.00	\$0.38**
Conv. chips	52	2.16	0.45	7.99	1.87	0	6.00	−\$0.30**
C. Environmental-only perspective								
GM potatoes	52	3.48	0.25	11.00	2.22	0	5.00	−\$1.26***
Conv. potatoes	53	3.19	0.10	15.00	3.03	0.10	6.00	−\$0.17
GM French fries	52	3.10	0.15	12.00	2.02	0	8.00	−\$1.08***
Conv. fries	52	3.02	0.02	12.00	2.75	0.10	9.00	−\$0.27
GM chips	53	2.54	0.10	8.00	1.75	0	5.00	−\$0.79***
Conv. chips	54	2.46	0.20	9.00	2.29	0	5.00	−\$0.17
D. Industry and scientific perspectives								
GM potatoes	51	2.54	0.25	9.00	2.98	0.05	9.00	\$0.43***
Conv. potatoes	51	2.57	0	6.00	2.33	0.01	6.00	−\$0.24**
GM French fries	50	2.30	0.10	8.00	2.68	0.05	7.00	\$0.38***
Conv. fries	49	2.36	0.05	5.00	2.13	0.05	5.00	−\$0.23*
GM chips	51	1.93	0.15	6.00	2.40	0.05	6.00	\$0.47***
Conv. chips	50	2.06	0.15	4.00	1.85	0.10	5.00	−\$0.21**
E. Industry and environmental perspectives								
GM potatoes	44	2.35	0.50	5.50	2.07	0	5.99	−\$0.28*
Conv. potatoes	44	2.28	0.25	5.00	2.68	0	9.00	\$0.40
GM French fries	43	2.40	0	6.99	1.95	0	5.00	−\$0.45**
Conv. fries	44	2.36	0.50	5.25	2.41	0	9.00	\$0.04
GM chips	43	1.87	0.25	4.00	1.65	0	5.00	−\$0.22*
Conv. chips	45	1.89	0	4.25	2.09	0	9.00	\$0.21

Table 2. (*Continued*)

Commodity Type	N	Mean Before	Min Information	Max	Mean After	Min Information	Max	Average Treatment Effect (ATE)
F. Scientific and environmental perspectives								
GM potatoes	38	2.25	0.01	5.00	2.05	0	10.00	−\$0.02
Conv. potatoes	39	2.21	0.01	5.00	2.47	0	8.10	\$0.26
GM French fries	37	2.12	0.01	6.00	1.89	0	6.00	−\$0.24
Conv. fries	38	2.02	0.01	7.00	2.03	0	5.00	\$0.02
GM chips	37	1.89	0.01	4.69	1.61	0	5.19	−\$0.28
Conv. chips	38	1.78	0.01	4.29	1.83	0	4.79	\$0.06

[a]For the right-hand column of Average Treatment Effects (ATEs), we perform a one-sample, two sided t-test of the null hypothesis of a zero ATE. For each of the six types of products across the six treatments, we test whether or not the mean of the bids pre-information minus the mean of the bids post-information equals zero. Significance is denoted as $^{*}p < 0.1$, $^{**}p < 0.05$, and $^{***}p < 0.01$.

systematically different average bids for the biotech products relative to the conventional products before introduction of information perspectives. Since the only perceptible difference among the goods is the package labeling, this suggests there is no labeling effect on average WTP.

For subjects receiving the treatment with only the industry perspective (panel A, Table 2), the mean bid on the conventional potatoes pre-information is \$2.49 and \$2.24 post-information, leading to a −\$0.25 ATE (significant at the 1% level). Similarly, the mean bid for conventional French fries is \$2.19 pre-information and \$1.98 post-information, which gives a −\$0.21 ATE (significant at the 5% level). All other ATEs for those receiving the industry-perspective-only treatment are insignificant.

All ATEs for those receiving the treatment with only the scientific perspective (panel B, Table 2) are significant at the 1% level or better. Among the biotech products, ATEs range in \$0.32 to \$0.55, with WTP for biotech potatoes increasing from \$2.34 to \$2.88, WTP for biotech French fries increasing from \$2.41 to \$2.73, and WTP for biotech chips increasing from \$2.09 to \$2.47. In contrast, ATEs for conventional products are negative, though somewhat smaller in magnitude, ranging between −$0.25 and −$0.30. After receiving the scientific-perspective-only treatment, WTP for conventional potatoes decreases from $2.68 to $2.38, WTP for conventional French fries decreases from $2.40 to $2.15, and WTP for chips decreases from $2.16 to $1.87.

Subjects receiving the treatment with only the environmental group's perspective (panel C, Table 2) have ATEs that are relatively large, negative, and significant at the 1% level for the three biotech products. For biotech potatoes, average WTP decreases from $3.48 to $2.22 (−$1.26 ATE), while for biotech French fries and chips, the average WTP decreases from $3.10 to $2.02 (−$1.08 ATE) and from $2.54 to $1.75 (−$0.79 ATE), respectively. The ATEs for conventional products among subjects receiving this treatment are insignificant. ATEs for those receiving the industry and scientific perspectives (panel D, Table 2) are all significant at the 10% level or better. For the biotech products, the ATEs range in $0.38 to $0.47 (all significant at the 1% level), and represent average WTP increases from $2.54 to $2.98, $2.30 to $2.68, and $1.93 to $2.40 for biotech potatoes, French fries, and chips, respectively. Average WTP for conventional products decrease by $0.21 to $0.24 after subjects receive this treatment. That is, average WTP decreases from $2.57 to $2.33 (potatoes), $2.36 to $2.13 (French fries), and $2.06 to $1.85 (chips).

For subjects receiving the industry and environmental group's perspectives, only ATEs for the biotech products are significant. For biotech potatoes, average WTP decreases from $2.35 to $2.07, an ATE of −$0.28 (significant at the 10% level). For biotech French fries, average WTP decreases from $2.40 to $1.95, an ATE of −$0.45 (significant at the 5% level). For biotech chips, the −$0.22 ATE (significant at the 10% level) reflects a decrease in average WTP from $1.87 to $1.65. Last, note that none of the ATEs for those receiving the scientific and environmental group's perspective are significant. This suggests that the positive information about bioengineering in the scientific perspective cancels out the negative information about biotech crops in the environmental group's perspective.

7. Regression Analysis of Bid Differences and Effects of Information

Our regression results of WTP differences corroborate our above finding of statistically significant ATEs. Importantly, these results extend the analysis from the previous subsection by accounting for potentially significant subject-specific controls and geographically-varying information effects, while removing unobserved subject heterogeneity.

Turning to the results in column 1 of Table 3, we find generally significant impacts of information treatments on WTP for biotech potatoes post-information less conventional potatoes pre-information, either in level terms or through their interactions with city fixed effects. Those receiving the treatment with only the scientific perspective had a WTP difference $1.16 higher relative to those receiving only the environmental group's perspective. Those receiving both the industry and scientific perspectives had a higher relative WTP difference for potatoes of $1.07, with those in Woburn experiencing a combined effect of $2.78 ($1.07 + $1.71). The treatment with only the industry perspective and the treatment with the scientific and environmental group's perspectives are only significant when interacted with the Woburn fixed effect. Among these subjects, the WTP difference for potatoes (relative to those receiving only the environmental group's perspective) is $1.61 and $2.67 larger, respectively. However, subjects in Woburn have a reduced WTP difference for potatoes of $1.17 relative to subjects in Des Moines.

For fresh potatoes, the fixed effect for Los Angeles is not statistically significant. The impacts of information treatments also do not significantly interact with the Los Angeles location. All demographic variables are insignificant at the 10% level. However, subjects who exercise regularly have a WTP difference for potatoes that is decreased by $0.50 relative to those who do not exercise regularly.

In the regression of WTP differences for French fries, we find that subjects who receive only the scientific perspective have an increased WTP difference of $1.22 relative to those receiving only the environmental group's perspective. By itself, the treatment with only

Table 3. Estimates for model of censored WTP differences for potatoes, French fries, and chips.[a]

Variable	(1) GM Pot WTP Post-Info – Conv Pot WTP Pre-Info	(2) GM Fry WTP Post-Info – Conv Fry WTP Pre-info	(3) GM Chip WTP Post-Info – Conv Chip WTP Pre-Info
Industry Information	0.119	0.203	0.027
Scientific Information	1.157***	1.223***	1.025***
Industry and Scientific Information	1.068*	0.982**	1.031***
Industry and Environmental Information	0.468	−0.145	0.193
Scientific and Environmental Information	0.040	−0.036	−0.331
LA site	0.111	−0.162	−0.092
Wobum site	−1.165**	−0.977**	−0.566
Industry Information × LA	1.186	1.192*	1.283**
Scientific Information × LA	−0.205	−0.231	−0.037
Industry and Scientific Information × LA	−0.152	−0.009	−0.375
Industry and Environmental Information × LA	0.372	0.805	0.253
Scientific and Environmental Information × LA	−0.132	0.373	0.572
Industry Information × Wobum	1.610**	1.278*	0.834
Scientific Information × Wobum	0.939	0.655	0.427
Industry and Scientific Information × Wobum	1.705**	1.438**	0.891*
Industry and Environmental Information × Wobum	1.019	1.341*	0.495
Scientific and Environmental Information × Wobum	2.672***	1.982***	1.916***
Age	0.028	0.064	0.076**
Age^2	−0.000	−0.001	−0.001**
Per-capita income	−0.003	0.001	0.000
Adults in household	0.127	0.064	0.166*
Previously informed about biotech food products	0.229	−0.007	−0.095
Previously informed about non-biotech food products	−0.328	−0.324	−0.159
Exercise regularly	−0.496**	−0.374**	−0.346**
Constant	−1.286	−1.856*	−2.165***
Observations	297	291	294

[a]Significance is denoted as $^{*}p < 0.1$, $^{**}p < 0.05$, and $^{***}p < 0.01$. The treatment containing only the environmental group's perspective is the excluded information treatment. The Des Moines site is the excluded location.

the industry perspective is insignificant. However, those receiving this treatment have an increased WTP difference of \$1.19 or \$1.28 if they are in Los Angeles or Woburn, respectively. Further, the relative impact of the treatment with the industry and scientific perspectives is \$0.98 in level terms, but those subjects in Woburn receiving the treatment have a combined relative WTP difference for fries that is \$2.42 (\$0.98 + \$1.44). The effects of the remaining two treatments are insignificant in level terms but have positive interactions with the Woburn location. Subjects in Woburn have WTP differences for French fries that are \$0.98 lower than subjects in the Des Moines area. Among these subjects, the information treatments with the industry and environmental group's perspectives and the scientific and environmental group's perspectives increase the WTP difference by \$1.34 and \$1.98, respectively.

Similar to the regression of the WTP differences for potatoes, many of the information treatments interacted with the Los Angeles fixed effects are insignificant. Demographic variables also tend to be insignificant, though subjects who regularly exercise have a WTP difference for French fries that is \$0.37 larger relative to subjects with less active lifestyles.

In contrast to the analyses for potatoes and French fries, regression results explaining WTP differences for chips contain information treatments effects that are small or not significantly different from zero. In level terms, the impacts of the scientific perspective and combined industry and scientific perspective are \$1.03, and the latter interacts significantly with the Woburn fixed effect (\$0.89). As in the analysis of WTP differences for French fries, the industry information is only significant among subjects in Los Angeles, whose relative WTP difference for chips increases by \$1.28. Similarly, the treatment containing the scientific and environmental group's perspectives has a significant Woburn fixed effect.

Interestingly, subject demographics play a more important role in explaining WTP differences for chips. The post-information GM premium for chips increases with subjects' age up to 38 years. Moreover, an additional adult in the household increases the WTP difference by \$0.17. Similar to results for potatoes and French fries, subjects who exercise frequently have a decreased WTP difference for chips of \$0.35. One reason why information treatments have relatively less influence and subject demographics have relatively more influence is that chips are a pre-processed snack food eaten primarily for convenience. Distinct from raw potatoes and frozen (uncooked) French fries, acrylamide levels in chips cannot be influenced by consumers.

8. Conclusion

Acrylamide is a probable carcinogen in humans and forms naturally in potatoes and potato products that have been cooked at high temperatures. This has been a concern of U.S. potato researchers and potato industry participants in recent years (Bethke and Bussan 2013). Proposition 65 in California requires a public notice to be displayed at businesses selling foods with acrylamide. Thus, there is substantial interest in the introduction of potato products with decreased acrylamide concentrations. Using experimental methods, we test the effects of labeling and information on subjects' WTP differences for conventional potato products and potato products using biotechnology to reduce acrylamide.

We find that consumers' purchasing behaviors of potatoes and potato products are strongly influenced by information. From our analysis of ATEs, average WTP for biotech potatoes, French fries, and chips are not statistically significantly different from average WTP for conventional products prior to subjects viewing their randomly-assigned information. In other words, there is no evidence of a pure labeling effect on WTP for biotech products relative to conventional products. However, WTP for conventional products tends to decrease after information, on average. For the biotech products, ATEs are negative for those receiving only the environmental group's perspective, as well as for those receiving the treatment with the industry and environmental group's perspective. There are positive ATEs for the biotech products for subjects receiving the scientific perspective and the treatment with both the industry and scientific perspectives.

From our regression analysis, we find those receiving the industry perspective, scientific perspective, industry and scientific perspectives, industry and environmental group's perspectives, and scientific and environmental group's perspectives have sizeable and significantly larger WTP relative to those who only receive the environmental group's perspective. By itself, the largest increase in relative WTP for all potato products occur after consumers receive only the scientific perspective. However, we find significant interaction effects of information with city fixed effects. On the whole, subjects in Woburn have relatively lower WTP differences for potatoes and French fries, but this is mediated by type of information received. Subjects' demographics have low power in explaining relative WTP for potatoes and French fries, though they are somewhat more important in explaining WTP differences for chips.

In late July, 2016, a national bioengineered food disclosure standard became legally binding in the U.S. The provisions of this national law direct the USDA to decide what food manufacturers must label and which ingredients will be considered GE foods within two years of enactment. Importantly, the law does not stipulate use of stringent wording, such as "made with GE ingredients." Rather, the disclosure is a text, symbol, or electronic/digital link with a disclosure option chosen by the food manufacturer. There are also reasonable accommodation options for manufacturers of foods with small or very small packages. The national standard supersedes all state labeling laws, including labeling legislation in Vermont that recently went into effect, but it does not address current voluntary labeling efforts by some food manufacturers (Public Law 114–216 2016). Although we found no evidence of a simple labeling effect on subjects' WTP for biotech potato products relative to conventional products, combinations of labels and information are likely to substantially influence consumers' valuations for foods using biotechnology to reduce dietary acrylamide exposure.

There are two primary limitations of our study. First, subjects in our experimental auctions did not have the option to win and take home low-acrylamide biotech potatoes, though they were informed of this prior to the conclusion of each session. This would have been a desirable feature of the experimental design, especially if some subjects were inadvertently disappointed with not being able to consume the new potato products. Second, our experimental design did not permit us to retain direct contact with participants after the experiments. Given the recent deregulation and commercialization of low-acrylamide biotech potatoes, it would have been desirable to follow up with subjects to determine if (or to what extent) their valuations for these products has been influenced by recent market events.

Our research methodology and results offer several opportunities for future research. First, it would be interesting to determine if consumers' objective level of knowledge about the use of GE methods in foods (rather than subjective understanding) has a significant impact on WTP for such foods. Developing a more objective measure of consumers' biotech understanding could be a potentially useful innovation. Second, how well consumers understand and incorporate new information, such as those contained in our information treatments, may correlate strongly with WTP for foods made from various processes. That is, in-depth understanding of information, rather than mere exposure to information, could have substantial and long-lasting effects on WTP. To the extent that this can be tested in an experimental setting, it could be a fruitful area for further research. Third, it would be useful to examine the effects of "authority" on consumer WTP for biotech foods, holding information content fixed. Some consumers may be more willing to believe and incorporate information from objective sources or perhaps from interested parties that confirm their previously-held beliefs.

Acknowledgements

We acknowledge funding and assistance from the USDA through the University of Wisconsin, the J.R. Simplot Company, and the Iowa Agriculture and Home Economics Experiment Station. The experiments were conducted while McFadden was at Iowa State

University and were approved by the ISU Institutional Review Board (IRB) in ISU IRB #14-115.

Appendix

Please take a few minutes to read the following.

An industry perspective on low acrylamide potatoes using biotechnology

General Statement:

- In 2002, acrylamide was discovered in foods containing certain natural sugars when cooked at high temperatures (above roughly 250 ° F), such as French fries, hash browns and potato chips. They are also formed in the baking of bread (crust) and cookies and roasting of coffee beans.
- Acrylamide, a chemical compound, is formed in foods from naturally occurring sugars and asparagine (a building block of a protein). It is a toxin and possible carcinogen in humans.
- Conventional plant breeding efforts have resulted in only small reductions of acrylamide in processed potato products.
- In contrast, biotech methods have significantly reduced the acrylamide level in processed potato products, thereby improving food safety.
- Biotech plant breeding methods take genes from one organism and transfer them to another. In new biotech potatoes, all genes come from wild and domesticated potato varieties. Hence, this biotech method has much in common with conventional plant breeding.

Nutrition and Health:

- Long-term, low-level intake of acrylamide by lab animals has been shown to create serious health problems. One study found that consuming conventional potato chips regularly for a month caused some health changes in humans.
- Under Proposition 65, California requires certain restaurants, including Applebee's, Chili's, McDonald's, Burger King, Wendy's, and KFC, to post a warning stating that "cooked potatoes that have been browned, such as French fries, hash browns and baked potatoes, contain acrylamide, a chemical known to the State of California to cause cancer."
- A major biotech accomplishment has been the development of new potato varieties that have approximately a 70% reduction in acrylamide levels in processed potato products relative to conventional.
- Biotech methods were also used to develop golden rice, which enhanced vitamin A content. Other opportunities exist for enhancing consumer attributes, such as, antioxidants and vitamins in food.
- All Food and Drug Administration (FDA)-approved biotech-foods have been assessed to be substantially equivalent in nutrient content to conventional foods. As such, labeling in the U.S. is voluntary.
- Biotech foods, relative to conventional foods, have similar low allergy potential.

Environmental Impacts and Food Security:

- Acrylamide is used in industrial processes to make paper, dyes, and plastics, but food and cigarette smoke are the two main sources of human exposure to acrylamide.
- Commercial potatoes are grown from pieces of whole potato and not seed. In addition, many commercial potatoes are either sterile or not sexually compatible with wild potatoes. Hence, there is very low risk of biotech potatoes crossing with other potato varieties or other plants.
- There are no adverse impacts on the environment of the new biotech potatoes.
- In general, biotech crops have reduced the use of toxic insecticides and increased the use of other environmentally friendly farming practices.
- Food security can be greatly improved with continued advancement and adoption of genetically modified crops.
- Approval is being sought for biotech potato exports to Canada, Mexico, South Korea, and Japan.

Fig. A.1. Industry perspective on low acrylamide potatoes using biotechnology.

Please take a few minutes to read the following.

A scientific perspective on acrylamide exposure

General Statement:

- In 2002, acrylamide was discovered in foods containing certain natural sugars when cooked at high temperatures (above roughly 250 ° F), such as French fries, hash browns and potato chips. Acrylamide is also formed in the baking of bread (crust) and cookies and roasting of coffee beans.
- Acrylamide, a chemical compound, is formed in these foods from naturally occurring sugars and asparagine (a building block of a protein).
- Compared to other foods in the American diet, French fries and potato chips are a major source of acrylamide.
- The amount of acrylamide increases with longer frying or baking times and higher cooking temperatures. However, boiling, steaming, and microwaving potatoes produces negligible acrylamide.
- Under Proposition 65, California requires certain restaurants, including Applebee's, Chili's, McDonald's, Burger King, Wendy's, and KFC, to post a warning stating that "cooked potatoes that have been browned, such as French fries, hash browns and baked potatoes, contain acrylamide, a chemical known to the State of California to cause cancer."
- However, the FDA has not advised people to stop eating these potato products.

Nutrition and Health:

- The International Agency for Research on Cancer (IARC) designates acrylamide as a "probable human carcinogen," and the US National Toxicology Program classifies acrylamide as "reasonably anticipated" to be capable of causing cancer in humans.
- The first health studies of acrylamide occurred in the 1960s and explored health hazards with on-the-job contact. Chronic exposure produces toxicity that weakens muscles and reflexes and leads to sensory loss.
- Evidence from studies in the 1970s and 1980s advanced the linkages between cancer and acrylamide.
- In a recent study, lab animals that consumed low doses of acrylamide placed in drinking water over a long period of time (roughly equal to annual acrylamide consumption of an adult) experienced gene damage.
- Another study found that humans who consumed conventional potato chips daily for four weeks had higher incidence of early development of cardio-vascular diseases.
- Additional studies suggest linkages between high acrylamide diets and development of breast cancer. High levels of acrylamide consumption by pregnant women have been associated with increased frequency of low birth weight babies.

Environmental Impacts and Food Security:

- Acrylamide is used in industrial processes to make paper, dyes, and plastics, but food and cigarette smoke are the two main sources of human exposure to acrylamide.
- If the occupational and health risks associated with acrylamide were to be substantially reduced or eliminated, food security would improve, especially in the United States and some other developed countries where a majority of the population regularly consumes foods that contain acrylamide.

Fig. A.2. Scientific perspective on acrylamide exposure.

Please take a few minutes to read the following.

An environmental group perspective on biotechnology

General Statement:

- Biotech plant breeding takes genes from one organism and places them into another. This process manipulates genes and alters genetic makeup and properties. The cutting of genetic material from one organism and inserting it into another is quite imprecise and can cause mutations. There has been inadequate testing of these products.
- Biotech methods frequently use antibiotic-resistant gene segments in soil bacteria or viruses and transfer them into plants. This process is risky, leading to unanticipated outcomes.
- Biotech plants are regulated by the federal government, but federal regulation relies heavily on data collected from field trials and other testing by the biotech industry.
- Biotech seeds were first marketed to U.S. farmers in the mid-1990s, and rapid farmer adoption occurred in field crops (corn, soybean, cotton and canola). Later developments have been in papaya and vegetables (sugar beet, squash and potato).
- In the United States, the sales of biotech foods have grown very rapidly. This growth is driven by self-interested producers and marketers seeking to maximize crop yields and minimize production costs.

Nutrition and Health:

- Of the laboratory plants that are successfully modified to express the "right" traits, genetic engineers select among those that look strong, healthy, and capable of further breeding. There is poor screening to eliminate varieties that produce harmful substances or low nutrient quality.
- New allergens are likely to be introduced into the food supply.
- Early research reported some health problems in laboratory animals consuming first-generation insect- resistant potatoes.
- Several scientific studies show that laboratory animals that have been fed biotech food developed one or more toxic effects on vital and/or reproductive organ functioning, relative to a control group.
- A recent Canadian study found a common protein from insect-resistant corn in the bloodstream of pregnant women and their fetuses. Another recent study found that high concentrations of this protein resulted in severe damage to human embryonic kidney cells.
- The nutritional content of biotech foods, relative to conventional foods, is variable.

Environmental Impacts and Food Security:

- Private companies are not capable of screening new biotech materials for every possible pathogen or environmental stress. Unnoticed and unsafe mutations could strike after the occurrence of extreme stress, such as plant disease outbreaks, droughts, floods, and heat waves.
- New biotech crops may cross-pollinate with other plants and are likely to cause super weeds.
- Some herbicides used on biotech crops diffuse into the air and leach into streams and waterways in some areas. These herbicides are toxic to amphibians (e.g., frogs, salamanders) and earthworms, which impacts bird populations.
- Biotech crops are doing little to help international food security or relieve hunger in poor countries. The major biotech crops, corn and soybeans, are mainly used for animal feed, biofuels, and processed human food in developed countries.

Fig. A.3. Environmental group perspective on biotechnology.

References

Antoniou, M., C. Robinson and J. Fagan (2012). *GMO myths and truths.* London: Earth Open Source.

Bethke, P.C. and A.J. Bussan (2013). Acrylamide in Processed Potato Products. *American Journal of Potato Research* 90: 403–424.

Colson, G.J., W.E. Huffman and M.C. Rousu (2011). Improving the Nutrient Content of Food Through Genetic Modification: Evidence from Experimental Auctions on Consumer Acceptance. *Journal of Agricultural and Resource Economics* 36, 343–364.

Garland, T. and M. Patterson (1967). Six Cases of Acrylamide Poisoning. *British Medical Journal* 4, 134–138.

Hayes, D.J., J.A. Fox and J.F. Shogren (2002). Experts and Activists: How Information Affects the Demand for Food Irradiation. *Food Policy* 27, 185–193.

Hayes, D.J., J.F. Shogren, S.S. Youll and J.B. Kliebenstein (1995). Valuing Food Safety in Experimental Auction Markets. *American Journal of Agricultural Economics* 77, 40–53.

Hoffman, E., D.J. Menkhaus, D. Chakravarti, R.A. Field and G.D. Whipple (1992). Using Laboratory Experimental Auctions In Marketing Research: A Case Study of New Packaging for Fresh Beef. *Marketing Science* 12, 318–338.

Huffman, W.E., M.C. Rousu, J.F. Shogren and A. Tegene (2007). The Effects of Prior Beliefs and Learning on Consumers' Acceptance of Genetically Modified Foods. *Journal of Economic Behavior and Organization* 63, 193–206.

Huffman, W.E., J.F. Shogren, M.C. Rousu and A. Tegene (2003). Consumer Willingness to Pay for Genetically Modified Food Labels in a Market with Diverse Information: Evidence from Experimental Auctions. *Journal of Agricultural and Resource Economics* 28, 481–502.

Institute of Medicine (2015). *Review of WIC food packages: An evaluation of white potatoes in the cash value voucher: Letter report.* Washington, DC: The National Academies Press.

JECFA (2011). *Safety evaluation of certain contaminants in food.* Rome: Technical Report Series.

Lacy, K. and W.E. Huffman (2016). Consumer Demand for Potato Products and Willingness-To-Pay for Low-Acrylamide, Sulfite-Free Fresh Potatoes and Dices: Evidence from Lab Auctions. *Journal of Agricultural and Resource Economics* 4, 116–137.

Lineback, D.R., J.R. Coughlin and R.H. Stadler (2012). Acrylamide in Food: A Review of the Science and Future Considerations. *Annual Review of Food Science and Technology* 3, 15–35.

List, J.A (2009). An Introduction to Field Experiments in Economics. *Journal of Economic Behavior and Organization* 70, 439–442.

Lusk, J.L. and J.F. Shogren (2007). *Experimental auctions: Methods and applications in economic and marketing research.* Cambridge: Cambridge University Press.

Maddala, G. (1983). *Limited dependent and qualitative variables in econometrics. In Econometric Society Monographs in Quantitative Economics*, Vol. 3. Cambridge: Cambridge University Press.

Menkhaus, D.J., G.W. Borden, G.D. Whipple, E. Hoffman and R.A. Field (1992). An Empirical Application of Experimental Economics in Marketing Research. *Journal of Agricultural and Resource Economics* 17, 44–55.

Mottram, D.S., B.L. Wedzicha and A.T. Dodson (2002). Acrylamide is Formed in the Maillard Reaction. *Nature* 419, 448–449.

Mucci, L.A. and H.O. Adami (2009). The Plight of the Potato: Is Dietary Acrylamide a Risk Factor for Human Cancer? *Journal of the National Cancer Institute* 101, 618–621.

Naruszewicz, M., D. Zapolska-Downar, A. Kosmider, G. Nowicka, M. Kozlowska-Wojciechowska, A. Vikstrom and M. Tornqvist (2009). Chronic Intake of Potato Chips in Humans Increases the Production of Reactive Oxygen Radicals by Leukocytes and Increases Plasma C-Reactive Protein: A Pilot Study. *American Journal of Clinical Nutrition* 89, 773–777.

Nixon, B.J., S.J. Stanger, B. Nixon and S.D. Roman (2012). Chronic Exposure to Acrylamide Induces DNA Damage in Male Germ Cells of Mice. *Toxicological Sciences* 129, 135–145.

O'Connell, J (2014). USDA deregulates biotech potato. Capital Press website: http://www.capitalpress.com/Nation_World/Nation/20141107/usda-deregulates-biotechpotato.

OAG (2008). Atty. gen. brown settles potato chip lawsuit with Heinz, Frito-Lay & Kettle Foods website: http://oag.ca.gov/news/pressreleases/atty-gen-brown-settles-potato-chip-lawsuit-heinz-frito-laykettle-foods.

OEHHA (2014). Proposition 65 website: http://www.oehha.ca.gov/prop65.html.

Public Law 114–216 (2016). An act to reauthorize and amend the National Sea Grant College Program Act, and for other purposes. *National Bioengineered Food Disclosure Standard.* 130 Stat. 834.

Rasmussen, J., K. Bradley and H. Baker (2013). *Innate varieties: 1.0 and beyond.* Working paper. J.R. Simplot company.

Rousu, M.C., R.H. Beach and J.R. Corrigan (2008). The Effects of Selling Complements and Substitutes on Consumer Willingness to Pay: Evidence from a Laboratory Experiment. *Canadian Journal of Agricultural Economics* 56, 179–194.

Rousu, M.C., W.E. Huffman, J.F. Shogren and A. Tegene (2007). Effects and Value of Verifiable Information in a Controversial Market: Evidence from Lab Auctions of Genetically Modified Food. *Economic Inquiry* 45: 409–432.

Schmitz, J.D., D.J. Menkhaus, G.D. Whipple, E. Hoffman and R.A. Field (1993). Impact of Changing Consumer Preferences on Willingness-To-Pay for Beef Steaks in Alternative Retail Packaging. *Journal of Food Distribution Research* 24, 23–35.

Shiraishi, Y (1978). Chromosome Aberrations Induced by Monomeric Acrylamide in Bone Marrow and Germ Cells of Mice. *Mutation Research* 57, 313–324.

Shogren, J., M. Margolis, C. Koo and J.A. List (2001). A Random nth-Price Auction. *Journal of Economic Behavior and Organization* 46, 409–421.

Tareke, E., P. Rydberg, P. Karlsson, S. Eriksson and M. Tornqvist (2002). Analysis of Acrylamide, a Carcinogen Formed in Heated Foodstuffs. *Journal of Agricultural and Food Chemistry* 50, 4998–5006.

Tran, N.L., L.M. Barraj and S. Collinge (2016). Reductions in Dietary Acrylamide Exposure — Impact of Potatoes with Low Acrylamide Exposure. *Risk Analysis* (forthcoming).

USFDA (2016). *Guidance for industry: Acrylamide in food.* Washington, DC: U.S. Government Printing Office.

Zellner, A (1962). An Efficient Method of Estimating Seemingly Unrelated Regressions and Tests for Aggregation Bias. *Journal of the American Statistical Association* 57, 348–368.

Chapter 14

Consumer Demand for Potato Products and Willingness-to-Pay for Low-Acrylamide, Sulfite-Free Fresh Potatoes and Dices: Evidence from Lab Auctions*

Katie Lacy and Wallace E. Huffman†

†*Agriculture and Life Science and Professor of Economics at Iowa State University*

We assess consumer demand for traditional fresh potatoes and processed potato products and willingness to pay for new experimental low-acrylamide and sulfite-free potato products. Demand for fresh potatoes, potato chips, and fries is unaffected by household income or education, but demand for chips and fries is affected by consumer age and exercise habits. Subjects display increased willingness to pay for new potato products after receiving a private company perspective about the technology and risks associated with exposure to acrylamide, a carcinogen, in fried conventional potatoes and a new product, potato dices. We find that consumers are willing to pay for enhanced food safety in fresh potato products achieved using biotechnology.

*This article originally appeared in the **Journal of Agricultural and Resource Economics (2016), 41 (1), 116–137**. This is published by permission of **Western Agricultural Economics Association**.

Jonathan McFadden, Nadia Huffman, Sonya Huffman, Zac Martin, Jared Martin, Nicole Caviris, and Jesse Strzok provided valuable assistance. Matt Rousu, Kyrre Rickertsen and two anonymous reviewers provided helpful comments on an earlier draft. The J.R. Simplot Company provided technical support, supervision of regulated products, experimental biotech products, and fresh-cut potato dices treated with sodium bisulfite for the lab displays. The University of Wisconsin-Madison, through a USDA-NIFA grant, the J.R. Simplot Company, and the Iowa Agricultural and Home Economics Experiment Station also provided financial support. The project was approved by the IRB for Human Subjects at Iowa State University.

Review coordinated by Hayley Chouinard.

Keywords: consumer-oriented GMOs, experimental auctions, food labels, fresh-cut potato dices, information effects, low-acrylamide potatoes, potato products, sulfite free, willingness-to-pay

1. Introduction

White potatoes remain a key vegetable in the American diet, whether consumed boiled, steamed, baked, or as hash browns, French fries, and potato chips. Per capita consumption of white potatoes in 2013 was approximately 115 lbs per year, about the same as in 1980, but the share going to processed foods increased from 53% to 69% during that time. However, since some of the potato is lost to waste during processing, per capita consumption of white potatoes has actually declined over this period (Rasmussen, Latulippe, and Yaktine, 2015). Richards, Kagan, and Gao (1997) summarize factors affecting the aggregate demand for potatoes, but very limited up-to-date research exists on factors that affect individual household demand for fresh or processed potatoes in the United States.

In 2002, acrylamide was first identified in starchy foods cooked at high temperatures (Tareke *et al.*, 2002), including high-temperature cooking of traditional white potatoes, such as frying, baking, or roasting to make French fries, hash browns, and chips.[1] Acrylamide occurs naturally when the amino acid asparagine and reducing sugars (fructose and glucose) are heated to above 250°F, as in frying, baking, and roasting. Once formed, acrylamide is a stable compound. The Maillard reaction, which produces acrylamide, also produces the dark-colored pigments or browning of French fries, chips, and hash browns (Bethke and Bussan, 2013). In general, the acrylamide content rises as these pigments darken. In addition, retail fresh-cut white potatoes are treated with sodium bisulfite to retard bruising and blackening when exposed to oxygen.

Both acrylamide and sulfites raise food safety concerns. Based largely on animal studies, acrylamide is a neuro-toxin and potential carcinogen in humans (Bethke and Bussan, 2013). Moreover, as a

[1] Acrylamide does not exist in fresh, boiled, or steamed potatoes.

result of a broad 2005 lawsuit brought by the State of California under Proposition 65 against the U.S. potato industry, many California restaurants are required to post signs that potato products that have browned in the cooking process contain acrylamide, a cancer-causing agent (California Office of Environmental Health Hazard Assessment, 2014). In addition, the U.S. potato industry has a mandate to largely eliminate acrylamide from potato products sold in California (California Department of Justice, 2008) and is working to lower acrylamide formation in processed potato products. Sodium bisulfite is a controversial preservative because some people are allergic to sulfides (Rangan, 2010).

Biotech methods have been used to eliminate these two health risks in white potatoes.[2] Changes in potato growing and storage practices and conventional breeding of potatoes have been unsuccessful in significantly reducing acrylamide content in high-temperature cooked potato products or darkening of fresh-cut potatoes. However, scientists have successfully used genetic engineering to significantly reduce acrylamide formation in potato products (Bethke and Bussan, 2013). In addition, the new low-acrylamide potatoes have the advantage of low bruising and blackening when cut, which reduces potato waste associated with manufacturing processed potato products. As a result, treatment with sodium bisulfate is no longer needed to reduce bruising and blackening in the new fresh-cut biotech potatoes. Hence, the new biotech potatoes reduce two types of food safety concerns — a major accomplishment.

In earlier lab auctions of genetically modified (GM) foods potentially carrying traits for herbicide tolerance and/or insect resistance, Huffman *et al.* (2003) and Rousu *et al.* (2007) found significant labeling and information treatment effects on willingness-to-pay (WTP) for GM foods. Huffman *et al.* (2003) found that most participants consistently bid less for products carrying GM food labels compared to products with conventional labels. However, the authors did not test for information treatment effects. In a later

[2]No food safety problems have arisen in two decades of human consumption of GM foods.

study, Rousu *et al.* (2007) examined both the effects of food labels and prepackaged information on WTP for conventional GM types and found positive effects of pro-biotech information and negative effects of anti-biotech information on WTP differences. Colson, Huffman, and Rousu (2011) found that subjects were willing to pay more for fresh vegetables enhanced with antioxidants and vitamin C (consumer-oriented traits) using biotech methods.

This study assesses consumer demand for traditional whole, fresh white potatoes and processed potato products and WTP for new, experimental whole, fresh potatoes and potato dices with low acrylamide-forming and browning/bruising potential due to biotech advances. In addition, we examine the effects of food labeling and pre-packaged information on WTP for new potato products. To do this, we design a random nth-price experimental lab auction (Shogren *et al.*, 2001) and recruit a sample of 102 adult consumers from three different regions of the United States to participate in experimental auctions.

Our empirical results show that the demand for traditional fresh potatoes and processed potato products is not responsive to subjects' household income. However, households with more adults are more likely to consume fresh potatoes and French fries but not chips. This result may be driven by falling per capita household income as the number of adults per household increases. Education has no significant effect on potato demand, and age has a significant quadratic effect on demand for chips and fries but not for fresh potatoes. *Ceteris paribus*, Boston and Los Angeles households are more likely to regularly consume processed potato products (chips and fries) than Des Moines households.

Our empirical results show that WTP differences for new experimental biotech fresh white potatoes and dices are also not significantly affected by the income of subjects' households. Male subjects, relative to females, are willing to pay less for biotech potato products after receiving an industry perspective on using biotechnology to reduce acrylamide and sulfide exposure relative to WTP pre-information. Other sociodemographic factors do not explain these WTP differences. However, subjects who are informed about biotech foods before the experiment are willing to pay more for the new experimental potato products than others. Additionally, subjects who receive the company perspective on using biotechnology to create low-acrylamide, sulfite-free fresh potatoes and fresh dices are willing to pay significantly more than those that receive the environmental group perspective. Hence, subjects are willing to pay for improved food safety achieved using biotechnology to improve the consumer attributes of fresh potatoes, traditionally consumed in large quantities in the United States, and on fresh potato dices, a totally new test food product.

2. Experiment Design

Experiments were conducted in the Boston, Masschusetts; Des Moines, Iowa; and Los Angeles, California in April and May 2014. Survey and Behavioral Research Services (SBRS) at Iowa State University developed the recruitment protocol. The target subjects were eighteen to sixty-five years of age.[3] Potential participants were told that an Iowa State University project was recruiting subjects to participate in sessions on how households select food and household products. In particular, they were told that the sessions involved a food preference experiment set up like an auction, the sessions would take about seventy-five minutes of their time, and they would be paid $65 for participating in the project. Also, they needed to be able to follow instructions and write in English and be willing to come to a common location in their area at a specified time.[4]

At each location, the lab was laid out classroom-style with a display table in the front of the room. Practice-round and experimental

[3]The IRB requires special procedures for subjects younger than eighteen or older than sixty-five.

[4]To recruit subjects in the Boston and Los Angeles areas, we worked with Answer Quest and Focus & Testing, food marketing and testing companies with databases of more than 50,000 and 120,000 individuals, respectively, who had participated in earlier marketing research projects and agreed to be contacted for future projects. Individuals from these databases were called by employees of Answer Quest or Focus & Testing and read the common protocol for recruitment and follow up.

products were placed on the table before subjects were admitted to the lab. Experimental potato products were placed in clear plastic bags with experimental food labels. No trademark, brand, or other information was on the bags. Experimental products were covered by blue bins placed on the table in the front of the lab.

Two concurrent sessions took place at a time, and subjects were alternately assigned to either Session A or Session B. Each session was conducted by a session monitor (who remained the same for each location) and one assistant. The sessions consisted of ten steps. See Figure 1 for a quick summary and sequence of the steps.

2.1. *Step one*

The subjects were greeted and asked to read and sign an informed consent form. Subjects were then assigned an ID number, handed a packet of project materials, and told to enter the lab and take a seat.[5] While waiting for the experiment to start, subjects were asked to complete a pre-auction questionnaire that collected socioeconomic information and information about their use of labels while shopping for food products. In addition, subjects were asked to rate their knowledge about biotech foods, non-biotech foods, and acrylamide before the experiments.[6] To relieve subjects' concerns about a credit constraint in the experiment, each subject was paid $65 for participating in the experiments at this point.

2.2. *Step two*

Subjects were informed that they would be participating in an auction and that there would be two practice rounds of bidding to

[5]Individuals who arrived together (e.g., a husband and wife, mother and daughter, etc.) were assigned to different sessions.

[6]By asking about knowledge in more than one area, we believe that any one of these questions is unlikely to bias our experimental results. Also, those subjects that indicate that biotechnology and GMs are different are anticipated to be more informed about this area of science than others.

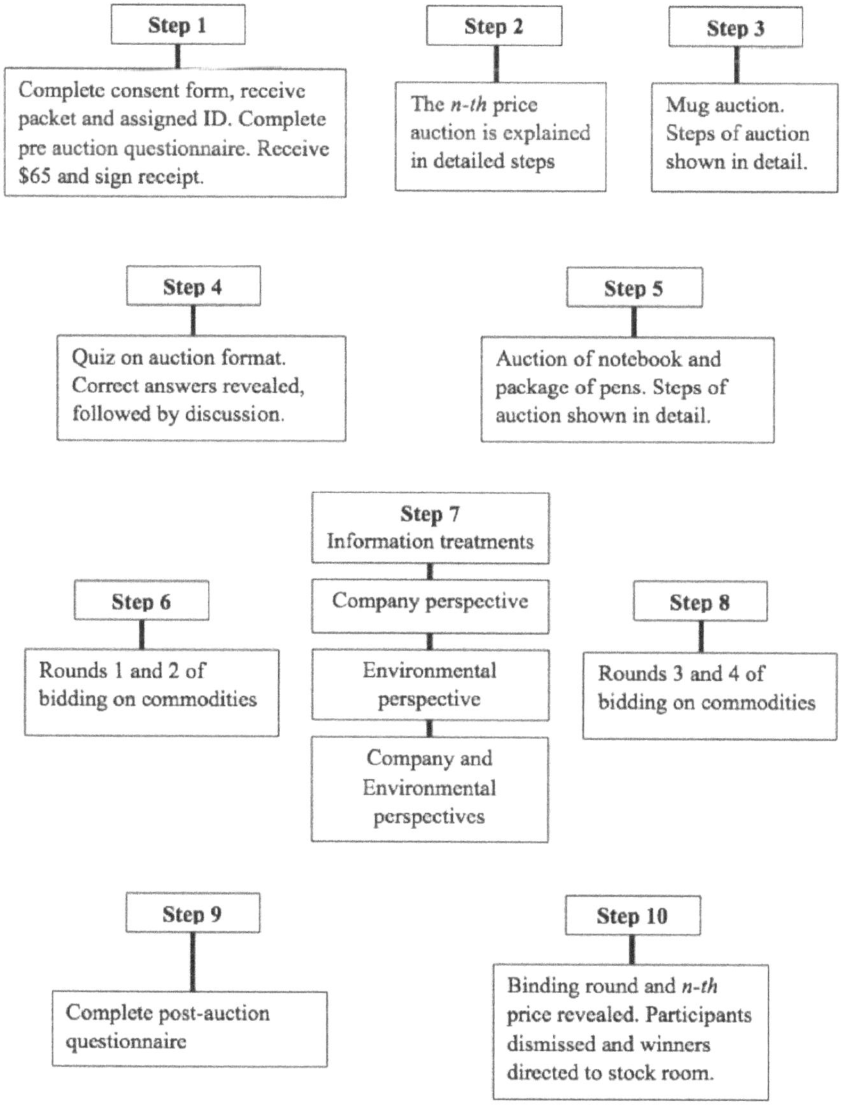

Fig. 1. Steps in the experiments.

allow them to become familiar with the auction mechanism. Subjects were told that it was always in their best interest to bid their true preferences. Subjects were asked to direct all questions to the monitor and refrain from communicating with other subjects.

Next, the session monitor read through an explanation of the random nth-price auction and addressed any questions raised by the subjects. The auction was explained to the participants.

2.3. *Step three*

Bidding on a single product, a generic ceramic coffee mug, was conducted to familiarize the subjects with the auction mechanism. Subjects were asked to come up to the front of the room, one by one, to view the product and then return to their seats.[7] After making their bids, the monitor and assistant collected the bid sheets and recorded the bids and ID numbers. The bids were then ranked from highest to lowest. The rank of the winning bid was determined by randomly drawing a number from a uniform distribution over 2 to k, where k is the number of subjects in a session. The random number, n, determined the rank of the random price. For example, if $k = 16$ and the random n is 4, the 3 subjects that bid strictly higher than the random price were the winners, and they paid the random price (fourth-highest price). These ranked bids with ID numbers and random nth price were displaced on a screen in front of the room.

2.4. *Step four*

Subjects were asked to complete a short, four-question quiz on the auction format, and the session monitor reviewed the answers and addressed questions.

2.5. *Step five*

The second practice round helped participants become familiar with bidding on two products — a notepad and a package of pens — simultaneously.

[7]Following Nunes and Boatwright (2004), we used nonfood items in our practice rounds to reduce impacts on later bids on experimental products.

2.6. *Step six*

Subjects placed bids on real experimental products: two rounds of bidding on a five-pound bag of traditional and biotech Russet Potatoes and a twelve-ounce bad of fresh-cut potato dices. To ease budget constraint issues, subjects were told that winners of these auctions would purchase at most one unit of each product (a five-pound bag of potatoes, a twelve-ounce bag of dices). In round 1, the products carried either a conventional or biotech food label (see Figure 2).[8] The order in which subjects saw either conventional or biotech products was randomized to minimize order effects in bidding. Subjects were asked to come to the front of the room, view the products, return to their seats, and place their bids. The bid sheets were then collected. In round 2, the participants were asked to bid on the same type of products as round 1, but this time the

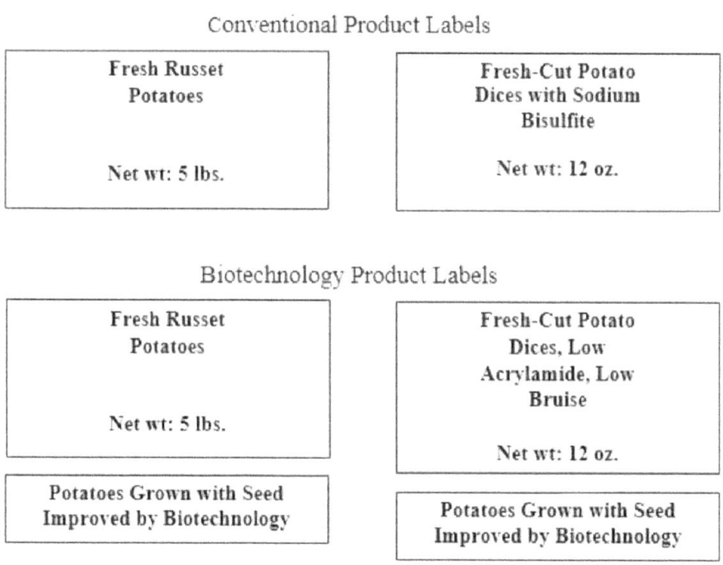

Fig. 2. Auction labels.

[8]A private food company supplied experimental products for the lab displays so that the product in the bag matched the attributes of the product stated on the food label.

products were conventional if they had been biotech in round 1 and vice versa.

2.7. *Step seven*

Subjects were given about ten minute to read the information statement(s) included in their packet. Each subject was randomly assigned to an information treatment group, which determined the information statements they received. The information treatments consisted of two different perspectives: a (positive) company perspective on using biotechnology to reduce bruising of potatoes in storage and processing and acrylamide levels in processed potato products and a (negative) environmental group perspective on biotechnology. See the appendices for both information treatments. A third information-treatment group was provided with both perspectives.[9] In the experiments, 35% of subjects received the company perspective, 33% received the environmental perspective, and 31% received both perspectives.

2.8. *Step eight*

Subjects placed bids for rounds 3 and 4 on experimental products. As in step six, the only difference between rounds 3 and 4 was whether the subjects saw food labels showing conventional products or biotechnology products first (which was randomized).

2.9. *Step nine*

After all bids were collected, subjects were asked to complete a short post-auction questionnaire that included questions about household consumption patterns for potatoes and potato products, how well the subject understood the information treatment that they received and read, and how well the subject was informed about biotechnology

[9]Greenpeace and Friends-of-the-Earth are two environmental groups that have active and strong negative perspectives on biotechnology and GMs.

and acrylamide after participating in the auction and reading the information treatment included in their packet.

2.10. *Step ten*

Subjects were informed that the biotechnology products were not currently available in grocery stores and that winners would receive conventional potato products obtained from a local grocery store. The monitor then randomly chose the binding round from the two rounds of bidding on the conventional products, displayed the ranked bids for this round for both products, and chose the random price or random n.[10] Winners were then identified. The winners were escorted to the stock room to purchase the products and the other subjects were dismissed.

3. Results

Table 1 presents the summary statistics for socioeconomic characteristics of the 102 subjects and potato-consumption frequency for subjects' households. The average subject was forty-three years old, and 39% of the subjects were male.[11] The mean education of subjects was fourteen years, which is equivalent to a two-year college degree. Eighty-seven percent of the subjects were white and 50% were married.[12] The average household had 2.2 adults 0.20 children under the age of eight, and 0.56 children eight years or older. Mean household income was $72,300, and 13% of subjects reported being blue-collar workers.[13] Also, 93% reported that they exercise weekly,

[10]Winners of fresh-cut potato dice auctions received a twelve-ounce package of a close substitute, Simply Potatoes™, a product available in the dairy case of major grocery stores and supermarkets.
[11]Approximately 60% of grocery-store shoppers are women.
[12]The unusually large share of white subjects, even for the Boston and Los Angeles areas, can be explained by the fact that recruiters screened for ability to communicate in English.
[13]A blue-collar worker was defined as anyone with an occupation of "building and grounds cleaning and maintenance," "construction/installation/repair," "farming/fishing/forestry," "production/manufacturing," or "transportation."

Table 1. Sample summary statistics.

Variables	Mean	St. Dev.
Potato Consumption Weekly (=1)	0.931	0.254
Potato Chip Consumption Weekly (=1)	0.824	0.383
French Fry Consumption Weekly (=1)	0.598	0.493
Household Income ($1,000)	72.30	47.04
Number of Adults	2.19	1.07
Number of Children $<$ 8 yrs	0.196	0.508
Number of Children $\geq$ 8 yrs	0.559	0.874
Rural (=1)	0.059	0.237
Anyone on Diet (=1)	0.353	0.480
Gender (1 = Male)	0.392	0.491
Age	43.32	13.13
Education (yrs)	14.38	2.00
Married (=1)	0.510	0.502
White (=1)	0.873	0.335
Reads Food Labels (=1)[a]	0.882	0.324
Exercises Weekly (=1)	0.931	0.254
Good Health (=1)[b]	0.794	0.406
Smokes Cigarettes (=1)	0.088	0.285
Blue Collar Occupation (=1)[c]	0.128	0.335
Boston (=1)	0.314	0.466
Los Angeles (=1)	0.333	0.474

Notes: [a]=1 if responded "Some of the time," "Often," or "Always" reads food labels when buying a food product for the first time and = 0 otherwise.

[b]=1 if respondent is in good or excellent physical health and = 0 otherwise.

[c]=1 if respondent is in an occupation of "building and grounds cleaning and maintenance," "construction/installation/repair," "farming/fishing/forestry," "production /manufacturing," or "transportation" and = 0 otherwise.

9% smoked cigarettes, and 79% indicated that they were in good to excellent health; 35% indicated that someone in their household was on a diet.

Regarding potato consumption, 93% of the subjects reported that their household consumed potatoes weekly, 82% reported consuming potato chips regularly, and 60% reported consuming French fries regularly. Given that we did not screen subjects for household potato product consumption, this high frequency of potato consumption indicates that our results are generalizable — at least to the population of people who can communicate in English. In the pre-auction questionnaire, 6% of the subjects reported that they were informed about acrylamide and 38% reported that they were informed about biotechnology.

Table 2 provides summary statistics of the subjects' bids by product and information treatment. In bidding rounds 1 and 2 (before the information treatment was included), the average bid for a five-pound bag of conventional potatoes was $2.71 and the average bid for a five-pound bag of biotechnology potatoes was $2.87, a difference of 16 cents. After each subject read his or her information treatment, the average bid for conventional potatoes was $2.69 per bag and the average bid for biotechnology potatoes was $2.50 per bag, a difference of 19 cents. After receiving an information treatment, the average bid for a bag of conventional potatoes was 2 cents lower and the average bid for a bag of biotech potatoes was 37 cents lower. These results imply that the negative information treatment had a stronger effect on bidding behavior than the positive information treatment.

The conventional fresh potato dices were treated with sodium bisulfide to prevent them from turning an unattractive brown color, and "sodium bisulfide" content was clearly presented on the label (see Figure 2). Before information treatment, the average subject's bid for a twelve-ounce bag of conventional potato dices treated with sulfites was $1.82 and the average bid for a twelve-ounce bag of biotechnology potato dices was $2.13, a difference of 31 cents. After information treatment, the average bid for a bag of conventional potato dices was $1.77 and the average bid for a bag of biotechnology potato dices was $1.78, a difference of 1 cent. After the information treatment, the average bid for a bag of conventional potato dices was 5 cents lower and the average bid for a bag of biotechnology dices was 35 cents lower.[14]

[14]We had very few zero bids and no cases where a subject bid zero on both products — whole potatoes and dices, GM or conventional.

Table 2. Bid (WTP) summary statistics for participants.

Commodity	N	Mean Bid	St. Dev.	Min.	Max.
A. All Bids: Pre-Information					
Conventional Potatoes	102	2.71	1.43	0.01	6.75
Conventional Potato Dices	102	1.82	1.26	0	6.00
Biotech Potatoes	102	2.87	1.52	0	6.95
Biotech Potato Dices	102	2.13	1.34	0	5.95
B. All Bids: Post Information					
Conventional Potatoes	102	2.69	1.45	0.01	6.00
Conventional Potato Dices	102	1.77	1.36	0	6.99
Biotech Potatoes	102	2.50	1.84	0	7.95
Biotech Potato Dices	102	1.78	1.31	0	6.00
A1. Bids Pre-Information: Before B1					
Conventional Potatoes	36	2.84	1.53	0.25	6.75
Conventional Potato Dices	36	1.94	1.12	0.10	4.89
Biotech Potatoes	36	2.92	1.60	0	6.00
Biotech Potato Dices	36	2.17	1.12	0	4.50
B1. Bids Post-Information: Company Perspective Only					
Conventional Potatoes	36	2.93	1.57	0.30	5.99
Conventional Potato Dices	36	1.98	1.14	0.10	5.00
Biotech Potatoes	36	3.11	1.77	0.01	7.00
Biotech Potato Dices	36	2.21	1.25	0.01	6.00
A2. Bids Pre-Information: Before B2					
Conventional Potatoes	34	2.75	1.29	0.89	6.00
Conventional Potato Dices	34	1.99	1.43	0.25	6.00
Biotech Potatoes	34	2.95	1.40	0.99	6.00
Biotech Potato Dices	34	2.19	1.42	0.49	5.95
B2. Bids Post-Information: Environmental Perspective Only					
Conventional Potatoes	34	2.73	1.29	0.75	6.00
Conventional Potato Dices	34	1.77	1.39	0	5.65
Biotech Potatoes	34	1.89	1.77	0	7.00
Biotech Potato Dices	34	1.46	1.29	0	5.00
A3. Bids Pre-Information: Before B3					
Conventional Potatoes	34	2.51	1.49	0.01	5.99
Conventional Potato Dices	34	1.51	1.19	0	4.25
Biotech Potatoes	34	2.74	1.59	0.01	6.95
Biotech Potato Dices	34	2.03	1.49	0	4.99
B3. Bids Post-Information: Company and Environmental Perspectives					
Conventional Potatoes	34	2.37	1.46	0.01	5.99
Conventional Potato Dices	34	1.56	1.54	0	6.99
Biotech Potatoes	34	2.45	1.82	0	7.95
Biotech Potato Dices	34	1.63	1.31	0	4.65

4. Regression Models

First, a model of a household's decision to consume potato product is presented, followed by a model of a household's WTP for low-acrylamide potato products. We specify a reference (no potato products consumed) household random indirect utility function:

$$U_{i0} = \boldsymbol{X}_{i0}\boldsymbol{\beta}_0 + \mu_{i0}, \tag{1}$$

where U_{i0} is the utility of the ith household when it does not consume a potato product, $\boldsymbol{X}_{i0}$ is a row vector and includes household income and prices of products purchased for consumption and socioeconomic variables that affect preferences, μ_{i0} represents other individually small effects on U_{i0} (and has a zero mean), and $\boldsymbol{\beta}_0$ is a column vector of unknown coefficients.

In contrast, the household's random indirect utility when its members choose to consume at least one unit of a potato product is

$$U_{i1} = \boldsymbol{X}_{i1}\boldsymbol{\beta}_1 + \mu_{i1}, \tag{2}$$

where U_{i1} is the utility of a household, given a decision to consume a potato product; $\boldsymbol{X}_{i1}$ is a row vector and includes household income and prices of potato products and other good purchased for consumption and socioeconomic attributes that affect tastes; μ_{i1} represents other individually small effects on U_{i1}, (and has a zero mean); and the coefficient vector $\boldsymbol{\beta}_1$ is a column vector of unknown coefficients.

A household consumes one unit of a potato product when its indirect utility is larger for consuming than for not consuming them. We define $D_i = 1$ if $U_{i0} < U_{i1}$, and 0 otherwise. Hence, the probability that $D_i = 1$ can be represented as

$$\begin{aligned} P_r(D_i = 1) &= P_r(\boldsymbol{X}_{i0}\boldsymbol{\beta}_0 + \mu_{i0} < \boldsymbol{X}_{i1}\boldsymbol{\beta}_1 + \mu_{i1}) \\ &= P_r(\mu_{i0} - \mu_{i1} < \boldsymbol{X}_{i1}\boldsymbol{\beta}_1 - \boldsymbol{X}_{i0}\boldsymbol{\beta}_0) \\ &= P_r(\mu_i < \boldsymbol{X}_i\boldsymbol{\beta}) \\ &= F(\boldsymbol{X}_i\boldsymbol{\beta}), \end{aligned} \tag{3}$$

where $F(\cdot)$ is a distribution function evaluated at $\boldsymbol{X}_i\boldsymbol{\beta}$. Equation (3) provides the conceptual framework for a model explaining the probability that a subject's household consumes a potato product over some time interval, say a week, which is one indication of the demand for a potato product. Equation (3) is fitted using the probit estimation routine, and then we calculate the marginal effect for each regressor.

In addition to household income, socioeconomic variables included as regressors are number of adults, children under age eight, and children eight years of age and older in the subject's household; whether anyone in the subject's household is on a diet; and the subject's gender, age, education, marital status, and racial-ethnic background. In addition, the regressors include dummy variables for whether the subject reads food labels on new purchases, exercises weekly, is in good health, and smokes cigarettes. We expect those subjects with more education and who read food labels when making new purchases will in general be more health conscious and less likely to consume processed potatoes and potato products. We don't have direct price data that the subject's household faces for potato products, but we expect prices of potatoes, chips, and fries to differ between (be higher in) rural and urban areas and between Boston and Los Angeles areas versus the Des Moines area. Hence, X also includes dummy variables for a subject's rural-urban residence and location in the Boston area (versus Des Moines) or Los Angeles area (versus Des Moines).

Next, we consider a model of willingness-to-pay for the gth commodity, vth variety, by the ith subject receiving the jth information treatment, W_{ij}^{gv}. The commodities are a five-pound bag of fresh whole potatoes and a twelve-ounce bag of fresh-cut potato dices. For fresh potatoes, the varieties (v) are conventional and low-acrylamide (achieved using potatoes grown with seed improved by biotechnology). For fresh-cut potato dices the varieties are conventional (with sodium bisulfite) and low-acrylamide/sulfite-free (achieved using potatoes grown with seed improved by biotechnology).[15] Each subject bid on a set of products before receiving a packaged

[15]Sodium bisulfite is a preservative used in conventional dices to limit product discoloration when exposed to air.

information treatment and then again after reading an environmental group perspective on biotechnology (antibiotech), a (positive) biotech company perspective on using biotechnology to lower bruising and acrylamide potential, or both perspectives.

We write WTP for the gth commodity, vth variety for the ith subject receiving the j-type of information as

$$W_{ij}^{gv} = \boldsymbol{X}_i \boldsymbol{\beta}_j^{gv} + \tau_1^{gv} I_{ij} + \varepsilon_{ij}^{gv}, \tag{4}$$

where $\boldsymbol{X}_i$ is a row vector and includes a set of socioeconomic attributes of the subject and his or her household, I_{ij} is 1 for information treatment of "a company perspective on low bruising and low acrylamide potential using biotechnology" (treatment 2) and 0 for receiving "an environmental group perspective on biotechnology" (treatment 1) or both perspectives (treatment 3). The last term of equation (4), ε_{ij}^{gv}, represents other individually small effects on W_{ij}^{gv} and has a zero mean. The baseline model is achieved when the ith individual engages in the first round of bidding (before the information treatment):

$$W_{i0}^{gv} = \boldsymbol{X}_i \boldsymbol{\beta}_0^{gv} + \varepsilon_{i0}^{gv}. \tag{5}$$

In our experiments, each of the i subjects bid first on commodity g of variety v before information treatment and then a second time after receiving an information treatment. Following earlier studies (e.g., Hoffman *et al.*, 1993; Rousu *et al.*, 2007), we convert the WTP model into one of WTP differences — WTP before relative to WTP after information treatment:

$$\begin{aligned} W_{i0}^{gv} - W_{i1}^{gv} &= \boldsymbol{X}_i(\boldsymbol{\beta}_0^{gv} - \boldsymbol{\beta}_1^{gv}) + \tau_1^{gv} I_{i1} + \varepsilon_{i0}^{gv} - \varepsilon_{i1}^{gv} \\ &= \boldsymbol{X}_i \boldsymbol{\beta}^{gv} + \tau_1^{gv} I_{i1} + \varepsilon_i^{gv}, \end{aligned} \tag{6}$$

where $\boldsymbol{\beta}^{gv} = \boldsymbol{\beta}_0^{gv} - \boldsymbol{\beta}_1^{gv}$ and $\varepsilon_i^{gv} = \varepsilon_{i0}^{gv} - \varepsilon_{i1}^{gv}$ and the last disturbance terms has a zero mean. The advantages to this specification include that the new dependent variable in equation (4) can be positive, zero, or negative; hence, the disturbance of the random disturbance term ε_i^{gv} is more likely to be normally distributed than for ε_{ij}^{gv}. Taking differences also removes any common individual fixed or random effects, including individual idiosyncrasies, which would otherwise be a possible source of biases in the estimated coefficients of the WTP equation (Greene, 2003).

Socioeconomic variables included in $\boldsymbol{X}_i$ are a subject's household income; number of children under age eight in subject's household; subject's gender, age, and years of completed schooling; whether subject indicated that they were informed about acrylamide and biotechnology in the preauction survey, read food labels when purchasing new food items, viewed biotech and GM foods as different, or indicated that their household regularly consumes potatoes; and a dummy variable for city location of subject (Des Moines, Boston or Los Angeles area). Subjects in households with children under eight years of age may be more concerned about the food safety of the food that they purchase in grocery stores, especially foods containing sulfites, than others. Because there is no scientific evidence of biotech foods being unsafe for human consumption, it is uncertain what effect the words "Made using potatoes grown with seed improved by biotechnology" in a food label will have. Although the perspective about genetically modified organisms (GMOs) presented by Greenpeace and Friends-of-the-Earth is negative, Colson, Huffman, and Rousu (2011) show that consumers are willing to pay more in experimental lab auctions of GM products with enhanced antioxidants and vitamin C when those traits were achieved using genes from within the species (intragenic). Women are more intensely involved in food shopping than men, and this difference in experience could affect the size of WTP differences, but we do not know in what direction. A subject's age is included to control for life-stage effects and to permit differences in WTP to occur by age. Individuals with more education are better able to read and digest consumer and food safety information (Schultz, 1975; Huffman, 1974) and hence may show larger responsiveness to information treatments. The effect of a subject being informed about biotechnology could have a positive or negative effect on how they bid, but Rousu *et al.* (2007) found that those who were informed bid more for GM products. Those that read food labels may be more conscious shoppers, but it is unclear how this attribute will affect bidding behavior. Households that regularly consume potatoes may behave differently in bidding on potato products than other households because they have more information about their tastes.

The locations of our food experiments are more than 1,300 miles apart and in vastly different parts of the country, where the daily attention to food production differs considerably. Iowa and California are large producers of food and agricultural products, while farmers in Massachusetts are engaged in small-scale farming, including agro-tourism. Massachusetts is heavily urbanized and far from centers of large-scale food production. Hence, we expect Boston subjects to be most strongly anti-biotech. In addition, three New England states have recently voted to require some type of mandatory labeling of GM foods. All state-level votes in other regions of the United States have failed (Huffman and McCluskey, 2014).

5. Regression Results for Potato Product Demand

Table 3 reports the results from fitting the empirical probit model explaining the probability of a subject's household consuming traditional potatoes and processed potato products weekly. A general model for equation (3) with fifteen regressors, including city fixed effects and intercept term, is first fitted. Then, variables having coefficients with unusually small z-values (implying that they are not significantly different from zero at even the 10% level) are deleted, except for a few core variables — household income, number of adults in household, and city dummies — and the model is re-fitted. The expectation is that the size of the estimated coefficients with large z-values at the first stage will be relatively unaffected when insignificant regressors are excluded but that their z-values may increase in size due to less near-multicollinearity among the remaining regressors.[16]

[16]Although we might test for information order effects, we do not expect them to be significant.

5.1. *Fresh potatoes*

Regression (1) of Table 3 includes fifteen regressors, largely socioeconomic variables, five of which are excluded from regression (2).[17] The probit results and marginal effects are in Table 3. In regression (2), a $1,000-per-year increase in household annual income reduces the probability of a subject's household consuming fresh potatoes slightly but not significantly. Adding one adult to a subject's household or a subject being married increases the probability of a subject's household consuming potatoes by 3.0 and 5.2 percentage points, respectively. Male subjects are 4.3 percentage points less likely to be in a household that consumes potatoes than female subjects. The marginal effects of a subject's age on the probability of the household consuming potatoes increases from age eighteen to forty, and then the probability of the household consuming potatoes decreases for subjects older than forty and continues to decrease as age increases. However, this age effect is statistically weak. If the subject reported that someone in the household was on a diet, then the probability of consuming potatoes decreases by 2.8 percentage points. Because most diets suggest reducing carbohydrate-intensive and high-starch foods, this result is not surprising. Households in the Boston and Los Angeles areas are slightly more likely to consume potatoes than Des Moines area households.

5.2. *Potato chips*

Regression (3) starts with the same fifteen regressors as regression (1), six of which are then excluded in regression (4).[18] In regression (4), the most significant factor affecting the probability of a subject's household consuming chips is his or her age.[19] The marginal effect

[17]The excluded regressors are whether a child age eight or older lives in the household, subject's gender, subject is white, exercises weekly, and in good health.
[18]These regressors are whether a child age eight or older lives in the household, anyone in the household is on a diet, subject's education, subject is married, white, and in good health.
[19]Additional household income has no economic effect on the probability of a subject's household consuming chips.

Table 3. Probit: Probability of household consuming commodities weekly ($N = 102$).

	Potatoes		Potato Chips		French Fries	
Regressors	(1)	(2)	(3)	(4)	(5)	(6)
Household Income ($1,000)	−0.00002	−0.00001	−0.00000	−0.00000	−0.00000	−0.00000
	(1.59)	(1.68)	(0.03)	(1.07)	(0.95)	(1.35)
	[−0.0000]	[−0.0000]	[0.0000]	[−0.0000]	[−0.0000]	[−0.0000]
Number of Adults	0.901	0.610	0.247	0.256	0.382	0.371
	(1.65)	(1.61)	(1.12)	(1.33)	(1.86)	(2.00)
	[0.017]	[0.030]	[0.047]	[0.053]	[0.144]	[0.140]
Number of Children ≥ 8 yrs	0.618	—	−0.227	—	0.034	—
	(0.92)		(1.04)		(0.19)	
	[0.011]		[−0.043]		[0.013]	
Anyone on Diet (=1)	−0.841	−0.474	−0.047	—	−0.235	—
	(1.13)	(0.94)	(0.12)		(0.73)	
	[−0.024]	[−0.028]	[−0.009]		[−0.090]	
Gender (1 = male)	−0.690	−0.706	−0.463	−0.420	0.731	0.699
	(1.14)	(1.40)	(1.31)	(1.23)	(2.33)	(2.33)
	[−0.017]	[−0.043]	[−0.094]	[−0.092]	[0.264]	[0.253]
Age	0.333	0.209	0.426	0.332	0.134	0.148
	(1.32)	(1.29)	(3.06)	(3.09)	(1.27)	(1.65)
	[0.006]	[0.010]	[0.081]	[0.069]	[0.051]	[0.056]
Age^2	−0.004	−0.002	−0.005	−0.004	−0.002	−0.002
	(1.37)	(1.40)	(3.03)	(3.02)	(1.30)	(1.66)
	[−0.0001]	[−0.0001]	[−0.001]	[−0.001]	[−0.001]	[−0.001]
Education (yrs)	−0.012	—	−0.146	—	0.003	—
	(0.07)		(1.38)		(0.03)	
	[−0.0002]		[−0.028]		[0.001]	

(*Continued*)

Table 3. (*Continued*)

Regressors	Potatoes (1)	Potatoes (2)	Potato Chips (3)	Potato Chips (4)	French Fries (5)	French Fries (6)
Married (=1)	1.030	0.928	−0.198	—	−0.124	—
	(1.34)	(1.40)	(0.49)		(0.34)	
	[0.024]	[0.052]	[−0.038]		[−0.047]	
White (=1)	0.755	—	−0.555	—	−0.067	—
	(0.76)		(0.84)		(0.14)	
	[0.029]		[−0.082]		[−0.025]	
Exercises Weekly (=1)	1.155	—	−0.829	0.668	−0.971	−1.102
	(1.12)		(1.57)	(1.38)	(1.81)	(2.15)
	[0.065]		[0.217]	[0.177]	[−0.301]	[−0.329]
Good Health (=1)	0.791	—	0.088	—	−0.389	—
	(0.88)		(0.17)		(0.95)	
	[0.028]		[0.017]		[−0.140]	
Boston (=1)	0.779	0.728	1.11	1.12	1.019	1.003
	(0.86)	(0.97)	(2.07)	(2.27)	(2.43)	(2.58)
	[0.011]	[0.029]	[0.169]	[0.186]	[0.345]	[0.340]
Los Angeles (=1)	0.751	0.701	0.756	0.830	1.047	1.020
	(0.97)	(1.03)	(1.75)	(2.04)	(2.60)	(2.77)
	[0.011]	[0.029]	[0.125]	[0.148]	[0.356]	[0.348]
Intercept	−7.587	−2.937	−6.583	−7.052	−2.525	−3.030
	(1.09)	(0.77)	(2.02)	(2.76)	(0.97)	(1.39)
Pseudo R^2	0.348	0.285	0.212	0.169	0.210	0.195

Notes: Absolute value of z-values in parentheses. t-values larger than 1.98 are significantly different from zero at the 5% level; larger than 1.65 are significant at 10% level. The marginal effects are in square brackets.

of a subject's age on the probability of his or her household consuming chips increases from eighteen to forty-three years; for subjects older than forty-three years, aging reduces the probability of the household consuming chips. Female subjects are 9.2 percentage point more likely to have a household that consumes potato chips than male subjects, and adding an additional adult to a subject's household increases the probability of the household consuming potato chips by 5.3 percentage points. If the subject reported he or she exercised regularly, then his or her household consuming chips is 17.7 percentage points more likely to consume chips than households with subjects reported not exercising regularly. Boston and Los Angeles households are 24.4 and 17.0 percentage points more likely to consume chips than Des Moines area households.

5.3. *French fries*

Regression (5) includes the same fifteen regressors as regression (1), six of which are excluded from regression (6).[20] In regression (6), an additional $1,000 in a subject's household income slightly reduces the probability of his or her household consuming fries, but the effects are not statistically significant). An additional adult in the subject's household increases the probability of the household consuming fries by 14 percentage points. The probability of a household consuming fries increases as the subject's age increases from eighteen to forty-six years of age and then declines as subject's age increases further. When the subject is a woman, the household is 14 percentage points more likely to consumes fries than if the subject is a man. Also, if the subject reported exercising regularly, then the household was 33 percentage points less likely to consume fries than when the subject reported not exercising regularly. Boston and Los Angeles area households are 34.3 and 29.6 percentage points more likely to consume fries that Des Moines area households.

[20]These regressors are whether a child age eight or older lives in the household, anyone in household is on a diet, subject's education, subject is married, white, and in good health.

There are major differences across the three potato products with regard to how the regressors affect the probability of a household consuming fresh versus processed potato products (i.e., chips and fries). A subject's marital status and whether someone in the household is on a diet are important variables for explaining fresh potato consumption but not chip or fry consumption. Since people who consider themselves to be on a diet are usually more conscious about the foods they eat as well as the food everyone else in the household eats, it is surprising that dieting is not an important explanatory variable for chips and fries. In contrast, regular reported exercise is important for explaining consumption of chips and fries but not fresh potatoes. A subject's household income has no significant effect on the probability of his or her household consuming potato products. In addition, our empirical model of the probability of a subject's household consuming potato products has the highest explanatory power for fresh potatoes, pseudo $R^2 = 0.285$, but is significantly lower for processed potatoes (fries and chips), pseudo R^2 of 0.195 and 0.169, respectively.

6. Regression Results for Willingness to Pay

Results from fitting equation (4) for the two commodities (a five-pound bag of fresh potatoes and a twelve-ounce bag of fresh potato dices) are reported. For a given commodity, within-subject WTP differences are expressed both within-variety and across-variety. Table 4 presents summary statistics for the dependent variables.

6.1. *Fresh potatoes*

In regressions (1) and (2) in Table 5, the variety of fresh potatoes is "biotech," (i.e., the food label on the front of the package displayed the phrase "Potatoes Grown with Seed Improved by Biotechnology"). Subjects bid twice on the biotech potatoes; first without information and a second time after receiving an information treatment. The dependent variable is a subject's WTP for biotech potatoes before information less WTP for biotech potatoes after

Table 4. WTP differences and other summary statistics.

Variables	Mean	St. Dev
WTP for improved biotechnology potatoes after information treatment less WTP for improvedbiotechnology potatoes before information treatment ($)	−0.38	$1.49
WTP for improved biotechnology potatoes after information treatment less WTP for conventional potatoes before information treatment ($)	−0.21	1.40
WTP for improved biotechnology dices after information treatment less WTP for improved biotechnology dices before information treatment ($)	−0.36	1.05
WTP for improved biotechnology dices after information treatment less WTP for conventional dices with sodium bisulfide before information treatment ($)	−0.05	0.98
Pre-auction,subject informed about:		
Acrylamide	0.059	0.236
Biotechnology	0.382	0.488
Subject received information treatment:		
Environmental group perspective	0.333	0.474
Company perspective	0.353	0.480
Company & environmental group perspectives	0.314	0.466

receiving an information treatment. Hence, the estimated coefficients on regressors reflect how a variable impacts the way treatment information changes WTP behavior of subjects.[21] In our regressions,

[21]One information treatment must be assigned to the intercept term to provide identification of the effects of the other treatment effects relative to the excluded one. Hence, in our results the intercept term absorbs the negative effect of the "environmental group perspective." This arrangement provides results that are easiest to interpret.

the dummy variable for a household being located in the Des Moines area is excluded and its effect is part of the intercept term.[22]

Regression (1) of Table 5 includes fifteen regressors, largely socioeconomic variables, information treatment effects, and city dummy variables; regression (2) excludes five of the insignificant regressors.[23] For regression (2), an increase of $1,000 in a subject's household income reduced subjects' WTP for a five-pound bag of biotech potatoes post-information treatment relative to their pre-information treatment by 0.00006, which is very small. Male subjects had a $0.62 per bag lower WTP for biotech potatoes post-information treatment relative to pre-information treatment; or women had larger WTP post-information. If a subject reported being informed about biotechnology pre-experiment, he or she had a $0.74 per bag higher WTP for biotech potatoes post-information treatment than pre-information treatment. If a subject indicated in the pre-auction survey that he or she reads food labels when buying new foods for the first time or views biotech and GM foods as being different, his or her WTP for a bag of biotech potatoes post-information treatment was $0.75 lower than for the pre-information treatment WTP. However, if the subject received the "company perspective" information treatment, his or her WTP for a bag of biotech potatoes was $1.43 higher post-information treatment than pre-information treatment. Hence, the "company perspective" is influential in modifying subject's WTP for improved food safety using biotechnology. If a subject received the third information treatment (the company perspective and the environmental group perspective), his or her WTP for a bag of biotech potatoes was $0.85 higher post-information treatment than pre-information treatment.

[22]There was no significant order effect (i.e., whether a subject first bid on biotech or conventional products).

[23]Excluded regressors are the number of children younger than eight in the subject's household, subject's age and education, whether the subject is informed about acrylamide, and an indicator for a subject's household consuming potatoes weekly. The null hypothesis that these five coefficients on the excluded variables are jointly zero yields a sample value of the F statistic of 0.001, but the tabled critical value of the F with 5 and 86 degrees of freedom at the 5% significance level is 2.33. Hence, we cannot reject the null hypothesis at the 5% significance level.

Table 5. Regression analysis of WTP differences: WTP after information treatment less WTP before treatment — fresh potatoes, 5lbs ($N = 102$).

	Biotech Potatoes		Biotech Potatoes vs. Conventional Potatoes	
Regressors	(1)	(2)	(3)	(4)
Household Income ($1,000)	−0.00006	−0.00006	−0.00004	−0.00003
	(1.69)	(1.73)	(1.06)	(1.16)
Number of Children < 8 yrs	−0.042	—	−0.164	—
	(0.15)		(0.58)	
Gender (1 = male)	−0.636	−0.621	−0.486	−0.461
	(2.32)	(2.34)	(1.79)	(1.75)
Age	−0.002	—	0.003	—
	(0.22)		(0.32)	
Education (yrs)	0.018	—	−0.002	—
	(0.23)		(0.02)	
Informed about Acrylamide (=1)	−0.192	—	0.498	—
	(0.32)		(0.82)	
Informed about Biotechnology (=1)	0.782	0.738	0.584	0.583
	(2.64)	(2.67)	(2.00)	(2.18)
Reads Food Labels (=1)	−0.774	−0.748	−0.398	—
	(1.77)	(1.78)	(0.92)	
Biotech and GM Foods Are Different (=1)	−0.727	−0.732	−0.536	−0.493
	(2.56)	(2.72)	(1.91)	(1.85)
Company Perspective	1.424	1.423	1.249	1.200
	(4.34)	(4.51)	(3.84)	(3.86)
Company Perspective × Environmental Group Perspective	0.808	0.845	0.874	0.898
	(2.36)	(2.61)	(2.59)	(2.82)
Household Consumes Potatoes Weekly	−0.358	—	−0.243	—
	(0.65)		(0.45)	
Boston (=1)	0.143	0.144	−0.99	−0.093
	(0.38)	(0.42)	(0.27)	(0.27)
Los Angeles (=1)	0.684	0.654	0.241	0.215
	(1.93)	(2.01)	(0.69)	(0.67)
Intercept	0.164	−0.090	0.058	−0.532
	(0.11)	(0.18)	(0.04)	(1.55)
R^2	0.318	0.312	0.241	0.223

Notes: Absolute value of t-values in parentheses. Coefficients with t-values larger than 1.98 are significantly different from zero at the 5% level; larger than 1.65 are significant at 10% level. Environmental group perspective on biotechnology is the excluded information treatment.

Hence, the positive "company perspective" continues to weigh heavily on WTP differences relative to the negative environmental group perspective. Subjects from the Los Angeles area were willing to pay \$0.65 more per bag of biotech potatoes post-information than pre-information treatment relative to Des Moines area subjects. There was no difference in WTP for a bag of biotech potatoes post- versus pre-information treatment for Boston area subjects than Des Moines subjects. Hence, Los Angeles subjects were most favorably impacted by the company information treatment.

Regressions (3) and (4) of Table 5 present a different perspective on WTP for biotech potatoes. It is a comparison of WTP for a five-pound bag of biotech potatoes post-information treatment with WTP for a five-pound bag of conventional potatoes pre-information treatment — a type of comparison that a consumer might face in a marketing campaign for new biotech potatoes appearing in grocery stores and supermarkets. Hence, both the variety type and information treatments are different in this comparison, and it is similar to the framing of WTP differences in the econometric analysis presented by Rousu *et al.* (2007). Regression (3) contains the same set of regressors as regression (1). The regressors that are much less significant in regression (3) than in regression (1) are a subject's household income, whether a subject reads food labels, and whether a subject resides in the Los Angeles area. Regression (4) excludes six regressors with small z-values.[24] For the remaining estimated coefficients, it is surprising that all of them are 30–50% smaller than for regression (2), except for the coefficient of the Boston area dummy variable, which is not significantly different from zero in both regressions.

[24]The excluded regressors are the number of children younger than eight in the subject's household, subject's age and education, whether subject was informed about acrylamide, whether subject reads food labels on new goods, and an indicator for a subject's household consuming fresh potatoes weekly. The null hypothesis that the estimated coefficients on these six regressors are jointly equal to zero yields a sample value of the F statistic of 0.34, but the tabled critical value of the F with 6 and 87 degrees of freedom at the 5% significance level is 2.20. Hence, we cannot reject the null hypothesis of no joint effects at the 5% significance level.

6.2. *Fresh potato dices*

In regressions (1) and (2) in Table 6, the variety of fresh potato dices is "biotech" (i.e., the food label on the front of the package displayed the phrase "Potatoes Grown with Seed Improved by Biotechnology"). Subjects bid twice on the biotech dices; first before receiving the information treatment and a second time after receiving an information treatment. The dependent variable is a subject's WTP for a twelve-ounce bag of biotech dices before information less WTP for biotech dices after receiving an information treatment.

Regression (1) includes fifteen regressors, largely socioeconomic variables, information treatment effects, and city dummy variables; regression (2) includes eight insignificant regressors.[25] In regression (2), if the subject indicated that he or she reads food labels, then his or her WTP post-information treatment declined by $0.43 per bag relative to the pre-information treatment WTP. Similarly, if the subject considered biotech and GM foods as being different, a subtle dimension, then his or her WTP was $0.44 less per bag post information than pre-information treatment. If the subject received the "company perspective" information treatments, his or her WTP for a bag of dices increased by $0.88 relative to his or her WTP pre-information treatment. As previously stated, this is not surprising since the "company perspective" provides positive information about biotechnology and biotech foods. If the subject received the "company perspective" and "environmental group perspective" in the information treatment, his or her WTP for a bag of dices increased by $0.38 relative to his or her WTP pre-information treatment. This is a decline of $0.50 per bag compared to just the "company perspective" and shows that the "environmental

[25]The excluded regressors are a subject's household income; the number of children younger than eight in subject's household; subject's gender, age, and education; whether subject was informed about acrylamide; whether subject was informed about biotechnology; and an indicator for a subject's household consuming potatoes weekly. The null hypothesis that the coefficients of these 8 coefficients variables are jointly zero yields a sample value of the F statistic of 0.43, but the tabled critical value of the F with 8 and 87 degrees of freedom at the 5% significance level is 2.20. Hence, we cannot reject the null hypothesis of no joint effects at the 5% significance level.

Table 6. Regression analysis of willingness-to-pay differences: WTP after information treatment less WTP before treatment — fresh cut potato dices, 12oz ($n = 102$).

	Low-Acrylamide Biotech Dices		Low-Acrylamide Biotech Dices vs. Dices w/Sodium Bisulfite	
Regressors	(1)	(2)	(3)	(4)
Household Income ($1,000)	−0.00001	—	−0.00001	—
	(0.56)		(0.53)	
Number of Children < 8 yrs	0.116	—	−0.112	—
	(0.56)		(0.55)	
Gender (1 = male)	−0.074	—	−0.017	—
	(0.37)		(0.09)	
Age	0.008	—	0.003	—
	(0.98)		(0.33)	
Education (yrs)	0.018	—	0.023	—
	(0.32)		(0.42)	
Informed about Acrylamide (=1)	−0.310	—	−0.017	—
	(0.70)		(0.04)	
Informed about Biotechnology (=1)	0.261	—	0.279	0.291
	(1.23)		(1.32)	(1.51)
Reads Food Labels (=1)	−0.550	−0.432	−0.400	−0.413
	(1.75)	(1.46)	(1.28)	(1.39)
Biotech and GM Foods Are Different (=1)	−0.475	−0.441	−0.107	—
	(2.33)	(2.29)	(0.52)	
Company Perspective	0.863	0.876	0.830	0.816
	(3.65)	(3.89)	(3.53)	(3.70)
Company Perspective × Environmental Group Perspective	0.341	0.376	0.613	0.614
	(1.38)	(1.61)	(2.51)	(2.70)
Household Consumes Potatoes Weekly	−0.033	—	0.180	—
	(0.08)		(0.46)	
Boston (=1)	−0.428	−0.522	−0.233	−0.299
	(1.58)	(2.29)	(0.87)	(1.34)
Los Angeles (=1)	0.417	0.365	0.244	0.148
	(1.63)	(1.60)	(0.96)	(0.67)
Intercept	−0.614	−0.169	−0.729	−0.229
	(0.50)	(0.50)	(0.70)	(0.70)
R^2	0.287	0.256	0.199	0.187

Notes: Absolute value of t-values in parentheses. Coefficients with t-values larger than 1.98 are significantly different from zero at the 5% level; larger than 1.65 are significant at 10% level. Environmental group perspective on biotechnology is the excluded information treatment.

group perspective," where packaged with the "company perspective," only slightly reduces the positive effect of the "company perspective" only treatment on WTP for biotech dices. Hence, the environmental group perspective does little to moderate the effects of the company perspective.

Regressions (3) and (4) in Table 6 present a different perspective on WTP for fresh-cut potato dices. The dependent variable in these regressions is a subject's WTP for fresh-cut potato dices made from "Potatoes Grown with Seed Improved by Biotechnology" post-information treatment relative to his or her WTP for a twelve-ounce bag of fresh cut potato dices made using sodium bisulfite. Hence, the "variety" of the commodity (dices) and available information are different in computing the WTP differences — fresh biotech dices without sulfites versus non-biotech dices with sodium bisulfite added. Regression (3) contains the same fifteen regressors as regression (1). The effect of a subject indicating in the pre-auction survey that he or she reads food labels when buying a food item for the first time is more significant in regression (3) than in regression (1), but the effect of a subject reporting in the pre-auction survey that biotech and GM foods are different is less significant in regression (3) than in regression (1). In regression (4), when a subject reports in the pre-auction survey that he or she is informed about biotechnology, his or her WTP for a bag of dices is $0.29 higher post-information treatment relative to pre-information treatment. When a subject indicates that he or she reads food labels, his or her WTP for a bag of biotech dices declines by $0.41 post-information treatment relative to a bag of dices containing sodium bisulfite pre-information treatment. If a subjects receives the "company perspective," he or she is willing to pay $0.82 more for a bag of biotech dices post-information treatment than for a bag of dices containing sodium bisulfite pre-information treatment. If a subject receives the "company perspective" and "environmental group perspective" information treatment, his or her WTP is $0.61 higher for a bag of biotech dices post-information than for a bag of sodium bisulfide dices pre-information treatment. Hence, the WTP difference per bag is only reduced by $0.20 when the "environmental group perspective" is added to the "company perspective." This is similar to the findings for fresh potatoes. Although subjects from the Boston area tend to pay less and for Los Angeles area to pay more for a bag of biotech dices relative to Des Moines residents, these city effects are not significantly different from zero at the 5% level.

The information treatment effects on WTP differences for fresh potatoes and dices are consistent with the information treatment effects obtained by Rousu *et al.* (2007) and Colson, Huffman, and Rousu (2011) for biotech foods. The explanatory power of the empirical models of WTP differences is largest for the comparisons of within-variety differences; R^2 being 0.312 for biotech potatoes and 0.256 for biotech dices. We lose about 0.10 from the R^2 in the comparisons of WTP for biotech product post-information treatment relative to conventional product pre-information. This suggests that there is additional unexplained noise in the cross-variety comparisons that does not exist in the within-variety comparisons.

7. Conclusion

This study provides new empirical evidence on household demand for traditional fresh and processed potato products and consumers' willingness to pay for a new variety of fresh biotech potato and new potato products — fresh potato dices, low-acrylamide potato products, and fresh potato dices that are also sulfite-free. The probability of a subject's household consuming traditional fresh potatoes, chips, and fries is not significantly related to household income. Hence, the results suggest that demand for potatoes in the United States will not change much as real household incomes rise over time. However, aging of the adult population is expected to decrease the probability of households consuming chips but not potatoes or fries. The probability of a household consuming potato products is unaffected by a subject's education. It is perhaps surprising that households in the Boston and Los Angeles areas were significantly more likely to consume processed potato products than households in the Des Moines area.

We find empirical evidence that some participants are willing to pay for food safety — fresh potatoes with low potential to produce acrylamide and white potato dices that are sulfite free — achieved using biotechnology, but not transgenic GMOs. However, WTP is conditioned by the information injected into the experimental auction. Retailers could segment their consumers into those that are receptive to GMOs and distribute information to them that shows the benefits of low acrylamide potato products. In this way, they can increase the demand for these products. Retailers can retain consumers who have non-GMO preferences by carrying both GM and conventional varieties of potato products. In the long run, biotech potato products that have improved food safety are expected to achieve higher rates of consumer acceptance in the United States than those with earlier traits for herbicide tolerance and/or insect resistance.

Limitations to our study include small sample size and under-representation of rural subjects and households. However, none of our preliminary results showed that subjects from rural households responded differently from subjects in urban households.

References

Bethke, P.C. and A.J. Bussan (2013). Acrylamide in Processed Potato Products. *American Journal of Potato Research* 90, 403–424. doi: 10.1007/s12230-013-9321-4.

California Department of Justice (2008). *Atty. Gen. Brown Settles Potato Chip Lawsuit With Heinz, Frito-Lay and Kettle Foods.* Sacramento, CA: Office of the Attorney General. Available online at http://oag.ca.gov/news/press-releases/atty-gen-brown-settles-potato-chip-lawsuit-heinz-frito-lay-kettle-foods.

California Office of Environmental Health Hazard Assessment (2014). *Proposition 65.* Sacramento, CA: California Office of Environmental Health Hazard Assessment. Available online at: http://oehha.ca.gov/prop65.html.

Colson, G.J., W.E. Huffman and M.C. Rousu (2011). Improving Nutrient Content Through Genetic Modification: Evidence from Experimental Auctions on Consumer Acceptance and Willingness to Pay for Intragenic Foods. *American Journal of Agricultural Economics* 93, 654–654. doi: 10.1093/ajae/aaq199.

Greene, W.H. (2003). *Econometric Analysis*, 5th Edn. Upper Saddle River, NJ: Prentice Hall.

Hoffman, E., D.J. Menkhaus, D. Chakravarti, R.A. Field and G.D. Whipple (1993). Using Laboratory Experimental Auctions in Marketing Research: A Case Study of New Packaging for Fresh Beef. *Marketing Science* 12, 318–338.

Huffman, W.E. (1974). Decision Making: The Role of Education. *American Journal of Agricultural Economics* 56, 85–97. doi: 10.2307/1239349.

Huffman, W.E. and J.J. McCluskey (2014). The Economics of Labeling GM Foods. *AgBioForum* 17, 156–160.

Huffman, W.E., J.F. Shogren, M.C. Rousu and A. Tegene (2003). Consumer Willingness to Pay for Genetically Modified Food Labels in a Market with Diverse Information: Evidence from Experimental Auctions. *Journal of Agricultural and Resource Economics* 28, 481–502.

Nunes, J.C. and P. Boatwright (2004). Incidental Prices and Their Effect on Willingness to Pay. *Journal of Marketing Research* 41, 457–466. doi: 10.1509/jmkr.41.4.457.47014.

Rangan, U. (2010). "The Real Risks of Sulfites." Available online at http://www.consumerreports.org/cro/news/2010/06/the-real-risks-of-sulfites/index.htm.

Rasmussen, K.M., M.E. Latulippe and A.L. Yaktine (2015). *Review of WIC Food Packages: An Evaluation of White Potatoes in the Cash Value Voucher.* Washington, DC: Institute of Medicine of the National Academies of Science.

Richards, T.J., A. Kagan and X. Gao (1997). Factors Influencing Changes In Potato And Potato Substitute Demand. *Agricultural and Resource Economics Review* 26, 52–66.

Rousu, M.C., W.E. Huffman, J.F. Shogren and A. Tegene (2007). Effects and Value of Verifiable Information in a Controversial Market: Evidence from Lab Auctions of Genetically Modified Food. *Economic Inquiry* 45, 409–432. doi: 10.1111/j.1465-7295.2007.00017.x.

Schultz, T.W. (1975). The Value of the Ability to Deal with Disequilibria. *Journal of Economic Literature* 13, 827–846.

Shogren, J.F., M. Margolis, C. Koo and J.A. List (2001). A Random *n*th-Price Auction. *Journal of Economic Behavior and Organization* 46, 409–421.

Tareke, E., P. Rydberg, P. Karlsson, S. Eriksson and M. TÃűrnqvist (2002). Analysis of Acrylamide, a Carcinogen Formed in Heated Foodstuffs. *Journal of Agricultural and Food Chemistry* 50, 4998–5006. doi: 10.1021/jf020302f.

Appendix A: Environmental Group Perspective on Biotechnology

General statement

- Biotech plant breeding takes genes from one organism and places them into another. This process manipulates genes and alters genetic makeup and properties. The cutting of genetic material from one organism and inserting it into another is quite imprecise and can cause mutations. There has been inadequate testing of these products.

- Biotech methods frequently use antibiotic-resistant gene segments in soil bacteria or viruses and transfer them into plants. This process is risky, leading to unanticipated outcomes.
- Biotech plants are regulated by the federal government, but federal regulation relies heavily on data collected from field trials and other testing by the biotech industry.
- Biotech seeds were first marketed to U.S. farmers in the mid-1990s, and rapid farmer adoption occurred in field crops (corn, soybean, cotton and canola). Later developments have been in papaya and vegetables (sugar beet, squash and potato).
- In the United States, the sales of biotech foods have grown very rapidly. This growth is driven by self-interested producers and marketers seeking to maximize crop yields and minimize production costs.

Nutrition and health

- Of the laboratory plants that are successfully modified to express the "right" traits, genetic engineers select among those that look strong, healthy, and capable of further breeding. There is poor screening to eliminate varieties that produce harmful substances or low nutrient quality.
- New allergens are likely to be introduced into the food supply.
- Early research reported some health problems in laboratory animals consuming first-generation insect-resistant potatoes.
- Several scientific studies show that laboratory animals that have been fed biotech food developed one or more toxic effects on vital and/or reproductive organ functioning, relative to a control group.
- A recent Canadian study found a common protein from insect-resistant corn in the bloodstream of pregnant women and their fetuses. Another recent study found that high concentrations of this protein resulted in severe damage to human embryonic kidney cells.
- The nutritional content of biotech foods, relative to conventional foods, is variable.

Environmental impacts and food security

- Private companies are not capable of screening new biotech materials for every possible pathogen or environmental stress. Unnoticed and unsafe mutations could strike after the occurrence of extreme stress, such as plant disease outbreaks, droughts, floods, and heat waves.
- New biotech crops may cross-pollinate with other plants and are likely to cause super weeds.
- Some herbicides used on biotech crops diffuse into the air and leach into streams and waterways in some areas. These herbicides are toxic to amphibians (e.g., frogs, salamanders) and earthworms, which impacts bird populations.
- Biotech crops are doing little to help international food security or relieve hunger in poor countries. The major biotech crops, corn and soybeans, are mainly used for animal feed, biofuels, and processed human food in developed countries.

Appendix B: Company Perspective on Low Bruising and Low Acrylamid Potential Using Biotechnology

General statement

- Potatoes have well-known vulnerability to bruising during harvest and while in storage. In addition, a fresh potato turns brown upon cutting when exposed to the air. Recently, it has been discovered that frying or broiling potatoes at high temperatures forms acrylamide.
- This tendency for bruising causes a large amount of potato wastage and costs potato growers and processors millions of dollars annually. Sodium bisulfite is sometimes used to preserve freshness.
- In 2002, acrylamide, a chemical compound, was discovered in foods containing certain natural sugars when cooked at high temperatures (above roughly 250°F), such as French fries, hash browns and potato chips. They are also formed in the baking of bread (crust) and cookies and roasting of coffee beans.

- Acrylamide is formed from naturally occurring sugars and asparagine (a building block of a protein) in food. It is a toxin and possible carcinogen in humans.
- The Simplot Company has developed improved versions of three popular potato varieties that have significantly reduced bruising and acrylamide forming potential than conventional potato varieties. These new potato varieties are slower to oxidize after peeling or cutting. Hence, these new potato varieties provide major advantages to potato producers, processors and consumers.
- The new varieties were made possible using InnateâĎć technology, which is a new biotech process for moving genes swiftly and precisely from wild and domesticated potatoes into popular varieties.

Nutrition and health

- Food and cigarette smoke are the two main sources of human exposure to acrylamide. In the American diet, processed potato products (potato chips and cooked French fries) are the leading source of acrylamide. Long-term, low-level intake of acrylamide by lab animals has been shown to create serious health problems. One study found that consuming conventional potato chips regularly for a month caused some health changes in humans.
- Under Proposition 65, California requires certain restaurants, including Applebee's, Chili's, McDonald's, Burger King, Wendy's, and KFC, to post a warning stating that "cooked potatoes that have been browned, such as French fries, hash browns and baked potatoes, contain acrylamide, a chemical known to the State of California to cause cancer."
- Simplot's latest Innate™ potato varieties reduce acrylamide formation by approximately 90% in processed potato products cooked at high temperatures. This reduction places them below the minimum amounts of California's Proposition 65, and removes acrylamide level health concerns.
- Biotech methods were also used to develop golden rice, which enhanced vitamin A content. Other opportunities exist for

enhancing consumer attributes such as antioxidants and vitamins in food.
- Biotech foods, relative to conventional foods, have similar low allergy potential.

Environmental impacts and food security

- Innate™ biotech modifies a plant's genes without incorporating foreign genes or antibiotic resistance. There are no adverse impacts on the environment or new food allergens.
- Commercial potatoes are grown from pieces of whole potato and not seed. Many commercial potatoes are either sterile or not sexually compatible with wild potatoes so there is very low risk of the Innate™ potato varieties crossing with other potato varieties or other plants.
- Innate™ potatoes have undergone extensive field testing and consumer taste testing by Simplot, and have been thoroughly evaluated by the USDA and FDA.
- Approval is being sought for Innate™ potato exports to Canada, Mexico, South Korea and Japan.

Chapter 15

Consumer Response to Information About a Functional Food Product: Apples Enriched with Antioxidants*

Armenak Markosyan[†,¶], Jill J. McCluskey[‡,‖], and Thomas I. Wahl[§,**]

[†]*Economist, Econ One Research, Inc., 601 West 5th Street, 5th Floor, Los Angeles, CA 90071*

[‡]*School of Economic Sciences and IMPACT Center, Washington State University, Pullman, WA 99164-6210*

[§]*Department of Agribusiness and Applied Economics, North Dakota State University, 217 Morrill Hal, Fargo, ND 58105-5636*

[¶]*amarkosyan@econone.com*

[‖]*mccluskey@wsu.edu*

[**]*tom.wahl@ndsu.edu*

Interest in functional foods has been growing as consumers become increasingly concerned with diet and nutrition. This article measures consumers' responses to apples enriched with an antioxidant coating. Antioxidant-enriched apples are believed to provide additional health benefits reducing the risk of cancer and heart diseases. We discuss the consumer's benefit-risk trade-offs. Although functional food products provide health benefits beyond basic nutrition, some consumers may reject them because they utilize new technology. Face-to-face consumer surveys with contingent valuation questions were conducted in Seattle and Spokane, Washington in 2006. Consumers who choose where to shop based on organic availability are less likely to pay a premium for

*This article originally appeared in **Canadian Journal of Agricultural Economics (2009), 57(3), 325–341**. This is published by permission of **John Wiley & Sons, Inc.**

apples enriched with antioxidants. Also, there is evidence that consumers in Spokane are more likely to pay a premium for the product than consumers in supermarkets in Seattle. Information regarding the potential health benefits of antioxidants has a positive significant effect on consumers' willingness to pay (WTP). The estimated mean WTP suggests that there is a small premium associated with this product in the mind of an average consumer.

L'intérêt pour les aliments fonctionnels ne cesse de croître étant donné que les consommateurs s'intéressent de plus en plus au régime alimentaire et à la nutrition. Le présent article évalue la réaction des consommateurs à la vente de pommes enrobées d'antioxydants. On croit que les pommes enrichies d'antioxydants procureraient des bienfaits supplémentaires pour la santé en diminuant le risque de cancer et de cardiopathies. Nous avons examiné les avantages et les risques pour le consommateur. Bien que les aliments fonctionnels apportent des bienfaits pour la santé en plus d'une nutrition de base, certains consommateurs peuvent les refuser parce qu'ils ont nécessité l'utilisation de nouvelles technologies. En 2006, nous avons effectué, à Seattle et à Spokane dans l'État de Washington, des sondages en personne à l'aide de l'approche des préférences exprimées. Les consommateurs qui choisissaient un magasin d'alimentation en fonction de la disponibilité de produits biologiques étaient moins enclins à payer une prime pour obtenir des pommes enrichies d'antioxydants. Les consommateurs de Spokane ont semblé plus enclins à payer une prime pour ce produit que les consommateurs qui fréquentaient les supermarchés de Seattle. L'information sur les bienfaits potentiels des antioxydants sur la santé a des répercussions positives considèrables sur la volonté de payer des consommateurs. La volonté de payer moyenne estimative laisse supposer que, dans l'esprit du consommateur moyen, ce produit procure un certain avantage.

1. Introduction

Technological advances in the food industry have resulted in a large variety of functional foods being sold in the marketplace. Interest in functional foods has been growing due to higher health-care costs; new legislation expanding the category of dietary supplements; and recent scientific discoveries linking dietary habits with the development of many diseases, including coronary heart disease and some cancers (Milner 2000). Functional food is broadly defined as any food or food components that provide health benefit beyond basic nutrition (Institute of Food Technologists 2005). Products, such as high-fiber breakfast cereals, orange juice with added calcium, vitamin-fortified milk, and vitamin-enriched diet coke and water, are now widely available in grocery stores. These foods promise to contribute to health maintenance through prevention of many illnesses.[1]

The functional food category has been growing rapidly over the last decade. U.S. sales of functional foods grew from $11.3 billion in 1995 to $18.5 billion in 2001. This accounts for 3.7% of the total food sales. The sales are projected to reach $49 billion by 2010 (U.S. General Accounting Office 2000). Despite the rapid growth, functional foods are not specifically defined under American law. The U.S. Food and Drug Administration (FDA) regulates functional foods under the same framework as conventional foods. The FDA has neither a definition nor a specific regulatory framework for foods marketed as functional foods. Apart from the FDA, the Federal Trade Commission regulates health claims made in advertising of food products. The regulation of functional foods provides policy challenges arising from information asymmetry. Issues considered in developing policies include factors influencing consumer preferences for these products and uncertainties in the markets for them (Veeman 2002). Hobbs (2002) points out regulatory uncertainty and credibility as important challenges that nutraceuticals and functional food industry faced in using labels as quality signals.

Although one might expect consumers to embrace the positive health benefits, some consumers may reject functional food products because they utilize new technology, which they feel is risky. Other consumers prefer to consume "natural" foods whenever possible. Although most consumers have positive feelings toward the health benefits associated with functional foods, many may still not be willing to purchase them, even with a significant discount. These consumers describe functional foods as "unnatural" and "potentially unsafe." This article examines the consumer response to a functional

[1]The impact of functional foods may be quite positive. As an example, Malla *et al.* (2007) estimate the value of potential health benefits of a functional food product, trans fat-free canola oil. They conclude that there is a significant impact of transfatty acids on coronary heart disease costs. Under different scenarios, they estimate the potential reduction in coronary heart disease cost to be between roughly CAD$54 million to CAD$440 million.

food product with consumer survey data collected for this purpose. One of the questions asked was, "How would you feel about apples with wax coatings enriched with antioxidants?" The responses ranged from, "Stop playing with natural food!" to "That sounds like a great idea!"

The functional food product that is the subject of this investigation is a protective coating that is applied to apples (and potentially other fruits). The coating is enhanced with specific flavonoids and stilbenes (antioxidants), which are believed to enhance the fruit's health benefits. In recent years, there has been a great increase in research and publicity devoted to the beneficial effects of antioxidants. Antioxidants are substances that may protect cells from the damage caused by unstable molecules known as free radicals. Free radical damage is believed to lead to cancer. Antioxidants interact with and stabilize free radicals and may prevent some of the damage free radicals otherwise might cause. Examples of antioxidants include beta-carotene, lycopene, vitamins C, E, and A, and other substances (National Cancer Institute 2004). These compounds, often called phytonutrients, are present in most fruits and vegetables naturally. The consumption of pills providing these compounds has increased in recent years (McKay 2007). Indeed, this trend suggests that consumers are much more aware of their health benefits. There is also information indicating that maximum benefits are achieved when these phytonutrients are consumed in natural products (e.g., fruits, wine) rather than in pill form. Significant evidence has been found that fruits and vegetables in combination have synergistic effects on antioxidant activities leading to greater reduction in risk of chronic diseases, specifically for cancer and heart diseases (International Food Information Council 2006).

The objective of this article is to measure consumers' responses to apples with enriched coatings and analyze the factors that affect consumers' choices. Specifically, we estimate confidence intervals for the possible premium consumers will pay for this product and analyze the exogenous variables that affect willingness to pay (WTP). In addition, we test whether the provision of information describing potential benefits of functional foods has a significant effect on consumers' WTP.

2. Previous Studies on Consumer Acceptance

There is a large literature on consumer acceptance of new technology. Several studies (e.g., Tegene *et al.* 2003; Li *et al.* 2004; Lusk *et al.* 2004; Hu *et al.* 2005) conclude that there is a significant positive shift in consumers' preferences toward new technology food products after being exposed to positive information. Although functional foods are relatively new to the market, there is an emerging literature on how the market is responding. Labrecque *et al.* (2006) find only minor differences among French, American, and French Canadian students' attitudes toward functional foods. Their results suggest that health-related and product-related benefits, credibility of information, and high knowledge have a positive effect on attitudes. Based on interviews with American mothers, Chema *et al.* (2006) find that functional attributes, such as higher protein, increased calcium, and lower cholesterol, are highly valued.

West *et al.* (2002) assess consumers' valuations of functional foods in Canada. They find that Canadian consumers have generally positive attitudes toward functional foods and may be willing to pay a premium for them. However, they indicate that a large proportion of respondents negatively perceive genetically modified (GM) relative to conventional foods, after controlling for price and health properties. Maynard and Franklin (2003) employ a survey with a sensory evaluation to assess the commercial potential of "cancer-fighting" dairy products. Households with children and health-conscious consumers were most willing to pay premiums for "cancer-fighting" dairy products.

Peng *et al.* (2006) examine the acceptance of conjugate linolic acid (CLA)-enriched dairy products among Canadian consumers and factors influencing the acceptance. The results suggest that gender and education do not significantly affect consumers' acceptances of CLA-enriched dairy products. Middle-aged consumers and consumers with teenagers in the household are more likely to buy this product. The authors argue that the target consumer segment for this product should be health-conscious, middle-aged consumers who believe in healthiness of conventional milk products.

This article measures consumers' responses to apples enriched with antioxidant coatings, a functional food product, while simultaneously considering the effect of information. To our knowledge, this is the first examination of preferences toward fresh produce marketed as functional food.

3. Data

The data utilized in this analysis were collected in face-to-face interviews that were conducted in September and October of 2006 in grocery stores in Seattle and Spokane, Washington. All of the stores offer a variety of fresh produce including both conventional and organic items. A total of 730 surveys were completed and used in the analysis. The questionnaire[2] included questions about awareness of antioxidants, attitudes toward nutritionally enriched foods and apples enriched with antioxidants, factors influencing apple purchases, and the choice of where to shop for food. Dichotomous choice contingent valuation (CV) questions with follow-up were included to elicit consumers' WTP for apples with enriched coatings. The questionnaire also included questions about the respondents' demographic characteristics, such as age, income, education, and the presence of children in the household.

The sample demographics and the U.S. Census Bureau 2000 data for Seattle and Spokane are presented in Table 1. We compare our sample to the Census data in order to assess how well it represents the population in both cities. The average age is 45 years, which is higher than the population average but expected since children were not approached. Sixty-one percent of the respondents were female. The average size of household was 2.75 members, and 35% of the respondents had children under 18 in the household. In terms of

[2]The survey instrument is available from the authors upon request.

Table 1. Summary statistics for demographic variables.

		Seattle		Spokane	
	Total Sample	Sample	Census 2000	Sample	Census 2000
Number of respondents	730	550	N/A	180	N/A
Median age (years)	45	44	35	46	35
Mean number of persons per household	2.75	2.55	2.1	3.35	2.0
Male	39%	32.27%	49.1%	40.56%	48.2%
Female	61%	62.73%	50.9%	59.44%	51.8%
Children under 18 present in the household	35%	33.10%	19.6%	41.67%	31.7%
Education (highest level)					
Some school	2%	1.64%	10.5%	4.44%	11.9%
High school diploma	12%	6.55%	15.3%	26.67%	26.3%
Some college	28%	24.55%	27.0%	34.44%	36.4%
Bachelor's degree	31%	34.36%	29.9%	21.11%	16.2%
Advanced degree or graduate degree	27%	31.82%	17.0%	11.11%	9.2%
Household income (in 2005)					
Less than $39,999	25%	19.27%	38.0%	40.00%	53.5%
$40,000–$79,999	37%	35.45%	34.8%	38.33%	33.3%
$80,000–$109,000	20%	20.36%	11.4%	13.89%	6.6%
$110,000–$149,000	11%	12.55%	9.4%	3.89%	4.3%
$150,000–$199,999	6%	6.73%	2.9%	1.67%	0.9%
Greater than $200,000	2%	2.36%	3.5%	0.00%	1.2%
Employment status					
Formally employed	62%	61.45%	66.3%	59.44%	58%
Self-employed	16%	16.36%	7%	13.33%	6.2%
Unemployed	3%	2.55%	3.6%	4.44%	5.7%
Retired	11%	9.82%	N/A	12.22%	N/A
Student	5%	4.36%	N/A	3.33%	N/A
Housewife	4%	3.45%	0.2%	5.56%	0.2%

Source: U.S. Census Bureau (2000).

education, 31% of the respondents had a bachelor's degree, 27% had advanced or graduate degrees, 28% attended some college, 12% had a high school diploma, and only 2% said they have had some school. The most common response to annual household income fell within $40,000 and $79,999. Also, a majority of the respondents were employed at the time of the survey. Sixty-two percent and 16% said they were formally employed and self-employed, respectively. Eleven percent reported to be retired, 5% were students, 4% were housewives, and 3% disclosed that they were unemployed.

Shopping habit and attitudinal variables are presented in Table 2. In terms of shopping behavior, 85% of the respondents were the primary shoppers in their household. The majority, 58%, of the respondents shop two to five times a week. Twenty-eight percent shops once a week, 8% shop daily, and 5% and 1% shop once every two weeks and once a month, respectively. The respondents were also asked about the most important factor in their choice of where to shop for food. Quality was the most common response with 65%. Price, variety, and location appeared to be roughly equally important to the consumers with 17%, 16%, and 12%, respectively. Four percent of the respondents listed "other" factors, out of which all but one said availability of organic food was the most important factor influencing their choice of grocery stores.

The survey responses suggest a general awareness of antioxidants and their health benefit. Twenty percent reported having no knowledge of antioxidants, 65% and 15% claimed to be "somewhat knowledgeable" and "very knowledgeable" of antioxidants, respectively. Twenty-seven percent of the respondents reported that they never take vitamin supplements, 47% take vitamins daily, and 27% take vitamins two to three times a week.

In response to a question about attitudes toward nutritionally enriched food, 63% of the respondents said they felt "very positive" or "somewhat positive." Fewer participants responded negatively to this question. Only 10% of the consumers surveyed said they felt "somewhat negative" or "very negative" toward nutritionally enriched food. If a respondent provided a "somewhat negative" or "very negative" response, then they were asked to provide a rationale. The most common explanations included, "It is unnatural," "It is better to get necessary nutrients naturally," "Additives are not good for one's health," and "Organic food is better."

Subsequently, the respondents were asked to rate their' feelings about apples that are "naturally wax coated with antioxidants."

Table 2. Response summary to selected questions.

Variable	Percentage of Respondents
How do you feel about...	
Nutritionally enriched food	
Very positive	25%
Somewhere positive	38%
Neutral	25%
Somewhat negative	8%
Very negative	2%
Don't know	2%
Apples with wax coatings, which are enriched with antioxidants	
Very positive	15%
Somewhat positive	27%
Neutral	28%
Somewhat negative	19%
Very negative	6%
Don't know	5%
Primary shopper in the household	85%
How often do you take vitamin or nutrient supplements?	
Never	27%
2-3 times a week	27%
Daily	47%
Most important factor in choosing where to shop for food	
Price	16%
Quality	65%
Variety	12%
Location	17%
Organic	4%
Most important factor in choosing apples	
Appearance	29%
Variety	32%
Nutrients	12%
Price	11%
Size	2%
Other:Taste	13%
Other:Organic	6%
Importance of higher nutrient content in food compared to buying food at the lowest price (1 = higher nutrient food most important... 10 = lower price most important)	5.4

There were fewer positive responses to this question. Forty-two percent of the overall customers surveyed said they had "very positive" or "somewhat positive" feelings, while 25% felt "somewhat negative," or "very negative" about apples enriched with antioxidants. Finally, 28% were neutral, and 5% said they "didn't know."

Again, customers who felt negatively about apples enriched with antioxidants were asked to explain why. The most common responses were, "I don't want to eat wax," "It is unnatural," "Additives to fruit are not necessary," "Washing apples removes the wax," "I prefer food with no additives," "I don't have enough information," "It is better to get nutrients naturally," "I prefer organic," and "It changes the taste."

There were two versions of the survey, with half of the respondents receiving each one. One version contained the following statement about potential health benefits of antioxidants, "Fruit enhanced with natural antioxidants will improve its health benefits by helping to prevent cancer, cardiovascular and other diseases." The other version of the survey contained no additional information. Including this positive information in half the surveys allows us to test for how positive information on consumers' affects attitudes toward naturally enriched apples and WTP.

4. Theory and Methods

CV method is a technique that is used to estimate WTP.[3] CV is based on a dichotomous choice model in which individual WTP can be estimated based on responses to market-type questions (Kanninen 1993). The survey for this study contained CV questions regarding the respondents' WTP a premium or accept a discount in order to buy enriched apples. We use the double-bounded model (Hanemann *et al.* 1991) to analyze the outcomes of the survey.[4] The respondents

[3]For further information, including recent reviews and comparison across models estimable from CV data with reiteration see, for example, Flachaire and Hollard (2006).
[4]There is a literature on the appropriate number of iterations to include in the bidding procedures used in the CV method. Cameron and Quiggin (1994) evidenced the problem of anchoring/starting point bias with iterations of bids. There is some bias with the double-bounded model, primarily due to inconsistencies which may be present between

Table 3. Distribution of bid responses.

	Premium				
	5%	10%	20%	30%	Total
Yes	66	76	60	56	258
No	35	43	66	69	213
Total	101	119	126	125	471
	Discount				
	5%	10%	20%	30%	Total
Yes	10	17	21	28	76
No	38	42	45	31	156
Total	48	59	66	59	232

were first asked if they were willing to pay an initial bid price for naturally enriched apples. If the answer to this question was "yes," a follow-up question was asked in which the respondents were asked if they were willing to pay the naturally enriched apples at a higher price. If the respondent's answer to the first question was "no," a follow-up question was asked in which the respondent was offered a discount on the naturally enriched apples.[5] The randomly assigned premium or discount levels had the following values: ± 5%, ± 10%, ± 20%, and ± 30%. Each level of premium or discount was used for one-fourth of the surveys. The distribution of responses to the premium and discount bids for apples enriched with antioxidants is presented in Table 3.

---

the consumers' first and subsequent bids (Hanemann and Kanninen 1999). Recent studies suggest that the loss from bias may outweigh the gain in efficiency relative to a single dichotomous choice question (Scarpa and Bateman 2000; Bateman *et al.* 2001; Cooper *et al.* 2002; Whitehead 2002, 2004; Aadland and Caplan 2004).

[5]Four outcomes are possible in the double-bounded dichotomous choice model. They are: (1) the respondent is neither willing to buy the enriched apples at *BI*, the initial bid price, nor is he or she willing to buy it at *BL*, the discounted price (i.e., "no" followed by "no"); (2) the respondent is not willing to buy the enriched apples at the initial bid price, but given the random discount, the respondent exhibits a willingness to buy it at that price (i.e., "no" followed by "yes"); (3) the respondent is willing to buy the enriched apples at the initial bid price but not at *BH*, the higher price (i.e., "yes" followed by "no"); (4) the consumer is willing to buy the enriched apples at the initial price and he or she is also willing to pay the random premium (i.e., "yes" followed by "yes").

Given the questioning sequence, one is able to isolate the range in which the respondents' true WTP lies, placing it into one of the following four intervals: $(-\infty, \text{BL})$, $[\text{BL}, \text{BI})$, $[\text{BI}, \text{BH})$, or $[\text{BH}, +\infty)$. The first and second bid combine to put an upper and a lower bound on the respondent's unobservable true WTP. When both the first and second decision are in the same direction (i.e., "yes, yes" or "no, no"), it raises the lower bound or lowers the upper bound, respectively. The WTP function is represented in general terms by[6]

$$\text{WTP} = \alpha + \rho B + Z'X + \varepsilon \tag{1}$$

where B is the ultimate bid amount each respondent faces, $X$is a column vector of observable characteristics of the individual respondent, ε is a random variable accounting for unobservable characteristics, and α, ρ, and Z are unknown parameters to be estimated. In the empirical implementation of the model, we define $G(\cdot)$ to be the standard logistic distribution with mean zero and standard deviation $\sigma = \pi/\sqrt{3}$. Maximum likelihood can be used to estimate the unknown parameters in the likelihood function. The empirical representation of Equation (1) for the enriched apples is defined as

$$\begin{aligned} \text{WTP} \equiv \alpha + \rho B + \sum_{i=1}^{2} z_i \times \text{Treatments}_i + \sum_{i=3}^{6} z_i \times \text{Demographics}_i \\ + \sum_{i=7}^{8} z_i \times \text{Vit. consumption}_i \\ + \sum_{i=9}^{10} z_i \times \text{Shopping preferences}_i + \varepsilon \end{aligned} \tag{2}$$

The information variable is included in the treatment group. The basis for the WTP analysis is the well-known Random Utility Model, in which respondents choose to purchase the good in question if

[6]We do not attempt to account for unobserved consumer heterogeneity. However, this is not a concern to the validity of the model and results. There is an extensive empirical literature that estimates preferences and WTP while not accounting for unobserved consumer heterogeneity (e.g., Hobbs *et al.* 2005; Loureiro *et al.* 2006; Peng *et al.* 2006; Labrecque *et al.* 2006).

Table 4. Description of explanatory variables.

Variable	Description
Treatments	
Seattle	1 = Seattle, 0 = Spokane
Info	1 = Presence of positive statement, 0 = no information
Demographics	
Gender	1 = Male, 0 = Female
Education	1 = Bachelor's degree or above, 0 = otherwise
Income	Midpoints of the income categories
Age	Reported age
Vitamin consumption	
Medium	1 = 2–3 times a week, 0 = otherwise
High	1 = Daily, 0 = otherwise
Shopping preferences	
Nutrition–Price	Trade-off between higher nutrition and lower price food, continuous scale of 1 = highest nutrition to 10 = highest value
Organic	1 = Organic most important factor in where to shop

the utility derived from the good is higher than the alternative. The information acts as a factor that affects consumers' utility. We hypothesize that the positive information makes the product attribute (i.e., additional antioxidants) more attractive to consumers, which increases the utility and subsequently increases the likelihood of switching to the functional food. As both opportunity cost and preferences for information vary across consumers, their use of information will vary as well (McCluskey and Swinnen 2004).

Table 4 presents a brief description and explanation of the coding for the explanatory variables used in the model. The B variable in Equation (2) represents the random bid offered to each consumer. Equation (2) was estimated via maximum likelihood using the MAXLIK module in GAUSS 8.0.

5. Estimation Results

The parameter estimation results are reported in Table 5, and the marginal effects of the estimated model with confidence intervals

Table 5. Maximum likelihood parameter estimates.

Variable	Coefficient	Standard Error	Z-statistic
Constant	−10.057***	0.536	−18.756
Bid	11.530***	0.700	16.461
Treatments			
Seattle	−0.385**	0.190	−2.027
Info	0.336**	0.154	2.179
Demographics			
Gender	0.134	0.161	0.838
Education	−0.101	0.169	−0.599
Income	0.002	0.002	0.715
Age	−0.007	0.006	−1.165
Vitamin consumption			
Medium	−0.055	0.185	−0.298
High	0.176	0.195	0.899
Shopping preferences			
Nutrition–Price	−0.053*	0.028	−1.876
Organic	−1.494***	0.344	−4.343
Observations	640		
LR-stat	40.41		
R^2 equivalent	0.736		

Note: *10% significance level, **5% significance level, ***1% significance level.

are reported in Table 6. To assess the significance of the model and the goodness of fit, we performed a likelihood ratio (LR) test to estimate an R^2 equivalent measure designed specifically for double-bounded logit models. From the LR test, we strongly reject the null hypothesis that the parameters in the model are jointly equal to zero and conclude that our model has a significant explanatory power. Further, we employ the sequential classification procedure to estimate a model fit measure. Kanninen and Khawaja (1995) show that the conventional R^2 measures, such as the McFadden pseudo R^2 and Pearson chi-square are not appropriate for double-bounded logit models.

As expected, the bid coefficient is significantly[7] negative, which means that as the hypothetical price increases, the probability that

[7]For purposes for this article, statistical significance is defined to be at the 5% level or better, unless otherwise stated.

Table 6. Marginal effects of the explanatory variables on mean WTP.

Variable	Marginal Effect	Standard Error	Z-statistic	90% Confidence Interval Lower Bound	90% Confidence Interval Upper Bound
Constant	—	—	—	—	—
Bid	—	—	—	—	—
Treatments					
Seattle	−0.038**	0.019	−2.023	−0.069	−0.007
Info	0.033**	0.015	2.186	0.008	0.059
Demographics					
Gender	0.013	0.016	0.837	−0.013	0.040
Education	−0.010	0.017	−0.599	−0.038	0.018
Income	0.000	0.0002	0.715	−0.000	0.001
Age	−0.001	0.0006	−1.165	−0.002	0.0003
Vitamin consumption					
Medium	−0.006	0.018	−0.298	−0.036	0.025
High	0.017	0.019	0.900	−0.014	0.049
Shopping preferences					
Nutrition–Price	−0.005*	0.003	−1.876	−0.010	−0.001
Organic	−0.149***	0.034	−4.334	−0.205	−0.092

Note: *10% significance level, **5% significance level, ***1% significance level.

the respondent chooses to purchase the product goes down. The negative and significant coefficient for the SEATTLE variable suggests that respondents from Seattle are less likely to pay a premium for apples enriched with antioxidants than those from Spokane. We offer the following explanation for this result. Spokane is located in the eastern part of the Washington, which is surrounded by large acreage of wheat crop land. Previous studies have found that consumers living in close proximity to conventional agriculture are more accepting of "innovative" value added agricultural production (McCluskey *et al.* 2003), and this effect may contribute to their WTP for antioxidant apples. This is related to Slovic's (1987) theories about the factors that contribute to risk perceptions: dread risk and unknown risk. Unknown risk is the primary factor associated with consumption of functional foods because many consumers are unfamiliar with these products. Since residents of agricultural areas are more familiar with agricultural processes and agricultural technologies, we would expect their levels of unknown risk to be lower.

The positive and significant coefficient on the health benefit information variable suggests that the statement regarding potential health benefits of antioxidant apples has a positive and significant effect on consumers' WTP for this product. On one hand, this suggests that most consumers need information in order to choose functional foods. On the other hand, this also suggests that consumers are easily influenced by information. It should be noted that we have provided only one strong health claim without referencing any source for reliability. The fact that consumers responded positively to this signal suggests significant vulnerability of consumers to potentially misleading (or false) health claims. In addition, empirical literature finds evidence that both positive and negative information affects preferences for GM food (e.g., Hu *et al.* 2005; Anand *et al.* 2007). It seems reasonable that negative information is likely to significantly reduce consumers' WTP for functional foods. However, having not tested this proposition explicitly, we do not make any conclusions in this regard.

None of the demographic variables are statistically significant in explaining consumers' WTP for apples enriched for antioxidants. Vitamin consumption is not significant either. Consumers' shopping preferences are significant. The nutrition-versus-price tradeoff variable is only significant at the 10% level and has a negative coefficient. This weakly suggests that consumers who are more price conscious relative to nutrition conscious are less likely to pay a premium for apples enriched with antioxidants. The ORGANIC variable has a negative and significant effect, which suggests that respondents for whom the availability of organic produce is important in choosing where to shop are willing to pay less for antioxidant-enriched apples.

5.1. *Mean WTP*

The mean WTP is calculated, following Hanemann (1984), as $WTP^{+} = \frac{1}{\hat{\rho}}(\hat{\alpha} + \hat{Z}'\bar{X})$. A 95% confidence interval can be calculated

Table 7. Estimates of mean WTP

Sample	Mean WTP	95% Confidence Interval
Full Sample	\$1.08	\$1.06–\$1.094
	(8% premium)	(6.4–9.4%premium)
Seattle	\$1.07	\$1.051–\$1.089
	(7% premium)	(5.1–8.9% premium)
Spokane	\$1.10	\$1.077–\$1.126
	(10% premium)	(7.7–12.6% premium)
No information	\$1.06	\$1.041–\$1.079
	(6% premium)	(4.1–7.9% premium)
Positive information	\$1.10	\$1.074–\$1.121
	(10% premium)	(7.4–12.1% premium)

for mean WTP using the delta method.[8] Mean WTP results are presented in Table 7. We use the initial bid of \$0.99 as a benchmark for estimating the mean WTP. The estimated mean premium for apples enriched with antioxidants suggests that average respondent is willing to pay a small premium for the enriched apples. This premium is estimated to fall between 6.4% and 9.4% in a 95% confidence interval.[9]

The estimated mean WTP for the consumers in Seattle is a 7% premium with a 95% confidence interval of (5.1%, 8.9%). The mean WTP in the Spokane grocery store was estimated to be approximately a 10% premium with a 95% confidence interval of

[8]A 95% confidence interval around estimated mean WTP for each case was calculated using the delta method, which approximates the asymptotic variance of a function of random variables as:

$$\text{asymptotic var } [f(b)] \cong \frac{\partial f(b)}{\partial b'}[\text{Var}(b)]\left[\frac{\partial f(b)}{\partial b'}\right]',$$

$$\text{where } f(b) = \frac{\alpha + Z'X}{\rho} \text{(Green 2003).}$$

[9]Previous literature suggests that the estimates of WTP for hypothetical private goods may be biased and inflated (e.g., List 2003), so we emphasize that these estimates suggest that the average respondent is willing to pay a small premium. Further, the upward bias may exist in the magnitude of the estimates, but it should not affect the ordinal rankings. Consequently, the WTP estimates are useful for comparing across categories, such as those respondents who received positive information with those who did not.

(7.7%, 12.6%). Since the confidence intervals overlap, we test this difference for statistical significance.[10] The null and alternative hypotheses for a lower-tailed test are

$$\begin{aligned} H_0 &: \mathrm{WTP}^{+}_{\text{Seattle}} = \mathrm{WTP}^{+}_{\text{Spokane}} \\ H_a &: \mathrm{WTP}^{+}_{\text{Seattle}} < \mathrm{WTP}^{+}_{\text{Spokane}} \end{aligned} \tag{3}$$

We reject the null hypothesis that the mean WTP for Seattle subsample is equal to the mean WTP for Spokane subsample. Our results provide support for the alternative hypothesis that Seattle residents are willing to pay less.

In addition, the mean WTP is estimated separately for the respondents who received the positive information statements and those who received no information. On average, respondents who received no information were willing to pay a 6% premium for the antioxidant-enriched apples with a 95% confidence interval of (4.1%, 7.9%). In contrast, respondents who received information were willing to pay a 10% premium with a 95% confidence interval of (7.7%, 12.6%). Again, we test the difference between the two estimated means for statistical significance. The null and alternative hypotheses for an upper-tailed test are

$$\begin{aligned} H_0 &: \mathrm{WTP}^{+}_{\text{Positive Information}} = \mathrm{WTP}^{+}_{\text{No Information}} \\ H_a &: \mathrm{WTP}^{+}_{\text{Positive Information}} > \mathrm{WTP}^{+}_{\text{No Information}} \end{aligned} \tag{4}$$

The null hypothesis can be rejected at 1% significance level, providing support for the conclusion that the mean WTP with positive information is greater than mean WTP with no information.

[10]We are testing for the difference in means of two separate samples. The fact that the confidence intervals of the means overlap raises suspicion but does not mean that the difference between the means is not significant (i.e., null hypothesis should not be rejected). A more common rule of thumb when making an inference regarding the difference in means using their confidence intervals is to check whether the confidence interval of one mean contains the other mean estimate. Neither of our subsample confidence intervals contains the mean of the other subsample. The most reliable way is to test directly for statistical significance using t-tests, which is the approach utilized in this article.

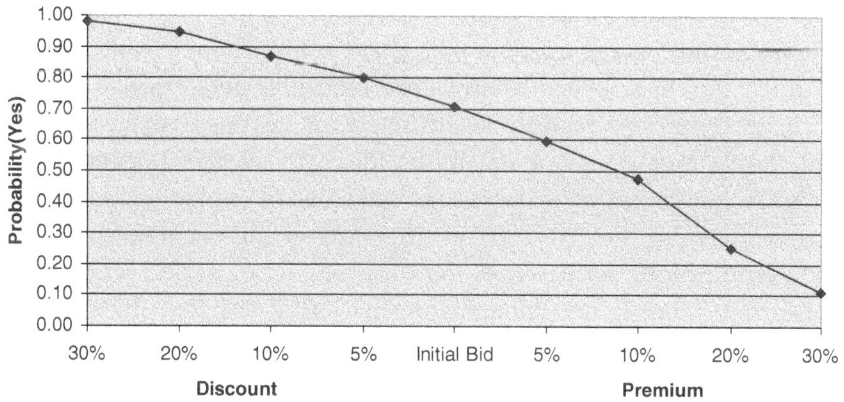

Fig. 1. Probability of choosing functional food as bid varies.

We estimate that the respondents from Seattle are, on average, willing to pay approximately 3.8% more than the respondents from Spokane. The respondents who received the positive information are willing to pay roughly 3.3% more than consumers who received no such information. Consumers who seek organic produce in choosing where to shop are, on average, willing to pay 14.8% less for antioxidant apples. Figure 1 shows the change in probability of saying "yes" to apples enriched with antioxidants given different bids. The probability drops from 0.98 for 30% discount to 0.71 for the initial offer down to 0.11 for 30% premium.

6. Conclusions

The purpose of the article was to examine consumers' attitudes toward functional foods, in general, and for apples enriched with antioxidant specifically. Face-to-face surveys were conducted in the State of Washington with a total of 730 responses. Although consumers generally have positive attitudes toward the health benefits associated with functional foods, some will reject functional food products because they utilize new technology, which they feel is risky. Other consumers prefer to consume "natural" foods whenever possible. When considering the specific functional food product, apples that are enriched with antioxidants, fewer of the respondents expressed positive feelings compared with functional foods in general. We argue that this is in some part due to the product being a fresh produce item as opposed to processed food products, such as orange juice or cereal. Attitudes, nevertheless, were positive in general, and the market does not reject the idea of this new product. The major reasons some consumers reject the idea of apples enriched with antioxidants are their perceptions that wax in general is not pleasant for consumption, additives in food are unnecessary and sometimes unhealthy, there is not enough information about the product and its safety, organic food is better and healthier, and additives in fruit specifically are unnatural.

The results of the WTP estimation are that, on average, our respondents are willing to pay a small premium for apples enriched with antioxidants. Considering factors affecting WTP, consumers who choose where to shop based on organic availability are less likely to pay a premium for apples enriched with antioxidants. Also, there is evidence that consumers in Spokane are more likely to pay a premium for the product than consumers in supermarkets in Seattle. The estimated confidence interval for the mean WTP in the Spokane area was approximately between 8% and 13%, while the estimated confidence interval for the mean WTP in the Seattle was approximately between 5% and 9%. We argue that the main reason for this may be that Seattle is a large growing market for organic produce. In contrast, Spokane is more closely aligned with traditional agriculture with large surrounding wheat acreage.

The statement providing positive information about antioxidants included in half the questionnaires had a significant and positive effect on consumers' WTP. The average premium with the positive information is estimated to be between about 4% more than without the statement. Therefore, there is evidence that similar statements on labels of functional foods may have positive effects on consumers' propensities for buying them.

This study contributes to the overall understanding of functional food markets with a specific emphasis on fresh-produce functional foods. Labrecque *et al.* (2006) suggest little differences among Canadian and American consumers' attitudes toward functional foods. Therefore, Canadian agribusiness firms and policy makers can learn important lessons from this study as well. The study guides agribusiness firms and retailers trying to enter fresh produce functional food markets in developing suitable marketing strategies. It appears that these products are most likely to be successful in areas with strong agricultural land and stores not catering to organic consumers.

Perhaps the most important lesson is that information is the key to success. Dissemination of positive information educating consumers about the benefits and safety of these products is vital for the growth of this industry. On the other hand, this study provides evidence of significant consumer sensitivity toward health claims, thus showing the importance of carefully developing and maintaining regulatory framework around such claims in order to protect consumers both in Canada and the United States. Finally, based on the response to the health benefit information, one can extrapolate that as more consumers recognize the potential health benefits, the market for functional foods is likely to grow.

Acknowledgments

The authors wish to thank without implicating Tom Marsh and two anonymous reviewers for helpful comments and Carolina Torres for scientific consultation. The authors gratefully acknowledge financial support from PACE International and the IMPACT Center at Washington State University.

References

Aadland, D. and A. Caplan (2004). Incentive Incompatibility and Starting-Point Bias in Iterative Valuation Questions: Comment. *Land Economics* 80, 312–315.

Anand, A., R.C. Mittelhammer and J.J. McCluskey (2007). Consumer Response to Information and Second-Generation Modified Food in India. *Journal of Agricultural and Food Industrial Organization* 5[Article 8], 145–172.

Bateman, I.J., I.H. Landford, A.P. Jones and G.N. Kerr (2001). Bound and Path Effects in Double Bounded and Triple Bounded Dichotomous Choice Contingent Valuation. *Resource and Energy Economics* 23, 191–213.

Cameron, T. and J. Quiggin (1994). Estimation Using Contingent Valuation Data from a 'Dichotomous Choice with Follow Up' Questionnaire. *Journal of Environmental Economics and Management* 27, 218–234.

Chema, S.K., L.A. Marks, J.L. Parcell and M. Bredahl (2006). Marketing Biotech Soybeans with Functional Health Attributes. *Canadian Journal of Agricultural Economics* 54, 685–703.

Cooper, J., W.M. Hanemann and G. Signorello (2002). One and One-Half Bound Dichotomous Choice Continent Valuation. *The Review of Economics and Statistics* 84, 742–750.

Flachaire, E. and G. Hollard (2006). Controlling Starting-Point Bias in Double-Bounded Contingent Valuation Surveys. *Land Economics* 82, 103–111.

Green, W.H. (2003). *Econometric Analysis*, 5th Edn. Upple Saddle River, NJ: Pearson Education.

Hanemann, M.W. (1984). Welfare Evaluations in Contingent Valuation Experiments with Discrete Responses. *American Journal of Agricultural Economics* 66(3), 332–341.

Hanemann, W.M. and B. Kanninen (1999). "The Statistical Analysis of Discrete-Response CV Data." In I.J. Bateman and K.G. Willis (Eds.), *Valuing Environmental Preferences: Theory and Practice of the Contingent Valuation Method in the US, EU, and Developing Countries*, pp. 302–442. Oxford: Oxford University Press.

Hanemann, M.W., J. Loomis and B.J. Kanninen (1991). Statistical Efficiency of Double-Bounded Dichotomous Choice Contingent Valuation. *American Journal of Agricultural Economics* 73, 1255–1263.

Hobbs, J.E. (2002). Evolving Supply Chains in the Nutraceuticals and Functional Foods Industry. *Canadian Journal of Agricultural Economics* 50, 559–568.

Hobbs, J.E., D. Von Bailey, D.L. Dickinson and M. Haghiri (2005). Traceability in the Canadian Red Meat Sector: Do Consumers Care? *Canadian Journal of Agricultural Economics* 53(1), 47–65.

Hu, W., F. Zhong and Y. Ding (2005). Actual Media Reports on GM Foods and Chinese Consumers' Willingness to Pay for GM Soybean Oil. *Journal of Agricultural and Resource Economics* 31, 376–390.

Institute of Food Technologists (2005). Functional Foods: Opportunities and Challenges. http://members.ift.org/IFT/Research/IFTExpertReports/functionalfoods_report.htm (accessed November 21, 2007).

International Food Information Council (2006). Functional foods fact sheet: Antioxidants. http://www.ific.org/publications/factsheets/antioxidantfs.cfm (accessed November 21, 2007).

Kanninen, B.J. (1993). Optimal Experimental Design for Double-Bounded Dichotomous Choice Contingent Valuation. *Land Economics* 69, 138–146.

Kanninen, B.J. and M.S. Khawaja (1995). Measuring Goodness of Fit for the Double-Bounded Logit Model. *American Journal of Agricultural Economics* 77, 885–890.

Labrecque, J., M. Doyon, F. Bellavance and J. Kolodinsky (2006). Acceptance of Functional Foods: A Comparison of French, American, and French Canadian consumers. *Canadian Journal of Agricultural Economics* 54, 647–661.

Li, Q., J.J. McCluskey and T.I. Wahl (2004). Effects of Information on Consumers' Willingness to Pay for GM-Corn-Fed Beef. *Journal of Agricultural and Food Industrial Organization* 2, Article 2. http://www.bepress.com/jafio/vol2/iss2/art9 (accessed June 18, 2009).

List, J.A. (2003). Does Market Experience Eliminate Market Anomalies? *Quarterly Journal of Economics* 118(1), 41–71.

Loureiro, M.L., A. Gracia and R.M. Nayga Jr. (2006). Do Consumers Value Nutrition Labels? *European Review of Agricultural Economics* 33, 249–268.

Lusk, J.L., L.O. House, C. Valli, S.R. Jaeger, M. Moore, J.L. Morrow and W.B. Traill (2004). Effect of Information About Benefits of Biotechnology on Consumer Acceptance of Genetically Modified Food: Evidence from Experimental Auctions in the United States, England, and France. *European Review of Agricultural Economics* 31, 179–204.

Malla, S., J.E. Hobbs and O. Perger (2007). Valuing the Health Benefits of a Novel Functional Food. *Canadian Journal of Agricultural Economics* 55, 115–136.

Maynard, L.J. and S.T. Franklin (2003). Functional Foods as a Value-Added Strategy: The Commercial Potential of "Cancer-Fighting" Dairy Products. *Review of Agricultural Economics* 25, 316–331.

McCluskey, J.J. and J.F.M. Swinnen (2004). Political Economy of the Media and Consumer Perceptions of Biotechnology. *American Journal of Agricultural Economics* 86, 1230–1237.

McCluskey, J.J., K.M. Grimsrud, H. Ouchi and T.I. Wahl (2003). Consumer Response to Genetically Modified Food Products in Japan. *Agricultural and Resource Economics Review* 32, 222–231.

McKay, T. (2007). Supplement Industry Growing. *The Denver Post*, June 3. http://www.denverpost.com/portlet/article/html/imageDisplay.jsp?contentItemRelationshipId=1454237 (accessed July 18, 2008).

Milner, J.A. (2000). Functional Foods: The U.S. Perspective. *American Journal of Clinical Nutrition* 71, 1654–1659.

National Cancer Institute (2004). Antioxidants and cancer prevention: Fact sheet. http://www.cancer.gov/cancertopics/factsheet/antioxidantsprevention (accessed June 18, 2009).

Peng, Y., G.E. West and C. Wang (2006). Consumer Attitudes and Acceptance of CLA-Enriched Dairy Products. *Canadian Journal of Agricultural Economics* 54, 663–684.

Scarpa, R. and I. Bateman (2000). Efficiency Gains Afforded by Improved Bid Design Versus Follow-Up Valuation Questions in Discrete Choice CV Studies. *Land Economics* 76, 299–311.

Slovic, P. (1987). Perception of Risk. *Science* 236, 280–285.

Tegene, A., W.E. Huffman, M. Rousu and J.F. Shogren (2003). *The Effects of Information on Consumer Demand for Biotech Foods, Evidence from Experimental Auctions.* U.S. Department of Agriculture, Economic Research Service, Technical Bulletin 1903.

U.S. Census Bureau (2000). Quick tables. http://factfinder.census.gov (accessed November 21, 2007).

U.S. General Accounting Office (2000). *Dietary Supplements and Functional Foods*. U.S. GAO/RCED-00-156.

Veeman, M. (2002). Policy Development for Novel Foods: Issues and Challenges for Functional Food. *Canadian Journal of Agricultural Economics* 50, 527–539.

West, G.E., C. Gendron, B. Larue and R. Lambert (2002). Consumers' Valuation of Functional Properties of Foods: Results from a Canada-Wide Survey. *Canadian Journal of Agricultural Economics* 50, 541–558.

Whitehead, J. (2002). Incentive Incompatibility and Starting-Point Bias in Iterative Valuation Questions. *Land Economics* 78, 285–297.

Whitehead, J. (2004). Incentive Incompatibility and Starting-Point Bias in Iterative Valuation Questions: Reply. *Land Economics* 80, 316–319.

Part IV

Consumer Response to New Technology in Other Cultures

Chapter 16

Russian Consumer Willingness to Pay for Genetically Modified Food*

Anthony R. Delmond[†], Jill J. McCluskey[†,§,¶],
Mirzobobo Yormirzoev[‡], and Maria A. Rogova[‡]

[†]*School of Economic Sciences,*
Washington State University, USA,
[‡]*School of Economics and Finance,*
National Research University Higher School of Economics,
Russia
[§]*mccluskey@wsu.edu*

Addressing a gap in the literature on genetically modified (GM) food preferences, we conducted a survey in Russia, where legislation forbids GM food production and importation. Based on in-person consumer surveys and choice experiments in Perm, Russia, we estimate the willingness to pay (WTP) for bread containing GMOs. In addition, we utilized a principal-component analysis (PCA) to construct variables that indicate types of consumers. Within this approach, we estimate how respondents' money- and health-consciousness affects their WTP for bread containing GMOs. Our findings are consistent with previous studies of Russian consumer preferences, who find that Russian consumers' food purchases are often motivated by health concerns and the "naturalness" of foods. Our results indicate that information-seeking health-conscious consumers tend to be less likely to purchase GM foods, while more money-conscious consumers are more likely to purchase the GM product. Overall, even with discounts ranging between 5% and 50%, only around 20% of respondents in our survey are willing to choose bread containing GMOs rather than bread made using conventional inputs.

*This article originally appeared in the **Food Policy (2018), 78, 91–100**. This is published by permission of **Elsevier**.
¶Corresponding author.

1. Introduction

In many parts of the developed world, there has been a consumer movement to reject modern agricultural technology such as genetically modified organisms (GMOs), which were introduced in the mid-1990s to improve product quality, yields, and disease and pest resistance (Pinstrup-Andersen and Schiøler, 2001). Despite a growing consensus in scientific research that genetically modified (GM) products are no more hazardous than conventional alternatives (Nicolia *et al.*, 2014; Tagliabue, 2016), the lack of universal consumer acceptance is well documented across many countries and has resulted in curbed demand for GM food products (see Huffman and McCluskey, 2014, for a discussion). This consumer skepticism is based on perceived risks over generally unknown long-run environmental and health consequences.

GM foods are not currently available in Russia, where government officials have taken a stance against the internal production and importation of GMOs in spite of significant cost-savings and increased yields generally associated with the technology. Russia's decision to ban GM foods could be an effective trade strategy given the growing demand for non-GM foods in many developed countries. In a speech given to Russian Parliament in December 2015, Russian President Vladimir Putin announced his intention to be the world's leading supplier of "ecologically clean and high-quality food" and criticized GMO food production in western countries even as demand for organic food continues to increase at high rates (Chow, 2015). Russia's position on GM crops is not unique; many European nations have, at some point, enacted (or attempted to enact) legislation restricting GM crop importation or production or banning it outright (LLC, 2014; GMO-free Europe, 2017).

This article investigates Russian consumer preferences for bread containing GMOs using data from in-person interviews in Perm, Russia, obtained for the purpose of this study. Given that GM foods are illegal to import or produce in Russia, market transactions data do not exist. Experimental auctions, though useful in many contexts, are not feasible here for three reasons. First, auctions use explicit prices for a hypothetical product with a direct relationship to an existing product for which consumers face heterogeneous prices. This introduces unnecessary starting-point bias into the WTP estimation. Second, auctions could potentially dissuade respondents from bidding for an illegal product if their personal information was to be maintained for ex post contact. Lastly, and sufficient on its own to eliminate the plausibility of auctions in this case, the product cannot be delivered to bidders.

A stated-preference (SP) approach is the best among our feasible options. With a SP approach, hypothetical bias (HB) is a concern. HB refers to situations in which the WTP estimates elicited from hypothetical formats diverge from the WTP elicited from non-hypothetical formats. In other words, there exists a major concern with SP surveys that stated WTP will exceed true WTP (List and Gallett, 2001; Lusk *et al.*, 2005). In the context of goods made with new technology, HB may take the form of an overstatement of the required discount for the food product made with new technology (Huffman and McCluskey, 2017). However, even with potential HB, useful information can be obtained from the responses. Since there is evidence of cross validation between SP approaches and revealed preference approaches, SP data can provide support for policy decisions (see Loureiro *et al.*, 2003; Chang *et al.*, 2009).

Specifically, we utilize a survey-based contingent valuation (CV) approach to elicit consumers' mean willingness to pay (WTP) for GM bread relative to the preexisting non-GM alternative.[1] In addition to basic bid and demographic survey questions, we obtained several categorical subjective variables. We employ principal-component analysis to combine subjective variables in respondents' questionnaires. Within this approach, we estimate how economic and social positions affect the mean consumer's relative WTP for bread containing GMOs. Though it is not possible to disentangle directly consumers' intrinsic preferences from those established by their environment, this article provides the groundwork for future research considering

[1]See Lusk (2003) for a detailed explanation of the merits and drawbacks of the CV approach.

regions with similar institutions and levels of economic development but different degrees of state intervention.

2. Background and Literature

Initially, GM technologies were employed to reduce crop susceptibility to disease and pests and to increase the efficiency of the agricultural sector (Pinstrup-Andersen and Schiøler, 2001). Second-generation GM products followed, such as functional foods fortified with nutritional supplements. Both first- and second-generation GM products have been subject to varying country-specific regulations (Veeman, 2002), but empirical evidence suggests that those GM products in the first generation have been more susceptible to negative consumer perception (Lusk *et al.*, 2005). The lack of unanimity in perception of GM foods may in part be due to their implicit classification as credence goods — goods for which the quality is unknown even after consumption (Darby and Karni, 1973). This temporally-persistent information asymmetry between sellers and buyers can deter the latter. In the case of GM foods, the information asymmetry can lead to consumers' fear about a product's creation or consumption, whether warranted or not.

Consumer perception and acceptance of GM foods varies across cultures and countries. Russia spans the continental landmass of Europe and Asia, so it is informative to examine studies of GM food perceptions across those continents as a point of comparison with Russian consumers. In Europe, broadly speaking, consumers are more critical toward food products utilizing new technology compared to their American counterparts (Lusk *et al.*, 2003; Lusk *et al.*, 2005; Huffman and McCluskey, 2014). Lusk *et al.* (2003) find that consumers in France, Germany, and the United Kingdom are willing to pay much higher premiums for beef not fed GM corn compared to US consumers. In a meta-analysis of 25 studies of GM food valuation, Lusk *et al.* (2005) find that European consumers, regardless of survey methodologies, hold higher valuations of non-GM food than U.S. consumers.

Research suggests that GM acceptance among Asia nations has been less consistent. China has been relatively accepting of GM foods, while Japan has rejected the technology. Li *et al.* (2002), using data from a consumer survey conducted in China, conclude that the Chinese consumers in their sample, on average, are willing to pay a 38% premium for Golden Rice, which is enhanced with additional vitamins. Lin *et al.* (2006) conduct a survey to estimate consumers' WTP for biotech foods, comparing biotech to non-biotech soybean oil and insect-resistant to non-biotech rice. They find that at least 60% of respondents are willing to purchase biotech foods without any discount. In contrast, McCluskey *et al.* (2003) find that only slightly over 3% of Japanese consumers are willing to purchase GM noodles at the same price as non-GM noodles, and only 17% are willing to purchase the former at a randomly assigned discount.

Some research has hinted at the impact of socioeconomic status on the biotechnology acceptance. With the exception of the United States and Canada, more developed countries tend to exhibit more intense aversion to GM foods than less developed countries. In Kenya, research suggests that a majority of consumers, though largely limited in knowledge about GM products, are willing to buy GM food products at the same price as their regular equivalent (Kimenju and De Groote, 2008). Curtis *et al.* (2004) directly address consumer acceptance of GM food in less developed countries, positing that the benefits of increased food availability that stems from biotechnological advances may outweigh any perceived potential long-run consequences.

Russia presents an interesting case because of its geographic size and location, its unique culture and political structure, and its intermediate level of economic development. Limited research has been conducted to investigate Russian consumer response to new food technology, and to our knowledge, there are no studies on the Russian consumer response to first-generation GM foods. Bruschi *et al.* (2015b) examine consumer acceptance and WTP for novel functional food bakery products among young, urban Russians, but their study focuses on an old wheat variety that is naturally rich in antioxidants. They find that with an information provision, their participants value the health-enhancing product over the base product and that the information type had a significant effect. For example, participants who were provided information that the purple wheat bakery products were made with an old variety of wheat are willing to pay more than those who were provided information about the specific antioxidants comprising the product. Also regarding functional foods, Dolgopolova *et al.* (2015) discuss the prevalence of food neophobia in Russia. They conclude that, in practice, novel foods are not purchased in Russia if traditional alternatives are available.

In nearby Uzbekistan, a former Soviet republic, Zaikin and McCluskey (2013) investigate the consumer response toward a new functional food product — apples enriched with an antioxidant coating. They find that the mean respondent was willing to purchase functional apples only at a discount, even when provided with positive health information. In contrast to the coated apples in Zaikin and McCluskey (2013), Bruschi *et al.*'s (2015b) product had a naturally occurring functional attribute, which may have made consumers more accepting of it. Comparing Zaikin and McCluskey's (2013) results with a U.S. study of the same product (Markosyan *et al.*, 2009), the Uzbek respondents were willing to pay less for apples with coatings that contain antioxidants (a mean discount of 6%) compared to the U.S. sample (a mean premium of 8%). In contrast to our research on first-generation GM products treated as a pure process attribute, the studies above explore consumer demand for second-generation GM products with decidedly positive nutritional attributes. And our results are correspondingly less optimistic about consumer acceptance of GM products.

On the more general topic of Russian food preferences, Honkanen and Frewer (2009) and Honkanen (2010) use survey data to examine Russian consumers' food-choice motives, consumption frequencies, health perceptions, and attitudes. Both studies suggest that sensory motives are the most important factors in food choice, followed by availability, naturalness, prices, consumers' moods, and state of health. The importance of "naturalness" suggests that Russians are unlikely to embrace GM foods. Russian consumers often cite food quality and health as a major factor influencing their food choices (Bruschi *et al.*, 2015a). Popova *et al.* (2010) found that

Russian consumers have doubts as to whether authorities can protect consumers from food risks. Using survey data, Bruschi *et al.* (2015a) study Russian consumer preferences for organic foods. Their results indicate that Russian consumers who purchase organic foods are mainly motivated by personal wellbeing and less by social or environmental concerns. Staudigel and Schröck (2015) estimate Russian food demand elasticities over time and across consumer segments, but their analysis is highly aggregated and does not extend to GM or organic food products.

To our knowledge, no other studies have evaluated consumer preferences for a GM food category in Russia. We contribute to the literature by studying consumer attitudes and WTP for a bread utilizing GM technology in Russia. The findings of this study provide insight into Russian consumer preferences for GM foods and may be useful in understanding how Russian consumers will response to other new food technologies.

3. Methods

3.1. *The contingent valuation approach*

We utilize the contingent valuation (CV) method to estimate consumers' WTP for bread[2] containing GMOs and examine which consumer characteristics affect choices. Following Hanemann (1984), our model uses a random utility framework to elicit information about consumers' mean WTP from close-ended, dichotomous-choice survey responses. Specifically, consumers were asked whether they would be willing to purchase bread containing GMOs at the same price as their usual non- GM bread. For those who answered in the affirmative, the line of questioning was terminated. For those who answered "no," we asked a follow-up question whether they would be willing to purchase bread containing GMOs at a discounted price. The discount was randomly selected, prior to the experiment, among

[2]Bruschi *et al.* (2015b) find that Russian consumers' knowledge of the base product is highly important. This supports using bread, a long-standing staple for Russian consumers, as our product of choice.

the following levels: 5%, 10%, 25%, 40%, and 50%. Each level of discount was used for one fifth of the surveys, not conditional on the first-stage response. The assignment of survey version (and thus, the discount) was random to the respondent. (See the Appendix A for an English translation of the contingent valuation questions.)

Our approach is identical to that in Grimsrud *et al.* (2004) and McCluskey *et al.* (2003), which the latter refers to as a "semi-double-bounded model." This is similar to the double-bounded dichotomous-choice contingent valuation survey technique described by Hanemann *et al.* (1991), but rather than extend the question conditionally based on either answer, we only asked a follow-up question to consumers whose WTP threshold was not met (i.e., the price was not low enough to induce the consumer to buy the GM bread). For the estimation of our model, we use an approach similar to the one-and-one-half-bound (OOHB) approach for our dichotomous-choice CVM as proposed by Cooper *et al.* (2002).[3] Kanninen (1995) shows that the double-bounded model is more efficient than the single-bounded model,[4] and the theoretical results in Cooper *et al.* (2002) indicate the OOHB model falls between the single- and double-bounded models in efficiency terms.

The rationale for excluding a follow-up to the affirmative response is that the GMOs used to make the bread have a process attribute, which reduces production costs — as opposed to a product-enhancing attribute. Proponents of this biotechnology claim that the GM products with process attributes are identical to non-GM products. Those opposed to genetic modification view it as a negative attribute. Therefore, it would not make economic sense after an initial "yes" response to follow up with a question that involves paying a premium for these GM products that only have cost-reducing attributes.

[3]Lusk *et al.* (2005) also refer to the survey technique used in our paper as a semi-double-bounded approach. To avoid further confusion, we draw a distinction between the survey technique and the estimation approach. Following Cooper *et al.* (2002), we will refer to the estimation approach as OOHB for the remainder of this paper.

[4]A single-bounded model uses a bid question without follow-up; a double-bounded model uses a first-stage-conditional follow-up question in a second stage for all respondents.

We did not discuss potential cost reductions that could be passed on to consumers with the respondents.

By not explicitly stating a value of the initial bid — replacing this with "the same price as the bread you usually buy" — we avoid the starting point bias inherent in contingent valuation surveys (Veronesi *et al.*, 2011). Instead of assigning specific values that can vary depending on respondents' heterogeneous WTP for non-GM bread, we consider relative WTP for GM bread as a percentage of the price each respondent pays for non-GM bread. In other words, since our questionnaire disregards specific prices paid to avoid any sort of innate anchoring, our second-round bids can be defined only as discounts from a normalized price of consumers' normal non-GM bread. Individual i's response probabilities π_i^y, π_i^{ny}, and π_i^{nn}(for responses of "yes," "no-yes," and "no-no," respectively) are as follows under the OOHB approach:

$$\pi_i^y(B_i^1) = \Pr\{B_i^1 \leq C_i\} = 1 - G(B_i^1;\theta) \tag{1}$$

$$\pi_i^{ny}(B_i^1, B_i^2) = \Pr\{B_i^1 \geq C_i \geq B_i^2\} = G(B_i^1;\theta) - G(B_i^2;\theta) \tag{2}$$

$$\pi_i^{nn}(B_i^1, B_i^2) = \Pr\{B_i^1 > C_i \cap B_i^2 > C_i\} = G(B_i^2;\theta) \tag{3}$$

where $B_i^1 = 1$ indicates the initial bid, $B_i^2 = \{0.5, 0.6, 0.75, 0.85, 0.95\}$ indicates the second bid (conditional on a negative response to the first bid), and C_i indicates maximum WTP. The corresponding log-likelihood function, therefore, is

$$\ln L^{\text{OOHB}}(\theta) = \sum_{i=1}^{N}\{d_i^y \ln \pi_i^y(B_i^1) + d_i^{ny} \ln \pi_i^{ny}(B_i^1, B_i^2) + d_i^{nn} \ln \pi_i^{nn}(B_i^1, B_i^2)\} \tag{4}$$

where d_i^y is a dummy variable taking value 1 when the initial response is "yes," and 0 otherwise; d_i^{ny} is a dummy variable taking value 1 when responses are "no-yes," and 0 otherwise; and d_i^{nn} is a dummy variable taking value 1 when responses are "no-no," and 0 otherwise.

Our model assumes a logistic cumulative distribution function (CDF) for $G(\cdot)$:

$$G(B) = [1 + e^{\alpha - \rho(B)}]^{-1}. \tag{5}$$

By assuming that functional form for the CDF and a linear relationship between WTP and its explanatory variables, we obtain

$$\ln L^{\text{OOHB}}(\theta) = \sum_{i=1}^{N} \left\{ d_i^y \ln \left[\frac{e^{\alpha-\rho B_i^1}}{1+e^{\alpha-\rho B_i^1}} \right] \right. + d_i^{ny} \ln \left[\frac{1}{1+e^{\alpha-\rho B_i^1}} - \frac{1}{1+e^{\alpha-\rho B_i^2}} \right] \left. + d_i^y \ln \left[\frac{1}{1+e^{\alpha-\rho B_i^2}} \right] \right\} \tag{6}$$

Since $\theta = \{\alpha, \rho\}$, we can use maximum likelihood estimation to determine the optimal parameters that solve $\frac{\partial \ln L^{OOHB}(\theta)}{\partial \theta} = 0$. We obtain

$$\frac{\partial \ln L^{OOHB}}{\partial \alpha} = \sum_{i=1}^{N} \left\{ d_i^y \left[\frac{1}{1+e^{\alpha-\rho B_i^1}} \right] \right. + d_i^{ny} \left[\frac{1-e^{2\alpha-\rho B_i^1-\rho B_t^2}}{\left(1+e^{\alpha-\rho B_i^1}\right)\left(1+e^{\alpha-\rho B_t^2}\right)} \right] \left. + d_i^{nn} \left[\frac{-e^{\alpha-\rho B_t^2}}{1+e^{\alpha-\rho B_t^2}} \right] \right\} = 0 \tag{7}$$

$$\frac{\partial \ln L^{OOHB}}{\partial \rho} = \sum_{i=1}^{N} \left\{ d_i^y \left[\frac{-B_i^1}{1+e^{\alpha-\rho B_i^1}} \right] \right. + d_i^{ny} \left[\frac{\frac{B_t^1 e^{\alpha-\rho B_i^1}}{(1+e^{\alpha-\rho B_i^1})^2} - \frac{B_i^2 e^{\alpha-\rho B_t^2}}{(1+e^{\alpha-\rho B_i^1})^2}}{\frac{1}{1+e^{\alpha 2-\rho B_i^1}} - \frac{1}{1+e^{\alpha-\rho B_t^2}}} \right] \left. + d_i^{nn} \left[\frac{B_i^2 e^{\alpha-\rho B_i^2}}{1+e^{\alpha-\rho B_i^2}} \right] \right\} = 0. \tag{8}$$

These equations implicitly define the optimal parameters, $\widehat{\alpha}^{MLE}$ and $\widehat{\rho}^{MLE}$, as functions of the bids. In the model without additional explanatory variables, mean and median relative WTP coincide at $(-\alpha/\rho)$.

Traditionally OOHB is used to form more consistent and efficient estimates than a double-bounded approach in which follow-up questions are only asked at random (i.e., when some respondents receive a follow-up question and others do not). In our case, we have a similar structure to OOHB, treating the random assignment of follow-up questions as if they directly coincide with those initially choosing not to purchase. This could potentially introduce a slight but unavoidable downward bias in our relative WTP estimates.

When we extend our baseline model to include additional exogenous explanatory variables, X, we use a standard logistic CDF with mean zero and standard deviation $\sigma = \pi/\sqrt{3}$:

$$G(B; X) = [1 + e^{\alpha - \rho B + \beta X}]^{-1}, \tag{9}$$

We can use these variables to make distinctions between the mean and median WTP values:

$$E[WTP] = (-1/\hat{\rho})(\hat{\alpha} + \hat{\beta}\bar{X}) \tag{10}$$

We can further use the explanatory variables to estimate individual respondents' relative WTP for the GM product. In order to gain insight into Russian consumer demand for GM food and to investigate the mechanisms driving WTP, we need these additional data.

In our survey we obtained both objective and subjective response variables, and to properly account for the latter we employ principal component analysis (PCA). To provide the most clear and complete explanation of our PCA, it is informative first to examine our data.

3.2. *Data and descriptive statistics*

We conducted in-person interviews of 300 respondents at several centrally-located grocery stores in Perm, Russia, which is a city of around one million residents located in the European part of Russia. It is about 1156 km (718 miles) east of Moscow. The rural population in our sample is 7.3%, which is significantly lower than the estimated rural population of 24.6% in Perm in 2014, but this is at least partially explained by the spatial limitations of our survey — as mentioned above, the grocery stores we selected for maximal survey participation were located near the city center. By using multiple locations and different store chains, we hoped to increase the diversity of the sample, both in terms of income and demography. Respondents were informed that their data would be treated anonymously and care was taken to ensure that multiple members of single households were not surveyed. The average age of respondents participating in the survey was approximately 35.8, which is only slightly younger than the average age of 38.7 in Perm in 2015. Average monthly income per capita in Perm was 32,053 Rubles in 2015, which lies in our mode range of household income (between 30,000 and 50,000 Rubles). For a description of the data, see Tables 1 and 2.

Given the potential overstatement of WTP inherent in SP surveys, bids among respondents in our study seem particularly low. Russian consumers' perceptions of GM foods are overwhelmingly negative with 79.67% of respondents stating that they are unwilling to purchase GM bread even up to a discount of up to 50%. Approximately 12% of respondents stated they would purchase GM bread for the same price as non-GM bread, and of the total sample, 8.33% were willing to purchase the GM bread if discounted relative to non-GM bread.

Examining the subjective questions in our survey, we find that the median respondent is relatively indifferent between saving the environment and saving jobs (though the frequency of individuals preferring to save the environment is higher). She prefers domestic products and is much more concerned with the safety of food than its price — a result that independently corroborates the findings of Bruschi *et al.* (2015a). The median respondent's perception of GMO consumption risk tends toward the more wary end of the spectrum, and she believes labels are very important for GM foods.

Cameron (1988) points out that traditional theoretical interpretations of dichotomous-choice survey responses, such as that of Hanemann (1984), and the resulting econometric methodologies for eliciting WTP do not appropriately account for the ranges indicated by a given response. Choices are discrete, but the space delineated by either response is continuous and potentially infinite. To ensure that our subsequent estimation follows from our true observations,

Table 1. Description and descriptive statistics for respondents' demographic variables.

Variable	Description	Descriptive Statistics	
Age	As of 2015	Mean:	35.76
		Std. dev.:	14.92
Gender	1 *if female*		65.7%
	0 *if male*		34.3%
Primary household shopper	1 *if primary shopper*		68.3%
	0 *otherwise*		31.7%
Highest education completed	*Secondary education*		4.7%
	Tertiary education		19.0%
	Some college		20.3%
	Bachelor's degree		17.0%
	Master's degree		38.3%
	Ph.D. or higher		0.7%
Children in household	1 *if children under* 18 *in household*		40.3%
	0 *otherwise*		59.7%
Household size	Total members of household	Mean:	2.87
		Std. dev.:	1.23
Living environment	*Urban dweller*		62.0%
	Sub-urban dweller		30.7%
	Rural dweller		7.3%
Employment	*Student*		11.0%
	Wageworker		60.0%
	Individual entrepreneur		14.3%
	Retired		6.7%
	Other		8.0%
Monthly household income[a,b]	*<20,000 RUB*		7.6%
	20,000–30,000 RUB		20.3%
	30,000–50,000 RUB		33.1%
	50,000–75,000 RUB		21.4%
	75,000–90,000 RUB		9.0%
	>90,000 RUB		8.6%

[a] Currency conversion as of December 31, 2015 was 73.41 Russian rubles per 1 US dollar.

[b]Ten respondents left the income question unanswered; income percentage calculations were based on 290 responses.

Table 2. Descriptions and frequencies for respondents' qualitative and subjective variables.

Variable	Description	Scaled Values	Response Frequency
Jobs vs. environment[a]	Quantifies respondents' choice between saving jobs and saving the environment	1 = saving jobs ⋮ 7 = environment (median: 4.00)	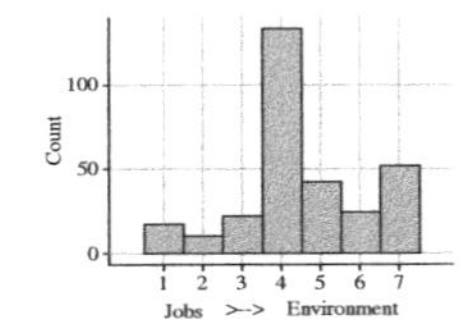
Food safety vs. price	Quantifies respondents' choice between food safety and product price	1 = food safety ⋮ 7 = lower price (median: 2.00)	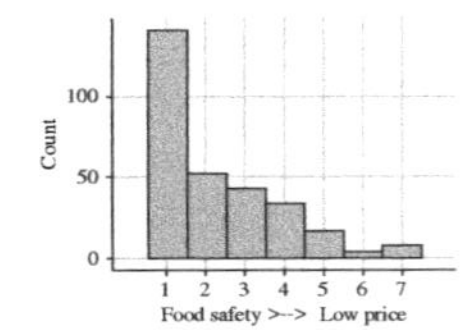
GMO risk	Quantifies respondents' perception of risk associated with GMOs	1 = high risk 2 = some risk 3 = no risk 4 = uncertain	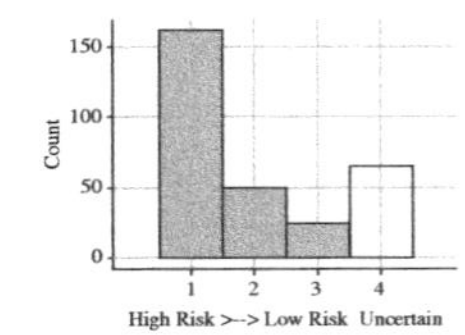
Biotech perception	Quantifies respondents' stated feeling about biotechnology use in food production	1 = very positive ⋮ 5 = very negative 6 = uncertain	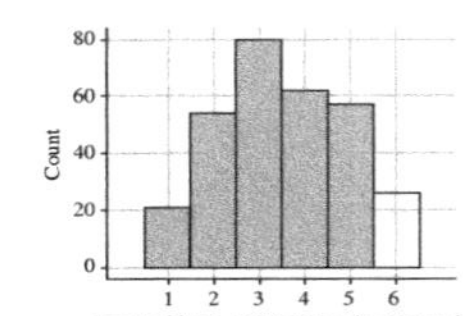

Table 2. (*Continued*)

Variable	Description	Scaled Values	Response Frequency
GMO knowledge	Quantifies respondents' self-reported knowledge about GMOs	1 = very knowledgeable 2 = some 3 = know little (median: 2.00)	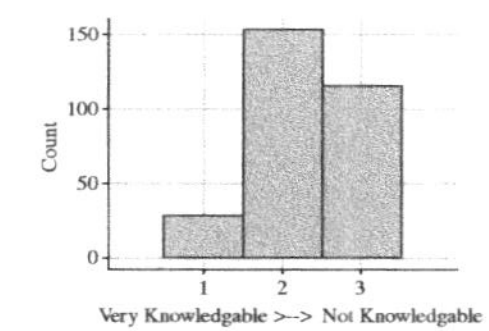
GM labeling	Quantifies respondents' views about the importance of labels on GM foods	1 = very important 2 = somewhat important 3 = not important (median: 1.00)	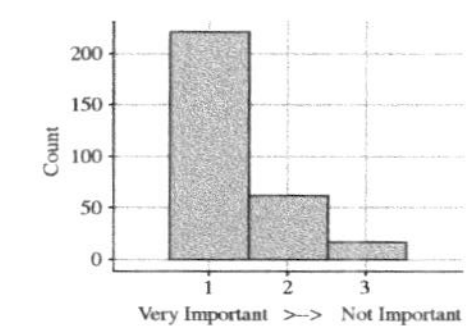
Preference for domestic	Binary preference for domestic or imported food	$\begin{cases} 1 & \text{if prefers domestic food} \\ 0 & \text{otherwise} \end{cases}$	84.33%

[a]One respondent left this unanswered; statistics reflect 299 remaining respondents.

Table 3. Empirical distribution of WTP derived using interval analysis.

WTP Interval	Probability
$(-\infty, 0.5)$	0.73333
(0.5, 0.6)	0.04704
(0.6, 0.75)	0.01501
(0.75, 0.85)	0.03385
(0.85, 0.95)	0.01756
(0.95, 1)	0.03321
$(1, \infty)$	0.12000

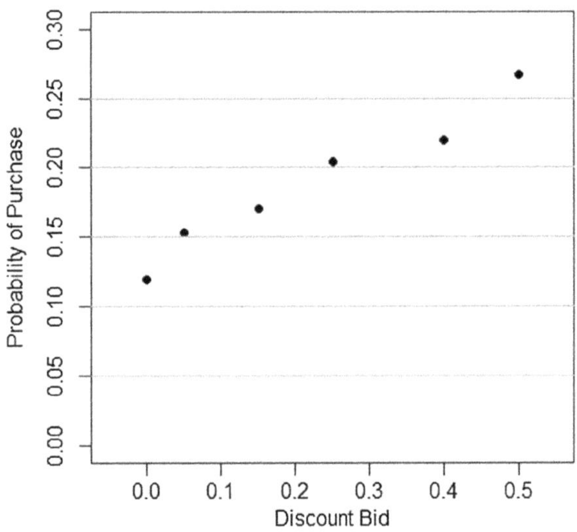

Fig. 1. Empirical distribution of WTP derived by interval analysis.

we first investigate the ranges of bid responses in the data, which are depicted in Table 3.

Performing interval tests to derive an empirical distribution function (Turnbull, 1976; Fay, 1996), we obtain estimated purchase probabilities by discount ranges, which we present in Fig. 1. Note that we only include those observations for which we have actual bids. As expected, fewer consumers said "yes" as the bid increased.

Interestingly, at the maximum discount offered (50%), the likelihood of consumers purchasing the GM bread is less than 30%. This indicates mean relative WTP is lower than our defined bid ranges and may in fact be negative (as our estimation results predict).

3.3. *Principal component analysis*

Given the questions asked in the survey, it is likely that responses across several questions may bear some correlation, especially those questions utilizing a Likert-type scale: comparing food safety to cost concerns, saving jobs versus saving the environment, and general feelings about the use of biotechnology in food production. This multicollinearity could skew our analysis; either ignoring the significance of one variable in favor of a highly correlated alternative or decreasing the significance of both simultaneously. Using PCA, we can reduce the dimensionality of our data and explore how certain types of consumers react without losing the information collected. This process essentially allows us to transform some of our explanatory variables into their principal components, identifying certain respondent profiles based on the data.

We can have as many principal components as we have explanatory variables used to construct them, but we restrict our analysis to those with the highest variance, as indicated by the highest eigenvalues in our covariance matrix of demeaned variables. Using our highest-variance, weighted linear combinations, we construct new variables indicating the relative influence of certain subjective values held by consumers. We define our principle components by the relative influence (size and direction) of each true variable on the component's construction. The principal components we retained for our analysis are presented in Table 4, where a "+" or "−" indicates a value above or below the sample mean, respectively.

The first principal component emphasizes a respondent's money-mindedness; i.e., how response to the food-safety-versus-low-price question will influence relative WTP. This principal component will be of a greater magnitude if a consumer is more interested in purchasing cheaper food than avoiding food-safety concerns.

Table 4. Description of three principal components included in the analysis.

Principal Component	Feeling about GM	Jobs vs. Environment	Food Safety vs. Price	Prefer Domestic	GMO Knowledge	GMO Consumption Risk	GMO Label Importance	Description
PC 1	−	−	+	−	+	+	+	Money-driven
PC 2	+	−	−	+	+	−	+	Biotech-driven, low information
PC 3	+	−	+	+	−	−	−	Biotech-driven, high information

Additionally, this component deemphasizes the influence of respondents' environmental concerns in favor of jobs, and the preference for domestic production is muted. Correspondingly, we will refer to this as the money-driven principal component. Stronger negative perceptions of biotechnology will produce lower values of this component. Other subjective responses that increase the money-driven principal component including knowing less about GMOs, perceiving lower risk, and being less interested in GMO labeling.

Compared with the first, the second and third principal components more strongly emphasize respondents' aversion to biotechnology. While both move positively with more negative biotech perceptions, the third principal component does so to a much larger degree. Like the money-driven component, these two deemphasize environmental concerns in favor of jobs, but unlike the first component these two move positively with domestic preference and inversely with perceived GMO consumption risk. The second and third components diverge from each other on food safety versus cheapness (the second emphasizes food safety while the third emphasizes cheapness) and knowledge and labeling (the second emphasizes knowing less and putting less importance on labeling while the third puts heavier weights on knowing more and preferring labels). Bearing in mind the component weights, we refer to the second and third principal components as *the biotech-driven, low-information* and *biotech-driven, high-information* principal components, respectively.

4. Estimation Results

4.1. *Analysis of factors that affect the choice of whether to purchase GM bread*

In Table 5, we present estimation results using Eqs. (4) and (9) for bids, general demographics, and respondents' preference for domestic foods to estimate willingness to pay for GM bread. With the exception of domestic food preference, the variables are nonsubjective. We observe low correlation between variables in this specification. As expected, the bid coefficient is significant to at least the 90% level, a result that holds across specifications including

Table 5. Parameter estimates for WTP model.

Parameters	Coeff. est.	St. err.	z-Score	p-Value	Marg. eff.	St. err.
Constant	1.502	(2.011)	0.747	0.455		
Bid	2.580*	(1.425)	1.810	0.070		
Prefer domestic	−1.134***	(0.408)	−2.779	0.005	−0.439***	(0.160)
Age	−0.020	(0.017)	−1.185	0.236	−0.008	(0.116)
Gender	−0.784**	(0.367)	−2.135	0.033	−0.304**	(0.144)
Children under 18	−0.341	(0.372)	−0.917	0.359	−0.132**	(0.065)
Highest education	*categ.*[a]					
Observations	300					
Pseudo-R^2	0.645					
ICCC	0.763					
FCCC	0.939					
Mean WTP	−1.206					

[a]Ordered variable treated as categorical dummies; individual results not reported.
*indicates 90%, **indicates 95%, and ***indicates 99% significance level.

additional socio-demographic variables tested but not presented in Table 5. This means that as the price of GM bread relative to non-GM bread increases, respondents are less likely to say they would purchase GM bread at the offered price.

Respondents' preference for domestic foods — a position held by 84.33% of respondents — exhibits a significant inverse relationship with relative WTP for GM bread. This high incidence of domestic preference is intuitive given Russian distrust of actors in the food chain, as described in Popova *et al.* (2010). A strictly domestic market should be expected to entail a shorter food supply chain and provide fewer opportunities for contamination. Additionally, Russia's position against the importation of GM foods dictates that domestic consumption is restricted to non-GM foods, so domestic preference coincides with a preference for non-GM foods. We cannot, however, disentangle any sort of causal order between consumer preference and government mandate from our research.

Gender accounts for a large and statistically significant amount of the variation in WTP, and its magnitude does not waiver considerably across alternate specifications. The coefficient estimate on age is negative but not statistically significant. Controlling for education level and whether children are present in the household does not significantly affect our results. Coefficient estimates for categorical levels of education and the presence of children in the household are all negative, but none is statistically significant.

4.2. *Estimates of the mean relative discounts for GM food*

In addition to coefficient estimates, Table 5 presents the mean estimated relative WTP for GM bread. As predicted by the responses limited by bid ranges in the questionnaire, our estimates show that few respondents were willing to pay any amount of money for the GM bread. Excluding explanatory variables from the model and focusing only on bids (and a constant), we estimate a mean relative WTP of $(-\alpha/\rho) = -1.034$, which indicates that the mean survey respondent must receive slightly more than the value of the non-GM alternative to be willing to assume the risk of consuming GM bread (i.e., a discount of 203.4%). To examine the validity of our estimates, we treated the initial bids as a single-bounded dichotomous choice and estimated mean WTP in the first stage as -0.809 based only on bids and a constant.

We should note that our estimator is biased since it is derived as a ratio of parameter estimates obtained through maximum likelihood estimation, but it is consistent by the law of large numbers (assuming that the asymptotic properties of our estimator are satisfied given our sample of 300 observations). The estimates themselves are reasonable given the distribution of bids and responses. Note that if we restrict the bids to be non-negative, our model estimates a mean relative WTP that is not significantly different from zero.

While we should not rely on the specific magnitudes of negative WTP estimates, we can interpret them as an indication that the mean Russian consumer has an intense aversion to purchasing first-generation GM food products. It is more appropriate to emphasize the estimated probabilities of purchase given varying discounts, presented above in Fig. 1. We observe that 12% of consumers were willing to purchase GM bread at the same price as conventional bread, but with a 50% discount the estimated probability of purchase only increases to 26.67%. In comparison to Grimsrud *et al.*'s (2004) study of Norwegian consumers' preferences for bread made with GM wheat, their Norwegian respondents were similarly resistant to GM bread, though not to as strong a degree as Russian respondents in the current study.

Though we analyzed specifications with subjective variables independently, we find weak relationships between most of these categorical variables and likelihood of purchase when treating them as continuous variables or dummies. Additionally we often found that the inclusion of one variable might have a strong and significant effect, but adding another would reduce the effect of both due to some degree of endogeneity between variables. Rather than present those and other consumer-perception variables independently, we construct principal components for a more inclusive analysis. Using this method, we are able to consider how different subjective-response profiles (or rather more pedantically, how emphasizing different linear combination weights of the subjective variables) are likely to influence our relative WTP estimates without sacrificing the broad scope of our data.

4.3. *Results of the principal component analysis*

Revisiting our OOHB model and incorporating our three principal components each in turn as explanatory variables and then all together, we report our results[5] in Table 6. We find that the money-driven and biotech-driven, high-information principal components are statistically significant in determining respondents' relative WTP for GM bread. The bid coefficient is statistically significant when we control for either biotech-driven component, but not when we control for money-driven component or when we run the model for all three principal components simultaneously. This is intuitive, since the bid coefficient is a monetary choice. Including a variable that emphasizes the influence of money on respondents' relative

[5]One observation with missing data was omitted in the principal components analysis.

Table 6. Estimation results of OOHB method using the logistic CDF and the three principal components.

	PC1 Model		PC2 Model		PC3 Model		All PC Model			
Variable	Coeff. est.	St. err.	Coeff. est.	St. err.	Coeff. est.	St. err.	Coeff. est.	St. err.	Marg. effects	St. err.
Constant	−1.245	(1.217)	−2.036	(1.582)	−1.866	(1.500)	−1.375	(1.323)		
Bid	1.870	(1.241)	3.072*	(1.605)	2.779**	(1.489)	1.984	(1.328)		
Money-driven PC	0.240**	(0.102)					0.258**	(0.104)	0.130	(0.235)
Biotech-driven, low-info PC			0.097	(0.126)			0.072	(0.124)	0.036	(0.053)
Biotech-driven, high-info PC					−0.274**	(0.128)	−0.292**	(0.129)	−0.147	(0.198)
Age	−0.022	(0.015)	−0.025	(0.015)	−0.023	(0.016)	−0.022	(0.015)	−0.011	(0.169)
Gender	−0.707**	(0.345)	−0.811**	(0.351)	−0.848**	(0.357)	−0.786**	(0.360)	−0.396**	(0.188)
Pseudo-R^2	0.092		0.119		0.112		0.068			
ICCC	0.809		0.732		0.776		0.779			
FCCC	0.947		0.962		0.955		0.924			
Mean WTP	−1.328		−1.128		−1.174		−1.357			

*indicates 90%, **indicates 95%, and ***indicates 99% significance level; $n = 299$.

WTP calculus, we may reduce the influence of the bid itself in determining WTP. Gender remains statistically significant across all specifications, still indicating a lower relative WTP for GM bread among female respondents.

The low pseudo-R^2 for each model does not indicate that these principal component models are less preferred to the model specifications above without principal components. The McFadden pseudo-R^2 compares the value of the maximized log-likelihood function for each model specification with the explanatory variables to the same model with only the intercept and bid coefficient. We interpret a low pseudo-R^2 to mean only that the maximized log-likelihood function's value is similar with or without the explanatory variables. Adding more variables will tend to increase this statistic. To further support the strength of our results, we present the ICCC and FCCC measures for goodness of fit for the initial and second-round probabilities of correct predictions, respectively (Kanninen and Khawaja, 1995). Higher values indicate stronger predictive power of the estimated model. Comparing these measures, the predictive power of the model is not significantly different when we use the principal components or the explanatory variables in the initial model specifications.

Our PCA provides some insight into how consumers' broad perceptions of biotechnology, including its far-reaching economic implications, influence their decisions about GM food. The principle component emphasizing consumers' economic incentives tends to increase consumers' WTP to pay for GM bread relative to non-GM bread. This suggests that if more consumers valued cheaper goods over food safety — contrary to what we observe in this data set — the mean relative WTP for GM bread may be higher. Alternatively, when biotech aversion and prevalence of GM information are more heavily emphasized, we estimate the opposite effect on relative WTP for GM food. In terms of policy, Russia's present stance against GM goods may be bolstered by keeping any debate in terms of potential health effects and steering clear of any economic impact analysis. This should not be difficult given Russian consumers' observed tendency to put more weight on nutritional and social concerns (food safety and the environment) than on economic ones (price and job safety), not mention their a priori distrust of GMOs.

5. Conclusions

A more nuanced examination of Russian consumers' attitudes and perceptions toward GM foods is essential in understanding whether any potential market could exist under a less stringent government policy. Even with a discount between 5% and 50%, only around 20% of respondents in our survey are willing to purchase bread containing GMOs rather than bread made using conventional inputs.

If consumers feel they are independently more reliable than food chain actors in promoting food safety as Popova *et al.* (2010) suggest, then it makes sense that they would strongly oppose a potential risk such as GMOs where direct regulation could be insufficient to avoid health hazards. We find that perceptions of GMOs in Russia are negative, and health concerns play a significant role in our sample's WTP for GM bread. Our findings are consistent with those of Bruschi *et al.* (2015a), who find that Russian consumers' food purchases are often motivated by health concerns and specifically the "naturalness" of foods. Our results indicate that placing a larger emphasis on information variables (i.e., higher GM knowledge and a stronger position regarding the importance of GM labeling) leads to a lower relative WTP for GM food. In contrast, when we emphasize economic concerns (high concern over job safety and the price of food products) we find that consumers are more likely to purchase the GM alternative.

The true long-run effects of GM food production and consumption remain unknown despite the broad consensus across scientific research indicating GM products are not significantly more hazardous than the conventional alternative (Nicolia *et al.*, 2014; Tagliabue, 2016). Current legislation in Russia opposing the production and import of GM foods undoubtedly influences individuals' perceptions about the safety of these goods. If the Russian government reduced its restrictions on GM food imports, we should not expect a large change since the majority of consumers prefer domestic products. However, if those reduced restrictions extended to domestic production and the cost savings were passed on to the consumer, it may be possible to design a profitable GM-food market strategy targeting money-conscious consumers, even as consumers continue to perceive GMOs as an undesirable alternative.

Appendix A. English Translation of Survey Questions

A.1. *Part I. Consumer information and perception variables*

1. How knowledgeable are you about biotechnology and genetically modified food?
   a. Very knowledgeable
   b. Somewhat knowledgeable
   c. Not informed
2. How do you feel overall about the use of biotechnology in foods?
   a. Very positive
   b. Somewhat positive
   c. Neutral
   d. Somewhat negative
   e. Very negative
   f. Don't know
3. Where would you place yourself on a scale from 1 to 7, if *food safety is the most important for you* is a 1 and, saving money, purchasing cheaper food products is a 7? [CIRCLE JUST ONE]

   1 2 3 4 5 6 7

4. Where would you place yourself on a scale from 1 to 7, if *savings jobs at all costs* is a 1 and, *saving the environment at all costs* is a 7? [CIRCLE JUST ONE]

   1 2 3 4 5 6 7

5. How important is labeling of genetically modified food?
    a. Very important
    b. Somewhat important
    c. Not very important
6. In your opinion, what risk entails the consumption of products containing GMOs?
    a. High level risk
    b. Low level risk
    c. No risk
    d. Don't know
7. Would you rather prefer foods produced in our country?
    a. Yes
    b. No

A.2. *Part II. Questions related to willingness to pay/accept*

Would you be willing to purchase bread containing GMOs at the same price as the usual bread?

a. Yes
b. No

Would you be willing to purchase bread containing GMOs at the price with [insert random discount%] discount rather than the usual bread?

a. Yes
b. No

A.3. *Part III. Social and demographic information*

Gender of respondent:

a. Female
b. Male

What year you were born: _____
Who is the primary shopper in your household?

a. I do shopping mostly for my household
b. Other family members do mostly shopping

How many members do you have in your household? _____
Do any children under 18 live in your household?

a. Yes
b. No

What kind of living environment are you in?

a. In down
b. In suburb
c. Rural

What is the highest level of education did you complete?

a. Secondary education
b. Secondary special or technical
c. undergraduate student
d. Bachelor's degree
e. Master's degree
f. PhD or higher

In which monthly income bracket would you household fall into in the year 2015?

a. Less than 30,000 Russian rubles
b. From 30,000 to 50,000 rubles
c. From 50,000 to 75,000 rubles
d. From 75,000 to 90,000 rubles
e. More than 90,000 rubles

Which one of the following categories best describes your employment status:

a. Student
b. Hired employee

c. Individual entrepreneur
d. Retired
e. Other

THANK YOU VERY MUCH FOR YOUR PARTICIPATION

Appendix B. Supplementary Material

Supplementary data associated with this article can be found, in the online version, at http://dx.doi.org/10.1016/j.foodpol.2018.02.004.

References

Bruschi, V., K. Shershneva, D. Dolgopolova, M. Canavari and R. Teuber (2015a). Consumer Perception of Organic Food in Emerging Markets: Evidence from Saint Petersburg, *Russia. Agribusiness* 31, 414–432.

Bruschi, V., R. Teuber and I. Dolgopolova (2015b). Acceptance and Willingness to Pay for Health-Enhancing Bakery Products: Empirical Evidence for Young Urban Russian Consumers. *Food Quality and Preference* 46, 79–91.

Cameron, T.A. (1988). A New Paradigm for Valuing Non-Market Goods Using Referendum Data: Maximum Likelihood Estimation by Censored Logistic Regression. *Journal of Environmental Economics and Management* 15, 355–379.

Chang, J.B., J.L. Lusk, F.B. Norwood (2009). How Closely Do Hypothetical Surveys and Laboratory Experiments Predict Field Behavior? *American Journal of Agricultural Economics* 91, 518–534.

Chow, L. (2015). Russia wants to be world's top exporter of non-GMO food. Organic Consumers Organization https://www.organicconsumers.org/news/russia-wants-be-worlds-top-exporter-non-gmo-food (accessed 13 March 2017).

Cooper, J.C., W.M. Hanemann and G. Signorello (2002). One-and-One-Half-Bound Dichotomous Choice Contingent Valuation. *Review of Economic Statistics* 84, 742–750.

Curtis, K.R., J.J. McCluskey and T.I. Wahl (2004). Consumer Acceptance of Genetically Modified Food Products in the Developing World. *AgBioForum* 7, 70–75.

Darby, M.R. and E. Karni (1973). Free Competition and the Optimal Amount of Fraud. *Journal of Law and Economics* 16, 67–88.

Dolgopolova, I., R. Teuber and V. Bruschi (2015). Consumers' Perceptions of Functional Foods: Trust and Food-Neophobia in a Cross-Cultural Context. *International Journal of Consumer Studies* 39, 708–715.

Fay, M.P. (1996). Rank Invariant Tests for Interval Censored Data Under the Grouped Continuous Model. *Biometrics* 52, 811–822.

GMO-free Europe (2017). GMO-free Regions by Country. http://www.gmo-free-regions.org/gmo-free-regions.html (accessed 24 September 2017).

Grimsrud, K.M., J.J. McCluskey, M.L. Loureiro and T.I. Wahl (2004). Consumer Attitudes to Genetically Modified Food in Norway. *Journal of Agricultural Economics* 55, 75–90.

Hanemann, W.M. (1984). Welfare Evaluations in Contingent Valuation Experiments with Discrete Responses. *American Journal of Agricultural Economics* 66, 332–341.

Hanemann, W.M., J. Loomis and B. Kanninen (1991). Statistical Efficiency of Double-Bounded Dichotomous Choice Valuation. *American Journal of Agricultural Economics* 73, 1255–1263.

Honkanen, P. (2010). Food Preference Based Segments in Russia. *Food Quality and Preference* 21, 65–74.

Honkanen, P. and L. Frewer (2009). Russian Consumers' Motives for Food Choice. *Appetite* 52, 363–371.

Huffman, W.E. and J.J. McCluskey (2014). "Labeling of genetically modified foods." In S.J. Smyth, P.W.B. Philips and D. Castle (Eds.), *Handbook on Agriculture, Biotechnology and Development*, pp. 467–487. Northampton, MA: Edward Elgar Publishing Inc.

Huffman, W.E. and J.J. McCluskey (2017). Using Stated Preference Techniques and Experimental Auction Methods: A Review of Advantages and Disadvantages for Each Method in Examining Consumer Preferences for New Technology. *International Review of Environmental and Resource Economics* 10, 269–297.

Kanninen, B.J. (1995). Bias in Discrete Response Contingent Valuation. *Journal of Environmental Economics and Management* 28, 114–125.

Kanninen, B.J. and M.S. Khawaja (1995). Measuring Goodness of Fit for the Double-Bounded Logit Model. *American Journal of Agricultural Economics* 77, 885–890.

Kimenju, S.C. and H. De Groote (2008). Consumer Willingness to Pay for Genetically Modified Food in Kenya. *Agriculture and Environment* 38, 35–46.

LLC (2014). Restrictions on Genetically Modified Organisms. U.S. Law Library of Congress. http://www.loc.gov/law/help/restrictions-on-gmos/restrictions-on-gmos.pdf (accessed 24 September 2017).

Li, Q., K.R. Curtis, J.J. McCluskey and T.I. Wahl (2002). Consumer Attitudes Toward Genetically Modified Foods in Beijing, China. *AgBioForum* 5, 145–152.

Lin, W., A. Somwaru, F. Tuan, J. Huang and J. Bai (2006). Consumers' Willingness to Pay for Biotech Foods in China: A Contingent Valuation Approach. *AgBioForum* 9, 166–179.

List, J.A. and C.A. Gallett (2001). What Experimental Protocol Influence Disparities Between Actual and Hypothetical Stated Values? *Environmental and Resource Economics* 20, 241–254.

Loureiro, M.L., J.J. McCluskey and R.C. Mittelhammer (2003). Are Stated Preferences Good Predictors of Market Behavior? *Land Economics* 79, 44–55.

Lusk, J.L. (2003). Using Experimental Auctions for Marketing Applications: A Discussion. *Journal of Agricultural and Applied Economics* 35, 349–360.

Lusk, J.L., M. Jamal, L. Kurlander, M. Roucan and L. Taulman (2005). A Meta-Analysis of Genetically Modified Food Valuation Studies. *Journal of Agricultural and Resource Economics* 30, 28–44.

Lusk, J.L., J. Roosen and J.A. Fox (2003). Demand for Beef from Cattle Administered Growth Hormones or Fed Genetically Modified Corn: A Comparison of Consumers in France, Germany, the United Kingdom, and the United States. *American Journal of Agricultural Economics* 85, 16–29.

Markosyan, A., J.J. McCluskey and T.I. Wahl (2009). Consumer Response to Information About a Functional Food: Apples Enriched with Antioxidants. *Canadian Journal of Agricultural Economics* 57, 325–341.

McCluskey, J.J., K.M. Grimsrud, H. Ouchi and T.I. Wahl (2003). Consumer Response to Genetically Modified Food Products in Japan. *Agricultural and Resource Economics Review* 32, 222–231.

Nicolia, A., A. Manzo, F. Veronesi and D. Rosellini (2014). An Overview of the Last 10 years of Genetically Engineered Crop Safety Research. *Critical Reviews in Biotechnology* 34, 77–88.

Pinstrup-Andersen, P. and E. Schiøler (2001). *Seeds of Contention: World Hunger and the Global Controversy over GM Crops*. Baltimore, MD: Johns Hopkins University Press.

Popova, K., L.J. Frewer, J. De Jonge, A. Fischer and E. Van Kleef (2010). Consumer Evaluations of Food Risk Management in Russia. *British Food Journal* 112, 934–948.

Staudigel, M. and R. Schröck (2015). Food Demand in Russia: Heterogeneous Consumer Segment Over Time. *Journal of Agricultural Economics* 66, 615–639.

Tagliabue, G. (2016). The Necessary "GMO" Denialism and Scientific Consensus. *Journal of Science Communication* 15, 1–11.

Turnbull, B.W. (1976). The Empirical Distribution Function with Arbitrarily Grouped, Censored, and Truncated Data. *Journal of the Royal Statistical Society Series B-Methodological* 38, 290–295.

Veeman, M. (2002). Policy Development for Novel Foods: Issues and Challenges for Functional Food. *Canadian Journal of Agricultural Economics* 50, 527–539.

Veronesi, M., A. Alberini and J.C. Cooper (2011). Implications of Bid Design and Willingness-To-Pay Distribution for Starting Point Bias in Double-Bounded Dichotomous Choice Contingent Valuation Surveys. *Environmental and Resource Economics* 49, 199–215.

Zaikin, A. and J.J. McCluskey (2013). Consumer Preferences for New Technology: Apples Enriched with Antioxidant Coatings in Uzbekistan. *Agricultural Economics* 44, 513–521.

Chapter 17

Consumer Attitudes to Genetically Modified Food in Norway*

Kristine M. Grimsrud[†], Jill J. McCluskey[‡,||], Maria L. Loureiro[§], and Thomas I. Wahl[¶]

[†] *Department of Economics, University of New Mexico, USA*
[‡] *Department of Agricultural and Resource Economics, Washington State University, USA*
[§] *Senior Researcher-Norwegian Agricultural Research Institute, Oslo, Norway*
[¶] *International Marketing Program for Agricultural Commodities and Trade (IMPACT) Center, Washington State University, USA*
[||] *mccluskey@wsu.edu*

There is a lack of public acceptance of genetically modified (GM) food products in Europe. Using a dichotomous choice contingent valuation methodology, we find that, on average, survey respondents with a positive attitude toward the use of biotechnology in food production, higher educated, younger, and male need less of a discount to choose to purchase bread made with GM wheat. Further, we estimate confidence intervals for the mean discount required for consumers to choose GM bread. The 95% confidence interval for the mean discount required for the respondents to choose the GM bread is a 37% to 63% discount relative to conventional bread.

1. Introduction

The introduction of genetically modified crops to world markets has created a new division between the crop trading countries.

*This article originally appeared in the **Journal of Agricultural Economics (2004), 55(1), 75–90**. This is published by permission of **John Wiley & Sons, Inc.**

The United States and Canada have great economic interests in exporting their transgenic crops. However, lack of public acceptance of genetically modified (GM) food products in the European Union (EU) and Japan have already resulted in reduced or curbed demand for GM food products. One suggested marketing strategy is consumer education about the safeness of GM foods. However, the results of the 1996 Eurobarometer survey suggested that more knowledgeable people do not necessarily have a more positive opinion; they just have a more definite opinion about biotechnology (Biotechnology and the European Public Concerted Action Group 1997). Furthermore, the 2001 Eurobarometer survey found that 70.9% of Europeans do not want genetically modified organisms (GMOs) in their food.

Consumer attitudes may be just as important as consumer knowledge. The Eurobarometer survey showed that moral doubts were more important than health risks in shaping public acceptance of gene technology (Biotechnology and the European Public Concerted Action Group 1997). Environmental concerns are also important — in the 2000 Eurobarometer survey, 59.4% of EU citizens interviewed said that they thought GMOs could have a negative impact on the environment. European consumer attitudes and behaviour toward genetically modified food products are complex, and a better understanding of them is essential for designing effective marketing strategies. This study investigates factors that affect consumer acceptance of GM food in Norway and the discount necessary for consumers to be willing to purchase GM food.

The Norwegian regulations regarding GM foods are somewhat stricter, but overall in accordance with the EU regulations. Differences with the EU are due to the fact that Norway is not an EU-member. The marketing of derived GM-foods is regulated by the "Act relating to the supervision and control of food and similar products."[1] The Norwegian Food Control Authority[2] is responsible for approval to use and market GM food material.[3] GM

[1]Næringsmiddelloven.
[2]Statens Næringsmiddel Tilsyn.
[3]GM-material includes food products that do not contain GMOs, but where GMOs are used in the production process.

foods may be approved if their use does not pose health risks to humans. Currently, the Food Control Authority has received several applications for approval of GM-foods, but the applications have been returned requesting more research concerning health risks. A second Law, The Gene Technology Act, regulates the import of living GM-organisms. The Norwegian Gene Technology Act requires that the production and use must be beneficial for society, ethically defendable, and contribute to sustainable development (Myrmæl and Tveito 2000). The Norwegian Directorate for Nature Management[4] evaluates potential environmental consequences of such imports.

The precautionary principle that dominates Norwegian and EU environmental and food safety policy, calls for preventive measures to be taken when an activity raises threats, even if direct cause-effect relationship has not been scientifically proven. In its strongest and most distinctive forms, the principle imposes a burden of proof on those who create potential risks, and it requires regulation of activities even if it cannot be shown that those activities are likely to produce significant harms (Sunstein 2002). The Precautionary Principle was applied regarding the use of antibiotic marker genes in transgenic crop varieties (Myrmæl and Tveito 2000) because there was a concern that such marker genes could lead to antibiotic resistance in humans. In 1997, all such varieties were banned in Norway, and consequently, Norway did not approve EU-approved genetically modified foods, which contained the marker genes. So far, only one transgenic tobacco plant and 11 transgenic carnation varieties are approved for importation (Myrmæl and Tveito 2000).

The Norwegian-government-appointed Walløe Commission, presented a report (NOU 2000), which argued that there are insufficient reasons to apply the Precautionary Principle to imported GM-food products. Their conclusions were partly based on the experiences from the United States and Canada where no health consequences have been discovered from use of GM food.

Since 1997, the requirements are that approved GM food products must be labeled as such if the food products/ingredients contain

[4] Direktoratet for Naturforvaltning.

more than 2% genetically modified materials.[5] The Food Control Authority recognizes that there inevitably are traces of GM material in batches of conventional crops, and they do not require approval in such instances. However, they are continuously evaluating what can be considered "unavoidable trace amounts" of GM material in conventional batches of imported crops.

The goals of this study are to analyse factors that induce Norwegian consumers to choose GM-food and to estimate their relative willingness to purchase GM-food with discounts compared to non-GM food. The paper is organised as follows: we first discuss previous literature on consumer preferences and attitudes towards GM foods with a focus on Norway; next, we introduce the survey data that is utilised in this study; we then present an empirical analysis of the data; and, finally, concluding remarks are offered.

2. Previous Studies

An analogous survey to the Eurobarometer survey was held in Norway[6] in 1993, 1996 and 1999, (Lund, Nielsen, and Skjåk, 2000). The results show that the scepticism in the Norwegian population toward gene technology is considerable. The percentage of people who thought that gene technology would make society better minus the percentage of people who thought it would make things worse was a negative 32% for Norway as opposed to a positive 9 for the EU in general. Possibly because of recognition of a potential for medical use, the percentage of Norwegian people who support biotechnology is higher than the percentage of people who are against it by 18 percentage points.

Grunert *et al.* (2000) investigated how perceptions of GM foods are related to basic motivators or life values in the Nordic countries of Denmark, Finland, Sweden and Norway. The GM food products

[5]The current EU threshold is 1% genetically modified materials. However, this regulation is expected to change in April 2004.

[6]Since Norway is not an EU member, Norway was not included in the Eurobarometer surveys.

considered were cheese, hard candy and salmon. Respondents were asked to rank the product varieties and were interviewed concerning their beliefs about each of the products. Grunert *et al.* (2000) found striking similarities in consumer preferences across the Nordic countries. The conventional product was associated with safety and good health, while the GM products gave negative associations including uncertainty, unnatural, diseases/ deformities, loss of species and ecological imbalance.

Consumers may be more positive about the use of biotechnology for health purposes. Grunert *et al.* (2000) found that GM confections with a low calorie content and that could be consumed by diabetics was valued by consumers. The food industry acknowledges the consumer scepticism and does not have any plans at present for developing GMO products for the Norwegian market (Abildgaard 2000).

Although most consumer studies concerning GM foods are qualitative in nature, there are a few published quantitative studies. For example, Burton *et al.* (2001) identified the willingness to pay to avoid GM foods in the United Kingdom. Of the quantitative studies, all but Boccaletti and Moro (2000) and Huffman *et al.* (2003) estimated willingness to pay (WTP) to avoid GM food. Boccaletti and Moro (2000) limit their study to GM foods with specific positive (hypothetical) attributes. Although one might expect for WTP for a non-GM product and the discount required for a consumer to choose the GM product to be roughly equivalent, this may not be the case empirically. Further, a non-GM claim implies that the product is not adulterated with GM ingredients, while the absence of a non-GM claim does not necessarily imply the presence of GM ingredients. This is different from stating positively that the product contains GM ingredients. Huffman *et al.* (2003) used a statistically based economic experiment to estimate the effect of a GM label on vegetable oil, tortilla chips, and potatoes. They found that their Midwestern U.S. participants required on average a 14% discount relative to the non-GM alternative. This paper attempts to contribute to this literature by quantifying the size of the GM discount in Norway.

3. Contingent Valuation and Willingness to Accept

The contingent valuation method (CVM) is a standard approach to elicit people's willingness to accept (WTA) or willingness to pay (WTP) through dichotomous choice, market-type questioning format with a direct survey such as via telephone, mail, and in-person questionnaire (Kanninen 1993). In the dichotomous choice CVM, each respondent is asked whether or not they would be willing to accept a particular discount for a particular good in a hypothetical market, letting them answer with "yes" or "no" along with the discounts "bid" amounts offered to each individual.

Our survey included contingent valuation questions regarding willingness to accept a discount to purchase bread made from genetically modified wheat. The hypothetical market for the good in question must be as close as possible to a real market in order to reveal people's true preferences if an actual market existed (Pearce and Turner 1990). The food product (bread) used in this study is appropriate for examination since bread is an everyday, staple food in Norway. All survey participants were familiar with bread as a food product and almost all consume it several times a week.

Consumers were first asked if they were willing to pay the same price for the GM- bread as the non-GM bread. If the respondent's answer to this question was "no," a follow-up question would be asked where the respondent was offered a percentage discount on the GM bread relative to the non-GM bread. (See the appendix for the English translation of the contingent valuation questions.) The discount was set at one of the following levels: 5%, 10%, 25%, 40%, and 50%.[7] Each level of discount was used for one fifth of the surveys. That is, eighty of the 400 surveys had a 5% discount for GM bread, another 80 surveys had a 10% discount for GM bread,

[7] A reviewer pointed out that the use of percentage discounts rather than absolute prices may affect the consumers' responses. This is an especially important point if consumers do not know the price of bread. However, discounts are widely used in sale offers, so consumers should be accustomed to responding to percentage discounts. Additionally, since only actual shoppers participated in the survey and because of the heavy consumption of bread in Norway, we believe that it is a reasonable expectation to think that our respondents know the price of bread.

and so on. The assignment of the survey version (and thus, discount) was random to the respondent. No follow-up question was asked if the customer's answer was "yes" to the initial question, and they were willing to purchase the GM bread at no discount.

The rationale for no follow up to a "yes" response is that the type of genetic modification associated with the GM wheat used to make the bread is a *process attribute*, which reduces production costs — as opposed to a *product-enhancing* attribute.[8] An example of a GM product with a product-enhancing attribute is the Flavr Savr Tomato. Proponents claim that the GM products with process attributes are identical to non-GM products.[9] Opponents view genetic modification as a negative attribute. Therefore, it would not make economic sense after an initial "yes", to respond with a follow-up question that involves paying a premium for these GM products that only have cost-reducing attributes. We did not discuss potential cost reductions that could be passed on to consumers with the respondents. However, if consumers perceive Roundup Ready wheat to reduce production costs, then in a "fairness" sense (Rabin, 1993) they may expect a discount.

Of the 400 respondents, more than one quarter of the sample would buy bread with GM wheat flour when offered no discount compared to conventional bread. Further, 12.4% of the sample replied that they would be willing to purchase the GM products if they were cheaper than the conventional products. For more specific statistics on the distribution of responses over the various discounts, see Table 1.

3.1. *Econometric model*

There are three possible outcomes in our methodology: (1) the respondent is willing to purchase the GM product at the same price

[8]An anonymous reviewer pointed out that although we stated that the wheat is Roundup Ready, respondents may not be aware that this is a process attribute. We acknowledge this is a potential drawback. Still, we did not mention any product-enhancing attributes to the respondents. The contingent valuation questions are presented in the appendix.
[9]Note that the GM bread is made with Roundup Ready wheat, so that reduced pesticide usage is not a benefit.

Table 1. Range and distribution of response rates to the randomly assigned discount offers.

	"Yes" to GM Food Product with No Discount	"No" If No Discount, But "Yes" with Discount Offered					No To Discount	Total N
Percent Discount	No Discount	5%	10%	25%	40%	50%	5% to 50%	
Respondents Who Chose GM Bread	107	7	7	12	10	14	243	400

as the non-GM product implying a "yes"; (2) the respondent is not willing to purchase the GM product neither at the same price but is willing to buy the GM food product at the random discount offered, *i.e.* a "no" followed by a "yes"; (3) the respondent is not willing to purchase the GM product neither at the same price nor at a discount relative to the non-GM product, *i.e.* "no" to both bids.

The model that is applicable to examine the outcomes of our survey can be considered a special case of the double-bounded logit model (Hanemann, Loomis, and Kanninen 1991). In the model, the initial bid B_0 equals zero and implies no price difference between GM and the corresponding, non-GM food products. The second bid (B_D), is the GM-food product offered at a random percentage discount relative to the corresponding, non-GM food product. This bid is only given to those individuals who answer that they would not buy GM food products at equal prices.

The sequence of questions isolates the range in which the respondents to the true WTA discounts for GM products relative to non-GM food products lie. The discount for the GM product relative to the non-GM product can be zero, B_0, or it can be located in one of the intervals (B_0, B_D), (B_D, $+\infty$) where B_D is the discount bid offered. The second bid, B_D, in conjunction with the response to the initial preference decision, allows bounds to be placed on the respondent's unobservable true WTA for GM food. The lower bound on the WTA discounts for GM food was determined *a priori* as no discount on GM-food in comparison to non-GM food because it is assumed that the genetic modification did not add any value to the product for the customer.

Let $B_D \leq 0$[10] denote the discount bid on the GM food product relative to the non-GM food product. The following discrete outcomes of the bidding process are observable:

$$D = \begin{cases} 1 & WTA \leq B_0 \\ 2 & B_0 < WTA \leq B_D \\ 3 & B_D < WTA \end{cases} \tag{1}$$

Respondents who indicate they require no discount fall into the first group, because they are willing to purchase the GM bread at zero discount compared to the non-GM bread. Those who fall into the second group require a discount that is positive but less than or equal to B_0 for GM bread. In order to choose the GM bread, the price of the GM bread must be discounted relative to the corresponding non-GM bread by an amount less than or equal to the discount bid. Finally, those who require the largest discount to choose GM bread fall into the third group. Customers in this group are not willing to purchase the GM-product at the discount offered. The WTA function for individual i is

$$WTA_i = \alpha + \rho B_i + \lambda' z_i + \epsilon_i \quad i = 1, \ldots, n \tag{2}$$

where B_i is the ultimate discount bid individual i faces, z_i is a column vector of observable characteristics of the individual, ε_i is a random variable accounting for random noise and possibly unobservable characteristics. Unknown parameters to be estimated are α, ρ, and λ. Linearity in z and ε is assumed. Furthermore, the distribution of the error term is assumed to follow $\varepsilon \sim G(0, \sigma^2)$, where $G(0,\sigma^2)$ denotes a cumulative distribution function with mean zero and variance σ^2. Under these assumptions, the choice probabilities individual i can be

[10]The randomly assigned discount bids were $B_D = \{0.05, 0.1, 0.25, 0.4, 0.5\}$, or expressed as percentages, 5%, 10%, 25%, 40%, and 50% of discount compared to the price of the conventional product.

characterized as[11]:

$$\text{Prob}\,(D_i = j) = \begin{bmatrix} G(\widetilde{\alpha} + \widetilde{\rho}B_{0i} + \widetilde{\lambda}'\mathbf{z}_i) \\ G(\widetilde{\alpha} + \widetilde{\rho}B_{Di} + \widetilde{\lambda}'\mathbf{z}_i) - G(\widetilde{\alpha} + \widetilde{\rho}B_{0_i} + \widetilde{\lambda}'\mathbf{z}_i) \\ 1 - G(\widetilde{\alpha} + \widetilde{\rho}B_{Di} + \widetilde{\lambda}'\mathbf{z}_i) \end{bmatrix}$$

$$\text{for} \quad j = \begin{bmatrix} 1 \\ 2 \\ 3 \end{bmatrix} \tag{3}$$

Thus, the log-likelihood function becomes:

$$L = \sum_{l} \begin{bmatrix} I_{D_i=1} \ln G(\tilde{\alpha} + \tilde{\rho}B_{0i} + \tilde{\lambda}'\mathbf{z}_i) \\ \quad + I_{D_i=2} \ln[G(\tilde{\alpha} + \tilde{\rho}B_{Di} + \tilde{\lambda}'\mathbf{z}_i) - G(\tilde{\alpha} + \tilde{\rho}B_{0i} + \tilde{\lambda}'\mathbf{z}_i)] \\ \quad + I_{D_i=3} \ln[1 - G(\tilde{\alpha} + \tilde{\rho}B_{Di} + \tilde{\lambda}'\mathbf{z}_i)] \end{bmatrix} \tag{4}$$

where I_K is an indicator function for the event K, and $D_i = j$ denotes that the jth alternative occurred. In the empirical implementation of the model, we define $G(\cdot)$ to be the standard logistic distribution function with mean zero and standard deviation $\sigma = \pi/\sqrt{3}$.

4. The Application

The survey was pre-tested with a small group of Norwegian subjects in December 2001 and conducted in January 2002, in the RIMI grocery store at Liertoppen in the Oslo-region of Norway. This region is the most populated part of Norway and lies in one of the Norwegian centers of economic activity. The survey data was collected at in-person interviews. By collecting data from consumers at the same time and place where actual purchase decisions are made, we hoped to better elicit consumers' true preferences about the products. Respondents were selected randomly with the criterion

[11]The condition of linearity on z is a simplifying assumption widely used in RUM (random utility models). This assumption implies that consumer's willingness to accept c dollars is generally represented as, $U(0, x_0, m) \leq U(1, x_1, m + c)\,\Pr[WTA \leq c] = \Pr[V_0 + \varepsilon_0 \leq V_1 + \varepsilon_1] = \Pr[\varepsilon_1 - \varepsilon_0 \leq V_0 - V_1]$ where $V_0 - V_1 = \alpha + \beta c$.

that the interviewer was to solicit every third customer who came in the survey area, and the turndown rate was about 5%. After the survey was completed, every respondent was given a gift certificate (worth approximately $5 in Norwegian Kroner).

In total, 400 consumers were surveyed. The majority of respondents are the primary food shoppers of the household (82%) and female (69%). Eighty-four percent of the shoppers in the survey shop for groceries daily or between two and five times a week. The respondents' average age is 41.6 years, which is close to the average age of 44 years for the general population of Norway in 1998. Fifty-seven percent of all respondents have children under the age of 18 years living in their household. The most frequent range of household income as a whole family is between 450,000 Norwegian Kroner (U.S. $50,300) and 600,000 Norwegian Kroner (U.S. $67,100)[12] for the 2001 fiscal year. This range is slightly higher than the average Norwegian income; in 1999, the average income for all households was Norwegian Kroner 381,500 which when inflated to 2001 represents an income of Norwegian Kroner 405,367. The most frequent level of education includes two to three years of college. This level of education may be somewhat higher than the national average; in 1998, the most frequent level of education completed was secondary school. The dataset contains 381 usable observations. Summary statistics and variable descriptions are presented in Table 2.

The survey asked for information relating to the respondents' attitudes concerning the environment, their self-reported knowledge and perceptions about biotechnology (see Table 3). Information about environmental attitudes was obtained by presenting a trade-off situation between environmental quality and economic growth (see the appendix for English translation of the environmental trade-off question). We find that 58% of the customers assign more importance to economic growth than preserving the environment.

[12]In order to obtain a high response rate, respondents were asked to place themselves in income intervals, rather than state their exact income amount. Survey respondents are typically reluctant to divulge incomes information. The exchange rate used is $1= 8.941 kroner.

Table 2. Summary statistics for the demographic variables.

Variable	Description	Descriptive Statistics
Age	Age of the consumer	Mean: 41.6 years
		St. Dev : 12.9 years
Female	1 if female,	69.3% females
	0 if male	30.8% males
Shopper	1 if main shopper	82.3% main shoppers
	0 otherwise	17.8% not main shopper
Education	compulsory school	15.5%
	HS diploma	29.3%
	2–3 year college	32.1%
	4-5 year degree	20.1%
	Adv./Prof. degree	2.3%
	refuse	0.5%
	0 = compulsory school, HS diploma, refuse 1 = 2 − 3, 4–5 year college, Adv./Prof. degree	
Children	1 if children <18 in the household	56.8%
	0 otherwise	43.3%
Household Size	Number of people in the household	SMean: 3.14
		St. Dev. 1.37
Income	1 = <150 NOK	3.6%
	2 = 150-300,000 NOK	19.5%
	3 = 300-450,000 NOK	23.6%
	4 = 450-600,000 NOK	27.7%
	5 = 600-750,000 NOK	13.2%
	6 = 750-900,000 NOK	6.9%
	7 = >900,000 NOK	5.6%

We acknowledge that a trade-off between economic growth and preserving the environment does not always exist. There can exist "win-win" opportunities, such as is suggested by the "Porter Hypothesis," in which environmental standards increase growth. Although an environmental trade-off question is not perfect, it can be an effective way of ensuring that the survey information is informative as well as useful in an empirical modeling context. For example, without the trade-off, most respondents will say that they value the environment highly. The resulting lack of variation in response can lead to a lack of statistical significance.

Table 3. Summary statistics for consumer information and perception variables.

Variable	Description	Scaled Values	Descriptive Statistics
Food Safety	Importance of food safety versus price	(1,2,. . . ,10) 1 = Food safety 10 = Cheaper food	Mean: 4.51 St. Dev.: 2.95
Risk	Risk Associated with GMOs. No risk, some risk, or high risk	1 = High risk 0 = Some, no risk, don't know	Mean: 0.48
Opinion	Favorable Opinion toward Use of Biotech in Food Production	1 = Very, somewhat positive, neutral, don't know 0 = Somewhat, very Negative	Mean: 0.50
Knowledge	Self-Reported knowledgeable about biotechnology	1 = Know a lot, know something 0 = Know little	Mean: 0.61
GM labelling	Views about the Importance of labeling GM foods	1 = Very important 0 = Somewhat, not Important	Mean: 0.87
Domestic	Preference of domestic vs. imported food products	1 = Yes 0 = No	Mean: 0.74

As in all surveys, sample representativeness is of concern. One possible sample bias is population choice, in which the population sampled does not adequately correspond to the population that would purchase genetically modified food products. The RIMI chain of grocery stores has chosen a low-price/limited-selection niche in the market and has in this way gained significant power in the Norwegian marketplace. Although the store-selection is wide enough to include all the most frequently consumed foods in Norwegian households, the chain does not focus on specialty products, such as organic and ecologically produced foods. This potential bias may be mitigated by the location of the grocery store at a mall where many different people drop by to pick up a few grocery store items. Further, the slightly above average income and education levels of our sample, indicates that the sample is not limited to low-income shoppers. Given the preceding concern, we acknowledge that the extent to which the findings can be fully generalised to broader populations is uncertain.

5. Empirical Analysis

The bid information and other information about the respondents were used to estimate the magnitude of factors that affect Norwegian consumers' choice of whether to purchase GM-foods and, specifically, how much of a relative discount Norwegian consumers will require to purchase GM bread. In the model, covariates to the bid information were self-reported knowledge about biotechnology (*Knowledge*), opinion regarding use of biotechnology in food production (*Opinion*), degree of preference for economic growth over environmental protection (*Enviro)*, education level (*Edu)*, gender (*Female*), age (*Age*) and whether there were children under 18 in the household (*Children*). The model (2) was estimated where

$$\mathbf{z}_i = \{Knowledge_i\ Opinion_i\ Enviro_i\ Edu_i\ Female_i\ Age_i\ Children_i\}.$$

Variable definitions and descriptions are given in Tables 2 and 3. Parameters to be estimated are $\lambda' = \{\lambda_1\lambda_2 \ldots \lambda_7\}$ in addition to α and ρ .

We include the knowledge and attitude variables in the empirical model because they have been statistically significant in previous studies. In their investigation of U.S. consumers' acceptance of GM corn flakes, Baker and Burnham (2001) found that cognitive variables (opinions, beliefs, knowledge) have a great influence on consumer preferences.

5.1. *Analysis of factors that affect the choice of whether to purchase GM-bread*

The parameter estimates are reported in Table 4. Not surprisingly, respondents who are favorable about use of biotechnology in food need a smaller discount to purchase GM-bread.

Table 4. Parameter estimates for WTA model.

Parameter	Variable Description	Estimate	t-value	p-value
α	*Intercept*	0.1799	0.322	0.7477
ρ	*Discount Bid*	2.2799	7.821	0.0000
λ_1	*Knowledge*	−0.1799	−0.802	0.4226
λ_2	*Opinion*	1.3347	5.898	0.0000
λ_3	*Environment*	−0.0829	−1.488	0.1367
λ_4	*Education*	0.4224	1.876	0.0606
λ_5	*Female*	−0.4412	−1.955	0.0506
λ_6	*Age*	−0.0283	−3.166	0.0015
λ_7	*Children*	−0.2032	−0.878	0.3798

The estimation results also indicate that on average, female respondents need higher discounts to choose the GM-bread. Baker and Burnham (2001) also find that women are more sceptical about GM-foods than men. The necessary discount also significantly increases with age. This means that older consumers are less accepting of GM foods than younger consumers. This may reflect that consumers generally become more health conscious as they age and become more susceptible to illness. On the other hand, it could also be interpreted to mean that younger consumers are more accepting of biotechnology. If the latter is the case, then it is promising for the future of biotechnology. Li *et al* (2003) also find that age is a significant factor for their Chinese sample. Interestingly, the estimation results show higher formal education levels decrease the required discount. The variable for self-reported knowledge about biotechnology is not statistically significant. The variables representing environmental attitudes and the presence of children under 18 both have the expected sign but are not statistically significant at conventional levels.

5.2. *Estimates of confidence intervals for the mean discount for GM-food*

The empirical mean discount necessary to choose GM-bread was estimated at the mean of the covariates and as $-(\widetilde{\alpha} + \overline{z}\widetilde{\lambda})/\widetilde{\rho}$. Our results indicate that, on average, the consumers in our sample require

Table 5. Required discount or WTA for GM bread (%) with confidence intervals.

Mean WTA Estimated	Average Discount Needed, $-(\widetilde{\alpha}+\widetilde{\bar{z}\lambda})/\widetilde{\rho}$	Bootstrap Standard Error	95% Confidence Interval
GM bread	0.4987	0.06539	(0.3705, 0.6269)
GM bread, below average age of the sample	0.3259	0.07162	(0.1856, 0.4663)
GM bread, above average age of the sample	0.6260	0.10204	(0.4260, 0.8260)[13]

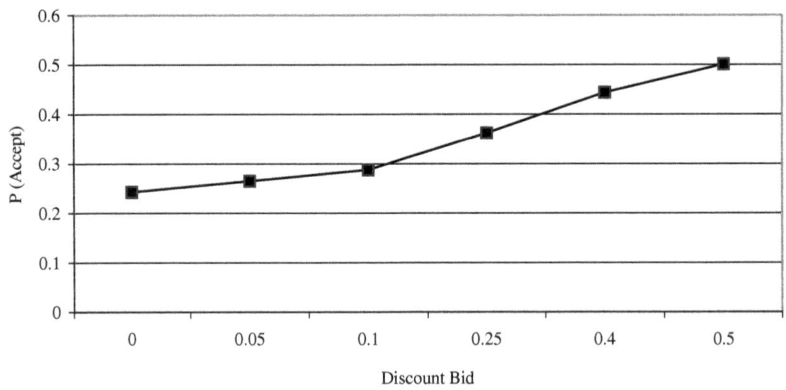

Fig. 1. Probability of choosing GM-bread as the discount varies.

a discount of 37% to 63% (this is a 95% confidence interval) relative to conventional bread to choose the GM bread (see Table 5). The distribution of the probability of accepting the discount bid offered over the range of bids is displayed in Figure 1. The confidence intervals (CIs) for the mean discount estimates are given in Table 5, and were calculated using 1000 bootstrap samples.[14]

Although the mean required discount (49.9%) is reasonable given the survey responses, a caveat is that it is just inside the highest

[13]As discussed previously, the mean WTA for respondents above the mean age that is outside the highest discount offered has a much wider confidence interval. Since there is great uncertainty surrounding this estimate, a safe interpretation of the mean WTA is as a required discount that is greater than 50%.

[14]$\mathrm{CI} = -(\tilde{\alpha}+\tilde{z}\tilde{\lambda})/\tilde{\rho} \pm \mathrm{Z}\ \mathrm{SE}_{\mathrm{Bootstrap}}$ for a 95% confidence interval Z = 1.96 and for a 90% confidence interval Z = 1.65.

discount offered (50%). When mean discount estimates are obtained from an extrapolation of the willingness-to-accept function, the level of uncertainty increases as the extrapolation gets further outside the highest discount. With an extrapolation, the confidence intervals around the mean discount estimate must be wider than otherwise would be necessary. An improvement for future work should include offering segments of respondents much larger discounts for GM foods, possibly even a 100% discount (given away for free).[15] Still, the estimation procedure is consistent, and the key issue in this survey calibration is the degree of precision in the final estimator. Since precision is the key issue from our survey design, we suggest that the confidence interval of 37% to 63% is the result that readers should focus on.

There is a strong age-discount relationship in the data indicating that the current scepticism towards GM food in the Norwegian population may be moderated over time. We are, therefore interested in estimating the difference in mean discount between younger and older respondents. In addition to including age as an integer variable in the previous model, we divided our sample into two groups: the first includes respondents older than the average age of the sample (41.6 years) and the second includes respondents younger than the average age. We then estimated the mean required discount for each of the two groups. To be willing to purchase GM bread, we find that respondents below the average age of 41.6 years on average only need a discount of 39.6% while for customers older than the average age the average discount needed is 62.6%. The confidence intervals are wider as we estimate the mean required discount at sub-samples and because the discount needed for the above-average age group is outside the range of the data.

6. Conclusions

In this paper, we analyse factors that induce Norwegian consumers in our sample to choose genetically modified bread and estimate the

[15] Note the pilot study did not detect the need for a much higher discount.

discount required for them to choose this GM product. Our data includes the results of in-person interviews with 400 shoppers in the Oslo-region of Norway conducted in January 2002. A better understanding of Norwegian consumers' attitudes and behaviour toward genetically modified food products and how these attitudes affect the purchasing choices for such food products is essential for marketing GM food products in Norway.

The results of this study suggests that our sample of Norwegian consumers are, on average, willing to purchase GM bread with a 49.9% discount compared to the corresponding non-GM product. This mean required discount to choose the GM bread is close to the highest discount offered. The consequence of high levels of rejection is that the willingness to accept function had to be extrapolated outside the data, so that the level of uncertainty is greater than it would have been if higher discounts had been offered. Still, we can infer from the results that a transformation of Norwegian consumers' perceptions and attitudes are needed for GM food products to successfully enter the Norwegian market.

To the extent that these findings apply more generally, this research highlights the effects of consumer attitudes, self-reported knowledge about biotechnology, socio-demographic characteristics, and price on Norwegian consumers' decision on GM bread. The Norwegian policy restricting GM food products and the results of surveys show that there is considerable scepticism in the Norwegian population toward GM foods, with women on average being more sceptical than men.

The strong scepticism in the Norwegian population may decrease over time, leaving a potential future market for GM foods. People younger than the average age of the sample were willing to purchase the GM bread with an average discount that is far less than the customers above the average age of the sample needed. The gap in between generations for WTA, indicate that younger people may be more open to GM foods and that it may be a question of time before the Norwegian market may be more open to GM food products.

Acknowledgment

The authors wish to thank without implicating Phil Wandschneider and Kyrre Rickertsen for helpful comments. The authors gratefully acknowledge financial support from the IMPACT Center at Washington State University.

References

Abildgaard, C. (2000). Forskningsrådets prioriteringer i forhold til forskning på genmodifisert mat og innvirkninger på helse og miljø. *Genialt* (4), 20–22.

Baker, G.A. and T.A. Burnham (2001). Consumer Response to Genetically Modified Foods: Market Segment Analysis and Implications for Producers and Policy Makers. *Journal of Agricultural and Resource Economics* 26(2), 387–403.

Biotechnology and the European Public Concerted Action Group (1997). Europe ambivalent on biotechnology. *Nature* (387), 845–847.

Boccaletti, S. and D. Moro (2000). Consumer Willingness-To-Pay For GM Food Products in Italy. *AgBioForum* 3, 259–267.

Grunert, K.G., L. Lähteenmäki, N.A. Nielsen, J.B. Poulsen, O. Ueland and A. Åström (2000). Consumer perception of food products involving genetic modification: results from a qualitative study in four Nordic countries. Working Paper no 72. MAPP — Centre for Market Surveillance, Research and Strategy for the Food Sector, Århus, Denmark.

Hanemann, W.M., J. Loomis and B.J. Kanninen (1991). Statistical Efficiency of Double-Bounded Dichotomous Choice Contingent Valuation. *American Journal of Agricultural Economics* 73, 1255–1263.

Huffman, W.E., J.F. Shogren, M. Rousu and A. Tegene (2003). Consumer Willingness to Pay for Genetically Modified Food Labels in a Market with Diverse Information. *Journal of Agricultural and Resource Economics* 28(3), 481–502.

Hviid Nielsen, T. (2000). Vitende Skepsis. *Genialt* (4), 35–37.

Kanninen, B.J. (1993). Optimal Experimental Design for Double-Bounded Dichotomous Choice Contingent Valuation. *Land Economics* 69(2), 138–146.

Li, Q., K.R. Curtis, J.J McCluskey and T.I. Wahl (2003). Consumer Attitudes toward Genetically Modified Foods in China. *AgBioForum* 5(4), 145–152.

Lund, M., T. Hviid Nielsen and K. Kalgraff Skjåk (2000). Norwegian attitudes to biotechnology. Norsk Samfunnsvitenskapelig Datatjeneste (NSD). Report no. 118.

Lusk, J.L., J. Roosen and J.A. Fox (2003). Demand for Beef from Cattle Administered Growth Hormones or Fed Genetically Modified Corn: A Comparison of Consumers in France, Germany, the United Kingdom, and the United States. *American Journal of Agricultural Economics* 85, 16–29.

Michael B., D. Rigby, T. Young and S. Sallie James (2001). Choice Modelling Consumer Attitudes to GMOs in Food. *European Review of Agricultural Economics* 28(4), 479–498.

Myrmæl, A. and G. Tveito (2000). Hvilken betydining har internasjonale avtaler for den politikken Norge kan føre når det gjelder genmat? *Genialt* (4), 24–26.

NOU [Norsk Offentlig Utredning] (2000). No. 29. GMO-mat. Helsemessige konsekvenser ved bruk av genmodifiserte naeringsmiddler og naeringsmiddelingredienser. Ministry of Health, Oslo, Norway.

Pearce, D.W. and R.K. Turner (1990). *Economics of Natural Resources and the Environment.* Harvester Wheatsheaf.

Rabin, M. (1993). Incorporating Fairness into Game Theory and Economics. *American Economic Review* 83, 1281–1302.

Sunstein, C.R. (2002). Beyond the Precautionary Principle. University of Chicago Law & Economics, Olin Working Paper No. 149.

Verbeke, W., R.W. Ward and J. Viaene (2000). Probit Analysis of Fresh Meat Consumption in Belgium: Exploring BSE and Television Communication Impact. *Agribusiness* 16, 215–234.

Appendix

Question used to elicit environmental attitude

Where would you place yourself on a scale from 1 to 10, if economic growth at all costs is a 1 and saving the environment at all costs is a 10. (CIRCLE JUST ONE)

1 2 3 4 5 6 7 8 9 10

Questions related to willingness to accept

Scientists at a U.S. university are developing Round-up Ready wheat, which is genetically modified. Would you be willing to purchase bread made with this wheat if it were offered at the same price as bread made without genetically modified wheat?

Yes (skip next question)
No

Would you be willing to purchase this bread if it were offered at a price that is [INSERT_RANDOM DISCOUNT] less than bread made without Round-up Ready wheat?

Yes
No

Chapter 18

Consumer Response to Genetically Modified Food Products in Japan*

Jill J. McCluskey*, Kristine M. Grimsrud†, Hiromi Ouchi‡, and Thomas I. Wahl§

**Department of Agricultural Economics, Washington State University*

†*University of New Mexico*

‡*Department of Agricultural Economics, Washington State University*

§*International Marketing Program for the Agricultural Commodities and Trade (IMPACT) Center, Washington State University*

In Japan, a large U.S. export market, there has been growing public opposition against genetically modified (GM) foods. Using a dichotomous choice contingent valuation method, findings show the discount needed for Japanese Seikyou consumers to purchase GM food products is positively affected (i.e., a greater discount is required) by higher levels of self-reported risk perceptions toward GM food, higher levels of concern about food safety and the environment, higher self-reported knowledge about biotechnology, education levels, and income. Interestingly, gender does not significantly affect the discount needed for GM food. Further, it can be inferred from the results that a transformation of Japanese consumers' perceptions and attitudes is needed for GM food products to successfully enter the Japanese market.

Keywords: Biotechnology, consumers, genetically modified foods, Japan, willingness to purchase

*This article originally appeared in the **Agricultural and Resource Economics Review (2003), 32(2), 222–231**. This is published by permission of **Cambridge University Press**.

Biotechnology has been touted as the future of agriculture. However, the business leaders, producers, and scientists involved in biotech did not foresee how controversial genetically modified (GM) foods would be with consumers, particularly for export markets such as Japan. A better understanding of consumer attitudes and behavior toward genetically modified food products in these export markets is essential for designing market strategies. This study focuses on consumers' choices and willingness to purchase GM food products with possible discounts.

In Japan, 34 local self-governing bodies are engaged in research and development of GM products (*Asahi Shimbun*, 2001). Yet, most of their products will not be marketed in Japan because of the growing public opposition to GM foods. Reflecting this trend, U.S. export orders have increased for soybeans and corn which are not genetically modified, even at premium prices.

In the United States in 1999, premiums of 8¢ to 15¢ per bushel were offered for non-GM corn, and premiums of 5¢ to 35¢ per bushel were offered for non-GM soybeans at the farm level (Nelson *et al.*, 1999). In response, some Japanese food companies and farmers have switched to non-GM ingredients and seeds. For example, Asahi and Kirin, the two leading Japanese beer companies, have announced they will switch entirely to non-GM ingredients, and soybean farmers who do not use GM seeds are enjoying a huge demand for their beans (Tolbert, 2000).

Labeling policies for GM foods are rapidly evolving worldwide. Japan's new system of labeling GM foods went into effect on April 1, 2001. Foods made from GM crops are required to be labeled as "genetically modified." Foods made from non-GM crops, which have been identity preserved, can be voluntarily labeled as "not genetically modified." Finally, foods that have not been identity preserved must be labeled as "not segregated from GM product." While few Japanese products carry labels identifying them as "genetically modified," many products claim to be free of genetically modified organisms (*Shinano Mainichi Shimbun*, 2001). GM status labels not only provide consumers with information, they also give the consumer the opportunity to choose. McCluskey (2000) argues that the appropriate labeling policy for GM foods depends on the size of the market and willingness to pay for non-GM foods relative to the costs of identity preservation.

The heated debate over labeling issues in Japan reflects increasing consumer concerns about food safety issues, including effects on human health and the environment. Several Japanese consumer groups actively campaigning against genetic modification of food are insisting on mandatory labeling of GM food products. These groups assert GM products are not being adequately tested for safety. Many school lunch programs in Japan serve non-GM foods to the extent possible, despite the efforts of the Japanese Ministry of Agriculture to convince consumers that GM products are safe to eat.

The objective of this study is to identify and analyze specific factors which induce Japanese consumers to choose GM foods. The remainder of the article is organized as follows. In the section below, we discuss the previous literature on consumer preferences and attitudes toward GM foods, with particular attention given to Japanese attitudes toward food safety. A description of the survey data utilized in this analysis is then provided. The next section presents the empirical analysis, and the factors that affect consumers' willingness to purchase discounted GM food products are analyzed. Concluding remarks are offered in the final section.

1. Related Studies

In recent years, the issue of GM labeling has received considerable attention. However, only a few published studies have analyzed consumer willingness to pay (WTP) or willingness to accept (WTA) compensation for food products containing GM ingredients. Lusk *et al.* (2001) estimated willingness to pay for corn chips made without genetically modified ingredients. In their experimental study, junior- and senior-level agricultural economics students at Kansas State University indicated their WTP by exchanging a bag of GM corn chips for a bag of GM-free corn chips. WTP findings revealed that an individual who is very concerned about GM foods would be 50% more likely to pay a premium to exchange GM chips for non-GM chips compared to an individual with little concern for GM foods. However, 70% of the study participants stated they were not willing to pay a premium for non-GM chips. The average bid to exchange GM chips for non-GM chips was $0.07/ounce. Still, 20% of participants were willing to pay at least $0.25/ounce for the exchange, and 2% offered bids of $0.50/ounce, suggesting there may be a potential niche market in the United States for non-GM food products.

Baker and Burnham (2001) investigated U.S. consumers' acceptance of GM corn flakes, and reported that 30% of those surveyed based their purchasing decision on GM content. Their findings show that cognitive variables (opinions, beliefs, knowledge) have a great influence on consumer preferences. The level of risk aversion, knowledge about genetic modification, and opinion about genetic modification are highly significant in explaining the purchasing decision. Earlier studies investigating the relationship between consumer characteristics and food safety concerns have generally found that sociodemographic variables (such as education and income) perform poorly as explanatory variables for purchasing decisions regarding GM food products. The exception is that women, in general, are more concerned with food safety.

Lusk, Roosen, and Fox (2003) estimate consumer willingness to pay for beef in France, Germany, the United Kingdom, and the United States. Their analysis uses a variety of quality variables, including whether the cattle were fed GM corn. Results suggest European consumers place a much higher value on beef from cattle that have not been fed genetically modified corn compared to their consumer counterparts in the United States. Yet, in a study of the discrepancy between European public opinion and consumer purchase behavior with regard to GM food products, Noussair, Robin, and Ruffieux (2002) found that consumers are typically unaware of the labeling indicating GM content.

A small number of survey studies have assessed consumer perceptions about food safety and bio technology in developed societies, including the United States and Japan. Jussaume and Judson (1992) studied the effects of sociological characteristics on consumer attitudes toward food safety. Using responses to mail surveys in Seattle, Washington, and Kobe, Japan, they examined differences between consumers in the two countries. Residents of Kobe were found to be significantly more concerned about food safety than residents in Seattle. For both countries, households more likely to care about food safety issues are those with children under the age of 18, who belong to consumer cooperatives, and with relatively high incomes.

Using telephone surveys conducted in Japan and the United States in 1995 and in 1998, Hoban (1999) studied consumer awareness and acceptance of biotechnology and willingness to purchase GM foods. Despite the recent controversy over the use of biotechnology, Hoban observed increasing acceptance of genetic modification for food production by Japanese and U.S. consumers between 1995 and 1998, and increasing numbers of consumers who are willing to buy GM foods in both countries. He concludes that the use of biotechnology does not negatively affect consumers' willingness to purchase those foods. However, Hoban cautions that consumer awareness and understanding of biotechnology is still relatively low in Japan compared with the United States.

Consumer attitudes toward biotechnology have been changing since Hoban's study, as more information has been provided to the pubic — primarily through the media. Macer and Chen Ng (2000) report that support for biotechnology and genetic engineering in Japan is decreasing, especially for agricultural applications. Based on a mail survey spanning the years 1991, 1993, 1997, and 2000, they found Japanese interest in science and biotechnology increased from 30% in 1991 to 47% in 2000. From the 2000 survey, 97% of respondents reported familiarity with the term "biotechnology," implying awareness of biotechnology has increased significantly among the Japanese public. Also from the 2000 survey, only 31%

of respondents stated they are likely to support GM foods, and only 20% said they are willing to buy GM fruits. Macer and Chen Ng conclude that although the majority of Japanese consumers are optimistic about biotechnology, they hold increasingly negative views toward its application for agriculture.

Since these studies were conducted, recent events in Japan may have affected consumer sentiments. The disappointing economy, government scandals, and the discovery of bovine spongiform encephalopathy (BSE, also known as "mad cow disease") in Japan may have caused consumers to be less trusting of government reassurances and GM food products.

2. Survey Data

After pre-testing with Japanese subjects in Nagano and Matsumoto, Japan, the survey used in this study was conducted at the Seikatsu Club Consumer Cooperative (Seikyou), a grocery store-like setting, in Matsumoto City, Japan, during June 2001. The survey data were collected with in-person interviews carried out in Japanese. By collecting data from consumers at the same time and place where actual purchase decisions are made, we hoped to better elicit consumers' true preferences about the products. Respondents were selected randomly, with the criterion that the interviewer was to solicit every third customer who came into the survey area.

Every respondent was given a gift certificate valued in Japanese yen (worth approximately U.S.$12, where U.S.$1 = 121 yen) as an incentive in return for participation in the survey. The turndown rate was about 50%, as observed by interviewers when they asked for participation.

In total, 400 consumers were surveyed. The majority of respondents are the primary food shoppers of the household (77%) and female (78%). Sixty-seven percent reported they shop for groceries daily or between two and five times a week. Most of the respondents are in their 40's, which is the average age of the population in Matsumoto City, as well as in Japan. Fifty-four percent of all respondents reported having children under the age of 18 years living in their household. This percentage is higher than Japan's national average, where 28% of all households have children under the age of 18.

For the 2000 fiscal year, the average household income of survey respondents ranged between 5,100,000 yen (U.S.\$42,011) and 7,600,000 yen (U.S.\$62,604).[1] The Matsumoto average income of 6,178,884 yen falls within this average income range. However, the average income level reported by survey participants is somewhat below Japan's average yearly household income for fiscal year 2000, which was 7,100,000 yen (U.S.\$58,485). The average education level of survey respondents included some years of college, while the Japanese average includes only high school. Summary statistics and descriptions of variables are presented in Table 1.

The survey solicited information regarding respondents' attitudes about the environment and food safety, and their self-reported knowledge and perceptions about biotechnology (see Table 2). Information about environmental and food safety attitudes was obtained by presenting tradeoff situations between environmental quality and economic growth, and between food safety and low prices, respectively (refer to the appendix for an English translation of the tradeoff questions).

Eliciting these attitudes from tradeoff scenarios is an effective way of ensuring the survey information is informative as well as useful in an empirical modeling context. For example, without the trade-off, most respondents will state that they value the environment highly. The resulting lack of variation in response can lead to a lack of statistical significance of the effect of the environmental variable.

As in all surveys, a primary consideration is that the sample is representative of the population under study. The potential bias of

[1] In order to obtain a high response rate, respondents were asked to place themselves in income intervals, rather than state their exact income amount. Survey respondents are typically reluctant to divulge information about income. The exchange rate used is U.S.\$1 = 121.398 yen.

Table 1. Summary statistics for demographic variables: Matsumoto city, Japan, survey participants, June 2001 ($N = 400$).

Variable	Description (coding)	Distribution of Responses, with Related Information	
Age	30 years	12%	Mean = 45.5 years; Std. Dev. = 13.9 years
	30 to 50 years	50%	
	50 to 70 years	32%	
	>70 years	5%	
Female	1 if female	78%	
	0 if male	22%	
Shopper	1 if main shopper	77%	
	0 otherwise	24%	
Education	Compulsory school	3.75%	CODING FOR ESTIMATION:
	HS diploma	42.50%	0 if compulsory school, HS diploma, or refuse;
	2–3 years college	19.50%	1 if 2-year college, 4-year degree, or advanced/
	4–5 years college degree	17.25%	professional degree
	Advanced/professional degree	13.50%	
	Refuse to answer	3.50%	
Children	1 if children under 18 years in household	54%	
	0 otherwise	46%	
Income	<2,500 yen	3.75%	Mean = 6,350 yen; Std. Dev = 2,500 yen
(in 1,000 yen)	2,510–5,000 yen	22.25%	
	5,010–7,500 yen	26.00%	CODING FOR ESTIMATION
	7,510–10,000 yen	15.75%	0 if less than average income;
	>10,010 yen	11.50%	1 if above average income
	Refuse to answer	20.75%	
Household	Number of people in household	—	Mean = 3.53 people; Std. Dev. = 1.31 people

Table 2. Summary statistics for consumer information and perception variables: Matsumoto city, Japan, survey participants, June 2001 ($N = 400$).

Variable	Description (coding)	Distribution of Responses
Environment	Importance of environmental sensitivity vs. economic growth, based on a scale of 1 to 10 where:	
	1 = economic growth is all-important	Mean = 6.5
	10 = environment is all-important	Std. Dev. = 1.91
Safety	Importance of food price vs. food safety, based on a scale of 1 to 10 where:	
	1 = food price is all-important	Mean = 7.9
	10 = food safety is all-important	Std. Dev. = 1.94
Risk	Risk associated with GM foods:	
	1 if high or low risk	74%
	0 if no risk	26%
Opinion	Opinion about use of biotechnology:	
	1 if favorable or neutral opinion	12%
	0 if negative opinion	88%
Knowledge	Self-reported knowledge about biotechnology:	
	1 if high or little knowledge	82%
	0 if no knowledge	18%
Label	Importance of labeling GM foods:	
	1 if labeling very important	98%
	0 if labeling somewhat or not very important	2%
Import	Preference for domestic vs. imported food products:	
	1 if preference is for domestic food	92%
	0 if no preference	8%

greatest concern in this study is the population choice bias — i.e., the possibility that the population chosen does not adequately correspond to the population which will purchase genetically modified food products.

As a marketing strategy, consumer cooperatives usually focus on offering "safe foods," and target members who are more willing to purchase safe foods (Jussaume, 1998). Therefore, as members of a consumer cooperative, we expect Seikyou consumers to have stronger concerns about food safety than consumers who shop at other grocery stores. A mitigating factor is that the Seikyou has significant power in the Japanese marketplace. A potential offsetting bias is that Matsumoto is a relatively agricultural area. About 13% of Matsumoto's population is represented by farm households, compared to 2% for all of Japan (Japan Ministry of Public Management, 2002). Consequently, residents of Matsumoto may be more closely associated with agricultural production and have more knowledge about the production processes.

Slovic (1987) suggests risk is influenced by two major factors: dread risk and unknown risk. We suspect agricultural residents are less sensitive to GM products because their level of unknown risk is probably lower. Further, while our response rate of approximately 50% is respectable, it does suggest the possible presence of non-respondent bias.[2] Given the preceding concerns, we caution that the extent to which the findings can be fully generalized to broader populations is uncertain.

3. Empirical Analysis

The empirical analysis associated with this research is divided into three subsections. The first provides a discussion of the contingent valuation dichotomous choice methodology used in the study. The second describes the econometric model used in the quantitative analysis of the data. The final subsection presents an analysis of

[2]Although 50% is lower than the commonly accepted 70% CV standard, there is not enough evidence to determine whether 50% is low for intercept interviews in Japan. One concern with a lower response rate is that people with a lower value of time, such as retirees or unemployed people, may be overrepresented. A mitigating factor in this survey is that the summary statistics suggest the average age of the sample is representative of the general population. A second concern is that those individuals who are more concerned about genetically modified foods elected to participate in the survey. If this is the case, it would bias the WTA results upward.

the factors affecting consumers' willingness to accept discounts to purchase GM foods.

3.1. *Contingent valuation dichotomous choice methodology*

The contingent valuation method (CVM) is a standard approach to elicit people's willingness to accept (WTA) through a dichotomous choice, market-type questioning format with a direct survey such as via telephone, mail, and in-person questionnaire (Kanninen, 1993). In the dichotomous choice CVM, each respondent is asked whether he/she would be willing to accept a particular discount for a particular good in a hypothetical market. A respondent answers with "yes" or "no," to the discount "bid" amount offered.

There are typically two types of bidding procedures used in the CVM: the single-bounded and the double-bounded dichotomous choice, with the double-bounded model gaining in popularity (Kanninen, 1993). The single-bounded model approach recovers the bid amount as a threshold by asking only one dichotomous choice question (Hanemann, Loomis, and Kanninen, 1991). The statistical efficiency of this approach can be improved by use of the double-bounded model, which engages in two bids.

Recent works using the double-bounded approach include an analysis by Yoo and Yang (2001) who estimated the benefit of a tap water quality improvement in Korea, and Hutchinson *et al.* (2001) who estimated the benefits from outdoor recreation in Northern Ireland. In the latter study, the authors provide empirical evidence of the gains in the statistical efficiency of both benefit and parameter estimates obtained by analyzing follow-up responses with double-bounded interval data analysis.

However, the double-bounded approach has also been critically evaluated. Hanemann, Loomis, and Kanninen (1999) note: "... there is also some bias in going from a single- to a double-bounded format because there is some evidence that some of the responses to the second bid are inconsistent with the first bid" (p. 382). They conclude, even if the double-bounded approach produces some bias, "the experience to date suggests that the bias is in a conservative direction and is greatly outweighed by the gain in efficiency in terms of minimizing overall mean squared error" (p. 388).

Our survey included contingent valuation questions regarding willingness to accept a discount to purchase noodles made from genetically modified wheat.[3] The hypothetical market for the good in question must be as close as possible to a real market in order to reveal consumers' true preferences if an actual market existed (Pearce and Turner, 1990). The food product (noodles) chosen for this study is appropriate for examination, because noodles are a frequently consumed food product by most of the Japanese population.

Consumers were first asked if they were willing to pay the same price for the GM noodles as the non-GM noodles. If the respondent's answer to this question was "no," a follow-up question was asked where the respondent was offered a percentage discount on the GM noodles relative to the non-GM noodles. The discount was set at one of the following levels: 5%, 10%, 25%, 40%, or 50%. Each discount level was used for one-fifth of the surveys — i.e., 80 of the 400 surveys had a 5% discount for GM noodles, another 80 surveys had a 10% discount for GM noodles, and so on. The assignment of survey version (and thus discount) was random to the respondent.

No follow-up question was asked if the consumer's answer was "yes" to the initial question, indicating a willingness to purchase the GM noodles at no discount. The rationale for no follow-up to a "yes" response is that the type of genetic modification associated with GM noodles is a *process attribute*, which reduces production costs (as opposed to a *product-enhancing* attribute). An example of a GM product with a product-enhancing attribute is the Flavr Savr tomato. Proponents claim the GM products with process attributes are identical to non-GM products.[4] Opponents view genetic modification as a negative attribute. Therefore, it would not make economic sense, after an initial "yes" response, to pose a follow-up question involving

[3]In a second set of contingent valuation questions, survey respondents were asked about their willingness to purchase tofu made with genetically modified soybeans. The results are very similar, and are not presented here.

[4]Note that the GM noodles are made with Roundup Ready wheat; thus, reduced pesticide usage is not a benefit.

Table 3. Distribution of response rates to the randomly assigned discount offer for GM noodles.

Description	"Yes"	<"Yes" with Discount>					"No" with Discount	Total
Discount	0%	5%	10%	25%	40%	50%		
Response rate to each discount	3.5%	1.8%	2.5%	3.5%	5.3%	3.8%	79.8%	100%

payment of a premium for these GM products which have only cost-reducing attributes.

Of the 400 respondents, only 3% said they would be willing to purchase the genetically modified noodles at the same price as the non-GM noodles (the "yes" group). Further, only 17% of consumers in the sample stated they would be willing to purchase the GM noodles if they were less expensive than the non-GM noodles (the "no, yes" group). The remaining 80% of respondents were not willing to purchase the GM noodles, even with the discount (the "no, no" group). For more specific statistics on the distribution of responses over the various discounts, see Table 3.

3.2. *The econometric model*

There are three possible outcomes in our methodology, instead of the four possible outcomes in the standard double-bounded model: (a) the respondent is not willing to purchase the GM noodles at the same price as non-GM noodles or at a discount relative to the non-GM noodles — i.e., a "no" response to both bids; (b) the respondent is not willing to purchase the GM noodles at the same price, but is willing to buy the GM food product at the random discount offered — i.e., a "no" response followed by a "yes"; and (c) the respondent is willing to purchase the GM product at the same price as the non-GM product, implying a "yes" response.

The applicable model for examining the outcomes of our survey can be considered a special case of the double-bounded logit model (Hanemann, Loomis, and Kanninen, 1991). In this model, the initial bid (B_0) equals zero and implies no price difference between

GM noodles and non-GM noodles. The second bid (B_D) is the GM noodles offered at a random percentage discount relative to the non-GM noodles. This second bid is only given to individuals who state they would not buy GM noodles at the same price as non-GM noodles. We refer to this model as a "semi-double-bounded" model.

Modeling of this type can be found in the literature since at least the 1989 work of Cameron and Huppert, and is envisioned in the multiple-bounded discrete choice modeling considered in Welsh and Poe (1998).

The sequence of questions isolates the range in which the respondents' true WTA discounts lie for GM products relative to non-GM food products. The second bid (B_D), in conjunction with the response to the initial preference decision, allows bounds to be placed on respondents' unobservable true WTA for GM food. Note that a discount greater than 100% can be interpreted as the discount a respondent would need to be paid to choose the GM product.

Let *WTA* denote an individual's WTA compensation (or bid function) for GM noodles, relative to non-GM noodles, and let $B_D \cdot\cdot 0$ denote the discount bid[5] on GM noodles relative to non-GM noodles. The following three discrete outcomes of the bidding process are observable[6]:

$$D = \begin{cases} 1 & WTA \leq B_0, \\ 2 & B_0 < WTA \leq B_D, \\ 3 & B_D < WTA. \end{cases} \tag{1}$$

Respondents who indicate they require no discount fall into the first group of expression (1), because they are willing to purchase the GM

[5]The randomly assigned discount bids are $B_D = \{0.05, 0.1, 0.25, 0.4, 0.5\}$

[6]Although we previously argued that with process attributes, it is not rational to pay a premium, we do allow for it by not restricting WTA to equal zero for the first group. If it is the case that the WTA = 0 restriction should be imposed for the first group and it is not, then we admit that the likelihood function does not fully account for what amounts to censoring of the WTA value. However, in this particular application, the degree of such censoring is very small, where only 3.5% said "yes" to a zero discount for GM noodles. Thus, in that case, our specification of the likelihood function is only an approximation, but the censored component is a minor term in the overall likelihood specification.

noodles at zero discount compared to the non-GM noodles. Those who fall into the second group require a discount that is positive but less than or equal to B_0 for GM noodles. In order to choose the GM noodles, the price of the GM noodles must be discounted relative to the corresponding non-GM noodles by an amount less than or equal to the discount bid. Finally, those who require the largest discount to choose GM noodles fall into the third group. Customers in this group are not willing to purchase the GM product at the discount offered. The WTA function for noodles for individual i is specified as:

$$WTA_i = \alpha + \rho B_i + \boldsymbol{\lambda}'\mathbf{z}_i + \varepsilon_i, \quad i = 1, \ldots, n, \tag{2}$$

where B_i is the ultimate discount bid individual i faces, $\mathbf{z}_i$ is a column vector of observable characteristics of the individual, and i is a random variable accounting for random noise and possibly unobservable characteristics. Unknown parameters to be estimated are α, ρ, and $\boldsymbol{\lambda}$. Linearity in $\mathbf{z}$ and ... is assumed for all individuals. Furthermore, the distribution of the error term is assumed to follow $\varepsilon \sim G(0, \sigma^2)$, where $G(0, \sigma^2)$ denotes a cumulative distribution function with mean zero and variance σ^2.

Under these assumptions, the choice probabilities for individual i can be characterized as[7]:

$$\text{Prob}(D_i = j) = \begin{Bmatrix} G(\tilde{\alpha} + \tilde{\rho}B_{0i} + \widetilde{\boldsymbol{\lambda}}'\mathbf{z}_i) \\ G(\tilde{\alpha} + \tilde{\rho}B_{D_i} + \boldsymbol{\lambda}'\mathbf{z}_i) - G(\tilde{\alpha} + \tilde{\rho}B_{0i} + \widetilde{\boldsymbol{\lambda}}'\mathbf{z}_i) \\ 1 - G(\tilde{\alpha} + \tilde{\rho}B_{D_i} + \boldsymbol{\lambda}'\mathbf{z}_i) \end{Bmatrix}$$

$$\text{for } j = \begin{Bmatrix} 1 \\ 2 \\ 3 \end{Bmatrix}. \tag{3}$$

[7]The condition of linearity on $\mathbf{z}$ is a simplifying assumption widely used in random utility models (RUMs). This assumption implies that consumers' willingness to accept c dollars is generally represented as follows: $U(0, x_0, m) \leq U(1, x_1, m + c)$, $\Pr\{WTA \leq c\} = \Pr\{V_0 + \varepsilon_0 \leq V_1 + \varepsilon_1\} = \Pr\{\varepsilon_1 - \varepsilon_0 \leq V_0 - V_1\}$, where $V_0 - V_1 = \alpha + \beta c$.

Thus, the log-likelihood function becomes:

$$L = \sum_i \left\{ \begin{array}{l} I_{D_i=1} \ln G(\widetilde{\alpha} + \widetilde{\rho} B_{0i} + \widetilde{\boldsymbol{\lambda}}' \mathbf{z}_i) \\ \quad + I_{D_i=2} \ln \left[G(\widetilde{\alpha} + \widetilde{\rho} B_{Di} + \widetilde{\boldsymbol{\lambda}}' \mathbf{z}_i) - G(\widetilde{\alpha} + \widetilde{\rho} B_{0i} - \widetilde{\boldsymbol{\lambda}}' \mathbf{z}_i) \right. \\ \quad + I_{D_i=3} \ln \left[1 - G(\widetilde{\alpha} + \widetilde{\rho} B_{Di} + \widetilde{\boldsymbol{\lambda}}' \mathbf{z}_i) \right] \end{array} \right\}, \tag{4}$$

where $D_i = j$ denotes that the jth outcome occurred. In the empirical implementation of the model, we define $G(\cdot)$ to be the standard logistic distribution function with mean zero and standard deviation $\sigma = \pi/\sqrt{3}$.

The bid information and other demographic information were used to estimate the magnitude of factors affecting Japanese consumers' WTA for GM food products, and how much of a relative discount Japanese consumers will require to purchase GM food products.

3.3. *Analysis of factors affecting WTA for GM food*

The variables included in the model are food safety and environmental attitudes (see the appendix for the tradeoff question corresponding to this variable), self-reported knowledge and risk perceptions about biotechnology, gender, income, and education. The model in equation (2) is estimated where $\mathbf{z}_i = \{Safety_i\ Environment_i\ Knowledge_i\ Risk_i\ Female_i\ Income_i\ Education_i\}$. Variable definitions and descriptions are given in table 2. Estimated parameters are $\boldsymbol{\lambda} = \{\lambda_1\ \lambda_2\ \lambda_3\ \lambda_4\ \lambda_5\ \lambda_6\}$, in addition to α and ρ. Other variables are not included because of multicollinearity problems. The parameter estimates are presented in Table 4.

The estimation results show that variables representing food safety and environmental attitudes, self-reported knowledge about biotechnology, self-reported risk perceptions toward GM foods, income, and education all significantly increase the WTA compensation for choosing GM foods (i.e., increase the required discount). The variable representing gender is not statistically significant in explaining WTA compensation to purchase GM food.

Table 4. Parameter estimates for the WTA model.

Parameter	Variable	Estimate	Std. Error	t-Value	p-Value
$\tilde{\alpha}$	Intercept	1.3214	0.3375	3.915	0.0000
$\tilde{\rho}$	*Bid*	5.3704	0.4623	11.616	0.0000
λ_1	*Safety Environment*	0.0846	0.0335	2.523	0.0058
λ_2	*Knowledge*	0.6543	0.3024	2.164	0.0152
λ_3	*Risk*	1.7128	0.4229	4.050	0.0000
λ_4	*Female*	0.0368	0.3104	0.119	0.4528
λ_5	*Income*	0.4604	0.3240	1.421	0.0776
λ_6	*Education*	0.4965	0.2847	1.744	0.0406

Based on the data, the mean WTA (the required discount to be willing to purchase) for our sample is at least 50%. However, we do not present an estimated mean willingness-to-accept value because this estimate would be outside of the highest discount offered (50%). Mean WTA estimates are obtained from an extrapolation of the WTA function, so that the level of uncertainty increases as one gets further outside the data. With an extrapolation, the confidence intervals around the mean WTA estimate must be wider than otherwise would be necessary. The "no, no" group likely includes people who would not choose GM foods even if they were given away for free.

An improvement for future work in evaluating consumer preferences for GM foods in Japan should include offering respondents much larger discounts, possibly even offering to pay a segment of respondents to consume GM foods (i.e., a discount greater than 100%). Even without the mean WTA estimate, we can infer from the results that a transformation of Japanese consumers' perceptions and attitudes is needed for GM food products to successfully enter the Japanese market.

4. Conclusions

A better understanding of Japanese consumers' attitudes and behavior toward genetically modified food products will be essential for designing market strategies for Japan. This study analyzes factors affecting Japanese consumers' WTA for GM food products. Eighty percent of the Seikyou consumers who participated in our survey would not choose the GM noodles over non-GM noodles, even with a discount set at one of the following levels: 5%, 10%, 25%, 40%, and 50%. Each discount level was used for one-fifth of the 400 surveys. The econometric model developed for this analysis is a semi-double-bounded logit model. Although the double-bounded dichotomous choice model is commonly used for estimating willingness to pay/accept compensation, the semi-double-bounded model better suits the case of a product with a cost-reducing attribute.

The results show that attitudes toward food safety and the environment, self-reported knowledge about biotechnology, self-reported risk perceptions toward GM foods, income, and education all significantly increase the WTA compensation for choosing GM foods (i.e., increase the required discount). These results support the findings of Baker and Burnham (2001) that cognitive variables (opinions, beliefs, knowledge) are very important in consumer preferences for GM foods.

Although these findings may not represent the attitudes and preferences of the general Japanese population, this study is nevertheless useful for firms desiring to sell food products in Japan. An implication of this study is that there is an opportunity to market food segregated from GM products in Japan. For those firms who want to market GM products, the picture is not so rosy. Japanese consumers need to be convinced of the safety of GM foods if these products are to be marketed successfully there. Firms who sell GM products must engage in consumer education and risk communication about biotechnology.

Acknowledgment

The authors wish to thank, without implicating, the anonymous reviewers, Maria Loureiro, Ray Jussaume, Phil Wandschneider, Kyrre Rickertsen, Mario Teisl, and Brian Roe for helpful discussion. Maria Loureiro also provided assistance with statistical programming codes. The authors gratefully acknowledge financial support from the IMPACT Center at Washington State University

References

Asahi Shimbun (2001, June 14). "GMO Crops: Research and Development in Thirty-Four Provinces," pp. 1–8 [Japanese newspaper]. Online. Available at www.asahi.com.

Baker, G.A. and T.A. Burnham (2001, December). Consumer Response to Genetically Modified Foods: Market Segment Analysis and Implications for Producers and Policy Makers. *Journal of Agricultural and Resource Economics* 26(2), 387–403.

Cameron, T.A. and D.D. Huppert (1989). OLS versus ML Estimation of Non-market Resource Values with Payment Card Interval Data. *Journal of Environmental Economics and Management* 17, 230–246.

Hanemann, W.M., J. Loomis and B.J. Kanninen (1991). Statistical Efficiency of Double-Bounded Dichotomous Choice Contingent Valuation. *American Journal of Agricultural Economics* 73, 1255–1263.

Hanemann, W.M., J. Loomis and B.J. Kanninen (1999). "The Statistical Analysis of Discrete Response CV Data." In I.J. Bateman and K.G. Willis (Eds.), *Valuing Environmental Preferences*, pp. 302–442. New York: Oxford University Press.

Hoban, T.J. (1999). Consumer Acceptance of Biotechnology in the United States and Japan. *Food Technology* 53(5), 50–53.

Hutchinson, W.G., R. Scarpa, S. M. Chilton and T. McCallion (2001). Parametric and Non-parametric Estimates of Willingness to Pay for Forest Recreation in Northern Ireland: A Discrete Choice Contingent Valuation Study with Follow-Ups. *Journal of Agricultural Economics* 52, 104–122.

Japan Ministry of Public Management, Home Affairs, Posts, and Telecommunications (2002). *Japan Statistical Yearbook.* Various population statistics. Online. Available at http://www.stat.go.jp/english/data/nenkan/index.htm.

Jussaume, R.A., Jr. and L. Higgins (1998). Attitudes Towards Food Safety and the Environment: A Comparison of Consumers in Japan and the U.S. *Rural Sociology* 63(3), 394–411.

Jussaume, R.A., Jr. and D.H. Judson (1992). Public Perceptions About Food Safety in the United States and Japan. *Rural Sociology* 57(2), 235–249.

Kanninen, B.J. (1993). Optimal Experimental Design for Double-Bounded Dichotomous Choice Contingent Valuation. *Land Economics* 69(2), 138–146.

Lusk, J.L., M.S. Daniel, D.R. Mark and C.L. Lusk (2001, July). Alternative Calibration and Auction Institutions for Predicting Consumer Willingness to Pay for Nongenetically Modified Corn Chips. *Journal of Agricultural and Resource Economics* 26(1), 40–57.

Lusk, J.L., J. Roosen and J.A. Fox (2003). Demand for Beef from Cattle Administered Growth Hormones or Fed Genetically Modified Corn: A

Comparison of Consumers in France, Germany, the United Kingdom, and the United States. *American Journal of Agricultural Economics* 85, 16–29.
Macer, D. and M.A. Chen Ng (2000). Changing Attitudes to Biotechnology in Japan. *Nature Biotechnology* 18, 945–947.
McCluskey, J.J. (2000, 2nd Quarter). Read the Warning: This Product May Contain GMOs. *Choices*, pp. 39–42.
Nelson, G.C., A. De Pinto, D. Bullock, E.I. Nitsi, M. Rosegrant, T. Josling, J. Babinard, C. Cunningham, L. Unnevehr and L. Hill (1999). "The Economics and Politics of Genetically Modified Organisms in Agriculture: Implications for WTO 2000." Bulletin No. 809, University of Illinois, Urbana.
Noussair, C., S. Robin and B. Ruffieux (2002). Do Consumers Not Care About Biotech Food or Do They Just Not Read the Labels? *Economic Letters* 75, 47–53.
Pearce, D.W. and R.K. Turner (1990). *Economics of Natural Resources and the Environment.* Baltimore: Johns Hopkins University Press.
Shinano Mainichi Shimbun (2001, April 26). Article announcing Japan's new system of labeling GM foods, p. 2. [Regional Nagano newspaper]. Online. Available at www.Shinmai.co.jp/Shinmai/Shimle.html.
Slovic, P. (1987). Perception of Risk. *Science* 236, 280–285.
Tolbert, K. (2000, January 24). In Japan, It's Back to Nature; Consumers Add Non-Modified Products to Shopping Cart. *The Washington Post*, p. A8.
Welsh, M.P. and G.L. Poe (1998). Elicitation Effects in Contingent Valuation: Comparisons to a Multiple-Bounded Discrete Choice Approach. *Journal of Environmental Economics and Management* 36, 170–185.
Yoo, S.-H. and H.-J. Yang (2001). Application of Sample Selection Model to Double-Bounded Dichotomous Choice Contingent Valuation Studies. *Environmental and Resource Economics* 20, 147–163.

Appendix: English Translation of Selected Survey Questions

Questions used to elicit food safety and environmental attitudes

- Where would you place yourself on a scale from 1 to 10, *if economic growth at all costs is a* 1, and *saving the environment at all costs is a 10*?

[CIRCLE JUST ONE]

1 2 3 4 5 6 7 8 9 10

- When you are purchasing food, how important are *lower food safety* risks versus *lower cost food*? Please place yourself on a scale of

1 to 10, where 10 means *food safety is all-important*, and 1 means *lower food prices are all-important.*

[CIRCLE JUST ONE]

1 2 3 4 5 6 7 8 9 10

Questions about GMO knowledge and attitudes

- How knowledgable are you about biotechnology and genetically modified (GM) foods?
  1. Very knowledgeable
  2. Somewhat knowledgeable
  3. Not informed
- How do you feel overall about use of biotechnology in foods?
  1. Very positive
  2. Somewhat positive
  3. Neutral
  4. Somewhat negative
  5. Very negative
  6. Don't know
- How much risk, if any, do you associate with genetically modified foods?
  1. High level risk
  2. Low level risk
  3. No risk
  4. Don't know
- How important is it to you that foods with genetically modified ingredients are labeled?
  1. Very important
  2. Somewhat important
  3. Not very important
- Do you prefer domestic to imported food products?
  1. Yes
  2. No

Questions related to willingness to accept

- A U.S. university is developing genetically engineered wheat. Would you be willing to purchase noodles made with this wheat if they were offered at the same price as noodles without genetically engineered wheat?

  1. Yes
  2. No

- Would you be willing to purchase these noodles if they were offered at a price that is [*insert random discount %*] less than noodles without genetically engineered wheat?

  1. Yes
  2. No

Chapter 19

Consumer Attitudes Toward Genetically Modified Foods in Beijing, China*

Quan Li[†], Kynda R. Curtis[†], Jill J. McCluskey[‡], and Thomas I. Wahl[§]

[†] *Washington State University*

[‡] *Department of Agricultural and Resource Economics and Food Policy Fellow of the International Marketing Program for Agricultural Commodities and Trade (IMPACT) Center, Washington State University*

[§] *Department of Agricultural and Resource Economics and Director, IMPACT Center, Washington State University*

A consumer survey in Beijing, China, was conducted in August 2002. Although the majority of surveyed consumers reported that they had little or no knowledge of biotechnology, their attitudes toward genetically modified (GM) foods was generally positive, especially for GM foods with product-enhancing attributes. Using dichotomous choice contingent valuation methodology, Chinese consumers' willingness to pay (WTP) for GM rice and GM soybean oil in our sample was positively affected by respondents' positive opinion toward GM foods for both products and by higher levels of self-reported knowledge for soybean oil. However, for GM rice, WTP was negatively affected by the respondent's age — the older respondents were less likely to choose GM rice. These results imply that, unlike Europe and Japan, there is a potential market for GM foods in China. GM food producers and exporters can use this information to design effective marketing strategies.

Keywords: China, biotechnology, genetically modified foods, soybean oil, rice, contingent valuation

*This article originally appeared in the **Journal of Agrobiotechnology Management & Economics (2002), 5, 145–152**. This is published by permission of **AgBioForum**.

1. Introduction

According to a recent survey of attitudes towards biotechnology, Chinese consumers appear to favor the use of biotechnology to grow pest-resistant crops requiring fewer chemicals (Environics International, 1999). In contrast, European and Japanese consumers worry about potential human and environmental effects. As a result of consumer concerns, many processors and retailers do not accept genetically modified (GM) ingredients for their food products in order to maintain consumer confidence. Countries such as Japan, Russia, South Korea, and the European Union alert consumers of genetic modification by labeling food products. More than 160 countries, including China, have signed the 2000 Cartagena Protocol on Biosafety, which includes a labeling requirement for GM products.

Although China has had a strong commitment to biotech research since the early 1990s (Gale, Lin, Lomar, & Tuan, 2002), the country has imposed regulations. The "Regulation on the Safety Administration of Agricultural GM Organisms" (published June 6, 2001) requires all GM products entering China for research, production, or processing have safety certificates from the Chinese agricultural ministry to ensure that they are safe for human consumption, animals, and the environment. As of March 20, 2002, labeling is required by the Chinese agricultural ministry's "Regulations on Labeling Agricultural GM Biological Products" for all listed GM biological products, including soybean seed, soybeans, soybean powder, soybean oil, soybean dregs, corn seed, corn, corn oil, corn powder, rapeseed, rapeseed dregs, cottonseed, tomato seed, tomato, and tomato sauce.

Implementation of these regulations has been widely reported in China's state-run media, possibly affecting attitudes towards GM products. Understanding Chinese consumers' willingness to accept GM foods and the use of biotechnology in food development will be important to understanding the potential of the Chinese market for GM food products.

The paper is organized as follows: first, previous studies of consumer attitudes with respect to food safety and biotechnology and Chinese attitudes towards GM foods are reviewed. Second, survey data and the methodology applied in this study of Beijing consumers are introduced. The empirical analyses of the data follows, where those factors that affect consumers' willingness to accept discounts or premiums to purchase GM rice and soybean oil are analyzed, and mean estimates for the willingness to accept discounts or premiums to purchase GM rice and soybean oil are presented. Finally, the major implications of this research are reviewed and concluding statements offered.

2. Prior Research

There are few published studies that estimate consumer willingness to pay or willingness to accept compensation for food products that contain GM ingredients. Baker and Burnham (2001) investigated US consumers' acceptance of GM corn flakes and found that 30% of US consumers surveyed based their purchasing decisions on GM content. Their analysis shows that cognitive variables (e.g., opinions, beliefs, knowledge) have a great influence on consumer preferences. The level of risk aversion, knowledge, and opinions about genetic modification are statistically significant[1] in explaining purchasing decisions. Previous studies (McCluskey, Ouchi, Grimsrud, & Wahl, in press; Mendenhall & Evenson, 2002) investigating the relationship between consumer characteristics and food safety concerns generally found that sociodemographic variables (such as education and income) perform poorly as explanatory variables for purchasing decisions regarding GM food products. The exception is that women in general are more concerned with food safety.

Lusk, Roosen, and Fox (2001) estimated consumer willingness to pay for beef in France, Germany, the United Kingdom, and the United States using a variety of quality variables, including whether the cattle were fed GM corn. Their results suggested that compared with US consumers, European consumers placed a much higher value on beef from cattle that have not been fed genetically modified corn. Noussair, Robin, and Ruffieux (2002) examined the discrepancy between European public opinion and consumers' purchasing behavior with regard to GM food products. They found that consumers were typically unaware of labels indicating GM content.

[1]They are statistically significant at the 1%, 10%, and 1% levels of probability, respectively.

McCluskey *et al.* (2003), found that on average Japanese consumers were willing to purchase GM noodles at a 60% discount over non-GM noodles, and on average were willing to purchase GM tofu at a 64% discount over non-GM tofu. Consumers' attitudes toward safety, self-reported knowledge, and risk were found to be significant indicators in their overall willingness to accept GM products. Consumers in Japan — not unlike those in Europe — are highly skeptical of government regulations and responsibilities when it comes to food safety.

In 1999, Environics International surveyed consumers in 10 countries and found that China's consumers were among the world's strongest supporters of agricultural biotechnology research. In China, 79% of respondents were favorable about the use of biotechnology to grow pest-resistant crops requiring fewer chemical crop applications, which was the highest among the 10 countries under consideration (78% in the US, 63% in Japan, 36% in Britain).

3. Data

This study uses data collected from 599 in-person interviews in Beijing, China, in August 2002. The survey instrument was pretested with Chinese students in the United States and conducted by four Chinese nationals (in Chinese). The survey was performed in four separate locations, including a supermarket, two outdoor markets, and one shopping area. These locations were chosen to ensure a random sample encompassing a cross section of the Beijing population and to survey consumers at the same time and place where actual purchasing decisions were made, in an effort to better elicit their true preferences. Respondents were randomly selected with the criterion that the interviewer was to solicit every third consumer that came into the survey area. Every respondent was given a gift pack of green tea (worth approximately $.65 in Chinese yuan) or a bottled cola drink (worth $.50 in Chinese yuan) as a reward for participating in the survey.

The summary statistics of the demographic variables and information and perception variables for the survey are presented in Tables 1 and 2, respectively. The majority of the 599 Beijing respondents were

Table 1. Summary statistics for demographic variables.

Variable	Description (coding)	Distribution of Survey Responses	
Age	1 if ≤ 39 years	50%	Mean = 38.8
	0 if > 39 years	50%	Std. dev. = 13.9
Female	1 if female	63.6%	
	0 if male	36.4%	
Shopper	1 if main shopper	68.8%	
	0 otherwise	31.2%	
Education	primary school or illiteracy	7.2%	Coding for estimation:
	junior secondary school	26.4%	0 if primary school, illiteracy, secondary school, or refuse;
	senior secondary	38.7%	1 if 2-year college, 4-year university or graduate.
	2-year college	16.0%	
	4-year university	9.2%	
	graduate	1.7%	
	refuse	0.8%	
Children	1 if children in household	61.1%	
	0 otherwise	38.9%	
Income	<10,000 yuan	12.2%	Coding for estimation:
	10,000–25,000 yuan	38.2%	0 if less than average income;
	25,000–40,000 yuan	24.9%	1 if above average income.
	40,000–55,000 yuan	12.7%	
	55,000–70,000 yuan	4.7%	
	>70,000 yuan	3.2%	
	refuse	4.1%	
Employment status	Full time employed	49.7%	Coding for estimation:
	Part time employed	15.7%	1 if full time employed;
	Unemployed	5.2%	0 if otherwise
	Housemaker	2.8%	
	Retired	18.2%	
	Refused	2.1%	
	Student	6.3%	
Family size	Number of people in household		Mean = 3.946
			Std. dev. = 12.343

Table 2. Summary statistics for consumer information and perception variables.

Variable	Description (coding)	Distribution of Responses
Safety	Importance of food safety vs. food price (scale from 1 to 10 where 1 = food safety all important; 10 = food price all important)	Mean = 3.47 SD = 2.20
Risk	Risk associated with GM foods (1 if high risk; 0 if little risk, no risk, or don't know)	
	high risk	7.8%
	little risk	27.5%
	no risk	25.0%
	don't know	39.7%
Opinion	Opinion about use of biotechnology (1 if favorable or neutral opinion, 0 if negative opinion or don't know)	
	favorable opinion	61.6%
	neutral opinion	6.9%
	negative opinion	9.3%
	don't know	22.2%
Knowledge	Self-reported knowledge about biotechnology (1 if high or little knowledge, 0 if no knowledge)	
	high knowledge	1.0%
	little knowledge	44.9%
	no knowledge	54.1%
Label	1 if GM labeling very important, 0 if somewhat or not very important	
	very important	61.9%
	somewhat important	27.9%
	not very important	10.2%
Import	1 if prefers domestic to imported foods, 0 otherwise	
	prefers domestic to imported foods	84.5%
	otherwise	15.5%

the primary food shoppers of the household (69%) and female (63%). Seventy-four percent of those shoppers purchased groceries two or more times per week. The majority of the respondents were in their late 30s or early 40s, with a mean age of 38.8 years — slightly higher than the overall average age in China (30 years). The household income reported in the survey ranged between 10,000 yuan (US$1,200) and 25,000 yuan (US$3,038) for the 2001 fiscal year, compared to the Chinese average income in 2001 of 16,619 yuan (US$2,201).[2] The average education level of the respondents was equivalent to a high school graduate, which is above the Chinese average (middle school education).[3]

4. Empirical Analysis

The contingent valuation method (CVM) is currently the standard approach used to elicit consumers' willingness to pay (WTP) using a dichotomous choice questioning format conducted by direct survey via telephone, mail, or face-to-face. There are typically two types of bidding procedures used in the CVM: the single-bounded and double-bounded dichotomous choice models (Kanninen, 1993). The single-bounded model approach recovers the bid amount as a threshold by asking only one dichotomous choice question (Hanemann, Loomis, & Kanninen, 1991). The statistical efficiency of this approach can be improved by use of the double-bounded model, which engages in two bids. However, the double-bounded approach has also been critically evaluated. Hanemann, Loomis, and Kanninen (1999) note that "there is also some bias in going from a single- to a double-bounded format, because there is some evidence that some of the responses to the second bid are inconsistent with the first bid" (p. 382). They conclude that even if the double-bounded approach produces some bias, "the experience to date suggests that the bias is in a conservative direction and is greatly outweighed by the gain in efficiency in terms of minimizing overall mean squared error" (p. 388).

---

[2]Because survey respondents typically are reluctant to divulge information regarding income, respondents were asked to place themselves in income intervals, rather than state their exact income amount, in order to obtain a higher response rate. The exchange rate used is 8.23 yuan/US$.

[3]The average Chinese household income of 2001 is calculated from China Statistical Yearbook 2002, and the average education is from China Statistical Yearbook 2002.

This survey included contingent valuation questions regarding willingness to pay a premium or accept a discount to purchase soybean oil made from GM soy and GM rice. Soybean oil and rice products used in this study are appropriate for examination, because these products are consumed frequently.

The respondents were first asked if they were willing to pay the same price for GM soybean oil and GM rice as for the corresponding non-GM products. If the respondent's answer to this question was "no," a follow-up question would be asked, where the respondent was offered a percentage discount on the GM product relative to the non-GM product. If the respondent's answer to the first question was "yes," a follow-up question would be asked where the respondent was offered a percentage premium on the GM product relative to the non-GM product. The discount was set at one of the following levels: 10%, 20%, 25%, 50%, and 75%. The premium for the GM rice was set at one of the following levels: 10%, 20%, 25%, 50%, and 100%. The premium for the GM soybean oil was set at one of the following levels: 5%, 10%, 20%, 25%, and 50%. Each level of discount or premium was used for one fifth of the surveys. That is, 120 of the 599 surveys had a 10% discount/premium for GM rice; another 120 surveys had a 20% discount/ premium for GM rice; and so on. The assignment of survey version (and thus, discount or premium) was random to the respondent. The rationale for using different premium amounts for the GM rice versus the GM soybean oil is that the respondent was given information regarding the *product-enhancing* attribute of GM rice, but was given no information regarding either a *product-enhancing* or a *process-enhancing* attribute for the GM soybean oil. Hence, we expected that those respondents willing to pay a premium for the GM product would pay more for the *product-enhancing* product than for the other. (See survey questions in the Appendix for a complete description of the willingness to pay questions. A complete survey is available upon request from the authors.)

Of the 599 respondents, 80% said that they would be willing to purchase the genetically modified rice at the same price as the non-GM rice. Furthermore, 43.9% of consumers in the sample stated that they would be willing to purchase the GM rice at a premium over the price of the non-GM rice (the "yes, yes" group), while 37.4% of all respondents said they would not be willing to pay a premium (the "yes, no" group). Only 4.7% of all respondents were not willing to purchase GM rice at the same price as non-GM rice, but were in fact willing to purchase GM rice at a discount from the non-GM rice price (the "no, yes" group). The rest of the respondents (14.0%) were not willing to purchase the GM rice even with the discount (the "no, no" group). For additional statistics on the distribution of responses over various discounts and premiums, see Tables 3 and 4.

In the case of the GM soybean oil, 73% of the respondents said that they would be willing to purchase the genetically modified

Table 3. Range and distribution of response rates to the randomly assigned discount.

	Discount	GM Rice	GM Soybean Oil
Yes to discount	10%	1.2%	1.8%
	20%	0.8%	1.8%
	25%	0.7%	1.2%
	50%	1.2%	1.7%
	75%	0.8%	2.0%
No with premium		14.0%	16.7%
Total		18.7%	25.2%

Table 4. Range and distribution of response rates to the randomly assigned premium.

	GM Rice		GM Soybean Oil	
	Premium		Premium	
Yes to premium	10%	11.0%	5%	9.3%
	20%	9.5%	10%	9.7%
	25%	10.7%	20%	8.0%
	50%	6.2%	25%	8.3%
	100%	6.5%	50%	4.3%
No with premium		37.4%		34.4%
Total		81.3%		74.0%[a]

[a]*0.8% of the respondents did not respond for GM soybean oil.*

soybean oil at the same price as the non-GM soybean oil. Furthermore, 39.6% of all consumers in the sample stated that they would be willing to purchase the GM soybean oil at a premium over the price of the non-GM soybean oil (the "yes, yes" group), while 34.4 % would not pay a premium (the "yes, no" group). Alternatively, 8.5% of all respondents were not willing to purchase the GM soybean oil at the same price as the non-GM soybean oil and were willing to purchase the GM soybean oil at a discount from non-GM soybean oil (the "no, yes" group). And 16.7% of the respondents, were not willing to purchase the GM soybean oil even with the discount (the "no, no" group).[4]

5. Econometric Model

In our double-bounded model there are four possible outcomes: (a) the respondent is not willing to purchase the GM product at the same price as non-GM product, nor at a discount relative to the non-GM product (i.e., "no" to both bids); (b) the respondent is not willing to purchase the GM product at the same price as the non-GM product, but is willing to purchase the GM product at the random discount offered (i.e., a "no" followed by a "yes"); (c) the respondent is willing to purchase the GM product at the same price as non-GM product, but is not willing to purchase it at a premium (i.e., a "yes" followed by a "no"); or (d) the respondent is willing to purchase the GM product at the same price as non-GM product and also willing to purchase the GM product at a random premium offered relative to the non-GM product (i.e., "yes" to both bids).

The model most applicable to examine the outcomes of our survey is the standard double-bounded logit model (Hanemann *et al.*, 1991). In this model, the initial bid (B_0) equals zero and implies no price difference between the GM product and the non-GM product. The second bid is contingent upon the response to the first bid. It will be a discount bid (B_D) if the respondents answer that they would not buy GM product at the same price as non-GM product. If they

[4]For GM soybean oil, 0.8% of the respondents did not respond.

answer that they would buy GM product at the same price as non-GM product, it becomes a premium bid (B_P).

The sequence of questions isolates the range of respondents' true willingness to pay for GM products relative to non-GM products. The second bid (B_D or B_P), in conjunction with the response to the initial preference decision, allows an upper bound and a lower bound to be placed on the respondent's unobservable true WTP for GM food products.

Let WTP_i denote an individual's WTP (bid function) for GM food products. The following discrete outcomes (D_g) of the bidding process are observable:

$$D_g = \begin{Bmatrix} \text{group 1} & WTP_i < B_D \\ \text{group 2} & B_D \leq WTP_i < B_0 \\ \text{group 3} & B_0 \leq WTP_i < B_P \\ \text{group 4} & B_P \leq WTP_i \end{Bmatrix} \tag{1}$$

Respondents who indicated they would require no discount and accepted the highest premium fall into the fourth group (D_4). Those indicating no discount and premium less than B_P fall into the third group (D_3). Next, respondents who required a discount greater than or equal to B_D fall into the second group (D_2). Finally, the first group (D_1) contains respondents indicating the lowest WTP. Consumers in this group are not willing to purchase the GM product at the discount offered. The WTP function for GM food products for individual i is

$$WTP_i = \alpha - \rho B_i + \lambda' z_i + \varepsilon_i \quad \text{for } i = 1, \ldots, n \tag{2}$$

where B_i is the ultimate bid individual i faces, z_i is a column vector of observable characteristics of the individual, ε_i is a random variable accounting for random noise and possibly unobservable characteristics. Unknown parameters to be estimated are α, ρ, and λ. Linearity in z and ε is assumed for all individuals. Furthermore, the distribution of the error term is assumed to follow $\varepsilon \sim G(0, \sigma^2)$, where $G(0, \sigma^2)$ denotes a cumulative distribution function with mean zero and variance σ^2. Under these assumptions, the choice probabilities for individual i can be characterized as

$$prob(D=j)=\begin{Bmatrix} G(\tilde{\alpha}-\tilde{\rho}B_D+\tilde{\lambda}'z), \quad \text{for} j=1 \\ G(\tilde{\alpha}-\tilde{\rho}B_0+\tilde{\lambda}'z)-G(\tilde{\alpha}-\tilde{\rho}B_D+\tilde{\lambda}'z), \ \text{for}\, j=2 \\ G(\tilde{\alpha}-\tilde{\rho}B_P+\tilde{\lambda}'z)-G(\tilde{\alpha}-\tilde{\rho}B_0+\tilde{\lambda}'z), \ \text{for}\, j=3 \\ 1-G(\tilde{\alpha}-\tilde{\rho}B_P+\tilde{\lambda}'z), \ \text{for}\, j=4 \end{Bmatrix} \tag{3}$$

Thus, the log-likelihood function becomes:

$$L=\sum_i \begin{Bmatrix} I_{D_i=1}\ln G(\tilde{\alpha}-\tilde{\rho}B_{Di}+\tilde{\lambda}'z_i) \\ +I_{D_i=2}\ln[G(\tilde{\alpha}-\tilde{\rho}B_{0i}+\tilde{\lambda}'z_i)-G(\tilde{\alpha}-\tilde{\rho}B_{Di}+\tilde{\lambda}'z_i)] \\ +I_{D_i=3}\ln[G(\tilde{\alpha}-\tilde{\rho}B_{Pi}+\tilde{\lambda}'z_i)-G(\tilde{\alpha}-\tilde{\rho}B_{0i}+\tilde{\lambda}'z_i)] \\ +I_{D_i=4}\ln[1-G(\tilde{\alpha}-\tilde{\rho}B_{Pi}+\tilde{\lambda}'z_i)] \end{Bmatrix} \tag{4}$$

where I_K is an indicator function for the event K, and $D_i = j$ denotes that the jth alternative occurred. In the empirical implementation of the model, we define $G(\cdot)$ to be the standard logistic distribution function with mean zero and standard deviation $\sigma = \pi/\sqrt{3}$. The empirical representation of equation 2 is formulated as

$$\begin{aligned} WTP_i &= \alpha - \rho B_i + \lambda_1 Children_i + \lambda_2 Education_i + \lambda_3 Age_i \\ &\quad + \lambda_4 Knowledge_i + \lambda_5 Income_i + \lambda_6 Opinion_i + \varepsilon_i, \end{aligned} \tag{5}$$

where B_i represents the random bid offered to each consumer, $Children_i$ is an indicator variable representing the presence of children under 18 years old in the household, $Education_i$ is an indicator variable representing the respondent's level of education, $Knowledge_i$ is an indicator variable representing the respondent's self-reported knowledge regarding biotechnology, $Income_i$ represents the respondent's reported income level, Age_i is an indicator variable representing the respondent's reported age, and $Opinion_i$ is an indicator variable representing the respondent's opinion regarding the application of biotechnology to food products.

6. Analysis of Willingness to Pay

Estimation results are presented in Tables 5 and 6. Results show that positive opinions regarding biotechnology significantly increase the WTP (i.e., increase the premium) for both GM products — at the 5% level of statistical significance for GM rice and very close to the 5% level for GM soybean oil. For GM soybean oil, self-reported knowledge about biotechnology increases consumers' WTP at the 10% level of statistical significance. This may indicate that self-reported knowledge was obtained from sources that were supportive of agricultural biotechnology.

Consumer attitudes concerning biotechnology reflect the Chinese government's traditionally strong support. Thus far, the level of debate occurring in Europe and Japan has not been as evident in the Chinese media, but new regulations regarding labeling and safety testing are most likely leading to increased public awareness of the application of biotechnology to agricultural products.

For GM rice, age level significantly decreased the consumers' WTP (i.e., increased the required discount) at the 5% level of statistical significance. The survey questions regarding GM rice contained information about product-enhancing attributes — specifically, that GM rice contains additional vitamins. Younger people were more

Table 5. Parameter estimates for willingness to pay model (GM rice).

Parameter	Variable Description	Estimate	Standard Error	t-value	p-value
α	Intercept	0.2958	0.2873	1.0297	0.3031
ρ	Bid***	−1.6386	0.1873	−8.7501	0.0000
λ_1	Income	0.1858	0.1599	1.1620	0.2452
λ_2	Knowledge	0.1315	0.1668	0.7884	0.4304
λ_3	Children	−0.2312	0.1590	−1.4538	0.1460
λ_4	Age**	−0.0106	0.0051	−2.0842	0.0371
λ_5	Education	0.0988	0.1886	0.5238	0.6004
λ_6	Opinion**	0.4044	0.1743	2.3196	0.0203

Note: *Single (*), double (**) and triple asterisks (***) denote statistical significance at the 10%, 5% and 1% levels, respectively.*

Table 6. Parameter estimates for willingness to pay model (GM soybean oil).

Parameter	Variable Description	Estimate	Standard Error	t-value	p-value
α	Intercept	0.0270	0.2809	0.0961	0.9234
ρ	Bid***	−2.5569	0.2637	−9.6960	0.0000
λ_1	Income	0.1977	0.1581	1.2501	0.2113
λ_2	Knowledge*	0.2837	0.1665	1.7037	0.0884
λ_3	Children	−0.1364	0.1562	−0.8734	0.3825
λ_4	Age	−0.0056	0.0051	−1.1042	0.2695
λ_5	Education	−0.1175	0.1829	−0.6422	0.5208
λ_6	Opinion**	0.3373	0.1758	1.9188	0.0550

Note: *Single (*), double (**) and triple asterisks (***) denote statistical significance at the 10%, 5% and 1% levels, respectively.*

accepting of biotechnology, which is the same reaction found in previous studies (McCluskey *et al.*, 2003; Chern & Rickertsen, 2001) of consumer attitudes toward GM foods.

The level of education, income, and the presence of children under age 18 in the household were not statistically significant in either case. These results are consistent with studies by Baker and Burnham (2001), where cognitive variables performed better as explanatory variables than sociodemographic variables.

The mean willingness to pay, WTP, is commonly estimated by restricting $\lambda_i = 0$ (Hanemann *et al.*, 1991). The empirical mean WTP can then be calculated as $-\widetilde{\alpha}/\widetilde{\rho}$. When we use this methodology, our results indicate that Beijing consumers on average were willing to pay a 38.0% premium for GM rice over non-GM rice, and were willing to pay a 16.3% premium for GM soybean oil over non-GM soybean oil.[5] This is not surprising, given that 61.6% of

[5]95% confidence interval around this estimated mean WTP is (0.24, 0.52) for GM rice and (0.01, 0.31) for GM soybean oil. These confidence intervals were calculated using the delta method, which approximates the asymptotic variance of the ratio of two random variables as (Kanninen, 1993):

$$a\sigma^2\left(\frac{-\alpha}{\rho}\right) \cong \frac{1}{\rho^2}\left[\left[\frac{a}{\rho}\right]^2 \sigma^2(\alpha) - 2\left(\frac{\alpha}{\rho}\right)\text{cov}(\alpha\rho) + \sigma^2(\rho)\right].$$

our survey respondents had a favorable opinion about the use of biotechnology in foods (Table 2). Additionally, 52.5% of respondents felt there was little or no risk associated with genetically modified foods.

The positive mean WTP for GM foods is unexpected, given studies in other countries. However, prevailing positive opinions regarding biotechnology are behind these results. The survey results indicate that only 9.3% of the respondents had a negative opinion concerning the use of biotechnology in foods, and only 7.8% associated high risk with GM foods. For each of these respective categories, 22.2% and 39.7% of respondents answered "don't know."

Why are our Chinese results so different from studies of other countries? One possible answer lies in historical differences. The European countries and Japan gradually developed modern capitalist societies while taking great concern and pride in preserving cultural traditions. For the Chinese, history took another turn. A decade of Cultural Revolution from 1966 to 1976 systematically tore down historical and traditional structures in the society. The past was condemned as "feudal and superstitious," (Beech, 2002). The vacuum remaining was to some extent replaced by the communist state. Now, with a highly desired and incredibly rapid transition to capitalism and with much of the old Chinese tradition crushed by the Cultural Revolution, the Chinese are looking forward. Technological novelties from the rest of the world are often considered much-needed improvements and not reasons for concern.

A caveat one must consider when interpreting results from a stated preference model is that differences may exist between responses from real and hypothetical valuation questions. There are findings both for and against the consistency of revealed and stated preferences (see Loureiro, McCluskey, & Mittelhammer, 2003 and Cummings, Harrison, & Rutstrom, 1995).

7. Conclusions

The results of this research indicate respondents to our consumer survey in Beijing, China, generally have a favorable view towards

GM rice and GM soybean oil — two products that are staples of the Chinese diet. These results are contrary to those found in previous studies, particularly those conducted in Europe and Japan. However, this consumer survey is one of the first studies of GM foods conducted within China, a developing country. Results suggest that a positive opinion toward biotechnology significantly increased Beijing consumers' willingness to pay for both GM products, while for GM soybean oil solely, higher levels of consumers' self-reported knowledge also increased their willingness to pay. For GM rice, WTP was negatively affected by the respondent's age — older respondents were less likely to choose GM rice.

In general, our surveyed Beijing consumers have positive attitudes towards the use of biotechnology in agriculture, although they have little knowledge. In China, consumer attitudes towards GM foods are influenced by positive media coverage, which is controlled by the government. China is the world's fourth largest producer of biotech crops and has been a large supporter of biotech research since the 1980s. Recently, China has seen great success in their biotechnology research efforts, such as in the case of insect-resistant Bt corn.

The marketing outlook for GM foods in China is optimistic. Younger people are more willing to purchase GM food products with product-enhancing attributes, which indicates that the Chinese market may be even more open to GM foods in the future. Additionally, government investment in biotechnology remains strong, as China works to fulfill its food self-sufficiency policies. If positive government sentiment towards new food technologies continues, Chinese citizens are unlikely to change their preferences due to limited coverage of anti-GM debates abroad.

Acknowledgment

The authors gratefully acknowledge financial support from the IMPACT Center at Washington State University.

References

Baker, G.A., and T.A. Burnham (2001). Consumer Response to Genetically Modified Foods: Market Segment Analysis and Implications for Producers and Policy Makers. *Journal of Agricultural and Resource Economics* 26(2), 387–403.

Beech, H. (2002, November 4). China's Next Cultural Revolution. *TIME* (Asia Edition), 160(18).

Chern, W.S. and K. Rickertsen (2001). Consumer Acceptance of GMO: Survey Results from Japan, Norway, Taiwan and the United States. *Taiwanese Agricultural Economic Review* 7(1), 1–28.

Cummings, R.G., G.W. Harrison and E.E. Rutstrom (1995). Homegrown Values and Hypothetical Surveys: Is the Dichotomous Choice Approach Incentive-Compatible? *American Economic Review* 85(1), 260–266.

Environics International (October 16, 1999). Biotech: Yes or no? *The Washington Post*, p. A19.

Gale, F., W. Lin, B. Lomar and F. Tuan (2002). Is Biotechnology in China's future? In *China's Food and Agriculture: Issues for the 21st Century* (AIB-775, pp. 34–37). Washington, DC: United States Department of Agriculture Economic Research Service.

Hanemann, W.M., J. Loomis and B.J. Kanninen (1999). "The statistical analysis of discrete response CV data." In I.J. Bateman and K.G. Willis (Eds.), *Valuing Environmental Preferences*, pp. 302–442. New York: Oxford University Press.

Hanemann, W.M., J. Loomis and B.J. Kanninen (1991). Statistical Efficiency of Double-Bounded Dichotomous Choice Contingent Valuation. *American Journal of Agricultural Economics* 73, 1255–1263.

Kanninen, B.J. (1993). Optimal Experimental Design for Double-Bounded Dichotomous Choice Contingent Valuation. *Land Economics* 69(2), 138–146.

Loureiro, M.L., J.J. McCluskey and R.C. Mittelhammer (2003). Are Stated Preferences Good Predictors of Market Behavior? *Land Economics* 79(1), 44–55.

Lusk, J.L., J. Roosen and J.A. Fox (2003). Demand for Beef from Cattle Administered Growth Hormones or Fed Genetically Modified Corn: A Comparison of Consumers in France, Germany, the United Kingdom, and the United States. *American Journal of Agricultural Economics* 81(1), 16–29.

McCluskey, J.J., K.M. Grimsrud, H. Ouchi and T.I. Wahl (2003). Consumer Response to Genetically Modified Food Products in Japan. *Agricultural and Resource Economic Review* 32(2), 222–231.

Mendenhall, C.A. and R.E. Evenson (2002). "Estimates of willingness to pay a premium for non-GM foods: A survey." In V. Santaniello, R.E. Evenson and D. Zilberman (Eds.), *Market Development for Genetically Modified Foods.* Trowbridge, UK: CABI Publishing.

National Bureau of Statistics of China (2002). *China Statistical Yearbook 2002*, no. 21. Beijing: China Statistics Press.

Noussair, C., S. Robin and B. Ruffieux (2002). Do Consumers Not Care About Biotech Food or do they Just Not Read the Labels? *Economic Letters* 75, 47–53.

Appendix: Questions Used to Elicit Consumers' Willingness to Pay

1. Scientists at universities are developing genetically engineered rice, which contains additional vitamins. Would you be willing to purchase this rice if it were offered at the same price as non-genetically modified rice? [If yes, go to question 3.]
2. Would you be willing to purchase this rice if it were offered at a price that is XX less than non-genetically modified rice? [Go to question 4.]
3. Would you be willing to purchase this rice if it were offered at a price that is XX more than non-genetically modified rice?
4. Would you be willing to purchase soybean oil made with genetically modified soybeans if it were offered at the same price as oil with non-genetically modified soybeans? [If yes, go to question 6.]
5. Would you be willing to purchase genetically modified soybean oil if it were offered at a price that is XX less than soybean oil with non-genetically modified soybeans?
6. Would you be willing to purchase genetically modified soybean oil if it were offered at a price that is XX more than soybean oil with non-genetically modified soybeans?

Chapter 20

Consumer Response to Information and Second-Generation Genetically Modified Food in India*

Alok Anand[†], Ron C. Mittelhammer, and Jill J. McCluskey[‡]

[†] *The Consumer Analytics and Modeling Unit, Citigroup*
[‡] *School of Economic Sciences and Food Policy Fellow, International Marketing Program for Agricultural Commodities and Trade (IMPACT) Center, Washington State University, 99164-6210 USA*

This study investigates the Indian market for GM wheat. First, we address the obvious valuation and acceptance questions for Indian consumers. Then we investigate the extent to which the provision of information about GM wheat has any significant effect on consumers' willingness to pay and the degree to which the effect varies based on the information treatment. Finally, we consider whether the type of primary beneficiary of GM wheat production affects valuation. Specifically, we assess the conjecture that consumers' willingness to pay for GM wheat changes according to whether the crop is portrayed as producer-friendly or consumer-friendly.

Keywords: Biotechnology, India, second-generation, information effects

1. Introduction

The application of biotechnology to food is highly controversial among consumers and policy makers including in the developing

*This article originally appeared in the **Journal of Agricultural & Food Industrial Organization (2007), 5(1), 1–20**. This is published by permission of **De Gruyter**.

world (Pinstrup-Andersen and Schiøler, 2001). Despite claims from scientists and official institutions that genetically modified (GM) products are safe, consumer attitudes toward GM foods are largely negative in many developed countries, especially in the European Union and Japan (Burton *et al.*, 2001; Lusk *et al.*, 2005). Consumer skepticism is often attributed to concern over unknown environmental and health consequences, such as unanticipated allergic responses, the spread of pest resistance or herbicide tolerance to wild plants, and inadvertent toxicity to wildlife. The United States is an exception to this rule, where consumers are largely indifferent about GM foods (Nelson, 2001; and Chern *et al.*, 2002). However, studies conducted in lesser developed countries find that consumer attitudes toward GM foods are less negative and in many cases positive (Li *et al.*, 2002; Pachico and Wolf, 2002; and Subrahmanyan and Cheng, 2000).

The United States, Argentina and Canada are the major producers of GM crops, with the United State being the largest. The major GM crop producing countries cite numerous benefits of GM crops, including food availability, nutrition enhancement, and economic advantages. The current world population is approximately six billion and growing at a rapid pace. There is a need to ensure an adequate food supply for this burgeoning population in the future. GM crops have beneficial characteristics that can be exploited to meet the growing food demand. They can be pest resistant, disease resistant, can have drought tolerance/salinity tolerance qualities, and can also possess some medicinal qualities. They can also be used for phytoremediation. For example, plants such as poplar trees have been genetically modified to remove heavy metal pollution from contaminated soil.

Most available GM products contain "first-generation" genetic modifications, which address production considerations and have input traits that may result in public good benefits, such as reduced pesticide usage. As "second-generation" GM products, which include quality traits or product-enhancing attributes such as nutritional benefits, enter the market, consumers may be more accepting of biotechnology in food because of the inherent private benefits. With benefits directly accruing to the consumers, those consumer-friendly GM products could be accepted without discounts or could receive a premium. Golden Rice, which by genetic modification produces beta-carotene, a precursor to vitamin A, is one of the emblematic cases of second-generation GM products. Although second-generation genetically modified organisms are not yet widely available for consumers, they are expected to be soon available in the market.

Many countries have imposed mandatory GM labeling and have blocked the registration of new varieties of GM crops. A concern for many countries is the potential market loss due to new regulations requiring labeling and traceability on all foods with GM ingredients. In an effort to ensure GM-free exports, lesser developed countries may be discouraged from planting genetically modified crops due to the extraordinary expense involved in segregating GM and GM-free crops.

Asia, being home to the world's largest consumer base, is an obvious candidate for market expansion of GM crops. Two Asian countries, China and India, are at the center stage of the future of GM crops. These are the world's two most populous countries with fast growing economies. Even though China is the only country in Asia producing a significant amount of GM crops, China has joined a coalition of several countries favoring strict regulation of GM crops. The scenario in India is somewhat different. In China, it is essentially the government that makes all of the decisions regarding GM crops and the government-controlled media has the prime responsibility for disseminating information regarding GM crops. In India, there are many interest groups that hold different sets of beliefs regarding GM crops and the government reacts to political pressures exerted by these groups. The Indian media is not state-run and both positive and negative views of GM food are presented.

Increased crop yields and dietary supplements provided by GM foods could be of great benefit in enhancing food availability and alleviating nutritional problems in India. Producers of dry land crops face severe pest and disease problems as well as drought. Existing GM methods can provide pest, disease, and drought resistance. Many political leaders, scientists and technocrats have advocated wider use of GM food as a means of enhancing agricultural productivity and reducing poverty. However, critics of GM food have worked within the open and democratic political system of India to secure a more cautious approach towards widespread acceptance of GM food.

India is currently investing in research relating to GM varieties of cotton, maize, cabbage, potatoes, rice, tomatoes and some other plants. There are at least 20 academic and research institutions that are actively involved in plant biotech research. Companies such as Monsanto, who are researching the development of GM crops, have been active in India for the last 50 years. Monsanto introduced Roundup in 1991, which has become an important weed control product for farmers throughout India. In 1998, they launched Leader, a new generation herbicide, for post-emergent weed control in wheat.

There is widespread optimism within the scientific community in India that GM crops can lead to a second green revolution. Given India's growing knowledge-based economy, GM crops are expected to play a pivotal role in the future development of the country. Like the United States, India does not require additional safety testing of GM foods beyond what is required for conventional foods and does not require labeling of GM foods. However, India has made efforts to seek premiums for GM-free products in export markets by labeling them as such. The Indian government has moved slowly on biotech approvals, fearing criticism from anti-GM organizations.

This study investigates the Indian market for GM wheat. First, we address the obvious valuation and acceptance questions for Indian consumers. Then we investigate the extent to which the provision of information about GM wheat has any significant effect on consumers' willingness to pay and the degree to which the effect varies based on the information treatment. Finally, we consider whether the type of primary beneficiary of GM wheat production affects valuation. Specifically, we assess the conjecture that consumers' willingness to pay for GM wheat changes according to whether the crop is portrayed as producer-friendly or consumer-friendly.

2. Literature Review

Numerous studies have estimated consumers' willingness to pay for food products that contain GM ingredients.[1] In this section, we focus on research which considers information effects and consumer response to second generation attributes. In an analysis of information effects, Tegene *et al.* (2003) conducted an experimental auction to elicit consumers' willingness to pay for "genetically modified" labeled and standard-labeled foods under different information regimes in two Midwestern U.S. cities. Under all information treatments, consumers discounted food items labeled "GM" by an average of 14%. While gender, income, and other demographic characteristics appeared to have only a slight impact on consumers' willingness to pay for GM foods, information from interested parties and third-party (independent) sources was found to have a strong impact. The results are consistent with other studies that show that individuals place a greater weight on negative information than on positive information.

Lusk *et al.* (2004) conducted incentive-compatible auctions to examine the effects of benefit information (including environmental, health, and benefits to lesser developed countries) across locations in England, France, and the United States. They found that benefit information significantly reduced the compensation consumers demanded to consume GM foods, and the effect varied by type of information and location.

Li *et al.* (2004) found that positive scientific information about biotechnology increased significantly the consumer's willingness to pay for GM products in the U.S. Pacific Northwest. Moreover, they found there was interaction between information provision and environmental attitudes, and between information provision and the level of education, which could have the interpretation that information provision has greater effects for particular groups of consumers.

---

[1]See Lusk *et al.* (2005) for a comprehensive review and a meta-analysis of the work in this area at the time of publication.

Scholderer and Frewer (2003) tested communication strategies for changing consumers' attitudes towards GM foods in an experimental study involving consumer samples from Denmark, Germany, Italy, and the United Kingdom. Different information strategies were tested against a control group for their ability to change consumer attitudes. They found that no attitude change occurred. However, unlike attitudes, consumers' product choices were sensitive to provision of information. All information materials significantly decreased the probability of consumers' choosing GM products. Frewer *et al.* (1999) investigated British consumers' response to a number of messages that all stressed the benefits of GM foods but varied in persuasiveness. None of the messages had a significant impact on consumer attitudes.

A few researchers have focused on second-generation or consumer-friendly GM products, which could be accepted without discounts or could receive a premium. Li *et al.* (2002) concluded that the Chinese consumers in their sample, on average, are willing to pay a 38% premium for Golden Rice. Further, Boccaletti and Moro (2000) reported a decrease in the percentage of consumers who refuse to buy GM food products, from 17% to 12%, when confronted with products that directly affect them, instead of generic GM food products in Italy. They found that consumers better accept GM products with increased nutritional and organoleptic characteristics, and longer shelf life. The same tendency was reported in Colombia, where the willingness to buy significantly increased when GM products contained a characteristic desired by consumers, such as enhanced nutrition and taste (Pachico and Wolf, 2002). West *et al.* (2002) reported that respondents were willing to pay higher prices for functional food products such as anti-cancer tomatoes, and heart-healthy potatoes and chicken in Canada. Maynard and Franklin (2003) reported a similar tendency with "cancer-fighting" dairy products. Grunert *et al.* (2004) investigated the effect of sensory experience with GM cheeses on attitudes and purchase intentions of participants from the Nordic countries. They report that the functional benefits of the products (fewer calories, less saturated fat, more calcium, and more zinc) had generally small effects.

The current study adds to the literature on the economics of consumer response to a new technology in the following ways: first we study consumers in India, an important emerging market; second, we analyze the effect of both positive and negative information relative to a no-information benchmark; and finally, we study the responses to first-generation (public benefits) versus second-generation (private benefits) biotechnology.

3. Data

The data set used in this study was collected through 600 in-person surveys conducted in New Delhi and Patna, India, in June 2004. New Delhi is the capitol of India with a population of more than 10 million, while Patna is the capitol city in the Indian State of Bihar with a population of more than 1 million. Two-thirds of the surveys were completed in New Delhi and one third in Patna. The survey was pre-tested with shoppers in New Delhi in January 2004. The survey was conducted in four different grocery stores in each city. These particular grocery stores were chosen in order to ensure a cross section of each city's population. Respondents were randomly selected and asked to participate by interviewers. Each respondent was given a bottled soft drink as a modest reward for participation. The data summarizing the demographic features of all 600 respondents are provided in Table 1.

The mean age of the respondents is 43.6 years, which is above the Indian average, but expected since children were not approached to participate in the survey. More than three-fourths of the respondents' households have children under 18 year's old living in their households. The average family size is 5.5. If one observes the distribution of respondents across levels of education, one finds that the group with maximum concentration is those with bachelor's degrees. Our sample has a higher level of education than the average in India, which is primary education below middle school. Around 45% of the respondents had income exceeding 200,000 rupees, which is well above the average of India, and the majority of the respondents were employed full time.

Table 1. Summary for demographic variables.

Variable	Description	Distribution of Survey Responses
Age	The participant's age in years	Mean =43.63 Std. Dev. = 10.42
Children	If children under 18 in household	
	yes	77.67%
	no	22.33%
Family Size	Number of people shopped for	Mean = 5.48 Std. Dev. = 2.80
Education	primary education	12.67%
	high school	26.17%
	10+2 level	16.33%
	bachelors degree	33.17%
	advanced degree	5.17%
	refused	6.50%
Income	<Rs. 25,000	5.33%
	Rs. 25,000–50,000	13.83%
	Rs. 50,001–100,000	10.33%
	Rs. 100,001–200,001	19.33%
	>Rs. 200,000	44.67%
	refused	6.50%
Employment	full time employed	56.17%
	part time employed	12.33%
	unemployed	1.86%
	homemaker	20.00%
	retired	4.67%
	refused	4.97%

As in all surveys, it is of concern whether the sample is representative of the population under study. The potential bias of greatest concern in this study is population choice bias, in which the population chosen does not adequately correspond to the population that will purchase genetically modified food products. Our sample was selected from the Indian population that shops for food at grocery stores, and it is expected that the values of certain socio-demographic statistics, such as income and education, will lie above the average for the country. However, given the preceding concerns, we acknowledge that the extent to which the findings can be fully generalized to the broader population of India is uncertain.

Table 2 provides summary statistics for the consumer information and perception variables. Approximately half of the respondents stated that they did not have any opinion about the use of biotechnology in foods. More than two thirds of the respondents did not know whether there is any risk associated with GM foods. This is not surprising as the majority of the respondents stated that they were not at all informed about GM foods. Even so, more than three-fourths of the respondents consider labeling of GM foods to be extremely important. This may seem surprising, but one generally finds that Indian consumers desire knowledge about any product they consume and also want to avoid adulteration of the products they are consuming. One of the most direct ways of achieving these objectives is to label the products.

The vast majority (87.5%) of the respondents are willing to consider purchasing foods with GM ingredients if they contain additional nutrients or vitamins. Similarly, 91.5% of the respondents are willing to consider purchasing foods with GM ingredients if they reduce the amount of pesticides applied to crops. Almost all respondents surveyed eat wheat chapatti daily, which is the likely end use of GM wheat.

The respondents were also asked questions about their attitudes concerning the environment and food safety. This information was obtained by presenting trade-off situations between food safety and low prices and between economic growth at all costs and saving the environment at all costs. Eliciting attitudes from trade-off scenarios is an effective way of ensuring that the survey information was informative, as well as useful in an empirical modeling context. Respondents placed more importance on food safety compared to food price, with a mean of 2.8 on a scale where food safety at all costs was one extreme (1) and food price being all-important was the other extreme (10). Similarly, the mean response was approximately 7 when economic growth at all costs was one extreme (1) and saving the environment at all costs was the other (10). This indicates that respondents self-reported that they leaned towards being environmentally conscious. Also, more than three fourths of the respondents stated that they prefer domestic to imported food products.

Table 2. Summary statistics for consumer information and perception variables.

Variable	Description	Distribution of Survey Responses
Opinion	Opinion about the use of biotechnology in foods	
	very positive	14.17%
	somewhat positive	14.83%
	neutral	12.67%
	somewhat negative	8.83%
	very negative	2.00%
	don't know	47.50%
Environment	Importance of economic growth versus savings	
	environment (1–10)	
	1 economic growth at all costs is all important	Mean = 7.00
	10 saving the environment at all costs is important	Std. Dev. = 1.55
Safety	Food safety versus food price (1–10)	
	1 food safety is all important	Mean = 2.82
	10 food price is all important	Std. Dev. = 1.41
Knowledge	Self-reported knowledge about GM foods	
	very knowledgeable	5.00%
	somewhat knowledgeable	40.83%
	not informed	54.17%
Labeling	GM foods should be labeled	
	very important	78.50%
	somewhat important	17.83%
	not very important	3.67%
Risk	Risk associated with GM foods	
	high risk	2.67%
	low risk	13.83%
	no risk	15.00%
	don't know	68.50%
Domestic	prefer domestic to imported food products	
	yes	77.83%
	no	22.17%
Nutrients	Purchase foods with GM ingredients if they contain additional nutrients or vitamins	
	yes	87.50%
	no	12.50%
Pesticide	Purchase foods with GM ingredients if they reduce the amount of pesticides applied to crops	
	yes	91.50%
	no	8.50%

The entire sample was divided into three equal groups for the purpose of analyzing information effects. Before consumers were asked about their preferences for products developed with biotechnology, one third of the sample was provided with positive public benefit information regarding reduced herbicide usage and production costs associated with GM wheat,[2] one third was given negative private benefit information that there may be long-term negative health consequences associated with consuming GM wheat,[3] and the final group was not given any information. Following this, contingent valuation questions were asked about GM wheat for use in making chapatti.

Finally, all 600 respondents were also asked questions about their willingness to pay for a type of GM wheat that is "good for the heart."[4] This is an example of a "second-generation" GM product, meaning that it has an attribute that enhances the final product from a consumer's viewpoint. One can compare this case with the one where positive public benefit information about GM wheat was given, the latter being an example of a "producer friendly" or "first generation" GM product.

---

[2]The positive information was, "Scientists at some universities are developing GM wheat. Genetic modification of the wheat decreases the amount of herbicide applied and reduces production costs."

[3]The negative information was, "The long-term consequences of eating chapatti made of GM wheat can be negative for health." Post-elicitation, the subjects were debriefed that there is no scientifically demonstrable evidence to support any negative claims and that they were provided this assertion purely for experimental purposes.

[4]Although this claim is not as specific as it could be, it is appropriate in context. There is a tradeoff between a broad understanding that something is healthy and providing technical information which the average consumer in a developing country would not understand. Heart disease is a major cause of death in India, and there is significant public concern as well as awareness about this health problem. In fact, many companies are marketing successful products in India which are "good for the heart." For example, sunflower oil is popular as a source of cooking oil in India as it is healthier for the heart owing to its low cholesterol. Hence, this "good for heart" claim was clear to the respondents based on the interactions (responses and follow-up discussions) with interviewers.

4. Empirical Model

The contingent-valuation method (CVM) is a technique that is used to estimate willingness to pay (WTP).[5] CVM is based on a dichotomous choice model in which individual WTP can be estimated based on responses to market-type questions (Kanninen, 1993). The survey for this study contained contingent valuation questions regarding the respondents' willingness to pay a premium or accept a discount in order to buy GM wheat in the form of chapatti.

We use the double-bounded model (Hanemann *et al.*, 1991) to analyze the outcomes of the survey.[6] In one set of questions, the respondents were first asked if they were willing to pay the same price for chapatti made from GM wheat as for chapatti made from non-GM wheat. If the respondent's answer to this question was "yes," a follow-up question was asked where the respondent was offered a percentage premium on the GM wheat-based chapatti relative to that for non-GM wheat. If the respondent's answer to the first question was "no," a follow-up question was asked where the respondent was offered a percentage discount on the GM wheat-based chapatti relative to non-GM wheat. The randomly assigned premium levels had the following values: 5%, 5%, 10%, 25% and 50%, while the randomly assigned discount levels had following values: 10%, 25%, 40%, 60% and 90%. Each level of premium or discount was used for one fifth of the surveys. These different levels of premium or discount were calibrated based on the results of the pre-survey. In a subsequent set of questions, the same questioning pattern was followed but the GM wheat had a claim that it was "good for the heart." The randomly assigned premium and discount levels were as before.

---

[5]For further information, including recent reviews and comparison across models estimable from CVM data with reiteration see, for example, Flachaire and Hollard, (2006).

[6]There is a literature on the appropriate number of iterations to include in the bidding procedures used in the CVM. Cameron and Quiggin (1994) evidenced the problem of anchoring/starting point bias with iterations of bids. There is some bias with the double-bounded model, primarily due to inconsistencies which may be present between the consumers' first and subsequent bids (Hanemann and Kanninen, 1999). Recent studies suggest that the loss from bias outweighs by the gain in efficiency relative to a single dichotomous choice question (Aadland and Caplan, 2004; Scarpa and Bateman, 2000; Bateman *et al.*, 2001; Whitehead 2002, 2004; and Cooper *et al.*, 2002).

Four outcomes are possible in the double bounded dichotomous choice model. They are: (1) the respondent is neither willing to buy the GM product at the same price as the non-GM product, nor is he or she willing to buy it at a discounted price, (i.e., "no" followed by "no"); (2) the respondent is not willing to buy the GM product at the same price as the non-GM product, but when given the random discount, the respondent exhibits a willingness to buy it at that price (i.e., "no" followed by "yes"); (3) the respondent is willing to buy the GM product at the same price as the non-GM product but not at a higher price (i.e., "yes" followed by "no"); (4) the consumer is willing to buy the GM product at the same price and he or she is also willing to pay the random premium (i.e., "yes" followed by "yes").

Given the aforementioned questioning sequence, one is able to isolate the range in which the respondents' true WTP lies, placing it into one of the following four intervals: $(-\infty, \text{BL})$, $[\text{BL}, \text{BI})$, $[\text{BI}, \text{BH})$, or $[\text{BH}, +\infty)$. The first and second bid combine to put an upper and a lower bound on the respondent's unobservable true WTP. When both the first and second decision are in the same direction (i.e. "yes, yes" or "no, no"), it raises the lower bound or lowers the upper bound, respectively. The WTP function is represented in general terms by:

$$WTP = \alpha - \rho B + \lambda' z + \varepsilon \tag{1}$$

where B is the ultimate bid amount each respondent faces, z is a column vector of observable characteristics of the individual respondent, ε is a random variable accounting for unobservable characteristics, and $\alpha, \rho,$ and λ are unknown parameters to be estimated. In the empirical implementation of the model, we define $G(.)$ to be the standard logistic distribution with mean zero and standard deviation $\sigma = \pi/\sqrt{3}$. The maximum likelihood principle can be used to estimate the unknown parameters in the likelihood function. The empirical representation of equation (1) for GM wheat is defined as:

$$\begin{aligned} WTP_i = \alpha - \rho B_i &+ \lambda_1 City_i + \lambda_2 Knowledge_i + \lambda_3 Opinion_i \\ &+ \lambda_4 Nutrients_i + \lambda_5 Pesticide_i + \lambda_6 Children_i \\ &+ \lambda_7 Education_i + \lambda_8 Age_i + \lambda_9 Risk_i + \varepsilon_i. \end{aligned} \tag{2}$$

Equation (2) was estimated for each of the aforementioned four different states of consumer information: zero additional information, positive public benefit information (reduction in herbicide use and production costs), negative private benefit information (relating to long term health consequences), and positive private benefit information (the product is "good for the heart").

The B variable in (2) represents the random bid offered to each consumer, *City* is an indicator variable which is coded as a 1 if the respondent was interviewed in Patna; *Knowledge* is self-reported knowledge about GM foods, which is coded as a 1 if the respondent indicated that he or she was at least somewhat knowledgeable about GM foods; *Opinion* represents whether the respondent was at least somewhat positive regarding the use of biotechnology in foods, coded as a 1 if such was the case; *Nutrients* is an indicator variable representing whether the respondent would purchase foods with GM ingredients if they contain additional nutrients, with the value 1 indicating an affirmative response; *Pesticide* is an indicator variable, which is coded as a 1 if the respondent indicated he or she would purchase foods with GM ingredients if they reduce the amount of pesticides applied to crops; *Children* is the number of children in the respondent's household, *Education* is an indicator variable representing the respondent's education level with higher than high school education indicated as 1, *Age* is the age of the respondent in years, and *Risk* is an indicator variable where if the respondent perceives any type of risk associated with GM foods, this is indicated by the value 1.

5. Analysis of Willingness to Pay

Mean willingness to pay can be estimating by restricting $\lambda_i's$ to zero (Hanemann *et al.*, 1991), and then calculating the empirical mean WTP by using the formula $\hat{\alpha}/\hat{\rho}$. A 95% confidence interval can then be calculated for mean WTP.[7] Table 3 provides mean willingness

[7]A 95% confidence interval around estimated mean WTP for each case was calculated using the delta method, which approximates the asymptotic variance of the ratio of two

Table 3. Mean willingness to pay (premium or discount) estimates with confidence intervals.

	Mean WTP	95% Confidence Interval
No Information	7.3%	[4.4%, 10.1%]
Positive, Public Good Information	10.4%	[7.2%, 13.7%]
Negative Information	−139%	[−153.7%, −124.2%]
Positive Private Good Information	23.7%	[19.9%, 27.5%]

to pay estimates, expressed in terms of premium or discount levels relative to non-GM food, along with confidence intervals for each type of information.

Regarding the effect of different types of information on mean willingness to pay for GM wheat, with no information presented, respondents, on average, were willing to pay a premium of 7.3% on GM wheat. One might be surprised at the mean WTP premium because the actual GM wheat in question is a first-generation GM product that had process attributes, which reduce production costs — as opposed to product-enhancing attributes.[8] Proponents claim that the GM products with process attributes are identical to non-GM products. Opponents view genetic modification as a negative attribute. Therefore, one would not expect consumers to pay a premium for a GM product that only has cost-reducing attributes. However, this finding supports Li *et al.* (2002) findings that Chinese respondents were also willing to pay a premium for GM soybean oil, which also has only process attributes. In both India and China, the concerns of food availability and nutritional intake are much greater compared to the United States, Europe, and Japan. It follows that consumers in China and India may feel generally more positive about the use of biotechnology. On the other hand, it may be the case that many of the respondents in the no-information treatment group

random variables as: *asymptotic var* $\left(\frac{\alpha}{\rho}\right) \cong \frac{\mu_\alpha^2}{\mu_\rho^2}\left[\frac{\sigma^2(\alpha)}{\mu_\alpha^2} - 2\frac{cov(\alpha,\rho)}{\mu_\alpha\mu_\rho} + \frac{\sigma^2(\rho)}{\mu_\rho^2}\right]$, where μ_α *and* μ_ρ denote the means of α *and* ρ.

[8]An anonymous reviewer pointed out that without debriefing for their thought processes in formulating a WTP response, we only speculate about what was prominent in the respondents' minds in the no-information case.

already understood the public good benefit of reduced pesticide use with GM wheat.

When positive public-benefit information (reduced herbicide usage and production costs) was presented, respondents were willing to pay, on average, a premium of 10.4%. This is a small, but not statistically significant increase, in the mean premium that respondents were willing to pay relative to the zero-information treatment. It may be the case that some respondents already capitalized these process attributes into the no-information case. Alternatively, it may be that consumers were not willing to pay a significant premium for the public good benefits of reduced herbicide usage and reduced production costs.

Other researchers have found empirical evidence on consumers' willingness to pay a premium for a product that provides public-good benefit. In their analyses of UK consumer preferences for GM foods, Rigby and Burton (2005 and 2006) find that their respondents have preferences for the public-good benefit of lower chemical use. Loureiro *et al.* (2001) consider both public- and private-good benefits in their assessment of consumer choice from among eco-labeled, organic, and regular apples. They find that both perceived quality of the apples (private good) and strong environmental attitudes (public good) have positive and significant effects on the probability of choosing one of the environmentally friendly choices. Hamilton *et al.* (2003) analyze preferences for public and private good aspects of food safety regulation. They consider the case of a ban on pesticide use that will result in better environmental quality, a public good. Willingness to pay to avoid pesticide residues is positively associated with a willingness to support (vote for) pesticide regulation.

Next consider the effect of the positive private-benefit information on mean willingness to pay. The respondents were, on average, willing to pay a 23.7% premium for GM wheat that is "good for the heart." This characteristic of GM wheat is product-enhancing or "consumer friendly." Comparing this with the cases of no information and positive public-benefit information respectively, one finds that respondents were willing to pay a significant premium when a "consumer friendly" or second-generation trait is thought to be present but a notably lesser amount when a "producer friendly" trait is present. This is consistent with other findings (Boccaletti and Moro, 2000; and West *et al.*, 2002).

In contrast, when negative information about potential health effects was provided, the average respondent required a discount of 139%, meaning that he or she was not willing to buy at any price and in fact required a payment in order to consume the product. Thus the respondent was unwilling to pay anything for GM wheat if the long term consequences of eating chapatti made with GM wheat may have negative health consequences. An important observation here is that this negative information had substantially more of an effect on consumers' willingness to pay than the positive information. This is consistent with many studies, including Tegene *et al.* (2003), who observed that individuals place a greater weight on negative information than on positive information. However, we emphasize that there are two characteristics of this negative information that should be noted when interpreting the ultimate effect on willingness to pay. First, the information was indeed unequivocally negative, and second, the information was about a private good (personal health) rather than a public good (reduced herbicide usage). Consequently, the more direct comparison of effect magnitude is to the heart-healthy wheat (discussed below), which was positive information of a private good nature.

Parameter estimation results are presented in Table 4, and the marginal effects of explanatory variables are presented in Table 5. Four information treatments are presented: no information, positive public-good information, and negative information, and positive private-good information. Most variables reported are significant at the 5% level and all variables are significant at the 10% significance level.

Although the focus of this analysis is on comparing willingness to pay across information treatments and benefit types, we briefly discuss significant variables in the models. In the no-information case, the estimation results show that variables representing the city, self-reported knowledge about GM foods, positive opinions regarding biotechnology, whether the presence of additional nutrients or

Table 4. Parameter estimates for WTP models for GM wheat.

	No Information		Positive Information, Public Good		Negative Information		Positive Information, Private Good	
Variable	Est.	Std. Error	Est.	Std. Error	Est.	Std. Error	Est.	Std. Error
Intercept	10.309	1.612	10.394	0.912	−1.184	0.385	7.930	0.685
Bid	−12.475	1.113	−9.527	0.852	−1.850	0.307	−6.015	0.352
City	−1.051	0.445	—	—	—	—	−1.025	0.212
Knowledge	−0.904	0.461	−1.146	0.360	−0.724	0.439	—	—
Opinion	1.496	0.537	1.600	0.374	1.810	0.467	—	—
Nutrients	1.531	0.745	—	—	—	—	—	—
Pesticide	4.835	1.030	—	—	—	—	0.855	0.305
Children	−0.743	0.373	—	—	—	—	—	—
Education	−0.667	0.316	—	—	—	—	0.342	0.179
Age	−0.045	0.018	—	—	—	—	−0.023	0.010
Risk	—	—	—	—	—	—	−0.607	0.238

Table 5. Marginal effects of explanatory variables on WTP for GM wheat.

	No Information		Positive Information, Public Good		Negative Information		Positive Information, Private Good	
Variable	Est.	Std. Error	Est.	Std. Error	Est.	Std. Error	Est.	Std. Error
City	−0.084	0.035	—	—	—	—	−0.170	0.035
Knowledge	−0.072	0.037	−0.120	0.038	−0.391	0.237	—	—
Opinion	0.119	0.043	0.168	0.040	0.977	0.275	—	—
Nutrients	0.123	0.060	—	—	—	—	—	—
Pesticide	0.388	0.076	—	—	—	—	0.142	0.051
Children	−0.059	0.030	—	—	—	—	—	—
Education	−0.054	0.025	—	—	—	—	0.057	0.030
Age	−0.004	0.001	—	—	—	—	−0.004	0.002
Risk	—	—	—	—	—	—	−0.101	0.040

reduction in pesticide usage affects the purchase decision, education and age are all statistically significant, with the expected signs, except for education, which has a negative effect on willingness to pay. In other studies (for example, see Li *et al.*, 2004), education has a positive effect, which has been interpreted to mean that more educated respondents are less skeptical of the scientific consensus. Younger consumers are willing to pay more compared to older people. This may reflect that consumers generally become more health conscious as they age and become more susceptible to illness. On the other hand, one could also interpret this to mean that younger consumers are more accepting of biotechnology. If the latter is the case, then it is promising for the future of biotechnology. Li *et al.* (2002) and Grimsrud *et al.* (2004) also find that age is a significant factor for their Chinese and Norwegian samples, respectively.

Apart from the intercept and bid variables, only two variables, knowledge and opinion, are statistically significant in the cases where positive public-good information and negative private-good information was given. All other variables, which were significant when no information was given, become insignificant. Apparently, the information (positive or negative) dominated other factors. For the positive private-benefit information relating to GM wheat that is "good for the heart," again fewer variables were significant compared to the case of GM wheat with zero information. One of the important remaining significant variables was risk. It had a negative sign, which means that even if some GM food is thought to be "consumer friendly," respondents who perceive there is risk associated with GM foods are willing to pay less.

6. Conclusions

The overall goal of this study was to assess the degree to which consumers' acceptance of GM foods in India, and in particular willingness to pay, is affected by different consumer information treatments and product benefit types. We find that respondents were willing to pay a small premium for GM wheat when no information was provided to them. This is surprising because one would not expect consumers to pay a premium for a GM product that only has cost-reducing attributes, but it supports Li *et al.* (2002) findings that Chinese respondents were also willing to pay a premium for GM soybean oil, which also has only process attributes. We consider that in both India and China the concerns of food availability and nutritional intake are much greater compared to the United States, Europe, and Japan. It follows that consumers in China and India might feel more positive about biotechnology. Further, in both China and India there is a general positive attitude towards science and technology.

Second, when positive "producer friendly" information was provided to the respondents, their willingness to pay for the GM wheat increased, but only by a small amount. This might be attributable to the fact that the positive information was about process attributes, i.e., reduced herbicide usage and production costs. It may be the case that some respondents already capitalized these process attributes into their no information bid response. Alternatively, it may be the case that consumers were not willing to pay a significant premium for the public good benefits of reduced herbicide usage and reduced production costs.

The results were dramatically different when negative information about potential health effects was provided. There was a strong negative consumer reaction. The average respondent was not willing to buy the GM wheat at any price. Third, when positive "consumer friendly" information was provided to the respondents, the willingness to pay for the GM product increased significantly, but still the effect was far smaller in absolute magnitude than when negative information was given to them.

In this open-economy era of India, understanding consumers' sentiments towards GM food is important for anticipating sales and development of new markets. Increased crop yields and dietary supplements provided by GM foods are of greater benefit for lesser developed countries and may contribute to consumer attitudes toward GM foods that are less negative and in many cases positive. The generally positive response in the case of India supports the findings of other studies in countries with emerging economies.

If GM crop production is to expand in the future, then it is natural to look for alternative markets such as India with its massive population and fast growing economy.

Further research should examine a number of issues surrounding information provision, including what types of benefits would convince consumers to accept GM products, how the information should be communicated so that consumers could make an informed choice, and which organizations consumers trust to provide information.

Acknowledgments

The authors wish to thank Tom Wahl for helpful comments. The authors express appreciation to the IMPACT Center of Washington State University for financial support of this research.

References

Aadland, D. and A. Caplan (2004). Incentive incompatibility and Starting-Point Bias in Iterative Valuation Questions: Comment. *Land Economics* 80, 312–315.

Bateman, I.J., I.H. Landford, A.P. Jones and G.N. Kerr (2001). Bound and Path Effects in Double Bounded and Triple Bounded Dichotomous Choice Contingent Valuation. *Resource and Energy Economics* 23, 191–213.

Boccaletti, S. and D. Moro (2000). Consumer Willingness to Pay for GM Food Products in Italy. *AgBioForum* 3, 259–267.

Burton, M., D. Rigby, T. Young and S. James (2001). Consumer Attitudes to Genetically Modified Organisms in the U.K. *European Review of Agricultural Economics* 28, 479–498.

Cameron, T. and J. Quiggin (1994). Estimation Using Contingent Valuation Data from a 'Dichotomous Choice with Follow Up' Questionnaire. *Journal of Environmental Economics and Management* 27, 218–234.

Chern, W.S., K. Rickertsen, N. Tsuboi and T. Fu (2002). Consumer Acceptance And Willingness To Pay For Genetically Modified Vegetable Oil and Salmon: A Multiple-Country Assessment. *AgBioForum* 5, 105–112.

Cooper, J., W.M. Hanemann and G. Signorello (2002). One and One-Half Bound Dichotomous Choice Continent Valuation. *The Review of Economics and Statistics* 84, 742–750.

Flachaire, E. and G. Hollard (2006). Controlling Starting–Point Bias in Double-Bounded Contingent Valuation Surveys. *Land Economics* 82, 103–111.

Frewer, L.J., C. Howard, D. Hedderley and E. Shepherd (1999). Reactions to Information About Genetic Engineering: Impact of Source Characteristics,

Perceived Personal Relevance and Persuasiveness. *Public Understanding of Science* 8, 35–50.

Grimsrud, K.M., J.J. McCluskey, M.L. Loureiro and T.I. Wahl (2004). Consumer Attitudes to Genetically Modified Food in Norway. *Journal of Agricultural Economics* 55, 75–90.

Grunert, K.G., T. Bech-Larsen, L. Lähteenmäki, Ø. Ueland and A. Åström (2004). Attitudes towards the Use of GMOs in Food Protection and their Impact on Buying Intention: The Role of Positive Sensory Experience. *Agribusiness* 20, 95–107.

Hamilton, S.F., D.L. Sunding and D. Zilberman (2003). Public Goods and the Value of Product Quality Regulations: The Case of Food Safety. *Journal of Public Economics* 87, 799–817.

Hanemann, W.M., J. Loomis and B.J. Kanninen (1991). Statistical Efficiency of Double Bounded Dichotomous Choice Contingent Valuation. *American Journal of Agricultural Economics* 73, 1255–1263.

Hanemann, W.M. and B. Kanninen (1999). "The Statistical Analysis of Discrete-Response CV Data." In I.J. Bateman and K.G. Willis (Eds.), *Valuing Environmental Preferences: Theory and Practice of the Contingent Valuation Method in the US, EU, and Developing Countries*, pp. 302–442. Oxford: Oxford University Press.

Kanninen, B.J. (1993). Optimal Experimental Design for Double-Bounded Dichotomous Choice Contingent Valuation. *Land Economics* 69, 138–146.

Li, Q., J.J. McCluskey and T.I. Wahl (2004). Effects of Information on Consumers' Willingness to Pay for GM-Corn-Fed Beef. *Journal of Agricultural and Food Industrial Organization* 2(2).

Li, Q., K.R. Curtis, J.J. McCluskey and T.I. Wahl (2002). Consumer Attitudes toward Genetically Modified Foods in China. *AgBioForum* 5, 145–152.

Loureiro, M.L., J.J. McCluskey and R.C. Mittelhammer (2001). Assessing Consumers Preferences for Organic, Eco-Labeled And Regular Apples. *Journal of Agricultural and Resource Economics* 26, 404–416.

Lusk, J.L., L.O. House, C. Valli, S.R. Jaeger, M. Moore, J.L. Morrow and W.B. Traill (2004). Effect of information about Benefits of Biotechnology on Consumer Acceptance of Genetically Modified Food: Evidence from Experimental Auctions in the United States, England, and France. *European Review of Agricultural Economics* 31, 179–204.

Lusk, J.L., M. Jamal, L. Kurlander, M. Roucan and L. Taulman (2005). A Meta-Analysis of Genetically Modified Food Valuation Studies. *Journal of Agricultural and Resource Economics* 30, 28–44.

Maynard, L.J. and S.T. Franklin (2003). Functional Foods as A Value-Added Strategy: The Commercial Potential of 'Cancer-Fighting' Dairy Products. *Review of Agricultural Economics* 25, 316–331.

Nelson, C.H. (2001). Risk perception, Behavior, and Consumer Response to Genetically Modified Organisms. *American Behavioral Scientist* 44, 1371–1388.

Pachico, D. and M. Wolf (2002). "Attitudes toward Genetically Modified Food in Columbia." Paper presented at the 6th International ICABR Conference in Ravello, Italy.

Pinstrup-Andersen, P. and E. Schiøler (2001). *Seeds of Contention: World Hunger and the Global Controversy over GM Crops.* Baltimore, MD: Johns Hopkins University Press.

Rigby, D. and M. Burton (2005). "Preference Heterogeneity and GM Food in the UK." *European Review of Agricultural Economics* 32, 269–288.

Rigby, D. and M. Burton (2006). Modeling Disinterest and Dislike: A Bounded Bayesian Mixed Logit Model of the UK Market for GM Food. *Environmental and Resource Economics* 33, 485–509.

Scarpa, R. and I. Bateman (2006). Efficiency Gains Afforded By Improved Bid Design versus Follow-Up Valuation Questions in Discrete Choice CV Studies. *Land Economics* 76, 299–311.

Scholderer, J. and L.J. Frewer (2003). The Biotechnology Communication Paradox: Experimental Evidence and the Need for a New Strategy. *Journal of Consumer Policy* 26, 125–157.

Subrahmanyan, S. and P.S. Cheng (2000). Perceptions and Attitudes of Singaporeans Toward Genetically Modified Food. *The Journal of Consumer Affairs* 34, 269–290.

Tegene, A., W.E. Huffman, M. Rousu, and J.F. Shogren (2003). *The Effects of Information on Consumer Demand for Biotech Foods, Evidence from Experimental Auctions.* U.S. Department of Agriculture, Economic Research Service, Technical Bulletin 1903.

West, G.E., C. Gendron, B. Larue and R. Lambert (2002). Consumers' Valuation of Functional Properties of Foods: Results from a Canada-Wide Survey. *Canadian Journal of Agricultural Economics* 50, 541–558.

Whitehead, J. (2002). Incentive Incompatibility and Starting-Point Bias in Iterative Valuation Questions. *Land Economics* 78, 285–297.

Whitehead, J. (2004). Incentive incompatibility and Starting-Point Bias in Iterative Valuation Questions: Reply. *Land Economics* 80, 316–319.

Part V

Conclusion and Recommendations

Chapter 21

Conclusion and Recommendations

Wallace E. Huffman and Jill J. McCluskey

We reviewed two major research methods for assessing consumers' WTP for new goods: (i) stated preference surveys and (ii) lab experiments. In stated preference surveys, subjects complete a survey that includes questions that uncover their underlying preferences for a new good. These questions can take the form of contingent valuation, where the subjects answer "yes" or "no" to whether they are willing to pay a specific price for the good in question or they may respond to discrete choice experiments that consist of presenting each respondent choice scenarios. In each scenario, respondents face a hypothetical situation in which they have to choose to buy a product from multiple options or buy nothing. An advantage to stated preference surveys is that this good in question might still be in the conceptual stage, so that it would be impossible to include it in a non-hypothetical setting. Moreover, these surveys generally take little of a subject's time and they can be conducted in-person, over the phone, through mail, or online. Hence, the cash payment to these subjects is generally quite modest — usually less than $20 per subject. However, the investigators must keep in mind that preferences or WTP are hypothetical. Subjects do not execute winning bids.

Lab experiments are a second important method to collect information on consumers' WTP for new goods. These lab experiments require subjects to come to a central location and participate in the live experiments. The experimental products must be available at least in sufficient quantity and the products auctioned in the lab experiments are correctly labeled. To keep subject burden down, we recommend keeping the total number of rounds of bidding for each subject to approximately four rounds.

There is much debate about the role of deception of subjects in WTP experiments. Investigators in psychological research are not prevented from using deception in their experiments. However, the field of economics has been more demanding. Some economics journals do not accept papers for publication that have used deception in almost any form (e.g., the *Journal of Experimental Economics*). However, it is impossible to eliminate deception in all forms. For example, subjects in economics experiments are not totally informed about the objectives of the study they are participating in. Recently, the Health and Human Services Administration has relaxed the human subjects' requirements for social science research, including economics, business and education. The only remaining human resource requirements are those on privacy of use of information collected, i.e., an individual's name will not be connected with data that he or she reports.

McFadden and Huffman (2017) raise interesting research issues that may occur before new biological products have been completely deregulated. Low-acrylamide GM potato products were not deregulated for commercial sale at the time of McFadden and Huffman's experiment. Hence, it was infeasible for subjects to take these products home. Their challenge was then to understand consumer preference for this product while minimizing deception.

Rousu *et al.* (2015) examine the cases for and against deception in applied economics research. They offer examples of published papers by agricultural economists that have used some form of deception, and furthermore, they describe some of the ways in which deception might be used successfully in research. Colson *et al.* (2016) report rankings of severity of deception found in a survey of scholars engaged in applied economics research. They identify the most severe forms of deception as cases where: (i) subjects are deceived in a way that causes physical or physiological trauma, (ii) subjects are promised payments for participating in an experiment but receive nothing, or (iii) subjects leave an experiment with a different product than what they thought that they had won. Deception of lesser consequences are identified as: (i) providing subjects with no or misleading information about the main purpose of a study, (ii) failing to inform subjects that they are part of a study, (iii) using confederates or bogus devices, and (iv) providing subjects with incomplete information about a product for sale in an experiment.

List (2008) emphasizes that a strict adherence to a no-deception policy is contrary to the goal of conducting and publishing research on important topics. For example, in some cases, research can only be undertaken if deception is used (e.g., testing for hiring discrimination by using résumés of individuals differing by gender or ethnic-associated names). He concluded that the benefits and costs of deception (and informed consent) should be carefully considered in each situation, and blanket rules against deception in economic experiments should be avoided. Moreover, the U.S. Office of Human Research Protection does not forbid deception in research (US-HHS-FDA, 2015), and in some other fields, deception in experiments is common and widely accepted. Psychologists have regularly used deception in laboratory experiments for more than 100 years (e.g., Solomons *et al.*, 1897). Hertwig and Ortmann (2008) report that more than half of the papers published in top social psychology journals in 2002 were based on research that involves some form of deception. Deception is widely used and accepted in some other fields as well, including marketing, nutrition, and medicine (e.g., Nunes and Boatwright, 2004; Wansink *et al.*, 2005; Wansink and Cheney, 2005).

McFadden and Huffman (2017) designed the experiments to minimize deception at each step, given the circumstances. At the initiation of each session, they told subjects they would be participating in an auction of food products. Subjects were told that the winners would execute winning bids, but that they would at most be a winner of only one unit of each type of product. In rounds of bidding with GM potato products, all labels were accurate, i.e., the label on the product correctly identified the product. Each subject participated in four rounds of bidding and no information on bids in a session was revealed until all bids were placed and collected. The monitor then randomly selected one of the two rounds of bidding on the conventional products to be binding. Hence, our procedures insured that products on display in the auctions were correctly labeled (authentic), no winner paid a price different from the rules in the auction, and no subject left the auction with a product that was different from what he or she thought had been won. The experimental procedures resulted in a low level of risk to human subjects, given the objective of the research project and state of the biotech product deregulation at the time of the experiments.

Lusk (2019) argues that the historical justifications typically given for the prohibition against deception in economic experiments are less relevant for today's experiments that are often conducted in non-lab settings with non-student subjects. He describes a variety of research questions that might be most adequately answered with some form of deception and calls for a more nuanced view of the issue that requires a consideration of the importance of the research question relative to the potential costs of deception. For example, in the case of new food products that have not yet been developed or approved, does the damage from hypothetical bias of stated preference analysis outweigh the damage from deceiving subjects in a non-hypothetical experiment? It is important for journals or academic professions, which ban the use of deception, to actually define what practices fall under the ban.

It is doubtful many people would avidly embrace a "pro-deception" position for economic research involving human subjects. We will not do so here. However, this short note will make the case for a more reasoned and nuanced consideration of the issue than is often espoused in the economic literature. Deception in economic experiments became taboo because of the context in which early experiments were conducted.

Several decades ago, when experimental economics began gaining traction as a distinct method and sub-discipline, the typical experiment involved student subjects who interacted in an induced value or other abstractly framed decision environment in a computer lab on campus. In this environment, the ban on deception had a rationale explained in early texts on the subject (e.g., Davis and Holt, 1993). In particular, subjects who participated in one experiment were highly likely to be recruited to participate in future experiments, and being deceived in a prior experiment may cause a participant to distrust the rules and instructions in a future experiment. More broadly, the student subjects may talk to one another outside the experiment, and a lab may develop a reputation for deception, perhaps even tarnishing the credibility of all economic experiments. In short, deception can contaminate the subject pool and degrade trust, which is a public good shared by all experimental economists. The rub is that the taboo surrounding deception has come to envelop experiments for which these sorts of pool-contamination and public-good critiques are less applicable.

Many experimental economists, in their roles as reviewers or editors, seem to have uncritically applied a blanket methodological prohibition to include all experiments even where there is little conceptual or empirical reason for the prohibition. Even a casual perusal of recent literature shows that today experiments are routinely conducted outside a lab, in field settings, with non-standard subject pools (Harrison and List, 2004; Levitt and List, 2009), and randomized controlled field trials have become standard tools for development economists (Duflo *et al.*, 2007). In these contexts, there is a vanishingly small chance a given individual will be asked to participate in a subsequent experiment. Because the experiments are not conducted at a particular lab, the risks to reputation are far lower. In the case of "true" field experiments, the participants are not even aware they took part in an experiment.

Aside from deontological reasoning, we ask what the argument is against deception in these sorts of field experiments. Whatever the reason, the argument must be different than the typical ones given for student-based lab experiments. There are ethical considerations to be sure, but most university human subjects institutional review boards (IRBs), which presumably exist to protect the interests of research participants, allow deception in certain circumstances. IRBs may require researchers to justify the use of deception (showing that the research question cannot be adequately answered any other way), require quasi-consent from the participant (where they agree to participate acknowledging the full purpose of the research has not been disclosed), or provide debriefings to subjects following deception, but there is no blanket prohibition against the practice.

When, in food and agricultural economics research, might deception produce valuable insights that would be difficult or impossible to obtain without deception? A common example is the interest in estimating consumers' preferences for a new food or technology that has not yet been developed (or is unapproved for human consumption, or in short supply). In such instances, researchers may want consumers to bid on a product that claims to have particular attributes or traits, when in fact it does not. One could, of course, ask hypothetical willingness-to-pay questions, but which is greater — the sin of hypothetical bias or the sin of deception?

If a journal or professional association is to enact a blanket ban on deception in experiments, our recommendation is that an explicit policy and definition should be formulated. Yet, this is much trickier than first meets the eye. What, exactly, constitutes deception? The question has been tackled by, among others, Krawczyk (2013), Rousu *et al.* (2015), and Wilson (2016). Is a boldface lie to be treated the same as withholding information? All deceptions are not created equal, and there is ample disagreement among researchers about the seriousness of a given practice. For example, Colson *et al.* (2016) surveyed food and agricultural economists, who had previously conducted experiments, and asked whether a list of practices should be banned. Their results suggest about 84% thought *buying* a mislabeled product should be banned, but only 57% thought *bidding* on a mislabeled product (in a design that assured it was not purchased) should be banned.

As a general rule, researchers, editors, and reviewers do not place a blanket ban on other research methods (say, computable general equilibrium models, neural networks, time series forecasting models, etc.) that produce results that sometimes (hopefully infrequently) lead to incorrect inference and harm. Rather, we rely on reviewers and editors to thoughtfully consider the strengths and weaknesses of the methods used in a paper relative to the ability of the methods to provide accurate insight on the question at hand. To be sure, it is much more difficult to evaluate the benefits of a particular piece of research relative to the various downsides posed by deception in the context of a particular study environment than to simply adopt a blanket ban, but if economists cannot equate marginal benefit with marginal cost, we are not sure who can.

References

Colson, G.J., J.R. Corrigan, C. Grebitus, M.L. Loureiro and M.C. Rousu (2016). Which Deceptive Practices, if Any, Should be Allowed in Experimental Economics Research? Results from Surveys of Applied Experimental Economists and Students. *American Journal of Agricultural Economics* 98(2), 610–621.

Davis, D.D and C.A. Holt (1993). *Experimental Economics.* Princeton: Princeton University Press.

Duflo, E., R. Glennerster and M. Kremer (2007). "Using Randomization in Development Economics Research: A Toolkit." In T.P. Schultz and J.A. Strauss (Eds.), *Handbook of Development Economics*, Vol. 4. pp. 3895–3962. Amsterdam: North-Holland.

Harrison, G.W. and J.A. List (2004). Field Experiments. *Journal of Economic Literature* 42(4), 1009–1055.

Hertwig, R. and A. Ortmann (2008). Deception in Experiments: Revisiting the Arguments in Its Defense. *Ethics & Behavior* 18(1), 59–92.

Krawczyk, M. (2013). Delineating deception in experimental economics: Researchers' and subjects' views, Working Papers 2013-11, Faculty of Economic Sciences, University of Warsaw.

Levitt, S.D. and J.A. List (2009). Field Experiments in Economics: The Past, the Present, and the Future. *European Economic Review* 53(1), 1–18.

List, J.A. (2008). Informed Consent in Social Science. *Science* 322(5902), 672.

Lusk, J. (2019). Viewpoint: The Costs and Benefits of Deception in Economic Experiments. *Food Policy* 83, 2–4.

McFadden, J.R. and W.E. Huffman (2017). Willingness-To-Pay for Natural, Organic, and Conventional Foods: The Effects of Information and Meaningful Labels. *Food Policy* 68(C), 214–232.

Nunes, J.C. and Boatwright, P. (2004). Incidental Prices and their Effect on Willingness to Pay. *Journal of Marketing Research* XLI, 457–466.

Rousu, M.C., G. Colson, J.R. Corrigan, C. Grebitus and M.L. Loureiro (2015). Deception in Experiments: Towards Guidelines on Use in Applied Economics Research. *Applied Economic Perspectives and Policy* 37, 524–536.

Solomons, L.M., J.E.A. Singer and W. James (1897). Contributions from the Harvard Psychological Laboratory: Discrimination in Cutaneous Sensations; Studies in Sensation and Judgment. *Psychological Review* 4(3), 246–271.

US-HHS-FDA (2015). Use of the term "natural" in the labelling of human food products; request for information and comments. 80 Fed. Reg. 69,905, 69,909.

United States Department of Health and Human Services, Food and Drug Administration.

Wansink, B., K. Van Ittersum and J.E. Painter (2005). How Descriptive Food Names Bias Sensory Perceptions in Restaurants. *Food Quality and Preference* 16(5), 383–478.

Wansink, B. and M.M. Cheney (2005). Super Bowls: Serving Bowl Size and Food Consumption. *The Journal of the American Medical Association* 293(14), 1727–1728.

Wilson, B.J. (2016). "The Meaning of Deceive in Experimental Economic Science." In B.J. Wilson, G. DeMartino and D. McCloskey (Eds.), *The Oxford Handbook of Professional Economic Ethics*. Oxford: Oxford University Press.

Index

acrylamide, 344, 382, 414
adverse selection, 90
advertising, 111
agenda setting, 114, 125
amplification stations, 124
anchoring, 29
antioxidants, 454
apples, 262, 454
asymmetric information, 50
auction session, 352
average treatment effects, 346, 358, 393

Bayesian-SUR-Tobit model, 323
Beijing, 556
Belgium, 94
Bid Differences, 401
bidder's surplus, 222
bovine spongiform encephalopathy (BSE), 124
bread, 514, 522
broccoli, 314

California's Proposition 37, 126
censored regression model, 217
Censored regression, fixed-effects model, 162
censoring, 360
censoring of bids, 215
chapatti, 582
"cheap talk", 22
China, 556
CNN factor, 125
Commercial mass media, 105
commodity, within-subject WTP differences, 436
confidence intervals, 523
conjoint analysis, 29
consumer skepticism, 47, 478
consumer sovereignty, 74
consumer traits, 395
consumer's social capital, 177
contingent valuation (CV) method, 17, 27, 479, 483, 514, 541, 542, 559, 582
contracting, 60
corn chips, 141
Council Directive 90/220, 82
Council for Biotechnology Education, 255
Council on Biotechnology Information, 176
Cultural Revolution, 567

dairy, 94
deception, 36, 356
deliberate release, 95
delta method, 467
Des Moines, 144
Developing countries, 48
dichotomous choice (DC), 22
dietary recommendations, 309
difference-in-differences, 251
difference-in-differences estimator, 249
differenced Tobit model, 360, 394
dioxin contamination, 94
double-bounded model, 27, 460, 562, 582

dread risk, 121, 465

education, 120
effects of information, 401
english auction, 25
enriched coatings, 454
environmental group perspective, 149
environmental groups, 51
environmental nongovernmental organizations, 194
EPA, 177
Escherichia coli outbreak, 125
EU Directive on the Contained Use of Genetically Modified Organisms, 95
European Union, 82, 95
experiment design, 143, 199, 386, 417
experiment session, 390
experimental auction, 18, 23

FCCC, 500
FDA, 177
Federal Insecticide, Fungicide, and Rodenticide Act, 94
Federal Plant Pest Act (FPPA), 94
fifth price auction, 71
'first-generation' genetic modifications, 49, 63, 572
'first-mover' advantage, 90
food labels, 57, 269, 388
food safety, 445
Food Safety and Inspection Service (USDA-FSIS), 260
food scare, 125
food security, 48
food-safety concerns, 493
formats of information, 62
framing, 108
framing role, 116
Free-riding, 51
French fries, 346
fresh broccoli, 262
functional food, 55, 75, 452

Gender, 120, 496, 500
GM contamination, 59
GM labeling, 196
GM labeling policy, 60
GM wheat, 17, 581
Golden rice, 49, 573
goodness of fit, 500
Greenpeace, 81
Greenpeace, Friends of the Earth, Action id, 54

health claims, 55
health risks in food, 120
healthy attitude proxies, 329
Heckman's two-step estimator, 35
heptachlor, 199
herbicide-tolerant, 192
human subjects, 36
hypothetical bias (HB), 20, 62, 479

ICCC, 500
identification, 26
"identity preservation system", 197
ideological bias, 111
incentive compatible, 19, 24, 69
independent third-party sources, 181
independent, third-party perspective, 149
India, 574
industry perspective, 149
information factor, 143
information perspectives, 314
information statements, 269
information treatments, 71, 149, 204, 206, 242, 250, 262, 270, 271, 274, 350, 352, 388, 389, 422, 429
insect-resistant, 192
institutional review board (IRB), 36
Integrated Pest Management (IPM), 288
interest groups, 51
"intragenic", 311
issue attention cycles, 115
Italy, 111

Japan, 532

labeling-order factor, 143
laboratory auctions, 387
laboratory experiments, 35, 68
Lancaster's approach, 30
large brown eggs, 262
Likert-type scale, 493
logistic, 462, 485, 518, 583
loss aversion, status quo bias, Bayesian updating, 198
lotteries, 238
low-acrylamide, 350

Maillard reaction, 414
mandatory labeling, 61
Matsumoto City, 536
mean discount, 523
mean willingness to pay, 467, 566
meat, 94
media, 103
Media Bias and Framing, 110
meta-analysis, 63
Michael Pollan, 127
monarch butterfly controversy, 116
Monsanto, Syngenta, 176
multicollinearity, 493
multinomial logit model, 184

nth-price auction, 3, 25, 241, 267, 387
Nagano, 536
National Organic Program (NOP), 260
"natural" foods, 260
negative information about potential health, 587
negative news coverage, 112
Netherlands, 94
New Delhi, 577
newspapers, 113
noodles, 542
Norway, 510
Nutrition Facts Panel, 57
Nutrition Labeling and Education Act (NLEA), 57, 196

one-and-one-half-bound, 484
open-ended questions, 22
organic, 260, 466
organic producers, 51
Oslo, 518

Patna, 577
payment for participating, 354
perception of risk, 107
"pesticide-free", 288
Policy implications, 372
postexperiment survey, 181
potato chips, 346
potatoes, 144, 200, 314, 344, 414
poultry, 94
practice round, 203, 354
practice round of bidding, 148
Practice-round, 417
pre-auction questionnaire, 390
precautionary principle, 49, 55, 126, 511
preference for domestic foods, 496
pretest survey, 32
principal-component analysis, 479, 493
prior beliefs, 235, 239
private benefits, 572
Private goods, 19
private information, 50, 82
process attribute, 515, 542
product-enhancing attribute, 515, 542, 560
production standards, 271
propensity score matching, 34
proposition 65, 404, 415
prospect theory, 57
public goods, 19, 68
public opinion, 122
public-benefit, 586

questionnaire, 275

random nth price auction, 145, 201, 262, 420
random digital dialing, 144
random indirect utility, 183, 427
random utility framework, 483
random utility model, 462

Randomizing the labeling treatment, 162
rationally ignorant, 109
Religious, 183
representative sample, 24
reputation, 91
response rate, 201
revealed preference (RP), 18
revealed-preference-based studies, 68
rice, 560
risk communication, 128
risk tolerance for GM content, 197
Roundup Ready wheat, 515
Russia, 478, 487

Seattle, 456
Second-generation GM products, 49, 55, 572
segregation" system, 197
segregation, 59
Seikyou, 536
Selection bias, 34
semi-double-bounded logit model, 548
semi-double-bounded model, 484, 544
single-bounded model, 27
slanting of news, 113
social amplification of risk, 108
social desirability, 21
sodium bisulfite, 414, 415, 425, 443
Soybean oil, 560
soybeans, 144
Spokane, 456
St. Paul, Minnesota, 144
StarLink, 116
starting point bias, 29, 485
state ownership of media, 105
stated preference (SP), 17
statistical sampling methods, 144
subject activities, 271
sulfite-free, 417, 428
sulfites, 414
Survey-based studies, 62

Television, 125
third-party perspectives, 144
'Tolerances', 60
tomatoes, 314
tortilla chips, 144, 200
traceability, 197
treatment effects, 33
trust in information sources, 178
Trust in Public Institutions, 93

U.S. Food and Drug Administration (FDA), 453
uninformed prior beliefs, 248
unknown risk, 121, 465
US Food and Drug Administration, 58
utility function, 179

Value of verifiable information, 226
vegetable oil, 144, 200
verifiable information, 181, 194, 221, 331
verified information system, 89
Vickrey second-price auction, 3, 25
Vitamin consumption, 466
voluntary labeling, 295

White Russet^{TM} potato, 373
Wilcoxon signed-rank tests, 156
willingness to accept (WTA), 514, 541
willingness to pay, 236
wisdom-of-crowds, 112
within-individual, 359
within-individual WTP differences, 393
within-subject comparison, 34

yea saying, 29
yellow corn, 144

zero bids, 275, 322

Lightning Source UK Ltd.
Milton Keynes UK
UKHW020302211020
371867UK00004B/263

9 789811 216718